International Conference on Advancement in Science and Technology 2012 (iCAST): Contemporary Mathematics, Mathematical Physics and their Applications

Journal of Physics: Conference Series Volume 435

Kuantan, Malaysia
7-10 November 2012

Editors:

Nasir Ganikhodjaev
Pah Chin Hee
Farrukh Mukhamedov

ISBN: 978-1-62748-312-4
ISSN: 1742-6588

Printed from e-media with permission by:

Curran Associates, Inc.
57 Morehouse Lane
Red Hook, NY 12571

CURRAN ASSOCIATES INC. proceedings.com

Some format issues inherent in the e-media version may also appear in this print version.

Printed by Curran Associates, Inc. (2013)

Institute of Physics
Dirac House, Temple Back
Bristol BS1 6BE UK

Phone: 44 1 17 929 7481
Fax: 44 1 17 920 0979

techtracking@iop.org

Additional copies of this publication are available from:

Curran Associates, Inc.
57 Morehouse Lane
Red Hook, NY 12571 USA
Phone: 845-758-0400
Fax: 845-758-2634
Email: curran@proceedings.com
Web: www.proceedings.com

TABLE OF CONTENTS

Foreword

The 4th International Conference on the Advancement of Science and Technology 2012 (iCAST 2012), with theme "Contemporary Mathematics, Mathematical Physics and their Applications", took place in Kuantan, Malaysia, from Wednesday 7 to Friday 9 November 2012. The conference was attended by more than 100 participants, and hosted about 160 oral and poster papers by more than 140 pre-registered authors. The key topics of the 4th iCAST 2012 include Pure Mathematics, Applied Mathematics, Theoretical/Mathematical Physics, Dynamical Systems, Statistics and Financial Mathematics. The scientific program was rather heavy since after the Keynote and Invited Talks in the morning, four parallel sessions ran every day. However, according to all attendees, the program was excellent with high level of talks and the scientific environment was fruitful; thus all attendees had a creative time.

The conference was aimed to promote the knowledge and development of high-quality research in mathematical fields concerned with applications of other scientific fields as well as modern technological trends in physics, chemistry, biology, medicine, economics, sociology, environmental sciences.

We would like to thank the Keynote and the Invited Speakers for their significant contributions to 4th iCAST 2012. We would also like to thank the members of the International Scientific Committee and the members of the Organizing Committee. We cannot end without expressing our many thanks to International Islamic University Malaysia and our sponsors for their financial support .

This book presents selected papers which have been peer-reviewed. The editors hope that it may be useful and fruitful for scholars, researchers, and advanced technical members of the industrial laboratory facilities for developing new tools and products

Guest Editors

Nasir Ganikhodjaev, Farrukh Mukhamedov, Pah Chin Hee

Journal of Physics: Conference Series **435** (2013) 011001 doi:10.1088/1742-6596/435/1/011001

EDITORIAL INFORMATION

Patron

Prof. Dato' Sri Dr. Zaleha bt Kamaruddin, Rector of IIUM

Advisor

Prof. Dr.Kamaruzzaman Yunus, Dean of Kulliyyah of Science

Chairman

Prof Dr. Nasir Ganikhodjaev

Secretariat Committee

Dr. Nik Ruzni Nik Idris
Dr. Siti Marponga Tolos
Dr. Mimi Hafizah Abdullah
Sr. Rohanum Johari
Sr. Nurul Najihah Mohamad

Publicity /Technical Committee

Dr. Samsun Baharin Mohamad
Dr. Pah Chin Hee
Dr. Hazimin Mohd Salleh
Dr. Muhammad Salihi Abdul Hadi
Br. Mohd Hirzie Mohd Rodzhan
Sr. Sarah 'Afifah Abdullah Sani

Abstract Review and Publication Committee

Prof. Dr. Farrukh Mukhamedov
Prof. Dr. Nasir Ganikhodjaev
Dr. Pah Chin Hee
Dr. Bakhram Umarov
Br. Abduaziz Abduganiev
Sr. Wan Nur Fairuz Alwani Wan Rozali
Sr. Siti Fatimah Zakaria

Registration/ Finance Committee

Dr. Aminul Islam
Dr. Mansoor Saburov
Br. Nor Zaihan Mohd Noar
Br. Abdul Malek Abdul Hadi
Br. Termizi Ahmad
Sr. Norfaieqah Ahmad

Advisory Board

PLENARY SPEAKERS

Luigi Accardi

After getting a scholarship at the CNR (Consiglio Nazionale delle Ricerche), Accardi became a Professor for probability at the Universitµa di Roma II in 1980. His fields of interest are: quantum probability and its application to physics; the foundation of probability and of quantum mechanics, analysis of functions and mathematical economy. Accardi made important contributions in quantum probability.

Shavkat Alimov

Professor Dr. Shavkat Alimov is the Professor of Moscow State University and National University of Uzbekistan, presently a visiting scientists at Laboratory of Computational Sciences and Mathematical Physics. Professor Alimov received his MSc. in Physics and PhD in Mathematics from Moscow State University. He was Professor of Mathematics at the Moscow State University (Department of Computational Mathematics) from 1974 to 1984. Since 1984 he has been Professor of Mathematics at the National University of Uzbekistan and since 2006 he is also Professor of Mathematics at the Tashkent Campus of the Moscow State University. Professor Alimov is the author of more than 120 mathematical works in mathematical physics, partial differential equations, spectral theory.

Mohd Salmi Md Noorani

Prof Salmi's research area is Ergodic Theory, dynamical systems and analysis. Currently he is also president of Malaysian Mathematical Sciences Society.

Moussa Larbani

Prof. Dr. Moussa Larbani is specialized in operational research. He is currently working at IIUM.

Mohd Uzir B Mahidin

Dr. Uzir is currently working as a president of Statistics Division of Industrial Production and Construction in Department of Statistics Malaysia.

Peer review statement

All papers published in this volume of *Journal of Physics: Conference Series* have been peer reviewed through processes administered by the proceedings Editors. Reviews were conducted by expert referees to the professional and scientific standards expected of a proceedings journal published by IOP Publishing.

Coefficient Estimates for a Subclass of Univalent Functions with Respect to Symmetric Points

A Janteng[1] **and S A Halim**[2]
[1] School of Science & Technology, Universiti Malaysia Sabah, Jalan UMS, 88400 Kota Kinabalu, Sabah
[2] Institute of Mathematical Sciences, Universiti Malaya, 50603 Kuala Lumpur

E-mail: [1]aini_jg@ums.edu.my, [2]suzeini@um.edu.my

Abstract. Let A be the class of functions which are analytic in the open unit disc $D=\{z:|z|<1\}$ given by $f(z)=z+\sum_{n=2}^{\infty}a_n z^n$, $z\in D$. We denote by S the subclass of A consisting of all functions in A which are also univalent in D. In this paper, the subclasses of S denoted by $C_s(g)$ and $K_s^*(g)$ are introduced. We obtain coefficient bounds for $f\in C_s(g)$ and $f\in K_s^*(g)$. These results generalize many known results.

1. Introduction

Let A be the class of functions which are analytic in the open unit disc $D=\{z:|z|<1\}$ given by

$$f(z)=z+\sum_{n=2}^{\infty}a_n z^n, \qquad z\in D \tag{1}$$

We denote by S the subclass of A consisting of all functions in A which are also univalent in D. For two functions f and g, analytic in D, we say that f is subordinate to g in D (written $f\prec g$) if there exists a Schwarz function $w(z)$, analytic in D with

$$w(0)=0 \text{ and } |w(z)|<1 \qquad (z\in D)$$

such that $f(z)=g(w(z))$for all $z\in D$. Specially, if g is univalent in D, then $f\prec g$ if and only if $f(0)=g(0)$ and $f(D)\subseteq g(D)$.

In many earlier investigations, various interesting subclasses of A and S have been studied from a number of different viewpoints. We choose to recall here the investigations by Ravichandran *et al* [3], Xu *et al* [7] and Srivastava *et al* [5]. In particular, let C_s be the subclass of S consisting of functions given by (1) satisfying

[1] A Janteng.

Journal of Physics: Conference Series **435** (2013) 012001 doi:10.1088/1742-6596/435/1/012001

$$\text{Re}\left(\frac{(zf'(z))'}{(f(z)-f(-z))'}\right)>0, \quad z\in D.$$

These functions are called convex with respect to symmetric points and were introduced by Das and Singh in 1977. Xu and Wu (2010) introduced the following subclasses of analytic functions denoted by $S_s^*(g)$ and $K_s(g)$.

Definition 1.1. *Let $g:D\to C$ be a convex function such that $g(0)=1$, $g(\bar{z})=\overline{g(z)}$ for $z\in D$, $\text{Re}\{g(z)\}>0$ on $z\in D$. Let f be an analytic function in D defined by (1). We say that $f\in S_s^*(g)$ if and only if it satisfies the following condition*

$$\frac{2zf'(z)}{f(z)-f(-z)}\in g(D), \quad z\in D.$$

Definition 1.2. *Let g satisfy the conditions of Definition 1.1 and f be an analytic function in D defined by (1). We say that $f\in K_s(g)$ if there exists a function $h\in S_s^*(g)$ such that*

$$\frac{2zf'(z)}{h(z)-h(-z)}\in g(D), \quad z\in D.$$

This paper develops two new subclasses of S by adopting the original idea of Xu and Wu (2010). Now, we give the defintition of these classes which are denoted by $C_s(g)$ and $K_s^*(g)$.

Definition 1.3. *Let $g:D\to C$ be a convex function such that $g(0)=1$, $g(\bar{z})=\overline{g(z)}$ for $z\in D$, $\text{Re}\{g(z)\}>0$ on $z\in D$. Let f be an analytic function in D defined by (1). We say that $f\in C_s(g)$ if and only if it satisfies the following condition*

$$\frac{2(zf'(z))'}{(f(z)-f(-z))'}\in g(D), \quad z\in D.$$

Definition 1.4. *Let g satisfy the conditions of Definition 1.1 and f be an analytic function in D defined by (1). We say that $f\in K_s^*(g)$ if there exists a function $h\in S_s^*(g)$ such that*

$$\frac{2z(zf'(z))'}{h(z)-h(-z)}\in g(D), \quad z\in D.$$

In this paper, by using the principle of subordination, we obtain coefficient bounds for functions $f\in C_s(g)$ and $f\in K_s^*(g)$.

2. Preliminary Results

The following preliminary results are required for proving the main results.

Lemma 2. 1([4]) *Let the function* g *given by*

$$g(z)=\sum_{k=1}^{\infty} g_k z^k, \quad z\in D$$

be convex in D. *Suppose that the function* f *given by*

$$f(z)=\sum_{k=1}^{\infty} a_k z^k, \quad z\in D.$$

be holomorphic in D. *If* $f \prec g$, $z\in D$ *then*

$$|a_k| \le |g_1|, \qquad (k\in N)$$

Theorem 2.1([6]) *Let* $f\in S_s^*(g)$, *then for* $n\ge 1$,

$$|a_{2n}| \le \frac{|g'(0)|}{n!2^n}\prod_{j=1}^{n-1}\left(|g'(0)|+2j\right)$$

and

$$|a_{2n+1}| \le \frac{|g'(0)|}{n!2^n}\prod_{j=1}^{n-1}\left(|g'(0)|+2j\right)$$

3. Main Results

We now state and prove the main results of our present investigation.

Theorem 3. 1 *Let* $f\in C_s(g)$, *then for* $n\ge 1$,

$$|a_{2n}| \le \frac{|g'(0)|}{(2n)n!2^n}\prod_{j=1}^{n-1}\left(|g'(0)|+2j\right) \tag{2}$$

and

$$|a_{2n+1}| \le \frac{|g'(0)|}{(2n+1)n!2^n}\prod_{j=1}^{n-1}\left(|g'(0)|+2j\right) \tag{3}$$

The result obtained is sharp.

Proof. First, we prove (2) using the principle of mathematical induction. Let

$$p(z)=\frac{2(zf'(z))'}{f(z)-f(-z)} \tag{4}$$

Since $f\in C_s(g)$, it follows that

$$p(0)=g(0)=1 \text{ and } p(z)\in g(D),(z\in D).$$

Therefore, we have

$$p(z) \prec g(z), \qquad (z \in D),$$

where

$$p(z) = 1 + p_1 z + p_2 z^2 + \ldots$$

According to Lemma 2.1, we obtain

$$|p_i| \le |g'(0)| \qquad (i \in N).$$

From (4), we reduce that

$$2\left(1 + 2^2 a_2 z + 3^2 a_3 z^2 + \ldots + (2n)^2 a_{2n} z^{2n-1} + (2n+1)^2 a_{2n+1} z^{2n} + \ldots\right)$$
$$= 2\left(1 + 3a_3 z^2 + 4a_4 z^3 \ldots + (2n-1)a_{2n-1} z^{2n-2} + (2n+1)a_{2n+1} z^{2n} + \ldots\right)$$
$$\bullet \left(1 + p_1 z + p_2 z^2 + \ldots + p_{2n} z^{2n} + p_{2n+1} z^{2n+1} + ..\right)$$

Equating the coefficients of the same powers of z, we obtain that

$$4a_2 = p_1, \quad 6a_3 = p_2 \tag{5}$$
$$16a_4 = p_3 + 3a_3 p_1, \quad 20a_5 = p_4 + 3a_3 p_2 \tag{6}$$
$$(2n)^2 a_{2n} = p_{2n-1} + 3a_3 p_{2n-3} + 5a_5 p_{2n-5} + \ldots + (2n-1)a_{2n-1} p_1 \tag{7}$$
$$(2n)(2n+1) a_{2n+1} = p_{2n} + 3a_3 p_{2n-2} + 5a_5 p_{2n-4} + \ldots + (2n-1)a_{2n-1} p_2 \tag{8}$$

Easily by using Lemma 2.1 and (5), we get

$$|a_2| \le \frac{|g'(0)|}{2(2)}, \quad |a_3| \le \frac{(|g'(0)|)}{3(2)} \tag{9}$$

Again by applying (9) and followed by Lemma 2.1, we get from (6)

$$|a_4| \le \frac{(|g'(0)|)(|g'(0)| + 2)}{4(4)(2)}, \quad |a_5| \le \frac{(|g'(0)|)(|g'(0)| + 2)}{5(4)(2)}$$

It follows that (2) and (3) hold for $n = 1,2$. We now prove (2) using induction. Equation (7) in conjuction with Lemma 2.1 yields

$$|a_{2n}| \le \frac{|g'(0)|}{(2n)^2}\left(1 + \sum_{k=1}^{n-1}(2k+1)|a_{2k+1}|\right) \tag{10}$$

We assume that (2) holds for $k = 3,4,\ldots,(n-1)$. Then from (10), we obtain

$$|a_{2n}| \le \frac{|g'(0)|}{(2n)^2}\left(1 + \sum_{k=1}^{n-1} \frac{|g'(0)|}{k!2^k} \prod_{j=1}^{k-1}(|g'(0)| + 2j)\right) \tag{11}$$

In order to complete the proof, it is sufficient to show that

$$\frac{|g'(0)|}{(2m)^2}\left(1+\sum_{k=1}^{m-1}\frac{|g'(0)|}{k!2^k}\prod_{j=1}^{k-1}(|g'(0)|+2j)\right)=\frac{|g'(0)|}{(2m)m!2^m}\prod_{j=1}^{m-1}(|g'(0)|+2j),\quad (m=3,4,\ldots,n) \tag{12}$$

Equation (12) is valid for $m=3$.
Let us suppose that (12) is true for all $m,\ 3<m\le(n-1)$. Then from (11)

$$\frac{|g'(0)|}{(2n)^2}\left(1+\sum_{k=1}^{n-1}\frac{|g'(0)|}{k!2^k}\prod_{j=1}^{k-1}(|g'(0)|+2j)\right)$$

$$=\frac{|g'(0)|}{(2n)^2}\left(1+\sum_{k=1}^{n-2}\frac{|g'(0)|}{k!2^k}\prod_{j=1}^{k-1}(|g'(0)|+2j)+\frac{|g'(0)|}{(n-1)!2^{n-1}}\prod_{j=1}^{n-2}(|g'(0)|+2j)\right)$$

$$=\left(\frac{n-1}{n}\right)^2\left(\frac{|g'(0)|}{2(n-1)(n-1)!2^{n-1}}\prod_{j=1}^{n-2}(|g'(0)|+2j)\right)+\frac{|g'(0)|}{(2n)^2}\left(\frac{|g'(0)|}{(n-1)!2^{n-1}}\prod_{j=1}^{n-2}(|g'(0)|+2j)\right)$$

$$=\frac{|g'(0)|}{2n(n)(n-1)!2^{n-1}}\prod_{j=1}^{n-2}(|g'(0)|+2j)\left(\frac{|g'(0)|+2(n-1)}{2}\right)$$

$$=\frac{|g'(0)|}{(2n)n!2^n}\prod_{j=1}^{n-1}(|g'(0)|+2j)$$

Thus, (12) holds for $m=n$ and hence (2) follows. Similarly we can prove (3). Letting $|p_i|=|g'(0)|,(i\in N)$ shows that the result is sharp.

Theorem 3. 2 *Let $f\in K_s^*(g)$, then for $n\ge1$,*

$$|a_{2n}|\le\frac{|g'(0)|}{(2n)n!2^n}\prod_{j=1}^{n-1}(|g'(0)|+2j)$$

and

$$|a_{2n+1}|\le\frac{|g'(0)|}{(2n+1)n!2^n}\prod_{j=1}^{n-1}(|g'(0)|+2j)$$

Proof. Theorem 3.2 can be proven by using similar arguments as in the proof of Theorem 3.1, so we omit the details involved.

If we set

$$g(z)=\frac{1+Az}{1+Bz},\quad (z\in D;\ -1\le B<A\le1)$$

in Theorem 3.1 and Theorem 3.2, we obtain easily Corollary 3.1 and Corollary 3.2 respectively.

Corollary 3. 1 *Let $f\in C_s(A,B)$, then for $n\ge1$,*

$$|a_{2n}| \le \frac{(A-B)}{(2n)n!2^n} \prod_{j=1}^{n-1} (|g'(0)| + 2j)$$

and

$$|a_{2n+1}| \le \frac{(A-B)}{(2n+1)n!2^n} \prod_{j=1}^{n-1} (|g'(0)| + 2j)$$

Corollary 3. 2 *Let* $f \in K_s^*(A,B)$ *, then for* $n \ge 1$,

$$|a_{2n}| \le \frac{(A-B)}{(2n)n!2^n} \prod_{j=1}^{n-1} (|g'(0)| + 2j)$$

and

$$|a_{2n+1}| \le \frac{(A-B)}{(2n+1)n!2^n} \prod_{j=1}^{n-1} (|g'(0)| + 2j)$$

References

[1] Das R N and Singh P 1977 On subclasses of schlicht mapping *Indian J. Pure Appl. Math.* **8** 864-872
[2] Janteng A and Halim S A 2008 A subclass of convex functions with respect to symmetric points *Proceedings of The 16th National Symposium on Mathematical Sciences* **2** 2-6
[3] Ravichandran V, Polatoglu Y, Bolcal M and Sen A 2005 Certain subclasses of starlike and convex functions *Hacettepe Journal of Mathematics and Statistics* **34** 9-15
[4] Rogosinski W 1943 On the coefficients of subordinate functions *Proc. London Math. Soc.* **48** 48-82
[5] Srivastava H M, Xu Q H and Wu G P 2010 Coefficient estimates for certain subclasses of spiral like functions of complex order *Appl. Math. Lett.* **23** 763-768
[6] Xu Q H and Wu G P 2010 Coefficient estimate for a subclass of univalent functions with respect to symmetric point *European Journal of Pure and Applied Mathematics* **3** 1055-1061
[7] Xu Q H, Gui Y C and Srivastava H M 2012 Coefficient estimates for a certain subclass of analytic and bi-univalent functions *Applied. Mathematics Letters* **25** 990-994

On Dobrushin Ergodicity Coefficient and weak ergodicity of Markov Chains on Jordan Algebras

Farrukh Mukhamedov

Faculty of Science, International Islamic University Malaysia, P.O. Box, 141, 25710, Kuantan, Pahang, Malaysia

E-mail: farrukh_m@iium.edu.my

Abstract. In this paper we study certain properties of Dobrushin's ergodicity coefficient for stochastic operators defined on non-associative L^1-spaces associated with semi-finite JBW-algebras. Such results extends the well-known classical ones to a non-associative setting. This allows us to investigate the weak ergodicity of nonhomogeneous discrete Markov processes (NDMP) by means of the ergodicity coefficient. We provide a necessary and sufficient conditions for such processes to satisfy the weak ergodicity.

1. Introduction

It is known (see [23]) that the investigations of asymptotical behavior of iterations of Markov operators on commutative L^1-spaces are very important. On the other hand, these investigations are related with several notions of ergodicity of L^1-contractions of measure spaces. To the investigation of such ergodic properties of Markov operators were devoted lots of papers (see for example, [7, 23]). On the other hand, such kind of operators were studied in noncommutative settings. Since, the study of quantum dynamical systems has had an impetuous growth in the last years, in view of natural applications to various field of mathematics and physics. It is then of interest to understand among the various ergodic properties, which ones survive and are meaningful by passing from the classical to the quantum case. Due to noncommutativity, the latter situation is much more complicated than the former. The reader is referred e.g. to [2, 15, 16, 18, 29] for further details relative to some differences between the classical and the quantum situations. It is therefore natural to study the possible generalizations to quantum case of the various ergodic properties known for classical dynamical systems. One of the generalizations of noncommutative algebras is Jordan algebras. Note that Jordan Banach algebras [3, 6] are a non-associative real analogue of von Neumann algebras. The existence of exceptional JBW-algebras does not allow one to use the ideas and methods from von Neumann algebras. The motivation of these investigations arose in quantum statistical mechanics and quantum field theory (see, [9]). Mostly, in those investigations homogeneous Markov processes were considered. Many ergodic type theorems have been proved for Markov operators acting in nonassociative and noncommutative L^p-spaces (see for example, [4, 5, 20, 21, 8, 18])

On the other hand, nonhomogeneous Markov processes with general state space have become a subject of interest due to their applications in many branches of mathematics and natural sciences. In many papers (see for example, [24, 17, 31]) the weak ergodicity of nonhomogeneous Markov process are given in terms of Dobrushin's ergodicity coefficient [12]. In [33] some sufficient conditions for weak and strong ergodicity of nonhomogeneous Markov processes are given and estimates of the rate of convergence are proved. Lots of papers were devoted to the investigation of ergodicity of nonhomogeneous Markov chains (see, for example [12]-[19],[30]).

Until now a limited number of investigations are devoted to the ergodic properties of nonhomogeneous Markov processes defined on nonassociative and noncommutative spaces (see for example, [1, 25]). In this paper we are going to study ergodic properties of nonhomogeneous discrete Markov processes defined on nonassociative L^1-spaces. Note that in the context of inhomogeneous Markov chains, ergodicity refers to the asymptotic behavior of products of stochastic operators where the number of factors grows unbounded. In the simplest case, when all factors in the products are identical to the same stochastic operator T, ergodicity corresponds to the investigation of iterations of T. The Dobrushin's ergodicity coefficient is one of the effective tools to study a behavior of such products (see [17] for review) . Therefore, we will define such a ergodicity coefficient of a positive mapping defined on nonassociative L^1-space, and study its properties. In this direction we extend the results of [24] to a noncommutative setting. This allows us to investigate the weak ergodicity of nonhomogeneous discrete Markov processes by means of such ergodicity coefficient. We shall provide sufficient conditions for such processes to satisfy the weak ergodicity. Note that in [13] similar conditions were found for classical ones to satisfy weak ergodicity.

2. Preliminaries

In this section recall some well known facts concerning Jordan algebras. Let A be a Jordan algebra with unity $\mathbf{1}$ and a Banach space over the reals. Multiplication operation on A we will denote by $\circ$. If the norm on A respects multiplication so that $\|a^2\| = \|a\|^2$ and $\|a^2\| \leq \|a^2 + b^2\|$ for all $a,\ b \in A$, then A is called a JB-algebra (see [3],[6]). A Jordan subalgebra A_0 of A is said to be *strongly associative* if $(a \circ b) \circ c = a \circ (c \circ b)$ for all $a,\ b \in A_0$ and $c \in A$. A family M of elements of A is said to be consistent if it is contained in a strongly associative subalgebra.

Definition 2.1. A partial ordering $\geq$ of a Jordan algebra A is said to be *compatible* with the algebraic operations if

(1) $a \geq b \ \Rightarrow \ a + c \geq b + c$ for each $\in A$;

(2) $a \geq b \ \Rightarrow \ \lambda a \geq \lambda b$ for all $\lambda \in \mathbf{R},\ \lambda \geq 0$;

(3) $a \geq 0,\ b \geq 0$, a and b are consistent $\Rightarrow \ a \circ b \geq 0$;

(4) $a^2 \geq 0$ for each $a \in A$.

Note that in each JB-algebra A the set $A^+ = \{a^2 : a \in A\}$ is regular convex cone and defines in A a partial ordering compatible with the algebraic operations. A JB-algebra A is called a JBW-algebra if there exists a Banach space N, which is said to be pre-dual to A, such that A is isometrically isomorphic to the space N^* of continuous linear functionals on N. On the JBW-algebra A one can introduce the $\sigma(A, N)$-weak topology. A linear functional f is continuous in the $\sigma(A, N)$-weak topology. It is known that the pre-dual space N of a JBW-algebra A can be identified with the space of continuous linear functionals A_* on A.

An important role in the theory of Jordan algebras is played by the operator U_a defined for each $a \in A$ by the formula $U_a x = 2a \circ (a \circ x) - a^2 \circ x$. If A is a JBW-algebra, then U_a has the following properties:

(1) U_a is positive, that is, $U_a A^+ \subset A^+$;

(2) U_a is normal, that is, for each increasing net x_α bounded above in A one has $U_a(\sup x_\alpha) = \sup U_a(x_\alpha)$;

(3) if a and b are consistent elements, then $U_{a \circ b} = U_a U_b$.

Definition 2.2. A *trace* on a JBW-algebra is a map $\tau : A^+ \to [0,\ \infty]$ such that

(1) $\tau(a+\lambda b)=\tau(a)+\lambda\tau(b)$ for all $a,\ b\in A^+$ and $\lambda\in\mathbf{R}_+$, provided that $0.(\infty)=0$,
(2) $\tau(U_s a)=\tau(a)$ for all $a\in A^+$ and $s\in A,\ s^2=\mathbb{1}$.

Note that [6] the last condition is equivalent to

$$\tau(x\circ a)\geq 0 \quad \text{for every} \quad a,x\in A^+.$$

A trace τ is said to be *faithful* if $\tau(a)>0$ for all $a\in A^+$, $a\neq 0$; it is normal if for each increasing net x_α in A^+ that is bounded above one has $\tau(\sup x_\alpha)=\sup\tau(x_\alpha)$; it is semifinite if there exists a net $\{b_\alpha\}\subset A^+$ increasing to $\mathbf{1}$ such that $\tau(b_\alpha)<\infty$ for all α, and it is finite if $\tau(\mathbb{1})<\infty$.

Throughout the paper we will consider a JBW-algebra A with a faithful finite normal trace τ. Therefore we omit this condition from the formulation of theorems.

An element $p\in A$ is called an idempotent if $p^2=p$. Let ∇ be the set of idempotents: ∇ forms a logic. For $p\in\nabla$ we set $p^\perp=\mathbf{1}-p$.

The map $\|\cdot\|_1: A\to[0,\ \infty)$ defined by the formula $\|\cdot\|_1=\tau(|a|)$ is a norm (see [6]). The completion of A in the norm $\|\cdot\|_1$ will be denoted by $L^1(A,\tau)$. It is shown [?] that the spaces $L^1(A,\tau)$ and A_* isometrically isomorphic, therefore they can be indentified. Further we will use this fact without noting.

Theorem 2.1.[6] *The space $L^1(A,\tau)$ coincides with the set*

$$L^1=\{x=\int_{-\infty}^{\infty}\lambda de_\lambda\in A:\int_{-\infty}^{\infty}|\lambda|d\tau(e_\lambda)<\infty\}.$$

Moreover,

$$\|x\|_1=\int_{-\infty}^{\infty}|\lambda|d\tau(e_\lambda).$$

Besides, If $x,y\in L^1(A,\tau)$ such that $x\geq 0,y\geq 0$ and $x\circ y=0$ then $\|x+y\|_1=\|x\|_1+\|y\|_1$.

For more information about Jordan algebras we refer a reader to [3, 6].

Let $T: L^1(A,\tau)\to L^1(A,\tau)$ be linear bounded operator. We say that a linear operator T is *positive* is $Tx\geq 0$ whenever $x\geq 0$. A linear operator T is said to be a *contraction* if $\|T\|\leq 1$. A positive operator T is called *stochastic* if $\tau(Tx)=\tau(x)$, $x\geq 0$. It is clear that any stochastic operator is a contraction. For given $y\in L^1(A,\tau)$ and $z\in A$ define a linear operator $T_{u,z}: L^1(A,\tau)\to L^1(A,\tau)$ as follows

$$T_{y,z}(x)=\tau(x\circ z)y.$$

Put $T_y:=T_{y,\mathbb{1}}$.

Recall that a family of contractions $\{T^{m,n}: L^1(A,\tau)\to L^1(A,\tau)\}$ $(m\leq n,\ m,n\in\mathbf{N})$ is called a *nonhomogeneous discrete Markov process (NDMP)* if one satisfies

$$T^{m,n}=T^{k,n}T^{m,k}$$

for every $m\leq k\leq n$. A NDMP $\{T^{m,n}\}$ is called *nonhomogeneous discrete Markov chain (NDMC)*, if each $T^{m,n}$ is a stochastic operator. A NDMP $\{T^{m,n}\}$ is called *uniformly asymptotically stable* or *uniformly ergodic* if there exist an element $y\in L^1(A,\tau)$ such that

$$\lim_{n\to\infty}\|T^{m,n}-T_y\|=0$$

for any $m\geq 0$.

Recall that if for a NDMP $\{T^{k,m}\}$ one has $T^{k,m}=(T^{0,1})^{m-k}$, then such a process becomes *homogeneous*, and therefore, in what follows, by T, where $T:=T^{0,1}$, we denote the homogeneous Markov process.

Journal of Physics: Conference Series **435** (2013) 012002 doi:10.1088/1742-6596/435/1/012002

3. Dobrushin ergodicity coefficient

Let A be a JBW-algebra with faithful normal semifinite trace τ. Let $L^1(A,\tau)$ be a L^1-space. In the sequel by $\|\cdot\|$ we mean the norm $\|\cdot\|_1$. Let $T : L^1(A,\tau) \to L^1(A,\tau)$ be a linear bounded operator. Define

$$X = \{x \in L^1(A,\tau) : \ \tau(x) = 0\},$$

$$\delta(T) = \sup_{x\in X, x\neq 0} \frac{\|Tx\|_1}{\|x\|_1}, \qquad \alpha(T) = \|T\| - \delta(T). \tag{1}$$

The magnitude $\delta(T)$ is called the *Dobrushin ergodicity coefficient* of T.

Remark 3.1. We note that in a commutative case, the notion of the Dobrushin ergodicity coefficient was studied in [11],[12],[32].

We have the following theorem which extends the results of [11],[32],[27].

Theorem 3.1. *Let $T : L^1(A,\tau) \to L^1(A,\tau)$ be a linear bounded operator. Then the following inequality holds*

$$\|Tx\|_1 \le \delta(T)\|x\|_1 + \alpha(T)|\tau(x)| \tag{2}$$

for every $x \in L^1(A,\tau)$.

Note that the proved theorem extends the results of [11],[32],[27]. Now before formulating a main result of this section we need an auxiliary result. Next lemma has been proved in [27].

First denote

$$D = \{x \in L^1(A,\tau) : \ x \ge 0, \|x\|_1 = 1\}.$$

Lemma 3.2. *For every $x, y \in L^1(A,\tau)$ such that $x - y \in X$ there exist $u, v \in D$, such that*

$$x - y = \frac{\|x-y\|_1}{2}(u - v).$$

The next result establishes several properties of the Dobrushin ergodicity coefficient in a noncommutative setting. Note that when M is commutative and τ is finite, similar properties were studied in [24, 17].

Theorem 3.3. *Let $T, S : L^1(A,\tau) \to L^1(A,\tau)$ be stochastic operators. Then the following assertions hold true:*

(i) $0 \le \delta(T) \le 1$;

(ii) $|\delta(T) - \delta(S)| \le \delta(T - S) \le \|T - S\|$;

(iii) $\delta(TS) \le \delta(T)\delta(S)$;

(iv) *if $K : L^1(A,\tau) \to L^1(A,\tau)$ is a linear bounded operator with $\tau(Kx) = 0$ for all $x \in A$, then $\|TK\| \le \|K\|\delta(T)$.*

(v) *one has*

$$\delta(T) = \sup\left\{\frac{\|Tu - Tv\|_1}{2} : \ u, v \in D\right\}. \tag{3}$$

(vi) *if $\delta(T) = 0$, then there is $y \in L^1(A,\tau)$, $y \ge 0$ such that $T = T_y$.*

Proof (i) is obvious. Let us prove (ii). From (1) we immediately find that $\delta(T-S) \le \|T-S\|$. Now let us establish the first inequality. Without loss of generality, we may assume that $\delta(T) \ge \delta(S)$. For an arbitrary $\varepsilon > 0$ from (1) one can find $x_\varepsilon \in X$ with $\|x_\varepsilon\|_1 = 1$ such that

$$\delta(T) \le \|Tx_\varepsilon\|_1 + \varepsilon.$$

Then we have

$$\begin{aligned}\delta(T)-\delta(S) &\le \|Tx_\varepsilon\|_1+\varepsilon-\sup_{x\in X,\|x\|_1=1}\|Sx\|_1\\ &\le \|Tx_\varepsilon\|_1-\|Sx_\varepsilon\|_1+\varepsilon\\ &\le \|(T-S)x_\varepsilon\|_1+\varepsilon\\ &\le \sup_{x\in X,\|x\|_1=1}\|(T-S)x\|_1+\varepsilon\\ &= \delta(T-S)+\varepsilon,\end{aligned}$$

and the arbitrariness of ε implies the assertion.

(iii). Let $x\in X$, then the stochasticity of S implies $\tau(Sx)=0$, hence due to (2) one finds

$$\begin{aligned}\|TSx\|_1 &\le \delta(T)\|Sx\|_1+\alpha(T)|\tau(Sx)|\\ &\le \delta(T)\delta(S)\|x\|_1\end{aligned}$$

which yields $\delta(TS)\le\delta(T)\delta(S)$.

(iv). Using the same argument as (iii) one gets the desired assertion.

(v). For $x\in X$, $x\ne 0$ using Lemma 3.2 we have

$$\begin{aligned}\frac{\|Tx\|_1}{\|x\|_1} &= \frac{\|T(x^+-x^-)\|_1}{\|x^+-x^-\|_1}\\ &= \frac{\frac{\|x^+-x^-\|_1}{2}\|T(u-v)\|_1}{\|x^+-x^-\|_1}\\ &= \frac{\|Tu-Tv\|_1}{2}.\end{aligned}$$

The equality (1) with the last one implies (3).

(vi). Let $\delta(T)=0$, then from (3) one gets $Tu=Tv$ for all $u,v\in D$. Therefore, denote $y:=Tu$. It is clear that $y\in D$. Moreover, $Ty=y$. Let $x\in L^1(A,\tau)$, $x\ge 0$, then noting $\|x\|_1=\tau(x)$ we find

$$Tx=\|x\|_1T\Big(\frac{x}{\|x\|_1}\Big)=\tau(x)y.$$

If $z\in L^1(A,\tau)$, then $z=z_+-z_-$, where $z_+,z_-\ge 0$. Therefore

$$T(z)=T(z_+)-T(z_-)=\tau(z_+)y-\tau(z_-)y=\tau(z)y.$$

This completes the proof.

4. Uniform ergodicity

In this section, as an application of Theorem 3.3 we are going to prove the uniform ergodicity of homogeneous Markov chain.

Theorem 4.1. *Let $\{T^n\}$ be a homogeneous Markov chain generated by a stochastic operator $T: L^1(A,\tau)\to L^1(A,\tau)$. The following assertions are equivalent:*

(i) *there exists $\rho\in[0,1)$ and $n_0\in\mathbf{N}$ such that $\delta(T^{n_0})\le\rho$;*

(ii) *T is uniformly ergodic.*

Proof (i) $\Rightarrow$ (ii). Let $\rho \in [0,1)$ and $n_0 \in \mathbf{N}$ such that $\delta(T^{n_0}) \le \rho$. Now from (iii) and (i) of Theorem 3.3 one gets

$$\delta(T^n) \le \rho^{[n/n_0]} \to 0 \quad \text{as} \quad n \to \infty, \tag{4}$$

where $[a]$ stands for the integer part of a.

Let us show that $\{T^n\}$ is a Cauchy sequence w.r.t. to the norm. Indeed, using (iv) of Theorem 3.3 and (4) we have

$$\|T^n - T^{n+m}\| = \|T^{n-1}(T - T^{m+1})\| \le \delta(T^{n-1})\|T - T^{m+1}\| \to 0 \quad \text{as} \quad n \to \infty. \tag{5}$$

Hence, there is a stochastic operator Q such that $\|T^n - Q\| \to 0$. Let us show that $Q = T_y$, for some $y \in L^1(M,\tau)$. To do so, due to (vi) of Theorem 3.3 it is enough to establish $\delta(Q) = 0$.

So, using (ii) of Theorem 3.3 we have

$$|\delta(T^n) - \delta(Q)| \le \|T^n - Q\|.$$

Now passing to the limit $n \to \infty$ at the last inequality and taking into account (4), we obtain $\delta(Q) = 0$, which is the desired assertion.

(ii)$\Rightarrow$ (i). Let y be the element of $L^1(M_{sa},\tau)$ defined at (ii). Let $\eta \in (0,1/4)$ be given a fixed number. Then (ii) implies that there is a number $n_0 \in \mathbf{N}$ such that $\|T^n - T_y\| < \eta$ for every $n \ge n_0$. Since $Ty = y$ we get that

$$\|T^{n_0}u - T^{n_0}v\|_1 \le \|T^{n_0}u - y\|_1 + \|T^{n_0}v - y\|_1 < 2\eta, \tag{6}$$

for every $u, v \in D$.

Hence, using (v) of Theorem 3.3 (see (3)) we obtain $\delta(T^{n_0}) \le 2\eta$ which yields the desired assertion. The proof is complete.

Remark 4.1. Note that the proved theorem is a non-associative version Bartoszek's result [7]. A similar result has been obtained in [8, 27] without using Dobrushin ergodicity coefficient, when A is a self-adjoint part of von Neumman algebra with a finite trace.

By $\Sigma(A)_{ue}$ we denote the set of all uniformly ergodic stochastic operators. Again using Theorem 3.3 we can prove the following

Theorem 4.2. *The set $\Sigma(A)_{ue}$ is a norm dense and open subset of $\Sigma(M)$.*

5. Weak ergodicity of nonhomogeneous Markov chains

In this section we study weak ergodicity of nonhomogeneous discrete Markov chains defined on $L^1(A,\tau)$.

Let us recall that NDMP $\{T^{k,n}\}$ defined on $L^1(A,\tau)$ is *weakly ergodic* if for every $k \in \mathbf{N} \cup \{0\}$ one has

$$\lim_{n\to\infty} \delta(T^{k,n}) = 0.$$

Theorem 5.1*Let $\{T^{k,n}\}$ be a NDMC defined on $L^1(A,\tau)$. If for each $k \in \mathbf{N} \cup \{0\}$ there exist $\lambda_k \in [0,1]$, a number $n_k \in \mathbf{N}$ such that $\delta(T^{k,k+n_k}) \le \lambda_k$ with*

$$\sum_{j' \ge 0} (1 - \lambda_{j'}) = \infty \tag{7}$$

for every subsequence $\{j'\}$ of $\{j\}_{j\in\mathbf{N}}$. Then the process $\{T^{k,n}\}$ is weak ergodic.

Proof Take any $k \in \mathbf{N} \cup \{0\}$. Then due to the condition of Theorem there exist $\lambda_k \in [0,1]$, a number $n_1 \in \mathbf{N}$ such that $\delta(T^{k,k+n_k}) \leq \lambda_k$. For $\ell_1 := k + n_k$ we again apply the given condition, then one can find λ_{ℓ_1}, n_{ℓ_1} such that $\delta(T^{\ell_1,\ell_1+n_{\ell_1}}) \leq \lambda_{\ell_1}$. Now continuing this procedure one finds sequences $\{\ell_j\}$ and $\{\lambda_{\ell_j}\}$ such that

$$\ell_0 = k,\ \ell_1 = \ell_0 + n_k,\ \ell_2 = \ell_1 + n_{\ell_1}, \dots, \ell_m = \ell_{m-1} + n_{\ell_{m-1}}, \dots$$

and $\delta(T^{\ell_j,\ell_{j+1}}) \leq \lambda_{\ell_j}$.

Now for large enough n one can find M such that

$$M = \max\{j :\ \ell_j + n_j \leq n\}.$$

Then due to (iii) of Theorem 3.3 we get

$$\begin{aligned} \delta(T^{k,n}) &= \delta(T^{n,\ell_M} T^{\ell_{M-1},\ell_M} \cdots T^{\ell_0,\ell_1}) \\ &\leq \prod_{j=0}^{M-1} \delta(T^{\ell_{M-j},\ell_{M-j+1}}) \\ &\leq \prod_{j=0}^{M-1} \lambda_{\ell_j}. \end{aligned}$$

Now taking into account (7), the last inequality implies the weak ergodicity of $\{T^{k,n}\}$.

It is well-known [30] that one of the most significant conditions for weak ergodicity is the Doeblin's Condition. Now we are going to define some non-associative analogous of such a condition.

We say that a NDMP $\{T^{k,n}\}$ defied on $L^1(A,\tau)$ satisfies *condition* $\mathcal{D}$ if there exists $\mu \in D$ and for each k there exist a constant $\lambda_k \in [0,1]$, an integer $n_k \in \mathbf{N}$, and for every $\varphi \in D$, one can find $\sigma_{k,\varphi} \in L^1(A_+,\tau)$ with $\sup_{\varphi} \|\sigma_{k,\varphi}\|_1 \leq \frac{\lambda_k}{4}$ such that

$$T^{k,n_k}\varphi + \sigma_{k,\varphi} \geq \lambda_k \mu, \tag{8}$$

and

$$\sum_{j' \geq 0} \lambda_{j'} = \infty \tag{9}$$

for every subsequence $\{j'\}$ of $\{j\}_{j \in \mathbf{N}}$.

Using Theorem 5.1 one can proves the following

Theorem 5.2. *Assume that a NDMC* $\{T^{k,n}\}$ *defined on* $L^1(A,\tau)$ *satisfies eak ergodic if and only if the condition* $\mathcal{D}$ *is satisfied.*

Acknowledgments

The author acknowledges the MOHE grant FRGS11-022-0170 and the Junior Associate scheme of the Abdus Salam International Centre for Theoretical Physics, Trieste, Italy.

References

[1] Accardi L and Fidaleo F 2003 *J. Funct. Anal.* **200** 324
[2] Albeverio S and Høegh-Krohn R 1978 *Comm. Math. Phys.* **64** 83
[3] Alfsen E M, Shultz F W and Stormer E 1978*Adv. in Math.* **128** 11
[4] Ayupov Sh A 1982 *Izv. Akad. Nauk UzSSR Ser. Fiz.-Mat.* **5** 7 (Russian)

[5] Ayupov Sh A 1981 *Russian Math. Surveys* **36** 201
[6] Ayupov Sh, Rakhimov A and Usmanov Sh. 1997 *Jordan, Real and Lie Structures in Operator Algebras* (Dordrecht: Kluwer Acad. Publ.)
[7] Bartoszek W 1990 *Anal. Polon. Math.* **52** 165
[8] Bartoszek W and Kuna B 2006 *Colloq. Math.* **105** 311
[9] Bratteli O and Robertson D W 1979 *Operator algebras and quantum statistical mechanics* (Berlin:Springer)
[10] Bożejko M, Kummerer B and Speicher R 1997 *Commun. Math. Phys.* **185** 129
[11] Cohen J E, Iwasa Y, Rautu G, Ruskai M B, Seneta E and Zbaganu G 1993 *Linear Algebra Appl.* **179** 211
[12] Dobrushin R L 1956 *Theor. Probab. Appl.* **1** 65, 329
[13] Dorea C C Y and Pereira A G C 2006 *Acta Math. Hungar.* **110** 287
[14] Emel'yanov E Yu and Wolff M P H 2003 *Positivity* **7** 3
[15] Fagnola F and Rebolledo R 2001 *Jour. Math. Phys.* **42** 1296
[16] Fagnola F and Rebolledo R 2003 *Theory Relat. Fields* **126** 289
[17] Ipsen I C F and Salee T M 2011 *SIAM J. Matrix Anal. Appl.* **32** 153
[18] Jajte R 1984 *Strong linit theorems in non-commutative probability* (Berlin: Springer)
[19] Johnson J and Isaacson D 1988 *J. Appl. Probab.* **25** 34
[20] Karimov A K and Mukhamedov F M 2003 *Sbornik Math.* **194** 237
[21] Karimov A and Mukhamedov F 2010 *Probab. Math. Statist.* **30** 153
[22] Komorowski T and Tyrcha J 1989 *Bull. Polish Acad. Sci. Math.* **37** 220
[23] Krengel U 1985 *Ergodic Theorems* (Berlin: Walter de Gruyter)
[24] Madsen R W and Isaacson D L 1973 *Ann. Probab.* **1** 329
[25] Mukhamedov F M 2004 *Izvestiya Math.* **68** 1009
[26] Mukhamedov F M (in press) *Rev. Mat. Compult.*
[27] Mukhamedov F, Temir S and Akin H 2005 *Siberian Adv. Math.* **15**(3) 28
[28] Nelson E 1974 *J. Funct. Anal.* **15** 103
[29] Parthasarathy K R 1992 *An intoroduction to quantum stochastic calculus* (Basel: Brirkäuser)
[30] Seneta E 1973 *Proc. Cambridge Philos. Soc.* **74** 507
[31] Tan Ch P 1996 *Statis. & Probab. Lett.* **26** 293
[32] Zaharopol R and Zbaganu G 1999 *Jour. Theor. Probab.* **99** 885
[33] Zeifman A I and Isaacson D L 1994 *Stochast. Process. Appl.* **50** 263

Classification of $\xi^{(s)}$-Quadratic Stochastic Operators on 2D simplex

Farrukh Mukhamedov[1], **Mansoor Saburov**[2], **Izzat Qaralleh**[3]

[1,2,3] Department of Computational & Theoretical Sciences, Faculty of Science, International Islamic University Malaysia, Kuantan, Pahang, Malaysia

E-mail: [1]farrukh_m@iium.edu.my, [2]msaburov@gmail.com, [3]izzat@yahoo.com

Abstract. A quadratic stochastic operator (in short QSO) is usually used to present the time evolution of differing species in biology. Some QSO has been studied by Lotka and Volterra. The general problem in the nonlinear operator theory is to study the behavior of operators. This problem was not fully finished even for the quadratic stochastic operators. To study this problem it was investigated several classes of such QSO. In this paper we study $\xi^{(s)}$-QSO class of operators. We study such kind of operators on 2D simplex. We first classify these $\xi^{(s)}$-QSO into 20 classes. Further, we investigate the dynamics of one class of such operators.

1. Introduction

A quadratic stochastic operator (in short QSO) is usually used to present the time evolution of differing species in biology (see [7, 6]). Note that, in general, such operators are defined on $n-1$-dimensional simplex of $\mathbb{R}^n$. Some QSO has been studied by Lotka and Volterra [11]. Many natural phenomena are being modeled by Lotka-Volterra (in short LV) quadratic systems. The investigation of dynamical properties and modeling in various fields running from economy to population dynamics have been using the LV system to be the source of analysis. The fascinating applications of quadratic stochastic operators to population genetics were given in the book [7]. It describes a distribution of species in the next generation if the distribution of these species of the current generation is given. In [1], it was given along self-contained exposition of the recent achievements and open problems in the theory of quadratic stochastic operators. A general problem in the nonlinear operator theory is to study the behavior of operators. This problem was not fully finished even for the quadratic stochastic operators (for the simplest nonlinear operators). The asymptotical behavior of the QSO (even) on small dimensional simplex is complicated (see [5]). In order to solve this problem, many researchers always introduced a certain class of quadratic operators and studied their behaviors: Volterra QSO [2, 10, 12], ℓ-Volterra QSO [9], Non-Volterra QSO, Strictly non-Volterra QSO [13], F-QSO, Separable QSO, Quadratic doubly stochastic operators and so on. For more information, one may refer to [1]. However, all these class of operators together would not cover all QSO. Therefore, there are many classes of QSO which were not studied yet.

In this paper we are going to study $\xi^{(s)}$-QSO class of operators which has been introduced in [8]. This class of operators depend on a partition of the index set. In case of two dimensional simplex, the index set has five possible partitions. In [8] it has been investigated $\xi^{(s)}$-QSO related to the point partition. In the present paper, we are going to describe and classify such operators generated by other three partitions. Further, we also investigate the dynamics of one class of such operators.

2. Preliminaries

Recall that a quadratic stochastic operator (QSO) is a mapping of the simplex

$$S^{n-1} = \{x = (x_1, x_2, ..., x_m)\} \in \mathbb{R}^m : \ x_i \geq 0, \ \sum_{i=1}^{m} x_i = 1 \tag{2.1}$$

into itself, of the form

$$V : x_k = \sum_{i,j=1}^{m} P_{i,j,k} x_i x_j, \tag{2.2}$$

where $P_{i,j,k}$ are coefficient of heredity, which satisfy the following conditions

$$P_{i,j,k} \geq 0, \ \ P_{i,j,k} = P_{j,i,k}, \ \ \sum_{i,j=1}^{m} P_{i,j,k} = 1. \tag{2.3}$$

Thus, each quadratic stochastic operator V can be uniquely defined by a cubic matrix $P = \left(P_{i,j,k}\right)_{i,j,k=1}^{m}$ with conditions (2.3). For a given $x^{(0)} \in S^{m-1}$ the trajectory $\{x^{(n)}\}$, $n = 0, 1, 2, ...$ of $x^{(0)}$ under the action of QSO V is defined by $x^{(n+1)} = V(x^{(n)})$, where $n = 0, 1, 2, ...$ By $W(x^{(0)})$ we denote the set of the limiting points of the trajectory. Since $x^{(n)} \subset S^{m-1}$ and S^{m-1} is compact, it follows that $W(x^{(0)}) \neq \emptyset$. Obviously, if $W(x^{(0)})$ consists of a single point, then the trajectory converges, and $W(x^{(0)})$ is fixed point of QSO.

In order to introduce a new class QSO, we need some auxiliary notations.

Note that each element $x \in S^{m-1}$ is a probability distribution on $I = \{1, ..., m\}$. Therefore, take two $x = (x_1, ..., x_m)$ and $y = (y_1, ..., y_m)$ from S^{m-1}. We say that x *is equivalent to* y if $x_k = 0 \Leftrightarrow y_k = 0$.

Let us denote $supp(x) = \{i : x_i \neq 0\}$. We say that x *is singular to* y $(x \perp y)$ if $supp(x) \cap supp(x) = \phi$. Note that if $x, y \in S^{m-1}$ then $x \perp y$ if and only if $(x, y) = 0$, here $(\cdot, \cdot)$ stands usual inner product in $\mathbb{R}^m$.

By $\mathbf{P}$ we denote the set of ordered pairs of I, i.e, $\mathbf{P} = \{(i,j) : \ i, j \in I, \ i < j\}$. Let $\xi = \{A_i\}_{i=1}^{N}$ be a partition of $\mathbf{P}$, i.e. $A_i \bigcap A_j = \emptyset, \bigcup_{i=1}^{N} A_i = \mathbf{P}$.

Definition 2.1. A quadratic stochastic operator V given by (2.2),(2.3) is called $\xi^{(s)}$*-QSO* if the following conditions are satisfied:

(i) for each A_k and every $(i,j), (u,v) \in A_k$ one has

$$(P_{i,j,1}, P_{i,j,2}, ..., P_{i,j,m}) \sim (P_{u,v,1}, P_{u,v,2}, ..., P_{u,v,m});$$

(ii) for every $(i,j) \in A_k$, $(u,v) \in A_l$, $(k \neq l)$ one has

$$(P_{i,j,1}, P_{i,j,2}, ..., P_{i,j,m}) \perp (P_{u,v,1}, P_{u,v,2}, ..., P_{u,v,m});$$

(iii) for every $i, j \in I$ one has

$$(P_{i,j,1}, P_{i,j,2}, ..., P_{i,j,m}) \perp (P_{u,v,1}, P_{u,v,2}, ..., P_{u,v,m}).$$

Journal of Physics: Conference Series **435** (2013) 012003 doi:10.1088/1742-6596/435/1/012003

3. Classification of $\xi^{(s)}$-QSO on 2Dsimplex

In this section we are going to study $\xi^{(s)}$-QSO in two dimensional simplex, i.e. $m = 3$. In this case, we have the following possible partitions

$$\begin{aligned}
\xi_1 &= \{\{(1,2)\},\{(1,3)\},\{(2,3)\}\}, \ |\xi_1| = 3,\\
\xi_2 &= \{\{(1,2)\},\{(1,3),(2,3)\}\}, \ |\xi_2| = 2,\\
\xi_3 &= \{\{(1,3)\},\{(1,2),(2,3)\}\}, \ |\xi_3| = 2,\\
\xi_4 &= \{\{(2,3)\},\{(1,2),(1,3)\}\}, \ |\xi_4| = 2,\\
\xi_5 &= \{\{(1,2),(1,3),(2,3)\}, \ |\xi_5| = 1.
\end{aligned}$$

In [8] it has been investigated $\xi^{(s)}$-QSO related to the partition ξ_1. In the present paper we are going to described all $\xi^{(s)}$-QSO related to the partitions ξ_2, ξ_3 and ξ_4. One can calculate that there are 108 parmetrical $\xi^{(s)}$-QSOs related to the said partitions.

Let us recall that two QSO V_1, V_2 are *conjugate*, if one can find a permutation π of $\{1,2,3\}$ such that one has $\pi V_1 \pi^{-1} = V_2$. Note that the $\xi^{(s)}$-QSO related partitions ξ_3 and ξ_4 are conjugate to $\xi^{(s)}$-QSO generated by the partition ξ_2. More exactly, these operators are conjugate by means of the permutation $\pi_1 = \begin{pmatrix} x & y & z \\ y & z & x \end{pmatrix}$. Therefore, below we are listing $\xi^{(s)}$-QSO related to the partition ξ_2. Let us introduce some notations

$$\begin{aligned}
&h(x) = x(1-x), \quad b = 1-a,\\
&g(x,y,z;\lambda;\alpha,\beta,\gamma) = x^2 + 2\lambda x^\alpha y^\beta z^\gamma, \quad \alpha,\beta,\gamma = \{0,1\}.
\end{aligned}$$

Using these notation, we are ready to list the $\xi^{(s)}$-QSOs which are the following ones:

$$V_1 : \begin{cases} x' = x^2 + 2ah(x) \\ y' = y^2 + 2bh(x) \\ z' = g(z,x,y;1;1,0,1) \end{cases} \quad V_2 : \begin{cases} x' = y^2 + 2ah(x) \\ y' = x^2 + 2bh(x) \\ z' = g(z,x,y;1;1,0,1) \end{cases} \quad V_3 : \begin{cases} x' = z^2 + 2ah(x) \\ y' = y^2 + 2bh(x) \\ z' = g(x,y,z;1;0,1,1) \end{cases}$$

$$V_4 : \begin{cases} x' = x^2 + 2ah(x) \\ y' = z^2 + 2bh(x) \\ z' = g(y,x,z;1;1,0,1) \end{cases} \quad V_5 : \begin{cases} x' = y^2 + 2ah(x) \\ y' = z^2 + 2bh(x) \\ z' = g(x,y,z;1;0,1,1) \end{cases} \quad V_6 : \begin{cases} x' = z^2 + 2ah(x) \\ y' = x^2 + 2bh(x) \\ z' = g(y,x,z;1;1,0,1) \end{cases}$$

$$V_7 : \begin{cases} x' = g(x,y,z;1;0,1,1) \\ y' = y^2 + 2ah(x) \\ z' = z^2 + 2bh(x) \end{cases} \quad V_8 : \begin{cases} x' = g(y,x,z;1;1,0,1) \\ y' = x^2 + 2ah(x) \\ z' = z^2 + 2bh(x) \end{cases} \quad V_9 : \begin{cases} x' = g(z,x,y;1;1,0,1) \\ y' = y^2 + 2ah(x) \\ z' = x^2 + 2bh(x) \end{cases}$$

$$V_{10} : \begin{cases} x' = g(x,y,z;1;0,1,1) \\ y' = z^2 + 2ah(x) \\ z' = y^2 + 2bh(x) \end{cases} \quad V_{11} : \begin{cases} x' = g(y,x,z;1;1,0,1) \\ y' = z^2 + 2ah(x) \\ z' = x^2 + 2bh(x) \end{cases} \quad V_{12} : \begin{cases} x' = g(z,x,y;1;1,0,1) \\ y' = x^2 + 2ah(x) \\ z' = y^2 + 2bh(x) \end{cases}$$

$$V_{13} : \begin{cases} x' = x^2 + 2ah(x) \\ y' = g(y,x,z;1;1,0,1) \\ z' = z^2 + 2bh(x) \end{cases} \quad V_{14} : \begin{cases} x' = y^2 + 2ah(x) \\ y' = g(x,y,z;1;0,1,1) \\ z' = z^2 + 2bh(x) \end{cases} \quad V_{15} : \begin{cases} x' = z^2 + 2ah(x) \\ y' = g(y,x,z;1;1,0,1) \\ z' = x^2 + 2bh(x) \end{cases}$$

$$V_{16}: \begin{cases} x' = x^2 + 2ah(x) \\ y' = g(z,x,y;1;1,0,1) \\ z' = y^2 + 2bh(x) \end{cases} \quad V_{17}: \begin{cases} x' = y^2 + 2ah(x) \\ y' = g(z,x,y;1;1,0,1) \\ z' = x^2 + 2bh(x) \end{cases} \quad V_{18}: \begin{cases} x' = z^2 + 2ah(x) \\ y' = g(x,y,z;1;0,1,1) \\ z' = y^2 + 2bh(x) \end{cases}$$

$$V_{19}: \begin{cases} x' = g(x,y,z;a;0,1,1) \\ y' = g(y,x,z;b;1,0,1) \\ z' = z^2 + 2h(x) \end{cases} \quad V_{20}: \begin{cases} x' = g(y,x,z;a;1,0,1) \\ y' = g(x,y,z;b;0,1,1) \\ z' = z^2 + 2h(x) \end{cases} \quad V_{21}: \begin{cases} x' = g(z,x,y;a;1,0,1) \\ y' = g(y,x,z;b;1,0,1) \\ z' = x^2 + 2h(x) \end{cases}$$

$$V_{22}: \begin{cases} x' = g(x,y,z;a;0,1,1) \\ y' = g(z,x,y;b;1,0,1) \\ z' = y^2 + 2h(x) \end{cases} \quad V_{23}: \begin{cases} x' = g(y,x,z;a;1,0,1) \\ y' = g(z,x,y;b;1,0,1) \\ z' = x^2 + 2h(x) \end{cases} \quad V_{24}: \begin{cases} x' = g(z,x,y;a;1,0,1) \\ y' = g(x,y,z;b;0,1,1) \\ z' = y^2 + 2h(x) \end{cases}$$

$$V_{25}: \begin{cases} x' = x^2 + 2h(x) \\ y' = g(y,x,z;a;1,0,1) \\ z' = g(z,x,y;b;1,0,1) \end{cases} \quad V_{26}: \begin{cases} x' = y^2 + 2h(x) \\ y' = g(x,y,z;a;0,1,1) \\ z' = g(z,x,y;b;1,0,1) \end{cases} \quad V_{27}: \begin{cases} x' = z^2 + 2h(x) \\ y' = g(y,x,z;a;1,0,1) \\ z' = g(x,y,z;b;0,1,1) \end{cases}$$

$$V_{28}: \begin{cases} x' = x^2 + 2h(x) \\ y' = g(z,x,y;a;1,0,1) \\ z' = g(y,x,z;b;1,0,1) \end{cases} \quad V_{29}: \begin{cases} x' = y^2 + 2h(x) \\ y' = g(z,x,y;a;1,0,1) \\ z' = g(x,y,z;b;0,1,1) \end{cases} \quad V_{30}: \begin{cases} x' = z^2 + 2h(x) \\ y' = g(x,y,z;a;0,1,1) \\ z' = g(y,x,z;b;1,0,1) \end{cases}$$

$$V_{31}: \begin{cases} x' = g(x,y,z;a;0,1,1) \\ y' = y^2 + 2h(x) \\ z' = g(z,x,y;b;1,0,1) \end{cases} \quad V_{32}: \begin{cases} x' = g(y,x,z;a;1,0,1) \\ y' = x^2 + 2h(x) \\ z' = g(z,x,y;b;1,0,1) \end{cases} \quad V_{33}: \begin{cases} x' = g(z,x,y;a;1,0,1) \\ y' = y^2 + 2h(x) \\ z' = g(x,y,z;b;0,1,1) \end{cases}$$

$$V_{34}: \begin{cases} x' = g(x,y,z;a;0,1,1) \\ y' = z^2 + 2h(x) \\ z' = g(y,x,z;b;1,0,1) \end{cases} \quad V_{35}: \begin{cases} x' = g(y,x,z;a;1,0,1) \\ y' = z^2 + 2h(x) \\ z' = g(x,y,z;b;0,1,1) \end{cases} \quad V_{36}: \begin{cases} x' = g(z,x,y;a;1,0,1) \\ y' = x^2 + 2h(x) \\ z' = g(y,x,z;b;1,0,1) \end{cases}$$

Theorem 3.1 *All $\xi^{(s)}$-QSO defined with respect to the partition ξ_2 on 2D simplex are classified into 20 non conjugate classes and they are listed below*

$$\begin{array}{llll} K_1 = \{V_1, V_{13}\}, & K_2 = \{V_2, V_{15}\}, & K_3 = \{V_3, V_{14}\}, & K_4 = \{V_4, V_{16}\} \\ K_5 = \{V_5, V_{18}\}, & K_6 = \{V_6, V_{17}\}, & K_7 = \{V_7\}, & K_8 = \{V_8, V_9\} \\ K_9 = \{V_{10}\}, & K_{10} = \{V_{11},\ V_{12}\}, & K_{11} = \{V_{19}, V_{31}\}, & K_{12} = \{V_{20}, V_{33}\} \\ K_{13} = \{V_{21}, V_{32}\}, & K_{14} = \{V_{22}, V_{34}\}, & K_{15} = \{V_{23}, V_{36}\}, & K_{16} = \{V_{24}, V_{35}\} \\ K_{17} = \{V_{25}\}, & K_{18} = \{V_{26}, V_{27}\}, & K_{19} = \{V_{28}\}, & K_{20} = \{V_{29}, V_{30}\} \end{array}$$

Proof. We are going to classify the above given operators with respect to permutation $\pi_3 = \begin{pmatrix} x & y & z \\ x & z & y \end{pmatrix}$. One can see that

$$\begin{aligned} \pi_3 \circ V_1 \circ \pi_3^{-1} &= \pi_3(x^2 + 2ax(1-x), z^2 + 2(1-a)x(1-x), y(1-x+z)) \\ &= (x^2 + 2ax(1-x), y(1-x+z), z^2 + 2(1-a)x(1-x)) \\ &= V_{13} \end{aligned}$$

From $\pi_3 = \pi_3^{-1}$, we infer that V_1, V_{13} are in the same class, which is denoted by K_1. The other operators can be classified by the same way, so one gets the above given list. This completes the proof.

4. Dynamics of $\xi^{(s)}$-QSO from the class K_{17}

In this section, for the sake of simplicity, we are going to study dynamics of $\xi^{(s)}$ -QSO taken from the class K_{17}. Namely, we consider V_{25} which is given by

$$V_{25}: \begin{cases} x' = x^2 + 2x(1-x) \\ y' = z^2 + 2ayx \\ z' = y^2 + 2(1-a)yz \end{cases} \tag{4.1}$$

where $0 < a < 1$. In the sequel, by e_1, e_2, e_3 we denote the vertices of the simplex S^2. i.e. $e_1 = (1,0,0)$, $e_2 = (0,1,0)$, $e_3 = (0,0,1)$.

In what follows, for the sake of simplicity, by $\mathbf{V}_a$ we denote V_{25}.

Theorem 4.1 *Let $\mathbf{V}_a : S^2 \to S^2$ be a $\xi^{(s)}$-QSO given by (4.1) Then the following statements hold true:*

(i) if $a \neq \frac{1}{2}$ then

1. $Fix(\mathbf{V}_a) = \left\{ e_1, \left(0, \frac{3-2a-\sqrt{4+(2a-1)^2}}{2(1-2a)}, \frac{-1-2a+\sqrt{4+(2a-1)^2}}{2(1-2a)}\right)\right\}$
2. $Per((\mathbf{V}_a) = \{e_2, e_3\}$

(ii) if $a = \frac{1}{2}$ then

1. $Fix(\mathbf{V}_a) = \left\{\left(0, \frac{1}{2}, \frac{1}{2}\right)\right\}$.
2. $Per(\mathbf{V}_a) = \{x = 0\}$

(iii) Let $x^{(0)} = (x, y, z) \in S^2$ be an initial point with $x = 0$, then we have three cases:

1. *if $0 < a < \frac{1}{2}$, then $W(x^{(0)}) = \{e_3\}$*
2. *if $\frac{1}{2} < a < 1$, then $W(x^{(0)}) = \{e_2\}$*
3. *if $a = \frac{1}{2} <$, then $W(x^{(0)}) = \{1 - x^{(0)}\}$*

(iv) Let $x^{(0)} = (x, y, z) \in S^2$ be an initial point with $x \neq 0$, then $W(x^{(0)}) = \{e_1\}$

Proof. (i) Assume that $a \neq 1/2$. To find fixed points of $\mathbf{V}_a$, we need to solve the equation, $Vx = x$, namely

$$\begin{aligned} &x^2 + 2x(1-x) = x \\ &z^2 + 2ayz = y \\ &y^2 + 2(1-a)yz = z. \end{aligned}$$

Solutions of the first equation $x^2 + 2x(1-x) = x$ are $x = 0, x = 1$. One can easily check that if $x = 1$, we have $z = y = 0$, and if $x = 0$ then $z + y = 1$. So, we have to solve the

$$(1-y)^2 + 2ay(1-y) = y. \tag{4.2}$$

One can find that solutions of (4.2) are $y = \frac{3-2a\pm\sqrt{4+(2a-1)^2}}{2(1-2a)}$ when $a \neq \frac{1}{2}$. We can verify that the only solution $y = \frac{3-2a-\sqrt{4+(2a-1)^2}}{2(1-2a)}$ is positive. Therefore, we have $z = \frac{-1+2a+\sqrt{4+(2a-1)^2}}{2(1-2a)}$. Hence,

$$Fix(\mathbf{V}_a) = \left\{e_1, \left(0, \frac{3-2a-\sqrt{4+(2a-1)^2}}{2(1-2a)}, \frac{-1-2a+\sqrt{4+(2a-1)^2}}{2(1-2a)}\right)\right\}, \ a \neq \frac{1}{2}.$$

In order to find 2-periodic points of $\mathbf{V}_a$, we have to solve the following equation: $\mathbf{V}_a^2(x) = x$. Let us consider first component of the last equation:

$$(x^2 + 2x(1-x))^2 + 2(x^2 + 2x(1-x))(1 - (x^2 + 2x(1-x)) = x$$

One can find solutions of this equation are $\left\{0, 1, \frac{3\pm\sqrt{3}i}{2}\right\}$. So, a possibility of appearance of the periodic point is $x = 0$ or $x = 1$. Let us find the corresponding value of y when $x = 0$. To do that, we have to solve the following one

$$(y^2 + 2(1-a)y(1-y))^2 + 2a(1-y)^2 + 2ay(1-y)(y^2 + 2(1-a)y(1-y)) = y \qquad (4.3)$$

This equation has solutions $\left\{0, 1, \frac{3-2a\pm\sqrt{4+(2a-1)^2}}{2(1-2a)}\right\}$. So, the periodic points are $(0,0,1), (0,1,0)$. Therefore, we have $Per(\mathbf{V}_a) = \{e_2, e_3\}$, when $a \neq \frac{1}{2}$.

(ii) Now, let us consider the case when $a = \frac{1}{2}$. Then (4.2) has a solution $y = \frac{1}{2}$. This means that $z = \frac{1}{2}$. Therefore, one has

$$Fix(\mathbf{V}_a) = \left\{\left(0, \frac{1}{2}, \frac{1}{2}\right)\right\}, \ a = \frac{1}{2}.$$

It is clear that the equation (4.3) has infinitely many solutions. Therefore, all points in the line $\{x = 0\}$ are periodic.

(iii) First of all let us consider the case when $a \neq \frac{1}{2}$. One can easily check that the line $\{x = 0\}$ is invariant w.r.t. $\mathbf{V_a}$. We want to study the behavior of the operator over this line. So, assume that $x = 0$. Then $\mathbf{V_a}$ takes the following form:

$$V_a : \begin{cases} x' = 0 \\ y' = z^2 + 2ayz \\ z' = y^2 + 2(1-a)yz \end{cases}$$

Let us consider the function $f_a(y) = (1-y)^2 + 2ay(1-y)$. One can check easily that $f_a(y)$ is decreasing on $[0,1]$, when $a \neq \frac{1}{2}$. We immediately find that $f_a(f_a(y)) > y$, when $y > \frac{3-2a-\sqrt{4+(2a-1)^2}}{2(1-2a)}$, and $f_a(f_a(y)) < y$, when $y < \frac{3-2a-\sqrt{4+(2a-1)^2}}{2(1-2a)}$. Therefore, we have

(a) for any $n \in N$ one has $f_a^{(2n+2)}(y) \geq f_a^{(2n)}(y)$, when $y > \frac{3-2a-\sqrt{4+(2a-1)^2}}{2(1-2a)}$. So, the sequence $f_a^{(2n)}(y)$ is increasing and bounded, moreover $\{f_a^{(2n)}(y)\}$ converges to y^*. One can see that y^* is a fixed point of $f_a(f_a(y))$. The only possibility is $y^* = 1$. Hence, from $f_a^{(2n)}(y) \to 1$, one concludes that $z^{(2n)} \to 0$. Thus, we get $W(x^{(0)}) = (0,1,0)$ if $a \neq \frac{1}{2}$ and $y > \frac{3-2a-\sqrt{4+(2a-1)^2}}{2(1-2a)}$.

(b) Similarly, one has $f_a^{(2n+2)}(y) \leq f_a^{(2n)}(y)$, for any $n \in N$, when $y < \frac{3-2a-\sqrt{4+(2a-1)^2}}{2(1-2a)}$. So, the sequence $f_a^{(2n)}(y)$ is decreasing and bounded, moreover $\{f_a^{(2n)}(y)\}$ converges to y^*. One can see that y^* is a fixed point of $f_a(f_a(y))$. The only possibility is $y^* = 0$. Therefore, one finds $z^{(2n)} \to 1$. So, we get $W(x^{(0)}) = (0, 0, 1)$, if $a \neq \frac{1}{2}$ and $y < \frac{3-2a-\sqrt{4+(2a-1)^2}}{2(1-2a)}$.

Now assume that $a = \frac{1}{2}$, then $\mathbf{V_a}$ has the following form.

$$V_{\frac{1}{2}} : \begin{cases} x' = 0 \\ y' = z^2 + yz \\ z' = y^2 + yz \end{cases}$$

Obviously, one can see that $y' = 1 - y$. This means that all points in the line $x = 0$ are periodic points. So, $W(x^{(0)}) = \{1 - x^{(0)}\}$.

(iv) Assume that $x \neq 0$. Now, let us consider the function $f(x) = x^2 + 2x(1-x)$. One can show that $f^{(n+1)}(x) \geq f^{(n)}(x)$, for any $n \in N$. We find that if $x \geq \frac{1}{2}$, then $x' \geq \frac{1}{2}$. That means the region when $x \geq \frac{1}{2}$ is invariant. Let us study the behavior of $\mathbf{V_a}$ in the $int(S^2)$. In order to do that we shall divide $int(S^2)$ into two subregions. Namely,

$$S_1 = \left\{(x, y, z) : x \geq \frac{1}{2}\right\}, \quad S_2 = \left\{(x, y, z) : x < \frac{1}{2}\right\}.$$

From the above observation, one concludes S_1 is invariant w.r.t. $\mathbf{V_a}$. Now, if $x < \frac{1}{2}$, then there exists n such that $x^{(n)} \geq \frac{1}{2}$. Since $x^{(n+1)} = (x^{(n)})^2 + 2x^{(n)}(1 - x^{(n)}) \geq x^{(n)}, x \in [0, 1]$, and $x^{(n)}$ converge to 1. This means for us it is enough to study the dynamics of $\mathbf{V_a}$ on S_1. Now we are going to show that $W(x^{(0)}) = e_1$, when $x^{(0)} \in S_1$.

Now, we want to show that the Lyapunov function $\varphi(x, y) = x - y$ is increasing when $0 < a < \frac{1}{2}$, $x \geq \frac{1}{2}$. It is sufficient to prove that $x' - y' \geq x - y$, i.e.

$$2x - x^2 - (1 - x - y)^2 - 2ay(1 - x - y) - x + y \geq 0,$$

when $x \geq \frac{1}{2}$. So, we need to prove that the minimum value of the expression in the LHS is greater than or equal zero.

One can calculate that only critical point of this function is $\left(\frac{2a^2+2a}{2a^2+4a-2}, f\left(\frac{2a^2+2a}{2a^2+4a-2}\right)\right)$. We easily can check this point doesn't belong to the simplex, so the minimum value will be in the boundary of the region $\Omega = \left\{(x, y) : x \geq \frac{1}{2}\right\}$. We can find the minimum value is 0, and it is clear in figure 1.

So, the Lyapunov function $\varphi(x, y) = x - y$ is increasing, this means that $x^n - y^n \to \alpha$. Since the sequence $\{x^{(n)}\}$ converges, therefore, one finds that $y^{(n)}$ is also converging. So, $(x^n, y^n, z^n) \to (x^*, y^*, z^*)$ one can see that (x^*, y^*, z^*) is a fixed point, and the only possibility is $(1, 0, 0)$. So, $W(x^{(0)} = e_1$. By the same way, we can also show that the Lyapunov function $\varphi(x, z) = x - z$ is increasing, when $\frac{1}{2} \leq a \leq 1$ and $x \geq \frac{1}{2}$. Consequently, we obtain $W(x^{(0)}) = e_1$.

This completes the proof.

Acknowledgments
The authors acknowledge the MOHE grant FRGS11-022-0170 for the financial support.

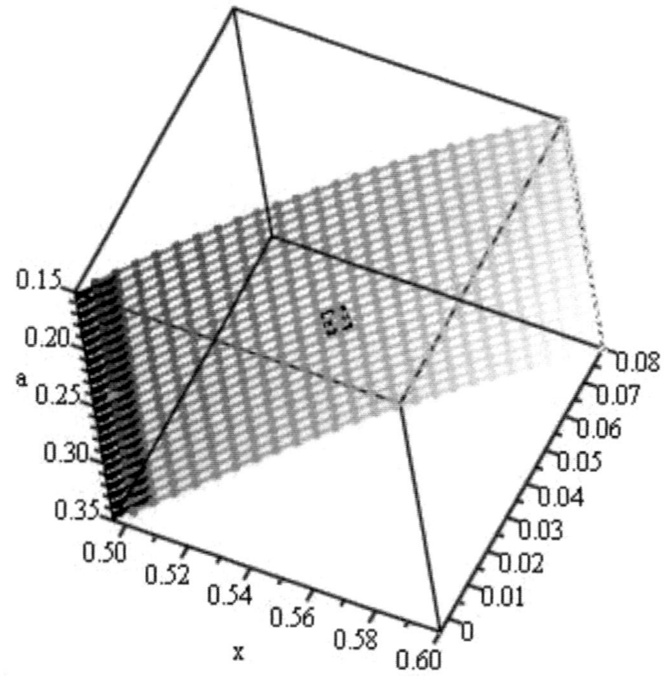

Figure 1.

References

[1] Ganikhodzhaev R N, Mukhamedov F and Rozikov U 2011 *Infin. Dim. Anal. Quant. Prob. Rel. Topics* **14** 279
[2] Ganikhodzhaev R 1993 *Russian Acad. Sci. Sbornik Math.* **76** 489
[3] Ganikhodzhaev R N and Eshmamatova D B 2006 *Vladikavkazskii Matematicheskii Zhurnal.* **8** 12
[4] Ganikhodzhaev N N and Zanin D V 2004 *Russian Mathematical Surveys* **59** 571
[5] Ganikhodzhaev R N and Saburov M K 2008*Journal of Siberian Federal University. Mathematics Physics* **1**188
[6] Kesten H 1970 *Adv. Appl. Probab.* **2** 1–82; 179–228
[7] Lyubich Yu I 1992 *Mathematical structures in population genetics* (Springer-Verlag: Berlin).
[8] Mukhamedov F, Saburov M, and Mohd Jamal A H 2012 *Inter. Jour. of Modr. Phys.: Conf. Series* **9** 299
[9] Rozikov U and A Zada 2010 *International Journal of Biomathematics* **3** 143
[10] Vallander S S 1972 *sov.Math.Dokl* **13** 123
[11] Volterra V 1927 *Association Franc. Lyon* **1926** 96
[12] Zakhrevich Z I 1978 *Russ. Math. Surv* **33** 207
[13] Zhamilov U U and Rozikov U A 2009 *Sbornik: Mathematics* **200** 1339

2012 iCAST: Contemporary Mathematics, Mathematical Physics and their Applications IOP Publishing
Journal of Physics: Conference Series **435** (2013) 012004 doi:10.1088/1742-6596/435/1/012004

Measurable bundles of C^*-algebras over ideals

Inomjon Ganiev[1], **Farrukh Mukhamedov**[2]

[1] Department of Science in Engineering, Faculty of Engineering, International Islamic University Malaysia, Kuala-Lumpur, Malaysia

[2] Department of Computational & Theoretical Sciences, Faculty of Science, International Islamic University Malaysia, Kuantan, Pahang, Malaysia

E-mail: [1]`inam@iium.edu.my`, [2]`farrukh_m@iium.edu.my`

Abstract. Let $(\Omega, \Sigma, \lambda)$ be a measurable space with a finite complete measure λ and $L_0 = L_0(\Omega)$ be an algebra of equivalence classes of complex measurable functions on Ω. In the present paper we define C^*-algebras over ideals of L_0. It is proved that any C^*-algebra over ideal can be represented as measurable bundle of C^*-algebras.

1. Introduction

It is known that the theory of Banach bundles stemming from paper [10], where it is was shown that such a theory had vast applications in analysis. For other applications of the measurable Banach bundles we refer the reader [4], [7], [8]. In this paper we are going to deal with C^*-algebras over a ring of measurable functions L_0. Such algebras are considered as modulus over L_0 and these C^*-algebras are examples of Banach–Kantorovich spaces over L_0. Note that in [5] these kind of algebras were introduced, in particular, it has been shown that any C^*-algebra over L_0 can be represented as measurable bundle of C^*-algebras over the field of complex number $\mathbb{C}$. Further, in [2] it was proved that a vector version of the Gelfand–Naimark's theorem for commutative C^*-algebras over L_0. In [1] a GNS-representation was obtained for noncommutative C^*-algebras over L_0.

In the present paper we are going to investigate a more general algebras, namely, we will define C^*-algebras over ideals of L_0. For such kind of algebra, we are going to prove that any C^*-algebra over ideal E can be represented as measurable bundle of C^*-algebras.

2. Preliminaries

In this section, we recall some notions which will be used in the paper. Let $(\Omega, \Sigma, \lambda)$ be a measurable space with a finite complete measure λ and $L_0 = L_0(\Omega)$ be an algebra of equivalence classes of complex measurable functions on Ω. Given functions $f, g \in L_0$, we set $f \leq g$ if $f(\omega) \leq g(\omega)$ almost everywhere. It is clear that L_0 becomes an ordered set in which two arbitrary functions $f, g \in L_0$ have a supremum $f \vee g$ and an infimum $f \wedge g$ in L_0 defined by the formulas

$$f \vee g = \max(f(\omega), g(\omega)), \quad f \wedge g = \min(f(\omega), g(\omega)).$$

A linear subspace E in L_0 is called *an ideal space* if for any $x \in L_0, y \in E$ with $|x| \leq |y|$ one gets $x \in E$, i.e. for any function taken from E, the set E contains its modulus and each function with smaller modulus. Denote by E^+ the cone of positive elements or the positive cone of an ideal space E, i.e. $E^+ = \{x \in E : \ x \geq 0\}$.

A sequence b_n is said to be *order convergent in* E (or (o)*-convergent in* E) to b if there is a sequence a_n in E satisfying an $a_n \downarrow 0$ and $|b_n - b| \leq a_n$ for all n, and written $b_n \stackrel{(o)}{\rightarrow} b$ or $b = (o) - \lim\limits_n b_n$.

In what follows, we will consider the ideal space E to be a unital algebra, i.e. E is an ideal subalgebra of L_0 with unit. Note that basic examples such kind of ideals are L_0, L^∞ and Arens algebrs L^ω.

A complex linear space $\mathcal{U}$ is said to be *normed* by E if there is a mapping $||\cdot|| : \mathcal{U} \to E$ such that the following conditions are satisfied: $||x|| \geq 0$; $||x|| = 0 \Leftrightarrow x = 0$; $||\lambda x|| = |\lambda|||x||$; $||x + y|| \leq ||x|| + ||y||$ for all $x, y \in \mathcal{U}, \lambda \in \mathbb{C}$. Then a pair $(\mathcal{U}, ||\cdot||)$ is said to be a *lattice normed* space over E. A lattice normed space is said to be *d-decomposable* if for every $x \in \mathcal{U}$ and the decomposition $||x|| = e_1 + e_2$ with disjoint positive elements e_1 and e_2 in E there exist $x_1, x_2 \in \mathcal{U}$ such that $x = x_1 + x_2$ with $||x_1|| = e_1, ||x_2|| = e_2$. A net $\{x_\alpha\} \subset \mathcal{U}$ is said to be (bo)-convergent to $x \in \mathcal{U}$ (in this case we write $x = (bo) - \lim x_\alpha$), if the net $\{||x_\alpha - x||\}$ (o)-converges to zero in E (in this case we write $(o) - \lim ||x_\alpha - x|| = 0$). A net $\{x_\alpha\}_{\alpha \in A}$ is called (bo)-fundamental if $\{x_\alpha - x_\beta\}_{(\alpha,\beta) \in A \times A}$ (bo)-converges to zero. A lattice normed space is called (bo)-complete if every (bo)-fundamental net (bo)-converges in it. A lattice normed space over E is called *Banach–Kantorovich space over* E if it is (bo)-complete and d-decomposable. It is well known [9] that every Banach–Kantorovich space $\mathcal{U}$ over E admits a E-module structure such that $||\lambda x|| = |\lambda|||x||$ for every $x \in \mathcal{U}, \lambda \in E$.

Let $\mathcal{U}$ be an arbitrary complex algebra, and at the same time, $\mathcal{U}$ be a module over E such that $(\lambda u)v = \lambda(uv) = u(\lambda v)$ for all $u, v \in \mathcal{U}, \lambda \in L_0$. We will consider in $\mathcal{U}$ some E-valued norm $||\cdot||$, which admits a E-module structure Banach–Kantorovich space in $\mathcal{U}$, in particular $||\lambda u|| = |\lambda|||u||$ for all $u \in \mathcal{U}, \lambda \in E$.

Definition 2.1 [1]. The algebra $\mathcal{U}$ is called a *Banach–Kantorovich algebra over* E if $||u \cdot v|| \leq ||u||||v||$ for all $u, v \in \mathcal{U}$.

If $\mathcal{U}$ is a Banach–Kantorovich algebra over E with a unit e such that $||e|| = \mathbf{1}$ then $\mathcal{U}$ is called an *unital Banach–Kantorovich algebra over* E, where $\mathbf{1}$ is the unit element in E.

Let $\mathcal{U}$ be an $*$-algebra and Banach–Kantorovich algebra over E such that $(\lambda u)^* = \overline{\lambda} u^*$ for all $u \in \mathcal{U}, \lambda \in L_0$. Then such an algebra (see [5]) is called a C^**-algebra over* E if $||u||^2 = ||u^* u||$ for all $u \in \mathcal{U}$.

Let X be a mapping, which sends every point $\omega \in \Omega$ to some C^*-algebra $(X(\omega), ||\cdot||_{X(\omega)})$. In what follows, we assume that $X(\omega) \neq \{0\}$ for all $\omega \in \Omega$. A function u is said to be a section of X, if it is defined almost everywhere in Ω and takes its value $u(\omega) \in X(\omega)$ for $\omega \in dom(u)$, where $\omega \in dom(u)$ is the domain of u.

Let L be some set of sections.

Definition 2.2 [5]. A pair (X, L) is said to be a *measurable bundle of* C^**-algebras* over Ω if

1. $\lambda_1 c_1 + \lambda_2 c_2 \in L$ for all $\lambda_1, \lambda_2 \in \mathbb{C}$ and $c_1, c_2 \in L$, where $\lambda_1 c_1 + \lambda_2 c_2 : \omega \in dom(c_1) \cap dom(c_2) \to \lambda_1 c_1(\omega) + \lambda_2 c_2(\omega)$;
2. the function $||c|| : \omega \in dom(c) \to ||c(\omega)||_{X(\omega)}$ is measurable for all $c \in L$;
3. for every $\omega \in \Omega$ the set $\{c(\omega) : c \in L, \omega \in dom(c)\}$ is dense in $X(\omega)$;
4. if $c \in L$ then $c^* \in L$, where $c^* : \omega \in dom(c) \to c(\omega)^*$;
5. if $c_1, c_2 \in L$ then $c_1 c_2 \in L$, where $c_1 c_2 : \omega \in dom(c_1) \cap dom(c_2) \to c_1(\omega) c_2(\omega)$.

A section s is said to be *simple*, if there are $c_i \in L, A_i \in \Sigma, i = \overline{1, n}$ such that $s(\omega) = \sum\limits_{i=1}^{n} \chi_{A_i}(\omega) c_i(\omega)$ for almost all $\omega \in \Omega$.

A section u is called *measurable* if there is a sequence $\{s_n\}$ of simple sections such that $s_n(\omega) \to u(\omega)$ almost everywhere on Ω.

The set of all measurable sections is denoted by $M(\Omega, X)$, and $L_0(\Omega, X)$ denotes the factorization of this set with respect to equality everywhere. We denote by $\hat{u}$ the class from $L_0(\Omega, X)$ containing a section $u \in M(\Omega, X)$, and by $||\hat{u}||$ the element of L_0 containing the function $||u(\omega)||_{X(\omega)}$. For $\hat{u} \in L_0(\Omega, X)$ we put $\hat{u}^* = \widehat{u(\omega)}^*$.

It is known [5] that $L_0(\Omega, X)$ is a C^*-algebra over L_0.

3. Main results

In this section we are going to prove our main results. Let E be an ideal space of L_0, as before, we assume that E is a unital subalgebra of L_0. Let us denote $E(X) = \{\hat{u} \in L_0(\Omega, X) : ||\hat{u}|| \in E\}$.

Theorem 1. *If X is a measurable bundle of C^*-algebras over Ω, then $E(X)$ is C^*-algebra over E.*

Proof. It is know that $E(X)$ is a Banach-Kantorovich space over E (see for example, [7]) Since $X(\omega)$ $(\omega \in \Omega)$ is a $*$-algebra, then $E(X)$ is $*$- algebra. As $X(\omega)$ $(\omega \in \Omega)$ is a Banach algebra, then we have

$$\begin{aligned} ||\hat{u} \cdot \hat{v}|| &= \|\widehat{u(\omega) \cdot v(\omega)}\|_{X(\omega)} \leq \|\widehat{u(\omega)\|_{X(\omega)} \cdot \|v(\omega)}\|_{X(\omega)} \\ &= \|\widehat{u(\omega)}\|_{X(\omega)} \cdot \|\widehat{v(\omega)}\|_{X(\omega)} = \|\hat{u}\| \cdot \|\hat{v}\|. \end{aligned}$$

By the same argument, one finds that $||\hat{u}^*|| = ||\hat{u}||$ and $\|\hat{u}^* \cdot \hat{v}\| = \|\hat{u}\|^2$. Therefore, $E(X)$ is a C^*-algebra over E. This completes the proof.

The set of all bounded measurable functions on Ω is denoted by $\mathcal{L}^\infty(\Omega)$ with the norm $||f||_{\mathcal{L}^\infty(\Omega)} = \sup\limits_{\omega\in\Omega} |f(\omega)|$. Let

$$L^\infty(\Omega) = \{\hat{f} \in L_0 : \exists a > 0, |\hat{f}| \leq a\mathbf{1}\}$$

with the norm $||\hat{f}|| = \inf\{a > 0 : |\hat{f}| \leq a\mathbf{1}\}$.

Denote

$$\mathcal{L}^\infty(\Omega, X) = \{u \in M(\Omega, X) : ||u(\omega)||_{X(\Omega)} \in \mathcal{L}^\infty(\Omega)\},$$

$$L^\infty(\Omega, X) = \{\hat{u} \in L_0(\Omega, X) : ||\hat{u}|| \in L^\infty(\Omega)\}.$$

We endow the defined sets with the following real-valued norms, respectively, by

$$||u||_{\mathcal{L}^\infty(\Omega,X)} = \sup_{\omega\in\Omega} |u(\omega)|_{X(\omega)}, \quad ||\hat{u}||_\infty = ||||\hat{u}||||_{L^\infty(\Omega)}.$$

It is known (see [7], [9]) that there is a homomorphism $p : L^\infty(\Omega) \to \mathcal{L}^\infty(\Omega)$ being a lifting such that

(i) $p(\hat{f}) \in \hat{f}$ and $domp(\hat{f}) = \Omega$;

(ii) $||p(\hat{f})||_{\mathcal{L}^\infty(\Omega)} = ||\hat{f}||_{L^\infty(\Omega)}$.

The homomorphism p is usually called a *lifting* from $L^\infty(\Omega)$ into $\mathcal{L}^\infty(\Omega)$.

Definition 2.3. The map $\ell : L^\infty(\Omega, X) \to \mathcal{L}^\infty(\Omega, X)$ is called a *vector-valued lifting* (associated with p), if for all $\hat{u}, \hat{v} \in L^\infty(\Omega, X)$ and $\lambda \in L^\infty(\Omega)$ the following conditions are valid:

(a) $\ell(\hat{u}) \in \hat{u}$, $dom\ell(\hat{u}) = \Omega$;
(b) $||\ell(\hat{u})(\omega)||_{X(\omega)} = p(||\hat{u}||)(\omega)$;
(c) $\ell(\hat{u} + \hat{v}) = \ell(\hat{u}) + \ell(\hat{v})$;
(d) $\ell(\lambda\hat{u}) = p(\lambda)\ell(\hat{u})$;
(e) $\ell(\hat{u}^*) = \ell(\hat{u})^*$;
(f) $\ell(\hat{u}\hat{v}) = \ell(\hat{u})\ell(\hat{v})$;
(g) for every $\omega \in \Omega$ the set $\{\ell(\hat{u})(\omega) : \hat{u} \in L^\infty(\Omega, X)\}$ is dense in $X(\omega)$.

Now we formulate the main result of the paper, which is a reverse to Theorem 1.

Theorem 2. *For any C^*-algebra $\mathcal{U}$ over E there exists a measurable bundle of C^*-algebras (X, L) with vector-valued lifting such that $\mathcal{U}$ is isometrically $*$-isomorphic with $E(X)$.*

Proof. We put

$$\Gamma = \{u \in \mathcal{U} : \|u\| \in L^\infty(\Omega)\}.$$

It is clear that Γ is an $L^\infty(\Omega)$-modulus and $(bo)-$ dense in $\mathcal{U}$. On the other hand, Γ is a $*$-algebra and one has $\|u * u\| = |\|u\|^2$ for all $u \in \Gamma$.

Define a seminorm α_ω on Γ by $\alpha_\omega(u) = p(\|u\|)(\omega)$ for all $\omega \in \Omega$, where p is the lifting on $L^\infty(\Omega)$.

Let $I_\omega^0 = \{u \in \Gamma : \alpha_\omega(u) = 0\}, \Gamma_\omega = \Gamma/I_\omega^0$, where $\|\cdot\|_\omega$ is a norm on Γ_ω, generated by the seminorm α_ω.

Let $\pi_\omega : \Gamma \to \Gamma_\omega$ be the standard projection from Γ into Γ_ω. Then $\pi_\omega(u \cdot v) = \pi_\omega(u) \cdot \pi_\omega(v)$ and $\pi_\omega(u^*) = \pi_\omega(u)^*$. As $\|\pi_\omega(u)\|_\omega = \alpha_\omega(u)$ for $u \in \Gamma$, then

$$\begin{aligned}\|\pi_\omega(u) \cdot \pi_\omega(v)\|_\omega &= \|\pi_\omega(u \cdot v)\|_\omega = \alpha_\omega(u \cdot v) = p(\|u \cdot v\|)(\omega) \leq p(\|u\| \cdot \|v\|)(\omega) \\ &= p(\|u\|)(\omega) \cdot p(\|v\|)(\omega) = \alpha_\omega(u) \cdot \alpha_\omega(v) = \|\pi_\omega(u)\|_\omega \cdot \|\pi_\omega(v)\|_\omega\end{aligned}$$

for all $\omega \in \Omega$ and $u, v \in \Gamma$. Similarly, one finds

$$\|\pi_\omega(u)^*\|_\omega = \|\pi_\omega(u^*)\|_\omega = \alpha_\omega(u^*) = p(\|u^*\|)(\omega) = p(\|u\|)(\omega) = \alpha_\omega(u) = \|\pi_\omega(u)\|_\omega.$$

Moreover, we have

$$\begin{aligned}\|\pi_\omega(u) \cdot \pi_\omega(u)^*\|_\omega &= \|\pi_\omega(u \cdot u^*)\|_\omega = \alpha_\omega(u \cdot u^*) = p(\|u \cdot u^*\|)(\omega) \\ &= p(\|u\|^2)(\omega) = p(\|u\|)^2(\omega) = \alpha_\omega(u)^2 = \|\pi_\omega(u)\|_\omega^2.\end{aligned}$$

So for each ω, the pair $(\Gamma_\omega, \|\cdot\|_\omega)$ satisfies all the axioms of C^*- algebra, without completeness. Let $X(\omega)$ be the C^*-algebra completion of the $*-$algebra Γ_ω [3].

Let $i_\omega : \Gamma_\omega \to X(\omega)$ be the canonical inclusion. It is known that $i_\omega(x \cdot y) = i_\omega(x) \cdot i_\omega(y)$ and $i_\omega(x^*) = i_\omega(x)^*$ for any $x, y \in \Gamma_\omega$. Therefore $\gamma_\omega = \pi_\omega \circ i_\omega$ is a $*-$homomorphism from Γ into $X(\omega)$.

Now let us define a mapping X which assigns for each $\omega \in \Omega$ the constructed C^*-algebra $X(\omega)$. By L we denote the set of all sections of the form $\omega \in \Omega : \omega \to \gamma_\omega(u)$, where $u \in \Gamma$. It is clear that (X, L) is a measurable bundle of C^* algebras.

Now consider C^*-algebra $E(X)$ over E with E-valued norm $\|\cdot\|$. Let us show that $\mathcal{U}$ is isometrically isomorphic to $E(X)$.

For each $u \in \Gamma$ define $\tau(u) = \widehat{\gamma_\omega(u)}$. Then for $u \in \Gamma$ and $\omega \in \Omega$ one has $\|\gamma_\omega(u)\|_\omega = \alpha_\omega(u) = p(\|u\|)(\omega)$. Hence, $\|\widehat{\gamma_\omega(u)}\|_\omega = \|u\|$, and therefore, τ is an isometry from Γ into $E(X)$. Since

Journal of Physics: Conference Series **435** (2013) 012004 doi:10.1088/1742-6596/435/1/012004

$\tau(\Gamma)$ contains the set of all simple sections then τ is isometry from Γ onto $E(X)$. Moreover, we have

$$\tau(u \cdot v) = \widehat{\gamma_\omega(u \cdot v)} = \widehat{\gamma_\omega(u)\gamma_\omega(v)} = \widehat{\gamma_\omega(u)}\widehat{\gamma_\omega(v)} = \tau(u)\tau(v)$$

and

$$\tau(u^*) = \widehat{\gamma_\omega(u^*)} = \widehat{\gamma_\omega(u)}^* = \tau(u)^*$$

So, τ is an $*$–isometrically homomorphism from Γ onto $E(X)$. The (bo)-completeness of Γ in $\mathfrak{U}$ implies that τ can be extended up to E-modular isometrically isomorphism from $\mathfrak{U}$ onto $E(X)$. Besides, it is clear that τ preserves the multiplication and involution, i.e. τ is an C^*–isomorphism of algebras $\mathfrak{U}$ and $E(X)$.

Now let us establish that (X, L) is a measurable bundle with a vector-valued lifting. First we will prove,that $\Gamma = L^\infty(\Omega, X)$. Due to $E(X) \subset L_0(\Omega, X)$ one gets that $\Gamma \subset L^\infty(\Omega, X)$. Let $u \in L^\infty(\Omega, X)$, then $\|u\| \in L^\infty(\Omega)$. The inclusion $L^\infty(\Omega) \subset E$ implies that $\|u\| \in E$ i.e. $u \in E(X)$ and $\|u\| \in L^\infty(\Omega)$. This yields that $\Gamma \supset L^\infty(\Omega, X)$. Therefore, $\Gamma = L^\infty(\Omega, X)$ (more exactly, Γ identified with $L^\infty(\Omega, X)$ with a help of C^*–isomorphism τ). From the following relations

$$\|\gamma_\omega(u)\|_{X(\omega)} = \|\pi_\omega(u)\|_{X(\omega)} = p(\|u\|)(\omega) \leq \|p(\|u\|)\|_{\mathcal{L}^\infty(\Omega)} = \|\|u\|\|_{L^\infty(\Omega)}$$

for any $u \in \Gamma = L^\infty(\Omega, X)$ and for all $\omega \in \Omega$, we find $\gamma_\omega(u) \in \mathcal{L}^\infty(\Omega, X)$. Now let us define a mapping $\ell : L^\infty(\Omega, X) \to \mathcal{L}^\infty(\Omega, X)$ by $\ell(\widehat{u})(\omega) = \gamma_\omega(\widehat{u})$.

Since Γ is identified with $L^\infty(\Omega, X)$, then an element $\widehat{u} \in \Gamma = L^\infty(\Omega, X)$ is identified with the element $\tau(\widehat{u}) = \widehat{\gamma_\omega(\widehat{u})}$. This means that $\ell(\widehat{u}) \in \widehat{u}$. Since $\gamma_\omega(\widehat{u})$ is defined for all $\omega \in \Omega$, then $dom(\ell(\hat{u})) = \Omega$. By the similar way, from the definition of ℓ it follows that $\|\ell(\hat{u})(\omega)\|_{X(\omega)} = p(\|\hat{u}\|)(\omega)$. It is clear that ℓ is linear, and from

$$\ell(\widehat{u} \cdot \widehat{v}) = \gamma_\omega(\widehat{u} \cdot \widehat{v}) = \gamma_\omega(\widehat{u})\gamma_\omega(\widehat{v}) = \gamma_\omega(\widehat{u})\gamma_\omega(\widehat{v}) = \ell(\widehat{u})\ell(\widehat{v})$$

and

$$\ell(\widehat{u}^*) = \gamma_\omega(\widehat{u}^*) = \gamma_\omega(\widehat{u})^* = \ell(\widehat{u})^*$$

we obtain (e) and (f) axioms of the definition of vector-valued lifting. The axiom (d) can be checked by the same argument.

According to the construction we conclude that $\ell(\widehat{u})(\omega) : \widehat{u} \in L^\infty(\Omega, X)\}$ is dense in $X(\omega)$ for all $\omega \in \Omega$.

Now we show that X is unique. Let X and Y be two measurable bundles of C^*-algebras with vector-valued lifting's ℓ_X and ℓ_Y for which $E(X)$ and $E(Y)$ are C^*–isomorphic onto $\mathfrak{U}$.

Let i be an isometrically isomorphism between $L^\infty(\Omega, X)$ and $L^\infty(\Omega, Y)$. Now we define a linear operator $H_\omega : X(\omega) \to Y(\omega)$ $(\omega \in \Omega)$ by

$$H_\omega(\ell_X(\hat{u})(\omega)) = \ell_Y(i(\hat{u}))(\omega),\ \hat{u} \in E(X).$$

Then for $\hat{u} \in E(X)$ we have

$$\|H_\omega(\ell_X(\hat{u})(\omega))\|_{Y(\omega)} = \|\ell_Y(i(\hat{u}))(\omega)\|_{Y(\omega)} = p(|i(\hat{u})|)(\omega) = p(|\hat{u}|)(\omega) = \|\ell_X(\hat{u})(\omega)\|_{X(\omega)},$$

i.e. H_ω is an isometry. By the same argument with properties of vector-valued lifting one yields that H_ω is a homomorphism and $\{H_\omega(u(\omega)) : u \in M(\Omega, X)\} = M(\Omega, Y)$. Hence, X and Y are isometrically isomorphic. This completes the proof.

Acknowledgments
The second named author (F.M.) acknowledges the MOHE Grant FRGS11-022-0170. He also thanks the Junior Associate scheme of the Abdus Salam International Centre for Theoretical Physics, Trieste, Italy.

References
[1] Chilin V I, Ganiev I G and Kudaybergenov K K 2007 *Vladikavkaz. Math. J.* **9** 33
[2] Chilin V I, Ganiev I G and Kudaybergenov K K 2008 *Russian Math.* **52** 58
[3] Dixmer J 1969 *Les algebres d' operateurs dand l' espase Hilbertian algebres de von Neumann*, (Paris).
[4] Ganiev I G 2006. In book: *Studies on Functional Analysis and its Applications*, (Moscow: Nauka) p. 9. (Russian).
[5] Ganiev I G and Chilin V I 2003 *Vladikavkaz. Math. J.* **5** 35
[6] Ganiev I G and Kudaybergenov K K 2004 *Vladikavkaz. Math. J.* **6** 21
[7] Gutman A E 1993 *Siberian Adv. Math.* **3** 8
[8] Kusraev,A G 2010 *Positivity* **12** 785
[9] Kusraev A G 2000. *Dominated operators* (Dordrecht: Kluwer Academic Publishers).
[10] Murray A and von Neumann J 1940. *Ann. Math.* **41** 94

On Non-Ergodic Transformations on S^3

Nasir N. Ganikhodjaev[1], **Uygun U. Jamilov**[1,2] **and Ramazon T. Mukhitdinov**[3]

[1]Department of Computational and Theoretical Sciences, Faculty of Science, IIUM, 25200 Kuantan, MALAYSIA.

[2]Institute of Mathematics,100125, Tashkent, UZBEKISTAN.

[3]Bukhara Engineering-Technical Institute of High Technologies, 105017 Bukhara, UZBEKISTAN.

E-mail: nasirgani@hotmail.com, jamilovu@yandex.ru, ramazon-mukhitdinov@rambler.ru

Abstract. In this paper we consider a set of all extremal Volterra quadratic stochastic operators on three dimensional simplex S^3, show that this set is parted into four equivalence classes with respect to group of transformations generated by permutations and describe the behaviour of trajectories extremal Volterra quadratic stochastic operators for each class. It is proved that the operators in some classes are non-ergodic transformation.

1. Introduction

Quadratic stochastic operators were first introduced by Bernstein [1]. Such operators frequently arises in many models of mathematical genetics, namely, theory of heredity [1],[7]-[11]. Consider a biological population, that is a community of organisms closed with respect to reproduction. Assume that each individual in this population belongs to precisely one species (genotype) $1, \cdots, m$. The scale of species is such that the species of the parents i and j unambiguously determines the probability of every species k for the first generation of direct descendants. Denote this probability, that is to be called the heredity coefficient, by $p_{ij,k}$. It is then obvious that $p_{ij,k} \geq 0$ for all i, j, k and that

$$\sum_{k=1}^{m} p_{ij,k} = 1 \quad (i, j, k = 1, \cdots, m).$$

The state of the population can be described by the $m-$tuple $(x_1, x_2, \cdots, x_m)$ of species probabilities, where x_k is the fraction of the species k in the total population. In the case of panmixia (random interbreeding) the parent pairs i and j arise for a fixed state $\mathbf{x} = (x_1, x_2, \cdots, x_m)$ with probability $x_i x_j$. Hence the total probability of the species k in the first generation of direct descendants is defined by

$$\sum_{i,j=1}^{m} p_{ij,k} x_i x_j, \quad (k = 1, \cdots, m)$$

Let

$$S^{m-1} = \{\mathbf{x} = (x_1, x_2, \cdots, x_m) \in R^m : \text{ for any } i \ x_i \geq 0, \text{ and } \sum_{i=1}^{m} x_i = 1\} \tag{1}$$

be the $(m-1)$-dimensional simplex. A map V of S^{m-1} into itself is called a quadratic stochastic operator (qso) if

$$(V\mathbf{x})_k = \sum_{i,j=1}^{m} p_{ij,k} x_i x_j \tag{2}$$

for any $\mathbf{x} \in S^{m-1}$ and for all $k = 1, \cdots, m$, where

$$a) p_{ij,k} \geq 0, \quad b) p_{ij,k} = p_{ji,k} \text{ for all } i,j,k; \quad c) \sum_{k=1}^{m} p_{ij,k} = 1.$$

Assume $\{\mathbf{x}^{(n)} \in S^{m-1} : n = 0,1,2,\cdots\}$ is the trajectory of the initial point $\mathbf{x} \in S^{m-1}$, where $\mathbf{x}^{(n+1)} = V(\mathbf{x}^{(n)})$ for all $n = 0,1,2,\cdots$, with $\mathbf{x}^{(0)} = \mathbf{x}$.

Definition 1.1 A point $\mathbf{a} \in S^{m-1}$ is called a fixed point of a qso V if $V(\mathbf{a}) = \mathbf{a}$.

Definition 1.2 A qso V is called regular if for any initial point $\mathbf{x} \in S^{m-1}$ the limit

$$\lim_{n\to\infty} V(\mathbf{x}^{(n)})$$

exists.

Note that the limit point be a fixed point of a qso V. Thus the fixed points of qso describe limit or long run behavior the trajectories of any initial point. Limit behavior of trajectories and fixed points of qso play important role in many applied problems [4],[5],[8],[9], [10]. The biological treatment of the regularity of qso is rather clear: in long run the distribution of species in next generation coincide with distribution of species in previous one, i.e., stable.

Definition 1.3 A qso V is said to be ergodic if the limit

$$\lim_{n\to\infty} \frac{1}{k} \sum_{n=0}^{k-1} V^n(x) \tag{3}$$

exists for any $\mathbf{x} \in S^{m-1}$.

On the basis of numerical calculations Ulam conjectured [12] that the ergodic theorem holds for any qso V. In 1977 Zakharevich [13] proved that this conjecture is false in general. Later in [3] was established necessary and sufficient condition to be non-ergodic transformation for qso defined on S^2. The biological treatment of non-ergodicity of qso V is the following: in long run the behavior of distributions of species is chaotic, i.e., unpredictable.

Note that a regular qso V is ergodic, but in generally from ergodicity does not follow regularity. Let π be a permutation of a set $\{1,2,\cdots,m\}$. Then one can define one-to-one transformation $T_\pi : S^{m-1} \to S^{m-1}$ as

$$T_\pi(x_1, x_2, \cdots, x_m) = (x_{\pi(1)}, x_{\pi(2)}, \cdots, x_{\pi(m)}) \tag{4}$$

It is evident that a transformation $T_\pi^{-1} V T_\pi$ is qso and limit behavior of trajectories of qso $T_\pi^{-1} V T_\pi$ coincide with limit behavior of trajectories of qso V [6]. Thus for any permutation π a qso $T_\pi^{-1} V T_\pi$ be regular (resp. ergodic) if and only if a qso V is regular (resp. ergodic).

The following notations will be used throughout this paper [8]. We let ∂S^{m-1} denote the boundary of S^{m-1}:

$$\partial S^{m-1} = \{\mathbf{x} \in S^{m-1} : x_i = 0 \text{ for at least one } i \in \{1,2,\cdots,m\}\},$$

and also for $k = 1,2,\cdots,m$:
$\mathbf{e}^k = (e_1^k, e_2^k, \cdots, e_m^k)$ be kth vertex of the simplex S^{m-1}, where $e_k^k = 1$ and $e_n^k = 0$ for $n \neq k$;
F_k be the kth face of S^{m-1}, where $F_k = \{\mathbf{x} \in S^{m-1} : x_k = 0\}$;
and interior of S^{m-1} be the set $intS^{m-1} = \{\mathbf{x} \in S^{m-1} : x_1 x_2 \cdots x_m > 0\}$.

Also denote by $\omega(x^0)$ ω-set of limit points of the trajectory $\{x^{(n)}, n = 0,1,2,...\}$.

2. Extremal Volterra Quadratic Stochastic Operators and Tournaments

Let $\Phi = \{1, \cdots, m\}$ be the scale of genotypes and V a quadratic stochastic operator .
Definition 2.1 The quadratic stochastic operator V is called Volterra, if $p_{ij,k} = 0$ for any $k \notin \{i, j\}$.
The biological treatment of such operators is rather clear: the offspring repeats one of its parents. Evidently for any Volterra qso $p_{ii,i} = 1$ for any $i = 1, \cdots, m$. A Volterra qso V defined on S^{m-1} has following form

$$(V\mathbf{x})_k = x_k^2 + 2 \sum_{i=1, i\neq k}^{m} p_{ik,k} x_i x_k, \tag{5}$$

where $k = 1, \cdots, m$.

Proposition 2.1 ([4] [8]) A qso V is a Volterra if and only if

$$(V\mathbf{x})_k = x_k(1 + \sum_{i=1}^{m} a_{ki} x_i) \tag{6}$$

where $A = (a_{ki})_{k,i=1}^m$ is a skew-symmetric matrix with $a_{ki} = 2p_{ik,k} - 1$, and $|a_{ki}| \leq 1$. Here $k, i \in \{1, 2, \cdots, m\}$.
Proposition 2.2 Let V be a Volterra qso. Then for any permutation $\pi \in S_m$ the transformation $T_\pi^{-1} V T_\pi$ is the Volterra qso.
Proof For qso V (5) and permutation $\pi \in S_m$ simple algebra gives

$$(T_\pi^{-1} V T_\pi \mathbf{x})_k = x_k^2 + 2 \sum_{i=1, i\neq k}^{m} p_{\pi^{-1}(i)\pi^{-1}(k), \pi^{-1}(k)} x_i x_k, \tag{7}$$

where $k = 1, \cdots, m$. It is evident that the transformation (7) is the Volterra qso.

In [4] and [5] the theory of Volterra qso was developed using theory of the Lyapunov functions and tournaments. Below we will consider a class of so-called extremal qso.
Definition 2.2 The quadratic stochastic operator V is called extremal Volterra, if $p_{ij,k} = 0$ or 1 for $k \in \{i, j\}$.
Let $\mathcal{E}$ be the set of all extremal Volterra qso on S^{m-1}. Since $p_{ik,i} = 1 - p_{ik,k}$, the total number of mixed products $x_i x_j$ with $i \neq j$ in (5) is equal to $m(m-1)/2$. Note that if V be an extremal Volterra qso then elements of matrix A take only two values ± 1. Thus there are $2^{\frac{m(m-1)}{2}}$ extremal Volterra qso.
Proposition 2.3 Let V be an extremal Volterra qso. Then for any permutation $\pi \in S_m$ the transformation $T_\pi^{-1} V T_\pi$ is the extremal Volterra qso.
Proof The proof immediately follows from Proposition 2.2.
Definition 2.3 Let $V_1, V_2 \in \mathcal{E}$ be two extremal Volterra qso. We will say that the qso V_1 is equivalent to V_2 and denote $V_1 \sim V_2$ if there exists a permutation $\pi \in S_m$ such that $V_2 = T_\pi^{-1} V_1 T_\pi$.
Since this relation is equivalence one we can part the set $\mathcal{E}$ into equivalence classes.

For $k = 2$ we have two extremal Volterra qso

$$\begin{aligned} x_1' &= x_1^2 + 2\alpha x_1 x_2 \\ x_2' &= x_2^2 + 2(1-\alpha) x_1 x_2 \end{aligned} \tag{8}$$

where $\alpha \in \{0, 1\}$ with single equivalence class. It is evident that the qso (8) has two fixed points $M_1 = (1, 0)$ and $M_2 = (0, 1)$. For $\alpha = 0$ (respectively $\alpha = 1$) any trajectory converges to M_2 (respectively M_1).

Let $k = 3$. Then we have 8 extremal Volterra operators

$$\begin{aligned} x_1' &= x_1^2 + 2\alpha x_1 x_2 + 2\beta x_1 x_3 \\ x_2' &= x_2^2 + 2(1-\alpha) x_1 x_2 + 2\gamma x_2 x_3 \\ x_3' &= x_3^2 + 2(1-\beta) x_1 x_3 + 2(1-\gamma) x_2 x_3 \end{aligned} \tag{9}$$

where $\alpha, \beta, \gamma \in \{0, 1\}$. This set consists of two equivalence classes: first class contain qso with $(\alpha, \beta, \gamma) = (0, 1, 0)$ and $(\alpha, \beta, \gamma) = (1, 0, 1)$ and second class contain all remaining 6 qso. The qso (9) with $(\alpha, \beta, \gamma) = (1, 0, 1)$ has following form:

$$\begin{aligned} x_1' &= x_1^2 + 2x_1 x_2 \\ x_2' &= x_2^2 + 2x_2 x_3 \\ x_3' &= x_3^2 + 2x_1 x_3 \end{aligned} \tag{10}$$

This operator was considered by Zakharevich [13] and was proven that it is non-ergodic transformation. Second operator with $(\alpha, \beta, \gamma) = (0, 1, 0)$ is reduced to (10) by a permutation of genotypes. For $(\alpha, \beta, \gamma) = (0, 0, 0)$ the corresponding qso has following form:

$$\begin{aligned} x_1' &= x_1^2 \\ x_2' &= x_2^2 + 2x_1 x_2 \\ x_3' &= x_3^2 + 2x_1 x_3 + 2x_2 x_3 \end{aligned} \tag{11}$$

and all others qso are reduced to (11) by some permutation of genotypes. Simple analysis shows that qso (11) is recurrent. One can rewrite qso (11) as following:

$$\begin{aligned} x_1' &= x_1^2 \\ x_2' &= x_2^2 + 2x_1 x_2 \\ x_3' &= x_3(2 - x_3). \end{aligned} \tag{12}$$

It is evident that one-dimensional transformation $\varphi(x) = x(2 - x)$ has two fixed points $x^* = 0$ and $x^{**} = 1$ with $\varphi'(0) > 1$ and $\varphi'(1) < 1$, such that x^* (resp. x^{**}) is repelling (resp. attracting) fixed point. Since $\varphi(x)$ is increasing function on segment $[0, 1]$ and x^{**} is fixed point, then $\lim_{n\to\infty} \varphi^{(n)}(x) = 1$, where $\varphi^{(n)}(x) = \varphi(\varphi^{(n-1)}(x))$, therefore any trajectory of qso (12) converges to vertex $(0, 0, 1)$, except fixed point $(1, 0, 0)$. In next section we consider the case $k = 4$.

3. Non ergodic QSO on S^3

For $k = 4$ the set $\mathcal{E}$ consists of 64 extremal Volterra operators

$$\begin{aligned} x_1' &= x_1^2 + 2\alpha x_1 x_2 + 2\beta x_1 x_3 + 2\gamma x_1 x_4 \\ x_2' &= x_2^2 + 2(1-\alpha) x_1 x_2 + 2\delta x_2 x_3 + 2\varepsilon x_2 x_4 \\ x_3' &= x_3^2 + 2(1-\beta) x_1 x_3 + 2(1-\delta) x_2 x_3 + 2\lambda x_3 x_4 \\ x_4' &= x_4^2 + 2(1-\gamma) x_1 x_4 + 2(1-\varepsilon) x_2 x_4 + 2(1-\lambda) x_3 x_4, \end{aligned} \tag{13}$$

where $\alpha, \beta, \gamma, \delta, \varepsilon, \lambda \in \{0, 1\}$. Simple but tedious algebra gives that the set $\mathcal{E}$ is parted into 4 equivalence classes.

Theorem 1.

1) First class containing QSO with $\alpha = \beta = \delta = \varepsilon = \lambda = 1$ and $\gamma = 0$ consists of 24 extremal Volterra operators and each operator from this class is non-ergodic;

2) Second class containing QSO with $\alpha = \beta = \gamma = \delta = \lambda = 1$ and $\varepsilon = 0$ consists of 8 extremal Volterra operators and each of them is ergodic;

3) Third class containing QSO with $\alpha = \beta = \delta = \varepsilon = 1$ and $\gamma = \lambda = 0$ consists of 8 extremal Volterra operators and each of them is non-ergodic;

4) Finally, fourth class containing QSO with $\alpha = \beta = \gamma = \delta = \varepsilon = \lambda = 1$ consists of 24 extremal Volterra operators and each of them is regular.

Below we consider separately each class.

3.1. First Equivalence Class

If $\alpha = \beta = \delta = \varepsilon = \lambda = 1$ and $\gamma = 0$ then corresponding qso (13) has following form

$$\begin{aligned} x_1' &= x_1^2 + 2x_1x_2 + 2x_1x_3 \\ x_2' &= x_2^2 + 2x_2x_3 + 2x_2x_4 \\ x_3' &= x_3^2 + 2x_3x_4 \\ x_4' &= x_4^2 + 2x_1x_4. \end{aligned} \tag{14}$$

The equivalence class containing this operator consist of 24 extremal Volterra qso. It is easy to see that the set of fixed points contains following points $M_1(1,0,0,0)$, $M_2(0,1,0,0)$, $M_3(0,0,1,0), M_4(0,0,0,1), M_5(1/3,0,1/3,1/3), M_6(1/3,1/3,0,1/3)$. Introducing new variables $y_1 = x_1, y_2 = x_2 + x_3, y_4 = x_4$, one can see that

$$y_1' = x_1', y_2' = x_2' + x_3', y_4' = x_4',$$

and qso (14) reduced to following

$$\begin{aligned} y_1' &= y_1^2 + 2y_1y_2 \\ y_2' &= y_2^2 + 2y_2y_3 \\ y_3' &= y_3^2 + 2y_1y_3. \end{aligned} \tag{15}$$

This qso is non-ergodic [13]. Thus we have proved following Proposition.

Proposition 3.1 Any qso from first class is non-ergodic operator.

3.2. Second Equivalence Class

If $\alpha = \beta = \gamma = \delta = \lambda = 1$ and $\varepsilon = 0$ then corresponding qso (13) has following form

$$\begin{aligned} x_1' &= x_1^2 + 2x_1x_2 + 2x_1x_3 + 2x_1x_4 \\ x_2' &= x_2^2 + 2x_2x_3 \\ x_3' &= x_3^2 + 2x_3x_4 \\ x_4' &= x_4^2 + 2x_2x_4. \end{aligned} \tag{16}$$

The equivalent class containing this operator consist of 8 extremal Volterra qso. It is easy to see that the set of fixed points consists of following points $M_1(1,0,0,0)$, $M_2(0,1,0,0)$, $M_3(0,0,1,0)$, $M_4(0,0,0,1), M_5(0,1/3,1/3,1/3)$. First equality in (16) one can rewrite as $x_1' = x_1(2 - x_1)$

similar (12). As proved above, if for initial $x^0 \in S^3$ we have $x_1^0 \neq 0$ then its trajectory converges to vertex $e_1 = (1,0,0,0)$. It is evident that the face $F_1 = \{x \in S^{m-1} : x_1 = 0\}$ is invariant subset and restriction of qso (16) on F_1 is reduced to Zakharevich's example (10).Thus we have proved following Proposition

Proposition 3.2 If extremal Volterra qso V belongs to the second class, then one of the faces is invariant subset and restriction of V on invariant face is non-ergodic qso and if initial point does not belong to invariant face then its trajectory converges to a vertex that does not belong this face.

3.3. Third Equivalence Class

If $\alpha = \beta = \delta = \varepsilon = 1$ and $\gamma = \lambda = 0$ then corresponding qso (13) has following form

$$\begin{aligned} x_1' &= x_1^2 + 2x_1x_2 + 2x_1x_3 \\ x_2' &= x_2^2 + 2x_2x_3 + 2x_2x_4 \\ x_3' &= x_3^2 \\ x_4' &= x_4^2 + 2x_1x_4 + 2x_3x_4. \end{aligned} \tag{17}$$

The equivalent class containing this operator consist of also 8 extremal Volterra qso. It is easy to see that the set of fixed points consists of following points $M_1(1,0,0,0)$, $M_2(0,1,0,0)$, $M_3(0,0,1,0)$, $M_4(0,0,0,1)$, $M_5(1/3,1/3,0,1/3)$. It is evident that $\lim_{n\to\infty} x_3^{(n)} = 0$. If for initial $x^0 \in S^{m-1}$ we have $x_1^0 \neq 0$ then $\omega(x^0) \subset F_3 = \{x \in S^{m-1} : x_3 = 0\}$. It is evident that the face $F_3 = \{x \in S^{m-1} : x_3 = 0\}$ is invariant subset and restriction of qso (17) on F_3 is reduced to Zakharevich's example (10). Thus we have proved following Proposition

Proposition 3.3 If extremal Volterra qso V belongs to the third class, then one of the faces is invariant subset and restriction of V on invariant face is non-ergodic qso and if initial point does not belong to invariant face then the set $\omega(x^0)$ of trajectory will be subset of this face.

3.4. Fourth Equivalence Class

If $\alpha = \beta = \gamma = \delta = \varepsilon = \lambda = 1$ corresponding qso (13) has following form

$$\begin{aligned} x_1' &= x_1^2 + 2x_1x_2 + 2x_1x_3 + 2x_1x_4 \\ x_2' &= x_2^2 + 2x_2x_3 + 2x_2x_4 \\ x_3' &= x_3^2 + 2x_3x_4 \\ x_4' &= x_4^2. \end{aligned} \tag{18}$$

The equivalent class containing this operator consist of 24 extremal Volterra qso. It is easy to see that the set of fixed points consists of following points $M_1(1,0,0,0)$, $M_2(0,1,0,0)$, $M_3(0,0,1,0)$, $M_4(0,0,0,1)$ and the sets F_1 and $\Gamma_{34} = \{\mathbf{x} \in S^3 : x_1 = x_2 = 0\}$ are invariant. First equality (18) one can rewrite as $x_1' = x_1(2 - x_1)$ and using properties of the function $\varphi(x) = x(2-x)$ we have

1) If $x_1^0 \neq 0$ then any trajectory of qso (18) converges to the vertex $M_1 = (1,0,0,0)$;
2) if an initial point belongs to set $F_1 \setminus \Gamma_{34}$, i.e., $x_1^0 = 0$, then the operator (18) is reduced to qso (11), therefore any trajectory converges to the vertex $M_2 = (0,1,0,0)$;
3) if $\mathbf{x} \in \Gamma_{34} \setminus M_4$ then the trajectory converges to third vertex $M_3 = (0,0,1,0)$. Thus we have the following Proposition.

Proposition 3.4 Any extremal Volterra qso V belonging to the fourth class is regular transformation.

Collecting together all four Propositions we complete the proof of Theorem 3.1

3.5. Acknowledgments
The second-named author, Post Doctorate Researcher (U.U.J) thanks International Islamic University Malaysia for kind hospitality and providing all facilities.

4. References

[1] Bernstein S N 1924 *Uchn. Zapiski. NI Kaf. Ukr. Otd. Mat.* **1** 83 (Russian)
[2] Ganikhodzhaev N N 2001 *Dokl. Math.* **61** 321
[3] Ganikhodjaev N N, Zanin D V 2004 *Russian Math.Surveys* **59** 571
[4] Ganikhodzhaev R N 1993 *Acad. Sci. Sb.Math.* **76** 489
[5] Ganikhodzhaev R N 1994 *Math. Notes* **56** 1125
[6] Ganikhodzhaev R N, Eshmamatova D B 2006 *Vladikavkaz. Mat. Zh.* **8** 12 (Russian)
[7] Ganikhodzhaev R N, Mukhamedov F M and Rozikov U A 2011 *Infinite Dimensional Analysis, Quantum Probability and Related Topics* **14** 279
[8] Jenks R D 1969 *J.Diff.Eqs.* **5** 497
[9] Kesten H 1970 *Adv.Appl.Prob.,* **2** 1
[10] Lyubich Yu I 1992 *Mathematical structures in population genetics* (Berlin: Springer)
[11] Lyubich Yu I 1978 *Russian Math. Surveys* **26** 51
[12] Ulam S 1960 *A collection of mathematical problems* (New-York-London: Interscience Publishers)
[13] Zakharevich M I 1978 *Russian Math.Surveys* **33** 265

Best proximity point on nonlinear contractive condition

Wasfi Shatanawi

Dept. of Mathematics, Hashemite University,Zarqa 13115, Jordan

E-mail: swasfi@hu.edu.jo

Abstract. In this paper, we obtain some proximity points of two mappings in metric space (X, d) by using (ψ, ϕ) contractive condition, where ψ and ϕ are altering distance mappings. Our results generalized several well-known results in the literature. Also, we apply our results to get an application of common fixed point theorem of integral types.

1. Introduction

Fixed point theory plays an important tool for solving existence problems in mathematical analysis. Some of among popular topics in the fixed point theory are cyclic contraction and best proximity. W. A. Kirk et al. [10] established the first result in this interesting area. Then after, many authors obtained many results in this area (see [1, 4, 5, 8, 11, 6, 10, 13]. Recently, Erdal Karapınar and İnci Erhan [6] studied some proximity points by using different types cyclic contraction.

We first define the cyclic map and the best proximity point.

Definition 1.1 *Let A and B be non-empty subsets of a metric space (X, d) and $T : A \cup B \to A \cup B$. Then T is called cyclic map if $T(A) \subseteq B$ and $T(B) \subseteq A$.*

A point $x \in A \cup B$ is called a best proximity point if $d(x, Tx) = d(A, B)$, where

$$d(A, B) = \inf\{d(a, b) : a \in A, b \in B\}.$$

In 2003, Kirk et al. [10] gave the following fixed point theorem for a cyclic map.

Theorem 1.1 *[10] Let A and B be nonempty closed subsets of a complete metric space (X, d). Suppose that $T : A \cup B \to A \cup B$ is a cyclic map such that*

$$d(Tx, Ty) \leq kd(x, y) \quad \forall x \in A, \ \forall y \in B.$$

If $k \in [0, 1)$, then T has a unique fixed point in $A \cap B$.

While Karapınar and Erhan [6] introduced the following types of cyclic contractions:

Definition 1.2 *[6] Let A and B be nonempty subsets of a metric space (X, d). A cyclic map $T : A \cup B \to A \cup B$ is said to be a Kannan type cyclic contraction if there exists $k \in (0, \frac{1}{2})$ such that*

$$d(Tx, Ty) \leq k(d(Tx, x) + d(Ty, y)) \quad \forall x \in A, \ \forall y \in B.$$

Journal of Physics: Conference Series **435** (2013) 012006 doi:10.1088/1742-6596/435/1/012006

Definition 1.3 *[6] Let A and B be nonempty subsets of a metric space (X,d). A cyclic map $T : A \cup B \to A \cup B$ is said to be a Reich type cyclic contraction if there exists $k \in (0, \frac{1}{3})$ such that*

$$d(Tx,Ty) \leq k(d(x,y) + d(Tx,x) + d(Ty,y)) \quad \forall x \in A, \ \forall y \in B.$$

Definition 1.4 *[6] Let A and B be nonempty subsets of a metric space (X,d). A cyclic map $T : A \cup B \to A \cup B$ is said to be a Ćirić type cyclic contraction if there exists $k \in (0, \frac{1}{3})$ such that*

$$d(Tx,Ty) \leq k \max\{d(x,y), d(Tx,x), d(Ty,y)\} \quad \forall x \in A, \ \forall y \in B.$$

Moreover, Karapınar and Erhan [6] obtained the following results:

Theorem 1.2 *[6] Let A and B be non-empty closed subsets of a complete metric space (X,d) and let $T : A \cup B \to A \cup B$ be a Kannan type cyclic contraction. Then T has a unique fixed point in $A \cap B$.*

Theorem 1.3 *[6] Let A and B be non-empty closed subsets of a complete metric space (X,d) and let $T : A \cup B \to A \cup B$ be a Reich type cyclic contraction. Then T has a unique fixed point in $A \cap B$.*

Theorem 1.4 *[6] Let A and B be non-empty closed subsets of a complete metric space (X,d) and let $T : A \cup B \to A \cup B$ be a Ćirić type cyclic contraction. Then T has a unique fixed point in $A \cap B$.*

In particular, there has been a number of works involving altering distance functions. There are control functions which alter the distance between two points in a metric space. Such functions were introduced by Khan et al. [9].

Definition 1.5 (altering distance function,[9]) *The function $\phi : [0,+\infty) \to [0,+\infty)$ is called an altering distance function if the following properties are satisfied:*

(i) ϕ is continuous and nondecreasing.

(ii) $\phi(t) = 0$ if and only if $t = 0$.

For some works in altering distance functions we refer the reader to [2, 3, 7, 14, 15, 16].

The purpose of this paper is to study some proximity points by using (ψ, ϕ) contractive condition, where ψ and ϕ are altering distance mappings. Our results generalize Theorems 1.1-1.4.

2. Main Result

We start with the following definition.

Definition 2.1 *Let (X,d) be a metric space and A, B be nonempty closed subsets of X. Let $f, T : X \to X$ be two mappings. The pair (f,T) is called a cyclic (ψ,ϕ)-contraction with respect to the pair (A,B) if*

(i) ψ and ϕ are altering distance functions;

(ii) $A \cup B$ has a cyclic representation w.r.t to the pair (f,T); that is, $fA \subseteq B$, $TB \subseteq A$ and $X = A \cup B$;

(iii) There exist altering distance functions ψ and ϕ such that for $x, y \in X$ with $x \in A$ and $y \in B$ we have

$$\psi(d(fx,Ty)) \leq \psi(\max\{d(x,y), d(fx,x), d(Ty,y)\}) - \phi(\max\{d(x,y), d(fx,x), d(Ty,y)\}). \quad (1)$$

From now on, by ψ and ϕ we mean altering distance functions unless otherwise stated.

Theorem 2.1 *Let (X,d) be a complete metric space and A,B be nonempty closed subsets of X with $X=A\cup B$. Let $f,T: A\cup B\to A\cup B$ be two mappings such that the pair (f,T) is cyclic (ψ,ϕ)-contraction with respect to (A,B). Then f and T have a common unique fixed point (say u) in $A\cap B$. Moreover u is a best proximity point.*

Proof. Let $x_0\in A$. Since $fA\subseteq B$, we choose $x_1\in B$ such that $fx_0=x_1$. Also, since $TB\subseteq A$, we choose $x_2\in A$ such that $Tx_1=x_2$. Continuing this process, we can construct a sequences (x_n) in X such that $x_{2n}\in A$, $x_{2n+1}\in B$, $x_{2n+1}=fx_{2n}$ and $x_{2n+2}=Tx_{2n+1}$. If $x_{2n_0+1}=x_{2n_0+2}$ for some $n\in\mathbf{N}$, then $x_{2n_0+1}=Tx_{2n_0+1}$. Thus x_{2n_0+1} is a fixed point of T in $A\cap B$. To prove that x_{2n_0+1} is a fixed point of f, we shall prove that $x_{2n_0+1}=x_{2n_0+2}=x_{2n_0+3}$. Now by (1), we have

$$\begin{aligned}
&\psi(d(x_{2n_0+2},x_{2n_0+3}))\\
=\;&\psi(d(Tx_{2n_0+1},fx_{2n_0+2}))\\
=\;&\psi(d(fx_{2n_0+2},Tx_{2n_0+1}))\\
\le\;&\psi(\max\{d(x_{2n_0+2},x_{2n_0+1}),d(fx_{2n_0+2},x_{2n_0+2}),d(Tx_{2n_0+1},x_{2n_0+1})\})\\
&-\phi(\max\{d(x_{2n_0+2},x_{2n_0+1}),d(x_{2n_0+3},x_{2n_0+2}),d(x_{2n_0+2},x_{2n_0+1})\})\\
=\;&\psi(\max\{d(x_{2n_0+2},x_{2n_0+1}),d(x_{2n_0+3},x_{2n_0+2}),d(x_{2n_0+2},x_{2n_0+1})\})\\
&-\phi(\max\{d(x_{2n_0+2},x_{2n_0+1}),d(fx_{2n_0+2},x_{2n_0+2}),d(x_{2n_0+2},x_{2n_0+1})\})\\
=\;&\psi(\max\{d(x_{2n_0+2},x_{2n_0+2}),d(x_{2n_0+3},x_{2n_0+2}),d(x_{2n_0+2},x_{2n_0+2})\})\\
&-\phi(\max\{d(x_{2n_0+2},x_{2n_0+1}),d(fx_{2n_0+2},x_{2n_0+2}),d(x_{2n_0+2},x_{2n_0+2})\})\\
=\;&\psi(d(x_{2n_0+3},x_{2n_0+2}))-\phi(d(x_{2n_0+3},x_{2n_0+2}))\\
\le\;&\psi(d(x_{2n_0+3},x_{2n_0+2})).
\end{aligned}$$

Thus $\phi(d(x_{2n_0+3},x_{2n_0+2}))=0$ and hence $d(x_{2n_0+3},x_{2n_0+2})=0$. Therefore $x_{2n_0+3}=x_{2n_0+2}$. Thus $x_{2n_0+1}=x_{2n_0+2}=x_{2n_0+3}$. Therefore $x_{2n_0+2}=fx_{2n_0+2}$. Hence $x_{2n_0+1}=fx_{2n_0+1}$. Therefore x_{2n_0+1} is also a fixed point of f. Thus, we may assume that $x_{2n+1}\neq x_{2n+2}$ for all $n\in\mathbf{N}$.

Given $n\in\mathbf{N}$. If n is even, then $n=2t$ for some $t\in\mathbf{N}$. By (1), we have

$$\begin{aligned}
&\psi(d(x_{n+1},x_{n+2}))\\
=\;&\psi(d(x_{2t+1},x_{2t+2}))\\
=\;&\psi(d(fx_{2t},Tx_{2t+1}))\\
\le\;&\psi(\max\{d(x_{2t},x_{2t+1}),d(fx_{2t},x_{2t}),d(Tx_{2t+1},x_{2t+1})\})\\
&-\phi(\max\{d(x_{2t},x_{2t+1}),d(fx_{2t},x_{2t}),d(Tx_{2t+1},x_{2t+1})\})\\
=\;&\psi(\max\{d(x_{2t},x_{2t+1}),d(x_{2t+2},x_{2t+1})\})\\
&-\phi(\max\{d(x_{2t},x_{2t+1}),d(x_{2t+2},x_{2t+1})\})\\
\le\;&\psi(\max\{d(x_{2t},x_{2t+1}),d(x_{2t+2},x_{2t+1})\}).
\end{aligned}$$

If

$$\max\{d(x_{2t},x_{2t+1}),d(x_{2t+2},x_{2t+1})\}=d(x_{2t+2},x_{2t+1})$$

then

$$\psi(d(x_{2t+1},x_{2t+2}))\le\psi(d(x_{2t+2},x_{2t+1}))-\phi(d(x_{2t+2},x_{2t+1})).$$

Therefore $\psi(d(x_{2t+1},x_{2t+2}))=0$ and hence $d(x_{2t+1},x_{2t+2})=0$. Thus $x_{2t+1}=x_{2t+2}$ which is a contradiction. Therefore

$$\max\{d(x_{2t},x_{2t+1}),d(x_{2t+2},x_{2t+1})\}=d(x_{2t},x_{2t+1}).$$

Hence

$$d(x_{n+1},x_{n+2}) = d(x_{2t+2},x_{2t+1}) \le d(x_{2t},x_{2t+1}) = d(x_n,x_{n+1}) \tag{2}$$

and

$$\psi(d(x_{n+1},x_{n+2})) \le \psi(d(x_n,x_{n+1})) - \phi(d(x_n,x_{n+1})). \tag{3}$$

If n is odd, then $n = 2t+1$ for some $t \in \mathbf{N}$. By (1), we have

$$\begin{aligned}
& \psi(d(x_{n+1},x_{n+2}) \\
= \ & \psi(d(x_{2t+2},x_{2t+3})) \\
= \ & \psi(d(x_{2t+3},x_{2t+2})) \\
= \ & \psi(d(fx_{2t+2},Tx_{2t+1})) \\
\le \ & \psi(\max\{d(x_{2t+2},x_{2t+1}), d(fx_{2t+2},x_{2t+2}), d(Tx_{2t+1},x_{2t+1})\}) \\
& -\phi(\max\{d(x_{2t+2},x_{2t+1}), d(fx_{2t+2},x_{2t+2}), d(Tx_{2t+1},x_{2t+1})\}) \\
= \ & \psi(\max\{d(x_{2t+3},x_{2t+2}), d(x_{2t+2},x_{2t+1})\}) \\
& -\phi(\max\{d(x_{2t+3},x_{2t+2}), d(x_{2t+2},x_{2t+1})\}) \\
\le \ & \psi(\max\{d(x_{2t+3},x_{2t+2}), d(x_{2t+2},x_{2t+1})\}).
\end{aligned}$$

If

$$\max\{d(x_{2t+3},x_{2t+2}), d(x_{2t+2},x_{2t+1})\} = d(x_{2t+3},x_{2t+2})$$

then

$$\psi(d(x_{2t+3},x_{2t+2})) \le \psi(d(x_{2t+3},x_{2t+2})) - \phi(d(x_{2t+3},x_{2t+2})).$$

Therefore $\phi(d(x_{2t+3},x_{2t+2})) = 0$ and hence $d(x_{2t+3},x_{2t+2}) = 0$. Thus $x_{2t+3} = x_{2t+2}$ which is a contradiction. Therefore

$$\max\{d(x_{2t+3},x_{2t+2}), d(x_{2t+2},x_{2t+1})\} = d(x_{2t+2},x_{2t+1}).$$

Hence

$$d(x_{n+2},x_{n+1}) = d(x_{2t+3},x_{2t+2}) \le d(x_{2t+2},x_{2t+1}) = d(x_{n+1},x_n). \tag{4}$$

$$\psi(d(x_{n+2},x_{n+1})) \le \psi(d(x_n,x_{n+1})) - \phi(d(x_n,x_{n+1})) \tag{5}$$

From (2) and (4), we have $\{d(x_{n+1},x_n) : n \in \mathbf{N}\}$ is a non increasing sequence. And from (3) and (5), we have

$$\psi(d(x_{n+2},x_{n+1})) \le \psi(d(x_n,x_{n+1})) - \phi(d(x_n,x_{n+1})) \quad \forall \ n \in \mathbf{N}. \tag{6}$$

Hence there is $r \ge 0$ such that

$$\lim_{n\to+\infty} d(x_n,x_{n+1}) = r.$$

Letting $n \to +\infty$ in (6) and using the fact that ψ and ϕ are continuous, we get that

$$\psi(r) \le \psi(r) - \phi(r).$$

Therefore $\phi(r) = 0$ and hence $r = 0$. Thus

$$\lim_{n\to+\infty} d(x_n, x_{n+1}) = 0. \tag{7}$$

Now, we prove that (x_n) is a Cauchy sequence in X. It is sufficient to show that (x_{2n}) is a Cauchy sequence. Suppose to the contrary; that is, (x_{2n}) is not a Cauchy sequence in X. Then there exists $\epsilon > 0$ for which we can find two subsequences of positive integers $(x_{2m(i)})$ and $(x_{2n(i)})$ such that $n(i)$ is the smallest index for which

$$n(i) > m(i) > i, \;\; d(x_{2m(i)}, x_{2n(i)}) \geq \epsilon. \tag{8}$$

This means that

$$d(x_{2m(i)}, x_{2n(i)-2}) < \epsilon. \tag{9}$$

From (8), (9) and triangular inequality, we get that

$$\begin{aligned}
\epsilon &\leq d(x_{2m(i)}, x_{2n(i)}) \\
&\leq d(x_{2m(i)}, x_{2n(i)-2}) + d(x_{2n(i)-2}, x_{2n(i)-1}) + d(x_{2n(i)-1}, x_{2n(i)}) \\
&< \epsilon + d(x_{2n(i)-2}, x_{2n(i)-1}) + d(x_{2n(i)-1}, x_{2n(i)}).
\end{aligned}$$

On letting $i \to +\infty$ in above inequalities and using (7), we have

$$\lim_{i\to+\infty} d(x_{2m(i)}, x_{2n(i)}) = \epsilon. \tag{10}$$

Also,

$$\begin{aligned}
\epsilon &\leq d(x_{2m(i)}, x_{2n(i)}) \\
&\leq d(x_{2m(i)}, x_{2m(i)-1}) + d(x_{2m(i)-1}, x_{2n(i)}) \\
&\leq 2d(x_{2m(i)}, x_{2m(i)-1}) + d(x_{2m(i)}, x_{2n(i)}).
\end{aligned}$$

Using (7), (10) and letting $i \to +\infty$, we get that

$$\lim_{i\to+\infty} d(x_{2m(i)}, x_{2n(i)}) = \lim_{i\to+\infty} d(x_{2m(i)-1}, x_{2n(i)}) = \epsilon. \tag{11}$$

By triangular inequality, we have

$$d(x_{2m(i)}, x_{2n(i)}) \leq d(x_{2m(i)}, x_{2n(i)+1}) + d(x_{2n(i)+1}, x_{2n(i)}). \tag{12}$$

Letting $i \to +\infty$ in (12), we obtain

$$\epsilon \leq \lim_{i\to+\infty} d(x_{2m(i)}, x_{2n(i)+1}). \tag{13}$$

From (1), we have

$$\begin{aligned}
&\psi(d(x_{2n(i)+1}), x_{2m(i)}) \\
= \;&\psi(d(fx_{2n(i)}, Tx_{2m(i)-1}) \\
\leq \;&\psi(\max\{d(x_{2n(i)}, x_{2m(i)-1}), d(fx_{2n(i)}, x_{2n(i)}), d(Tx_{2m(i)-1}, x_{2m(i)-1})\}) \\
&-\phi(\max\{d(x_{2n(i)}, x_{2m(i)-1}), d(fx_{2n(i)}, x_{2n(i)}), d(Tx_{2m(i)-1}, x_{2m(i)-1})\}) \\
= \;&\psi(\max\{d(x_{2n(i)}, x_{2m(i)-1}), d(x_{2n(i)+1}, x_{2n(i)}), d(x_{2m(i)}, x_{2m(i)-1})\}) \\
&-\phi(\max\{d(x_{2n(i)}, x_{2m(i)-1}), d(x_{2n(i)+1}, x_{2n(i)}), d(x_{2m(i)}, x_{2m(i)-1})\}).
\end{aligned}$$

Letting $i \to +\infty$ and using (7), (11), (13), and the continuity of ψ and ϕ, we get that

$$\psi(\epsilon) \leq \psi(\epsilon) - \phi(\epsilon).$$

Therefore, we get that $\phi(\epsilon) = 0$. Hence $\epsilon = 0$ a contradiction. Thus (x_n) is a Cauchy sequence in X. Since (X, d) is a complete metric space, then there exists $u \in X$ such that

$$\lim_{n \to +\infty} x_n = u.$$

Since $x_{2n} \to u$, $x_{2n+1} \to u$ and A, B are closed subset of X, we have $u \in A \cap B$. Thus by (1), we have

$$\begin{aligned} & \psi(d(x_{2n+1}, Tu)) \\ = \; & \psi(d(fx_{2n}, Tu)) \\ \leq \; & \psi(\max\{d(x_{2n}, u), d(fx_{2n}, x_{2n}), d(Tu, u)\}) - \phi(\max\{d(x_{2n}, u), d(fx_{2n}, x_{2n}), d(Tu, u)\}) \\ = \; & \psi(\max\{d(x_{2n}, u), d(x_{2n+1}, x_{2n}), d(Tu, u)\}) - \phi(\max\{d(x_{2n}, u), d(x_{2n+1}, x_{2n}), d(Tu, u)\}). \end{aligned}$$

Letting $n \to +\infty$ in above inequalities, we get that

$$\psi(d(u, Tu) \leq \psi(d(Tu, u)) - \phi(d(Tu, u)).$$

Therefore $\phi(d(u, Tu)) = 0$ and hence $d(u, Tu) = 0$. Thus u is a fixed point of T and $u \in A \cap B$. Also, by (1), we have

$$\begin{aligned} & \psi(d(fu, x_{2n+2})) \\ = \; & \psi(d(fu, Tx_{2n+1})) \\ \leq \; & \psi(\max\{d(u, x_{2n+1}), d(fu, u), d(Tx_{2n+1}, x_{2n+1})\}) - \phi(\max\{d(u, x_{2n+1}), d(fu, u), d(Tx_{2n+1}, x_{2n+1})\}) \\ = \; & \psi(\max\{d(u, x_{2n+1}), d(fu, u), d(x_{2n+2}, x_{2n+1})\}) - \phi(\max\{d(u, x_{2n+1}), d(fu, u), d(x_{2n+2}, x_{2n+1})\}). \end{aligned}$$

Letting $n \to +\infty$ in above inequalities, we get that

$$\psi(d(fu, u) \leq \psi(d(fu, u)) - \phi(d(fu, u)).$$

Therefore $\phi(d(fu, u)) = 0$ and hence $d(fu, u) = 0$. Thus u is a fixed point of f and $u \in A \cap B$.

To prove the uniqueness of u, assume there is $z \in A \cap B$ such that $fz = Tz = z$. Then by (1), we have

$$\psi(d(u, z)) = \psi(d(fu, Tz)) \leq \psi(d(u, z)) - \phi(d(u, z)).$$

Therefore $\phi(d(u, z)) = 0$ and hence $u = z$. Thus f and T have unique fixed point in $A \cap B$.

Notice that u is a best proximity point. Indeed $d(A, B) = 0$ since $u \in A \cap B$. Also, since u is a fixed point of T, we have $d(Tu, u) = 0$. Thus $d(Tu, u) = d(A, B)$. ■

As a consequent result of Theorem 2.1, we have the following results:

Corollary 2.1 *Let A and B be a nonempty closed subsets of a metric space (X, d). Let $f, T : A \cup B \to A \cup B$ be two mappings such that $A \cup B$ has a cyclic representation with respect to (f, T). Suppose there exists an altering distance function ϕ such that for $x \in A$ and $y \in B$, we have*

$$d(fx, Ty) \leq \max\{d(x, y), dx, fx), d(y, Ty)\} - \phi(\max\{d(x, y), d(x, fx), d(y, Ty)\}).$$

Then f and T have a unique common fixed point (say u) in $A \cap B$. Moreover, u is a best proximity point.

Proof. Defining $\psi : [0, +\infty) \to [0, +\infty)$ by $\psi(t) = t$. Then the pair (f, T) is a cyclic (ψ, ϕ) with respect to (A, B). The results follows from Theorem 2.1 ■

Corollary 2.2 *Let A and B be a nonempty closed subsets of a metric space (X, d). Let $f, T : A \cup B \to A \cup B$ be two mappings such that $A \cup B$ has a cyclic contraction with respect to (f, T). Suppose there exists $k < 1$ such that for $x \in A$ and $y \in B$, we have*

$$d(fx, Ty) \leq k \max\{d(x, y), d(x, fx), d(y, Ty)\}.$$

Then f and T have a unique common fixed point (say u) in $A \cap B$. Moreover, u is a best proximity point.

Proof. Follows from Corollary 2.1, by defining $\psi, \phi : [0, +\infty) \to [0, +\infty)$ by $\psi(t) = t$ and $\psi(t) = (1 - k)t$ and noting that the pair (f, T) is a cyclic (ψ, ϕ)-contraction with respect to (A, B). ■

Corollary 2.3 *Let A and B be a nonempty closed subsets of a metric space (X, d). Let $f, T : A \cup B \to A \cup B$ be two mappings such that $A \cup B$ has a cyclic contraction with respect to (f, T). Suppose that there exists $a, b, c \in \mathbf{R}^+$ such that for $x \in A$ and $y \in B$, we have*

$$d(fx, Ty) \leq ad(x, y) + bd(x, fx) + cd(y, Ty).$$

If $a + b + c \in [0, 1)$, then f and T have a unique common fixed point (say u) in $A \cap B$. Moreover, u is a best proximity point.

Proof. Follows from Corollary 2.2 by noting that

$$ad(x, y) + bd(x, fx) + cd(y, Ty) \leq (a + b + c) \max\{d(x, y), d(x, fx), d(y, Ty)\}. \quad ■$$

By taking $f = T$ in Theorem 2.1 we have the following result.

Corollary 2.4 *Let A and B be a nonempty closed subsets of a metric space (X, d). Let $T : A \cup B \to A \cup B$ be a cyclic map such that for $x \in A$ and $y \in B$, we have*

$$\psi(d(Tx, Ty)) \leq \psi(\max\{d(x, y), dx, Tx), d(y, Ty)\}) - \phi(\max\{d(x, y), d(x, Tx), d(y, Ty)\})$$

where ϕ is an altering distance function. Then T has a unique fixed point (say u) in $A \cap B$. Moreover, u is a best proximity point.

Proof Follows from Theorem 2.1 by noting that the pair (f, f) is a cyclic (ψ, ϕ)-contraction with respect to (A, B). ■

By taking $f = T$ in corollaries 2.1, 2.2 and 2.3 we have the following results:

Corollary 2.5 *Let A and B be a nonempty closed subsets of a metric space (X, d) and $T : A \cup B \to A \cup B$ be a cyclic mappings. Suppose there exists an altering distance function ϕ such that for $x \in A$ and $y \in B$, we have*

$$d(Tx, Ty) \leq \max\{d(x, y), dx, Tx), d(y, Ty)\} - \phi(\max\{d(x, y), d(x, Tx), d(y, Ty)\}).$$

Then T has a unique common fixed point (say u) in $A \cap B$. Moreover, u is a best proximity point.

Corollary 2.6 *Let A and B be a nonempty closed subsets of a metric space (X, d) and $T : A \cup B \to A \cup B$ be a cyclic map such that for $x \in A$ and $y \in B$, we have*

$$d(Tx, Ty) \leq k \max\{d(x, y), d(Tx, x), d(Ty, y)\}.$$

If $k \in [0, 1)$, then T has a unique fixed point (say u) in $A \cap B$. Moreover, u is a best proximity point.

Corollary 2.7 *Let A and B be a nonempty closed subsets of a metric space (X, d) and $T : A \cup B \to A \cup B$ be a cyclic map such that for $x \in A$ and $y \in B$, we have*

$$d(Tx, Ty) \leq ad(x, y) + bd(Tx, x) + cd(Ty, y).$$

If $a + b + c \in [0, 1)$, then T has a unique fixed point (say u) in $A \cap B$. Moreover, u is a best proximity point.

Remark 1 *Theorem 2.1 of [6] is a special case of Corollary 2.2.*

Remark 2 *Theorem 2.3 of [6] is a special case of Corollary 2.2.*

Remark 3 *Theorem 2.4 of [6] is a special case of Corollary 2.2.*

Remark 4 *Theorem 1.1 of [10] is a special case of Corollary 2.2.*

3. Application

Denote by Λ the set of functions $\mu : [0, +\infty) \to [0, +\infty)$ satisfying the following hypotheses:
(h1) μ is a Lebesgue-integrable mapping on each compact of $[0, +\infty)$.
(h2) For every $\epsilon > 0$, we have

$$\int_0^\epsilon \mu(t)\, dt > 0.$$

Theorem 3.1 *Let (X, d) be a complete metric space and A, B be nonempty closed subsets of X. Let $f, T : X \to X$ be two mappings such that $A \cup B$ has a cyclic representation w.r.t to the pair (f, T). Suppose that there exist $\mu, \gamma \in \Lambda$ such that for any two comparable elements $x, y \in X$ we have*

$$\begin{aligned} & \int_0^{d(fx,Ty)} \mu(s)\, ds \\ \leq & \int_0^{\max\{d(x,y),d(x,fx),d(y,Ty)\}} \mu(s) ds \\ & - \int_0^{\max\{d(x,y),d(x,fx),d(y,Ty)\}} \gamma(s) ds. \end{aligned}$$

Then f and T have a unique common fixed point.

Proof. Define the two functions $\psi, \phi : [0, +\infty) \to [0, +\infty)$ by

$$\psi(t) = \int_0^t \mu(s)\, ds$$

and

$$\phi(t) = \int_0^t \gamma(s)\, ds.$$

Then its easy to check that ψ and ϕ are altering distance functions. Also, we see that the pair (f, T) is a cyclic (ψ, ϕ)-contraction with respect to (A, B). Thus the result follows from Theorem 2.1. ■

Acknowledgment:
The author thanks the editor and the referee for their valuable comments and suggestions.

References

[1] M. A. Al-Thafai and N. Shahzas: Convergence and existence for best proximity points, Nonlinear Analysis 70 (2009) 3665-3671.

[2] H. Aydi, M. Postolache, W. Shatanawi, Coupled fixed point results for (ψ, ϕ)-weakly contractive mappings in ordered G-metric spaces, Computers and Mathematics with Applications (2012) 63, Pages 298–309.

[3] H. Aydi, E. Karapinar, W. Shatanawi, Coupled fixed point results for (ψ, φ)-weakly contractive condition in ordered partial metric spaces, Computer and Mathematics with Applications 62 (2011), 4449–4460.

[4] A. A. Eldered and P. Veeramani: Proximal pointwise contraction, Topology and its Aplications 156 (2009) 2942-2948.

[5] A. A. Eldered and P. Veeramani: Convergence and existence for best proximity points, J. Math. Anal. Appl. 323 (2006) 1001-1006.

[6] E. Karapinar and İ M. Erhan: Best proximity point on different type contractions, Applied Mathematics and Information Sciences 5 (2011) 342-353.

[7] E Karapinar, W Shatanawi, On Weakly (C, ψ, ϕ)-Contractive Mappings in Ordered Partial Metric Spaces, Abstract and Applied Analysis Volume 2012, Article ID 495892, 17 pages doi:10.1155/2012/495892

[8] S. Karpagam and S. Agrawal: Best proximity points for cyclic Meir-Keeler contraction maps, Nonlinear Analysis 74 (2011) 1040-1046.

[9] M. S. Khan, M. Swaleh and S. Sessa: Fixed point theorems by altering distancces between the points, Bull. Aust. Math. Soc. 30 (1984) 1-9.

[10] W. A. Kirk and P. S. Srinavasan and P. Veeramani: Fixed points for mapping satisfying cyclical contractive conditions, Fixed Point Theory 4 (2003) 79-89.

[11] G. Petruşhel: Cyclic representations and periodic points, Studia Uni. Babes-Bolyai Math. 50 (2005) 107-112.

[12] M. Păcurar and I. A. Rus: Fixed point theory for cyclic ϕ-contractions, Nonlinear Anal. 72 (2010) 1181-1187.

[13] W. Shatanawi and S Manro, Fixed point results for cyclic (ψ, ϕ, A, B)-contraction in partial metric spaces, Fixed Point Theory and Applications 2012, 2012:165.

[14] W. Shatanawi, A. Al-Rawashdeh, H. Aydi, and H.K. Nashine, On a fixed point for generalized contractions in generalized metric spaces, Abstract and Applied Analysis Volume 2012 (2012), Article ID 246085, 13 pages.

[15] W. Shatanawi, B. Samet, On (ψ, ϕ)-weakly contractive condition in partially ordered metric spaces, Computers and Mathematics with Applications 62 (2011), 3204–3214, doi:10.1016/j.camwa.2011.08.033.

[16] W. Shatanawi, Fixed point theorems for nonlinear weakly C-contractive mappings in metric spaces, Mathematical and Computer Modelling 54 (2011), 2816–2826, doi:10.1016/j.mcm.2011.06.069

An iteration problem

Ural Bekbaev

Department of Science in Engineering, Faculty of Engineering, International Islamic University Malaysia, 57528, P.O. 10, Kuala Lumpur, Malaysia

e-mail: bekbaev@iium, bekbaev2011@gmail.com

Abstract. Let F stand for the field of real or complex numbers, $\varphi : F^n \to F^n$ be any given polynomial map of the form $\varphi(x) = x+$"higher order terms". We attach to it the following operator $D : F[x] \to F[x]$ defined by $D(f) = f - f \circ \varphi$, where $F[x] = F[x_1, x_2, ..., x_n]$-the F-algebra of polynomials in variables $x_1, x_2, ..., x_n$, $f \in F[x]$ and $\circ$ stands for the composition(superposition) operation. It is shown that trajectory of any $f \in F[x]$ tends to zero, with respect to a metric, and stabilization of all trajectories is equivalent to the stabilization of trajectories of $x_1, x_2, ..., x_n$.

Let S^{n-1} be $n-1$ dimensional simplex in the real vector space and $D : S^{n-1} \to S^{n-1}$ be a(for example, polynomial or continuous) map. The behavior investigation of the trajectories $(D^m(x))_{m\in N}$, where $x \in S^{n-1}$, D^m means m times application of D, is one of the important problems in Discrete Dynamical Systems theory. In particular characterization of polynomial maps D, for which trajectories $(D^m(x))_{m\in N}$ converge at any $x \in S^{n-1}$, is one of the unsolved problems. One can consider stronger case: When does $(D^m(x))_{m\in N}$ stabilize, that is there exists $m_0 = m_0(x) \in N$ such that $D^m(x) = D^{m_0}(x)$ at any $m > m_0$, for any $x \in S^{n-1}$? This problem is not solved even for polynomial maps of degree ≤ 2. In this paper we consider a linear transformation D of the space of multivariate polynomials which appears as a regular pattern in the formal inversion formula for multivariate power series [1,2]. It will be shown that for this type transformations trajectories stabilization problem at all points can be reduced to the stabilization of the trajectories of a finite system of points. Moreover it is noted that for these transformations the stabilization problem is closely related to the Jacobian conjecture. It is interesting to note that the same regular pattern is the main object of investigation in [3], where it is considered by functional analysis's point of view with applications to problems of Dynamical System and Functional Analysis.

Let F stand for the field of real or complex numbers, $\varphi : F^n \to F^n$ be any given polynomial map. We attach to it the following transformation $D_\varphi = D : F[x] \to F[x]$ defined by $D(f) = f - f \circ \varphi$, where $F[x] = F[x_1, x_2, ..., x_n]$-the F-algebra of polynomials in variables $x_1, x_2, ..., x_n$, $f \in F[x]$ and $\circ$ stands for the composition(superposition) operation. The following Proposition shows that this transformation reminds a differential operator.

Proposition 1. *The following identities are true*

1. *D is a F-linear map.*
2. *$D(fg) = D(f)(g \circ \varphi) + fD(g)$ whenever $f, g \in F[x]$.*
3. *$D(f) \circ \varphi = D(f \circ \varphi)$*

Proof. Proof is an easy one, for example, here is a proof the second identity.

$$D(fg) = fg - (fg) \circ \varphi = (f - f \circ \varphi)(g \circ \varphi) + f(g - g \circ \varphi) = D(f)(g \circ \varphi) + fD(g)$$

In future it will be assumed that the polynomial map $\varphi(x)$ is of the form $\varphi(x) = x +$ "higher order terms". In this case one can check easily that $ord(D(f)) > ord(f)$ whenever $ord(f) < \infty$, where $ord(f)$ stands for the least degree of nonzero terms of f, $ord(0) = \infty$. Therefore D_φ has only one fixed point, namely zero polynomial. If one considers $F[x]$ with respect to the metric

$$d(f,g) = 2^{-ord(f-g)}$$

then every trajectory tends to this fixed point. But here we are interested in the following problem: When does every trajectory terminate(stabilize)? In other words when for every $f \in F[x]$ one can find $m = m(f)$ such that $D^m(f) = 0$, where $D^m_\varphi = D^m$ stands for $D^{\circ m}_\varphi = D^{\circ m}$.

Let $K_{\varphi,m} = K_m = Ker(D^m)$.

Proposition 2. *The following equalities are true.*

$$D^{m+1}(f) = D^m(f) - D^m(f) \circ \varphi$$

$$D^m(f) = \sum_{k=0}^{m} (-1)^k \binom{m}{k} f \circ \varphi^{\circ k}$$

$$D^m(fg) = \sum_{k=0}^{m} \binom{m}{k} D^k(f) D^{m-k}(g \circ \varphi^{\circ k})$$

$$K_m K_l \subset K_{m+l}$$

Proof. A proof of this Proposition can be done by induction on m. Here is a proof of the last but one and the last properties. At $m = 0$ and $m = 1$ the equality is true. Due to Proposition 1 and induction

$$D^{m+1}(fg) = D^1(D^m(fg)) = D^1(\sum_{k=0}^{m} \binom{m}{k} D^k(f) D^{m-k}(g \circ \varphi^{\circ k})) =$$

$$\sum_{k=0}^{m} \binom{m}{k} (D^{k+1}(f) D^{m-k}(g \circ \varphi^{\circ k}) \circ \varphi + D^k(f) D^{m-k+1}(g \circ \varphi^{\circ k})) = \sum_{k=0}^{m+1} \binom{m+1}{k} D^k(f) D^{m+1-k}(g \circ \varphi^{\circ k})$$

To prove the last property let $f \in K_m$, $g \in K_l$ and consider

$$D^{m+l}(fg) = \sum_{k=0}^{m+l} \binom{m+l}{k} D^k(f) D^{m+l-k}(g \circ \varphi^{\circ k})$$

As far as at least one of $D^k(f)$, $D^{m+l-k}(g \circ \varphi^{\circ k})$ is zero whenever $0 \le k \le m+l$ the above sum has to be zero that is $fg \in K_{k+l}$..

If $K_\varphi = K$ stands for $\bigcup_{m=0}^{\infty} K_{\varphi,m}$ then due to the Propositions it is clear that K is a subalgebra of $F[x]$, $K \circ \varphi \subset K$, $D(K) \subset K$ and $D^{-1}(K) \subset K$.

Here is the main result of this paper.

Theorem. *All trajectories stabilize if and only if trajectories of all $x_1, x_2, ..., x_n$ stabilize. In other words $K = F[x]$if and only if $\{x_1, x_2, ..., x_n\} \subset K$.*

Proof. If all trajectories stabilize then, in particular, trajectories of $x_1, x_2, ..., x_n$ stabilize. If trajectories of $x_1, x_2, ..., x_n$ stabilize then due to finiteness of this system one can find natural m such that all $x_1, x_2, ..., x_n$ are in K_m. For any polynomial f of degree l due to Proposition 2 one has $f \in K_{ml}$ that is trajectory of f stabilizes. This the end of the proof.

Let id stand for the identity map $id : F[x] \to F[x]$, $id(x) = x$. Termination of trajectories of all $x_1, x_2, ..., x_n$ means that there exists m such that

$$D^m(id) = \sum_{k=0}^{m} (-1)^k \binom{m}{k} \varphi^{\circ k} = 0$$

, where $D^m(id)$ stands for $(D^m(x_1), D^m(x_2), ..., D^m(x_n))$. But due to [1,2] it implies that the inverse map φ^{-1} is also a polynomial map. Therefore one has the following Corollary.

Corollary. *If all trajectories stabilize then φ^{-1} is also a polynomial map.*

If one knows that φ^{-1} is also a polynomial map can he be sure that all trajectories stabilize? It is also an open problem.

Though termination of trajectories of all elements $F[x]$ is reduced to the termination of trajectories of the finite number elements $x_1, x_2, ..., x_n$ checking termination of trajectories of these elements is not an easy task. But termination of trajectories of all $x_1, x_2, ..., x_n$ can be written in the following equivalent form: There exists m such

$$(\partial\varphi)(x)(mE + \sum_{k=2}^{m}(-1)^{k-1}\begin{pmatrix} m \\ k \end{pmatrix}(\partial\varphi)|_{\varphi(x)}(\partial\varphi)|_{\varphi^{\circ 2}(x)}...(\partial\varphi)|_{\varphi^{\circ(k-1)}(x)}) = E \qquad (1)$$

, where $\partial_k = \frac{\partial}{\partial x_k}$, $(\partial\varphi(x))^i_j = \partial_i\varphi_j(x)$, $(\partial\varphi)|_{\varphi^{\circ(k-1)}(x)} = (\partial\varphi)(\varphi^{\circ(k-1)}(x))$, E is n order identity matrix. By taking determinant of both sides (1) one can see that for termination of every trajectory the condition $det((\partial\varphi(x))^i_j) = 1$ is a necessary condition. Checking of this necessary condition can be considered as an easy thing. Therefore one can ask the following natural question.

Question. *Does $det((\partial\varphi(x))^i_j) = 1$ imply that every trajectory stabilizes?*

In common case the answer to this question seems to be negative. But we have no corresponding example.

Here we would like to offer an approach to construct such an example. It has been noted that equality (1) guarantees that the inverse map φ^{-1} is also a polynomial map. Therefore the above theorem is closely related to the well known unsolved Jacobian Conjecture [4]. It is known ([5]) that to prove the Jacobian Conjecture it is enough to prove it for polynomials of the form

$$\varphi(x) = x + ((xA_1)^{*3}, (xA_2)^{*3}, ..., (xA_n)^{*3}) = x + (xA)^{*3}$$

, where x^{*3} stands for $(x_1^3, x_2^3, ..., x_n^3)$, $A_i = (a_i^1, a_i^2, ..., a_i^n)$ is a column vector consisting of numbers, $A = (A_1, A_2, ..., A_n)$ is the corresponding matrix. Therefore one can look for a possible counterexample to our question among such polynomial maps. For such polynomial $\varphi(x)$ the equality $det((\partial\varphi(x))^i_j) = 1$ is equivalent to the system of equalities

$$\sum_{1\leq i_1<i_2<...<i_k\leq n}(xA_{i_1})^2(xA_{i_2})^2...(xA_{i_k})^2 det((a_{i_t}^{i_s})_{s,t=1,2,...,k}) = 0$$

, where $k = 1, 2, ..., n$. In particular $detA = 0$. Let us consider only $rk(A) = n - 1$ case. In this case one can assume that $A_n = \lambda_1A_1 + \lambda_2A_2 + ... + \lambda_{n-1}A_{n-1}$ and therefore $y_n = \lambda_1y_1 + \lambda_2y_2 + ... + \lambda_{n-1}y_{n-1}$, where $y_i = xA_i$. For given n one can first consider ideal I generated by polynomials

$$y_n - \sum_{i=1}^{n-1}\lambda_iy_i, a_n^k - \sum_{i=1}^{n-1}\lambda_ia_i^k, \quad \sum_{1\leq i_1<i_2<...<i_k\leq n}(y_{i_1})^2(y_{i_2})^2...(y_{i_k})^2 det((a_{i_t}^{i_s})_{s,t=1,2,...,k})$$

, where $k = 1, 2, ..., n$ and $\lambda = (\lambda_1, \lambda_2, ..., \lambda_{n-1})$, in $F[y, \lambda, A]$ then, due to $D^1(x) = \Phi_1(y) = -y^{*3}$ and $\Phi_{m+1}(y) = \Phi_m(y) - \Phi_m(y + y^{*3}A)$, check if all components of $\Phi_m(y)$ are in I. If "Yes" consider next n, if, for example, for all $m = 1, 2, ..., 100$ not all components of $\Phi_m(y)$ are in I then he should consider the corresponding $\varphi(x)$ as a possible counterexample to our question. Of course after that the obtained $\varphi(x)$ should be checked wether it is a counterexample or not.

References

[1] Ural Bekbaev, *An inversion formula for multivariate power series*, arXiv:1203.3834v1 [math. AG] 17 Mar. 2012.

[2] Wenhua Zhao, *Exponential formulas for the Jacobians and Jacobian matrices of analytic maps* , *Journal of Pure and Applied Algebra*,166(2002) 321-336.

[3] Yu.I.Lyubich, *The cohomological equations in nonsmooth categories*, arXiv:1211.0229v1 [math. FA] 1 Nov.2012.

[4] H.Bass, E.Connell, D.Wtight, *The Jacobian conjecture, reduction of degree and formal expansionof the inverse*, *Bull.Amer.Soc.*, 7(1982) 287-330.

[5] Druzkowski L.M., *An effective approach to Keller's Jacobian Conjecture, Math. Ann.*, 264(1983),303-313.

Fuzzy θ-semi-generalized closed sets.

N A Abdul Wahab[1] **and Z Salleh**[2]
[1,2]Department of Mathematics, Faculty of Science and Technology, Universiti Malaysia Terengganu, 21030 Kuala Terengganu, Terengganu, MALAYSIA.

Email: [1]adilla_aw@yahoo.com, [2]zabidin@umt.edu.my

Abstract. In this paper, a new notion of the fuzzy generalized closed sets called fuzzy θ-semi-generalized closed sets in fuzzy topological spaces is introduced and studied their properties. Furthermore, we study these new sets in relation to some other types of already known fuzzy sets.

1. Introduction.

The concept of fuzzy sets due to Zadeh [16] naturally plays important role in the study of fuzzy topological spaces which has been introduced by Chang [4]. In the year of 1970, Levine [7] initiated the study of generalized closed set in general topology. Generalized closed (open) sets play an important role in general topology. After that, Balasubramanian and Sundaram [3] defined fuzzy generalized closed set in fuzzy topological spaces. Later, El-Shafei [5] introduced fuzzy semi-generalized closed and fuzzy generalized semi-closed set with some of their properties in year 2007.

In this paper, we introduced another new notion of fuzzy generalized closed set called fuzzy θ-semi-generalized closed set which is an alternative generalization of fuzzy semi-closed set by utilizing semi-θ-closure operator in fuzzy topological spaces. The relationships among some fuzzy generalized closed sets and this new notion are obtained. Some examples and their properties were discussed in details.

2. Preliminaries.

Throughout this paper, let X be a set and I the unit interval. A fuzzy set in X is an element of the set of all functions from X to I. The family of all fuzzy sets in X is denoted by I^X. A fuzzy singleton x_α is a fuzzy set in X define by $x_\alpha(x)=\alpha$, $x_\alpha(y)=0$ for all $y\neq x, \alpha(0<\alpha\leq 1)$. The set of all fuzzy singletons in X is denoted by $S(X)$. For every $x_\alpha \in S(X)$ and $\mu\in I^X$, we define $x_\alpha \in \mu$ if and only if $\alpha\leq\mu(x)$. Space (X,τ) (simply X) always means fuzzy topological spaces. By 1_X and 0_X, we mean the fuzzy sets with constant function 1 (unit function) and 0 (zero function) respectively. A fuzzy set μ is quasi-coincident with a fuzzy set η denoted by $\mu q\eta$ if and only if there exist $x\in X$ such that $\mu(x)+\eta(x)>1$. If μ and η are not quasi-coincident, it is denoted as $\mu\bar{q}\eta$. A fuzzy subset μ of X is called fuzzy semi-closed [2] if $int(\eta)\leq\mu\leq\eta$ where η is fuzzy closed while μ is fuzzy semi-open if $\eta\leq\mu\leq cl(\eta)$ where η is fuzzy open set. The family of all fuzzy semi-open and fuzzy semi-closed sets in (X,τ) will be denoted by $FSO(X,\tau)$ and $FSC(X,\tau)$, respectively. A fuzzy subset μ of X is called fuzzy semi-θ-closed if $\mu=scl_\theta(\mu)$ and fuzzy semi-θ-open if $\mu=sint_\theta(\mu)$. The operators fuzzy semi-closure and fuzzy semi-interior of μ are defined by $scl(\mu)=\wedge\{\gamma:\mu\leq\gamma,\gamma\in FSC(X,\tau)\}$ and $sint(\mu)=\vee\{\gamma:\mu\geq\gamma,\gamma\in FSO(X,\tau)\}$ respectively. Saeid Jafari and Miguel Caldas [12] introduced θ-semi-generalized closed set in topological spaces and now we proceed θ-semi-generalized closed set in fuzzy topological spaces.

Now we introduce some basic notions that were used in the sequel.

Definition 2.1. A fuzzy set μ in (X,τ) is called;

(i) Fuzzy generalized closed set [3] (briefly, fg-closed set) if $cl(\mu)\leq\eta$ whenever $\mu\leq\eta$ and η is fuzzy open;

(ii) Fuzzy semi-generalized closed set [5] (briefly, fsg-closed set) if $scl(\mu)\leq\eta$ whenever $\mu\leq\eta$ and η is fuzzy semi open;

(iii) Fuzzy θ-generalized closed set [6] (briefly, f-θg-closed set) if $cl_\theta(\mu)\leq\eta$ whenever $\mu\leq\eta$ and η is fuzzy open.

The complement of fuzzy generalized closed (resp. fuzzy semi-generalized closed, fuzzy θ-generalized closed) set is fuzzy generalized open (resp. fuzzy semi-generalized open, fuzzy θ-generalized open) set.

Lemma 2.1. [13] Let μ be a fuzzy set in (X,τ). Then,

$$\mu\leq scl(\mu)\leq scl_\theta(\mu)$$

and hence fuzzy semi-θ-closed set is a fuzzy semi-closed set.

3. Fuzzy θ-Semi-Generalized Closed Sets.

In this section, we introduce fuzzy θ-semi-generalized closed sets in fuzzy topological space and study some of their characteristics and their relationships with other notions.

Definition 3.1. Let (X,τ) be a fuzzy topological space and μ be a fuzzy set of X. The operators of fuzzy semi-θ-closure and fuzzy semi-θ-interior of μ are defined as follows,

$$scl_\theta(\mu)=\wedge\{scl(\gamma):\mu\leq\gamma,\gamma\in FSO(X,\tau)\},$$

$$sint_\theta(\mu)=\vee\{sint(\eta):\mu\geq\eta,\eta\in FSC(X,\tau)\}.$$

Definition 3.2. A fuzzy subset μ of (X,τ) is said to be fuzzy θ-semi generalized closed set (briefly f-θsg-closed set) if $scl_\theta(\mu)\leq\eta$ wherenever $\mu\leq\eta$ and $\eta\in FSO(X,\tau)$.

The complement of fuzzy θ-semi generalized closed set is fuzzy θ-semi generalized open set (briefly f-θsg-open set).

Lemma 3.1. Every fuzzy semi-θ-closed set in X is fuzzy θ-semi generalized closed.

Proof. Let μ be a fuzzy semi-θ-closed set in X, then $\mu=scl_\theta(\mu)$. Suppose that $\mu\leq\eta$ and $\eta\in FSO(X,\tau)$. It follows that $scl_\theta(\mu)\leq\eta$ and hence μ is fuzzy θ-semi generalized closed set in X. □

Example below shows that the converse of Lemma 3.1 does not true.

Example 3.1. Let $X=\{a,b,c\}$ and $\tau=\{0_X,\mu_1,\mu_2,\mu_3,1_X\}$ where $\mu_1=a_0\vee b_0\vee c_{0.4}$, $\mu_2=a_{0.9}\vee b_{0.6}\vee c_0$, and $\mu_3=a_{0.9}\vee b_{0.6}\vee c_{0.4}$. The family of all fuzzy semi-open sets is

$$FSO(X,\tau)=\left\{0_X,1_X,a_x\vee b_y\vee c_z \quad either \quad \begin{matrix}0\le x\le 0.1\\ 0\le y\le 0.4\\ 0.4\le z\le 0.6\end{matrix} \quad or \quad \begin{matrix}0.9\le x\le 1\\ 0.6\le y\le 1\\ 0\le z\le 0.6\end{matrix}\right\}.$$

Hence the family of all fuzzy semi-closed sets is

$$FSC(X,\tau)=\left\{0_X,1_X,a_x\vee b_y\vee c_z \quad either \quad \begin{matrix}0\le x\le 0.1\\ 0\le y\le 0.4\\ 0.4\le z\le 1\end{matrix} \quad or \quad \begin{matrix}0.9\le x\le 1\\ 0.6\le y\le 1\\ 0.4\le z\le 0.6\end{matrix}\right\}.$$

If $\mu=a_{0.7}\vee b_{0.7}\vee c_{0.6}$, then μ is fuzzy θ-semi-generalized closed set. But μ is not fuzzy semi-θ-closed set since $scl_\theta(\mu)\ne\mu$.

Lemma 3.2. Every fuzzy θ-semi-generalized closed set is fuzzy semi-generalized closed.

Proof. Let μ be a fuzzy θ-semi-generalized closed set of (X,τ). Let $\mu\le\eta$ and $\eta\in FSO(X,\tau)$. Since μ be a fuzzy θ-semi-generalized closed, $scl_\theta(\mu)\le\eta$. It follows that $scl(\mu)\le\eta$ since $scl(\mu)\le scl_\theta(\mu)$ by Lemma 2.1. Hence μ is fuzzy semi-generalized closed. □

The following example shows that the converse of Lemma 3.2 is not true.

Example 3.2. Let $X=\{y\}$ with fuzzy topology $\tau=\{0_X,y_{2/3},y_{3/4},1_X\}$. The family of all fuzzy semi-open sets is

$$FSO(X,\tau)=\left\{0_X,1_X,y_p \quad where \quad \frac{2}{3}\le p<1\right\},$$

and the complement is

$$FSC(X,\tau)=\left\{0_X,1_X,y_q \quad where \quad 0<q\le\frac{1}{3}\right\}.$$

Let $\mu=y_{1/3}$ then μ is fuzzy semi-generalized closed set. But $scl_\theta(\mu)=1_X\nleq y_{2/3}$ where $y_{2/3}\in FSO(X,\tau)$ hence μ is not fuzzy θ-semi-generalized closed.

Lemma 3.3. Every fuzzy θ-generalized closed set is fuzzy θ-semi-generalized closed but the converse may not be true in general.

Example 3.3. Let $X=\{a\}$ and $\tau=\{0_X,a_{0.3},1_X\}$. The family of all fuzzy semi-open sets is;

$$FSO(X,\tau)=\{0_X,1_X,a_x \quad where \quad 0.3\le x\le 0.7\}.$$

Hence the family of all fuzzy semi-closed sets is

$$FSC(X,\tau)=\{0_X,1_X,a_x \quad where \quad 0.3\le x\le 0.7\}.$$

If $\mu=a_{0.2}$, then μ is fuzzy θ-semi-generalized closed set but μ is not fuzzy θ-generalized closed since $cl_\theta(\mu)=a_{0.7}\nleq a_{0.3}$ where $a_{0.3}\in\tau$.

Base on the discussion above, every fuzzy semi-closed set is fuzzy semi-generalized closed set but the converse is not true (see [5]). Moreover, fuzzy semi-θ-closed implies fuzzy θ-semi-generalized closed but the converse may not be true as in Example 3.1. Lemma 3.1 shows that fuzzy θ-semi-generalized closed imply fuzzy semi-generalized closed but the reverse implication is not true in general as shown in Example 3.2. Example 3.3 above shows that fuzzy θ-semi-generalized closed set does not implies fuzzy θ-generalized closed. Furthermore, fuzzy θ-closed implies fuzzy θ-generalized closed but the converse is not true (see [6]). The Figure 1 below summarizes the relationships among some fuzzy generalized closed sets discussed above where none of these implications are reversible.

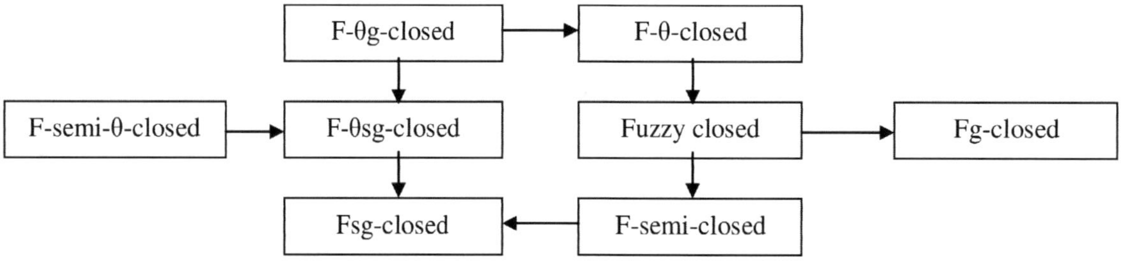

Figure 1. Relationship among some fuzzy generalized closed sets.

4. Some Properties of Fuzzy θ-Semi-Generalized Closed Sets.

In this section, we study some properties of fuzzy θ-semi-generalized closed set.

Theorem 4.1. If μ is fuzzy semi-open and fuzzy θ-semi-generalized closed set in X, then μ is fuzzy semi-θ-closed in X.

Proof. Let μ be fuzzy semi-open and f-θsg closed set in X. Obviously that $\mu\le scl_\theta(\mu)$. Since $\mu\le\mu$, by definition $scl_\theta(\mu)\le\mu$. This implies $\mu=scl_\theta(\mu)$. Hence μ is fuzzy semi-θ-closed in X.

□

Lemma 4.1. [11] $A\le B$ if and only if A and B^c are not quasi-coincident; particularly, $x_\alpha\in A$ if and only if x_α is not quasi-coincident with A^c.

Theorem 4.2. A fuzzy set μ is a fuzzy θ-semi-generalized closed set if and only if $\mu\bar{q}\eta$ implies $\mathrm{scl}_\theta(\mu)\bar{q}\eta$ for every fuzzy semi-closed set η in X.

Proof. Suppose that μ is a f-θsg closed subset of X. Let η be a fuzzy semi-closed set in X such that $\mu\bar{q}\eta$. Then by Lemma 4.1, $\mu\leq 1-\eta$ and $1-\eta$ is fuzzy semi-open set in X. Therefore, $scl_\theta(\mu)\leq 1-\eta$ as μ is f-θsg closed in X. Hence, $scl_\theta(\mu)\bar{q}\eta$.

Conversely, let η be a fuzzy semi-open set in X such that $\mu\leq\eta$ where μ is a fuzzy subset of X. By definition, $\mu\bar{q}(1-\eta)$ and $(1-\eta)$ is fuzzy semi-closed set in X. Then we have $scl_\theta(\mu)\bar{q}(1-\eta)$ which implies $scl_\theta(\mu)\leq\eta$. Hence μ is f-θsg closed in X. □

Lemma 4.2.[9] If μ is a fuzzy semi-open set in ***X*** then $scl(\mu)=scl_\theta(\mu)$.

Theorem 4.3. Let μ be a fuzzy semi-open set in a fuzzy topological space (X,τ). The fuzzy set μ is a fuzzy θ-semi-generalized closed if and only if μ is fuzzy semi-generalized closed.

Proof. (Necessity). It is obvious by Lemma 3.2.
(Sufficiency). Let μ be a fuzzy semi-generalized closed set and $\mu\leq\eta$ where $\eta\in FSO(X,\tau)$. Hence $scl(\mu)\leq\eta$ and since μ is fuzzy semi-open in X, $scl_\theta(\mu)\leq\eta$ by Lemma 4.2. Thus μ is fuzzy θ-semi-generalized closed set. □

Theorem 4.4. A fuzzy set μ is fuzzy θ-semi generalized open set if and only if $\eta\leq sint_\theta(\mu)$ whenever η is fuzzy semi-closed in X and $\eta\leq\mu$.

Proof. (Necessity) Let μ be f-θsg-open set in ***X*** and $\eta\leq\mu$ where η is fuzzy semi-closed. Obvious that μ^c is contained in η^c. Since μ^c is f-θsg-closed set then $scl_\theta\left(\mu^c\right)\leq\eta^c$ and hence $scl_\theta\left(\mu^c\right)=\left(sint_\theta(\mu)\right)^c\leq\eta^c$ such that $\eta\leq sint_\theta(\mu)$.
(Sufficiency) If μ is a fuzzy semi-closed set with $\eta\leq sint_\theta(\mu)$ whenever $\eta\leq\mu$, then it follows that $\mu^c\leq\eta^c$ and $\left(sint_\theta(\mu)\right)^c\leq\eta^c$ such that $scl_\theta\left(\mu^c\right)\leq\eta^c$. Hence μ^c is f-θsg-closed and therefore μ is f-θsg-open set. □

Lemma 4.3. If α and β are fuzzy subsets of a fuzzy topological space (X,τ), then:

(i) $scl_\theta(\alpha\vee\beta)=scl_\theta(\alpha)\vee scl_\theta(\beta)$;
(ii) $scl_\theta(\alpha\wedge\beta)\leq scl_\theta(\alpha)\wedge scl_\theta(\beta)$;
(iii) $sint_\theta(\alpha\wedge\beta)=sint_\theta(\alpha)\wedge sint_\theta(\beta)$;
(iv) $sint_\theta(\alpha\vee\beta)\geq sint_\theta(\alpha)\vee sint_\theta(\beta)$;
(v) $\left(scl_\theta(\alpha)\right)^c=sint_\theta\left(\alpha^c\right)$

Proof. (i) Let $scl_\theta(\alpha)$ and $scl_\theta(\beta)$ be both fuzzy semi-θ-closed sets and

$$scl_\theta(\alpha)\leq scl_\theta(\alpha\vee\beta), scl_\theta(\beta)\leq scl_\theta(\alpha\vee\beta).$$

By combining both of them,

$$scl_\theta(\alpha)\vee scl_\theta(\beta)\leq scl_\theta(\alpha\vee\beta).$$

Conversely,

$$\alpha\leq scl_\theta(\alpha) \text{ and } \beta\leq scl_\theta(\beta)$$

implies,

$$\alpha\vee\beta\leq scl_\theta(\alpha)\vee scl_\theta(\beta)$$

And $scl_\theta(\alpha)\vee scl_\theta(\beta)$ is a fuzzy semi-θ-closed. But since $scl_\theta(\alpha\vee\beta)$ is the smallest fuzzy semi-θ-closed set containing $\alpha\vee\beta$, hence

$$scl_\theta(\alpha\vee\beta)=scl_\theta(\alpha)\vee scl_\theta(\beta).$$

This gives the equality.

(ii) Let $\alpha\wedge\beta\leq\alpha$ and $\alpha\wedge\beta\leq\beta$. Therefore

$$scl_\theta(\alpha\wedge\beta)\leq scl_\theta(\alpha) \text{ and } scl_\theta(\alpha\wedge\beta)\leq scl_\theta(\beta).$$

Combining,

$$scl_\theta(\alpha\wedge\beta)\leq scl_\theta(\alpha)\wedge scl_\theta(\beta)$$

or,

$$\alpha\wedge\beta\leq scl_\theta(\alpha)\wedge scl_\theta(\beta)$$

implies that $scl_\theta(\alpha\wedge\beta)\leq scl_\theta(scl_\theta(\alpha)\wedge scl_\theta(\beta))=scl_\theta(\alpha)\wedge scl_\theta(\beta)$.

(iii) is the complement of (ii).
(iv) is the complement of (i).

(v) Observe that

$$\begin{aligned}(scl_\theta(\alpha))^c &= 1-scl_\theta(\alpha)\\ &= 1-\wedge\{scl(\varepsilon):\alpha\leq\varepsilon,\varepsilon\in FSO(X,\tau)\}\\ &=\vee\{sint(1-\varepsilon):1-\alpha\geq 1-\varepsilon,\varepsilon\in FSO(X,\tau)\}\end{aligned}$$

By letting $\eta=1-\varepsilon$, we have $\vee\{sint(\eta):1-\alpha\geq\eta,1-\eta\in FSO(X,\tau)\}=sint_\theta(1-\alpha)$. □

Theorem 4.5. In a fuzzy topological space (X,τ), the collection of all fuzzy semi-θ-open sets in (X,τ) is a fuzzy topological space.

Proof. Note that 0_X, $1_X\in\tau$ and obviously that 0_X and 1_X also fuzzy semi-θ-open set since $sint_\theta(0_X)=0_X$ and $sint_\theta(1_X)=1_X$.

Suppose that $\{\mu_i, i \in I\}$ be a collection of arbitrary fuzzy semi-θ-open sets in X. Then $\mu_i = sint_\theta(\mu_i)$ for each $i \in I$. Let $\mu = \vee\{\mu_i, i \in I\}$. It is obvious that $sint_\theta(\mu) \le \mu$. On the other hand, since $\mu_i \le \mu$ then $sint_\theta(\mu_i) \le sint_\theta(\mu)$ for each $i \in I$. So, $\vee\{sint_\theta(\mu) : i \in I\} \le sint_\theta(\mu)$. Thus we have $\mu = \vee\{\mu_i, i \in I\} \le sint_\theta(\mu)$. Hence we have $\mu = sint_\theta(\mu)$ and this shows that union of arbitrary fuzzy semi-θ-open sets is fuzzy semi-θ-open set.

On the other side, let μ_1 and μ_2 be two fuzzy semi-θ-open sets in X. Then $\mu_1 = sint_\theta(\mu_1)$ and $\mu_2 = sint_\theta(\mu_2)$. Let $\mu = \mu_1 \wedge \mu_2$. Obviously that $sint_\theta(\mu) \le \mu$. Since $sint_\theta(\mu_1) \wedge sint_\theta(\mu_2) \le \mu_1 \wedge \mu_2$ then by Lemma 4.3(iii),

$$\begin{aligned} sint_\theta(sint_\theta(\mu_1) \wedge sint_\theta(\mu_2)) &\le sint_\theta(\mu_1 \wedge \mu_2) \\ sint_\theta(sint_\theta(\mu_1)) \wedge sint_\theta(sint_\theta(\mu_2)) &\le sint_\theta(\mu) \\ sint_\theta(\mu_1) \wedge sint_\theta(\mu_2) &\le sint_\theta(\mu) \\ \mu_1 \wedge \mu_2 &\le sint_\theta(\mu) \\ \mu &\le sint_\theta(\mu) \end{aligned}$$

Hence we have $\mu = sint_\theta(\mu)$ and this shows that the intersection of two fuzzy semi-θ-open sets is also fuzzy semi-θ-open set. It's completes the proof. □

Remark. The collection of all fuzzy semi-θ-open sets in (X, τ) is a fuzzy topological spaces denoted by $\tau_{s\theta}$.

By Theorem 4.5, the collection of all fuzzy semi-θ-open sets in (X, τ) is a fuzzy topological space. Similar argument as previous discussion, we will have

Proposition 4.1. The union of two fuzzy θ-semi-generalized closed sets is always fuzzy θ-semi-generalized closed set.

Proof. Suppose that α and β are fuzzy θ-semi-generalized closed sets in X and let $\eta \in FSO(X, \tau)$ such that $\alpha \vee \beta \le \eta$. Since α and β are fuzzy θ-semi-generalized closed, then we have $scl_\theta(\alpha) \vee scl_\theta(\beta) \le \eta$ and by Lemma 4.3(i), $scl_\theta(\alpha \vee \beta) \le \eta$. Hence, $\alpha \vee \beta$ is fuzzy θ-semi-generalized closed. □

Intersection of two fuzzy θ-semi-generalized closed sets is not necessarily a fuzzy θ-semi-generalized closed set as the following example.

Example 4.1. Let $X = \{a, b\}$ and $\tau = \{0_X, a_{0.4}, b_{0.5}, a_{0.4} \vee b_{0.5}, 1_X\}$. The family of all fuzzy semi-open sets is;

$$FSO(X,\tau) = \left\{ 0_X, 1_X, a_x \vee b_y \quad either \quad \begin{matrix} 0.4 \le x \le 0.6 \\ 0 \le y \le 0.5 \end{matrix} \quad or \quad \begin{matrix} 0 \le x \le 0.6 \\ y = 0.5 \end{matrix} \right\}.$$

Hence the family of all fuzzy semi-closed sets is

$$FSC(X,\tau)=\left\{0_X,1_X,a_x \vee b_y \quad either \quad \begin{matrix} 0.4\le x\le 0.6 \\ 0.5\le y\le 1 \end{matrix} \quad or \quad \begin{matrix} 0.4\le x\le 1 \\ y=0.5 \end{matrix}\right\}.$$

If $\mu=a_{0.8}\vee b_{0.3}$ and $\eta=a_{0.5}\vee b_{0.7}$, then μ and η are fuzzy θ-semi-generalized closed sets since the only semi-open superset of μ and η is 1_X. But $\mu\wedge\eta=a_{0.5}\vee b_{0.3}$ is not fuzzy θ-semi-generalized closed set since $scl_\theta(\mu\wedge\eta)=a_{0.5}\vee b_{0.5}\nleq a_{0.1}\vee b_{0.5}$ where $a_{0.1}\vee b_{0.5}\in FSO(X,\tau)$.

Theorem 4.6. If μ is a fuzzy θ-semi-generalized closed set in X and $\mu\le\beta\le scl_\theta(\mu)$ then β is also a fuzzy θ-semi-generalized closed set in X.

Proof. Let η be a fuzzy semi-open subset of X such that $\beta\le\eta$. Then, $\mu\le\eta$. Since μ is fuzzy θ-semi-generalized closed, its follows that $scl_\theta(\mu)\le\eta$. Now, $\beta\le scl_\theta(\mu)$ implies $scl_\theta(\beta)\le scl_\theta(scl_\theta(\mu))=scl_\theta(\mu)$. Thus, $scl_\theta(\beta)\le\eta$. This proves that β is also fuzzy θ-semi-generalized closed subset of X. □

Theorem 4.7. Let μ be a f-θsg-closed subset of (X,τ). Then

(i) $scl_\theta(\mu)-\mu$ does not contain a nonzero fuzzy semi-closed set.
(ii) $scl_\theta(\mu)-\mu$ is f-θsg-open set

Proof. (i) Let μ be a fuzzy set of (X,τ) and suppose that there exists a nonzero fuzzy semi-closed subset ρ of X such that $\rho\le scl_\theta(\mu)-\mu$ and $\rho\ne 0_X$. Now, $\rho\le scl_\theta(\mu)-\mu$, i.e $\rho\le\mu^c$ which implies $\mu\le\rho^c$. Since ρ^c is fuzzy semi-open and μ is f-θsg-closed set, $scl_\theta(\mu)\le\rho^c$, i.e. $\rho\le(scl_\theta(\mu))^c$. Then $\rho\le scl_\theta(\mu)\wedge(scl_\theta(\mu))^c=0_X$ and hence $\rho=0_X$ which is a contradiction.
(ii) Suppose that μ be f-θsg-closed and h be a fuzzy semi-closed set such that $h\le scl_\theta(\mu)-\mu$, then by (i), h is zero and therefore $h\le sint_\theta(scl_\theta(\mu)-\mu)$. Hence, $scl_\theta(\mu)-\mu$ is f-θsg-open by Theorem 4.4. □

Recall that a fuzzy topological space (X,τ) is said to be fuzzy semi-$T_{1/2}$ [8] if and only if every fuzzy semi-generalized closed set is fuzzy semi-closed in X.

Theorem 4.8. A fuzzy topological space (X,τ) is said to be fuzzy semi-$T_{1/2}$-space if and only if;

(i) every fuzzy singleton is fuzzy semi-open or fuzzy semi-closed.
(ii) every fuzzy θ-semi-generalized closed set is fuzzy semi-closed.

Proof. (i) Let (X,τ) be fuzzy semi-$T_{1/2}$-space and for some $x_\alpha\in S(X)$, x_α is not fuzzy semi-closed. Then $1-x_\alpha$ is not fuzzy semi-open and hence 1_X is the only fuzzy semi-open set containing $1-x_\alpha$. Therefore, $1-x_\alpha$ is fsg-closed in (X,τ). Since X is a fuzzy semi-$T_{1/2}$-space, then $1-x_\alpha$ is fuzzy semi-closed set or equivalently x_α is fuzzy semi-open set.

Conversely, assume that every singleton set of (X,τ) is either fuzzy semi-closed or fuzzy semi-open set. Let μ be a fsg-closed set of (X,τ). Let $x_\alpha \in S(X)$ and we consider the following two cases:
Case I: Suppose that x_α is fuzzy semi-closed. If $x_\alpha \notin \mu$, then $x_\alpha \in scl(\mu)-\mu$. Now $scl(\mu)-\mu$ contains a nonzero fuzzy semi-closed set. Since μ is fsg-closed set, by Theorem 4.7(i), which is contradiction. Hence $x_\alpha \in \mu$.
Case II: Assume that x_α is fuzzy semi-open. Since $x_\alpha \in scl(\mu)$, then $x_\alpha \wedge \mu \neq 0_X$. So, $x_\alpha \in \mu$. Thus in both case $x_\alpha \in \mu$. So $scl(\mu) \leq \mu$. Therefore $\mu = scl(\mu)$ or equivalently μ is a fuzzy semi-closed set. Thus every fsg-closed is fuzzy semi-closed set. Hence, (X,τ) is fuzzy semi-$T_{1/2}$-spaces.
(ii) *Necessity*. Let μ be a f-θsg-closed set in (X,τ). By Lemma 3.2, μ is fsg-closed set. Since (X,τ) is a fuzzy semi-$T_{1/2}$-space, μ is fuzzy semi-closed set.
Sufficiency. Let $x_\alpha \in S(X)$. If x_α is not fuzzy semi-closed, then $1-x_\alpha$ is not fuzzy semi-open set and thus the only superset of $1-x_\alpha$ is 1_X. So, $1-x_\alpha$ is f-θsg-closed. By hypothesis, $1-x_\alpha$ is fuzzy semi-closed or equivalently x_α is fuzzy semi-open. Hence (X,τ) is a fuzzy semi-$T_{1/2}$-space. □

Conclusion.
In this paper, we introduce θ-semi-generalized closed set in fuzzy setting. We also investigate the relationship of some generalized closed sets that already known which related to fuzzy θ-semi-generalized closed set. Those give some new relationships which has been found to be useful in the study of generalized closed sets in fuzzy topological spaces.

Acknowledgement.
This research has been partially supported by the Universiti Malaysia Terengganu under the Geran Galakan Penyelidikan (GGP) Ref. No. : 68007/2013/80.

References

[1] Athar M and Ahmad B 2008 Fuzzy boundary and fuzzy semiboundary *Hindawi Publishing Corporation*. (doi: 10.1155/2008/586893)
[2] Azad K K. 1981 On fuzzy semi-continuity, fuzzy almost continuity and fuzzy weakly continuity, *J. Math. Anal. Appl.* 82 14-32.
[3] Balasubramanian G and Sundaram P 1997 On some Generalizations of fuzzy continuous functions *Fuzzy sets and systems*. 86 93-100.
[4] Chang C L 1968 Fuzzy topological spaces *J. Math. Anal. Appl.* 24 182-190.
[5] El-Shafei M E & Zakari A 2007 Semi-generalized continuous mappings in Fuzzy Topological Spaces *Journal of the Egyptian Mathematical Society*. 15 No.1 57-67.
[6] El-Shafei M E & Zakari A 2005 θ - generalized closed sets in Fuzzy Topological Spaces. *The Arabian Journal for Science and Engineering*. 31 No.2A 197-206.
[7] Levine N. 1970. Generalized closed sets in topology, *Rend. Circ. Mat. Palermo*. 19 89-96.
[8] Maki H et.al 1998 Generalized closed sets in fuzzy topological spaces I. *Meetings on Topological Spaces Theory and its Applications*. 23-36.
[9] Malakar S 1992 On fuzzy semi-irresolute and strongly irresolute functions *Fuzzy sets and Systems*. 45 239-244.
[10] Murugesan S & Thangavelu P 2008 Fuzzy Pre-semi-closed sets *Bulletin of the Malaysian Mathematical Sciences Society*. 31 No.2 223-232.

Journal of Physics: Conference Series **435** (2013) 012008 doi:10.1088/1742-6596/435/1/012008

[11] Palaniappan N 2007, Fuzzy Topology, *Second Edition, Oxford, Alpha Science International Ltd.,* 20.
[12] Saeid Jafari & Miguel Caldas 2003 On θ-semi-generalized closed set in Topology *Kyungpook Math. J.* 43 135-148.
[13] Seong Hoon Cho and Jae Keun Park 2003 A Note on Fuzzy Irresolute and Strongly Irresolute Functions *Commun. Korean Math. Soc.* 18 No.2 355-366.
[14] Sinha S P. On S*-Closedness in fuzzy setting, preprint.
[15] Yalvac T H 1988 Semi-interior and semi-closure of a fuzzy set *J. Math. Anal. Appl.* 132 No.2 356-364.
[16] Zadeh L A 1965 Fuzzy sets *Information and control.* 8 338-352.

Journal of Physics: Conference Series **435** (2013) 012009 doi:10.1088/1742-6596/435/1/012009

Polynomial Spline Approach for Double Integrals with Algebraic Singularity

[1,2]**Z. K. Eshkuvatov**, [1,2] N.M.A. Nik Long, [1,2]Habshah Midi, [3]Abduvali Khaldjigitov

[1]Department of Mathematics, Faculty of Science, and
[2]Institute for Mathematical Research, Universiti Putra Malaysia.
[3]Tashkent University of Information Technologies , Uzbekistan

Email: ezaini@science.upm.edu.my

Abstract. In this note, cubature formulas are constructed to evaluate the double integrals on the rectangle with algebraic singularity by replacing the density function $f(x, y)$ with the $S_\Delta(P)$ modified spline function interpolation of type (0,2). Rate of convergence are obtained in the classes of function $f(x, y) \in C^{2,\alpha}(D)$.

INTRODUCTION:

The issue of approximation with spline functions of several variables was studied by the number of mathematicians in the past and recent years ([1]-[3]) and some of its applications in the approximation of one dimensional integration problems are given in ([4]-[7]) and literatures cited therein. The spline functions of two variables of Hermitian type and modified spline function of two variables are much interesting in the approximation of two variables function [1].

For the calculation of multivariate singular integrals it is important to have efficient approximate formulas with different weight function [8]-[12]. Particularly, Diethelim [8] investigated the numerical approximations of double integrals with Cauchy kernel which is related to boundary integral approach for many type of partial differential equations. The integration region B is a polygon and it is divided into number of triangles. He used two possible approaches to this problem and particularly he discussed error estimates and showed that the methods were highly competitive. Nenad Ujevic [10] obtained upper and lower error bounds for an optimal 2-point quadrature rule of open type for two dimensional integral on the rectangle. He showed that these error bounds were sharp. Applications in numerical integration were also given.

Let us consider the problem of numerical evaluations of double integrals with the function of algebraic singularity

$$I(f) = \int_D f(\bar{x}) |\bar{x} - \bar{x}_0|^{\alpha-2} dA, \quad 0 < \alpha \leq 1, \tag{1}$$

where D is a closed bounded simple connected region in $\square^2$ and $\bar{x}=(x,y)\in D$ are variables and $\bar{x}_0=(x_0,y_0)$ is a fixed point in D.

Double integral in (1) has algebraic singularity at a point of the plane D where $|\bar{x}-\bar{x}_0|\to 0$. Therefore the numerical calculation of (1) requires a care about subtle point.

For the double integral (1) density function $f(x,y)$ to be existed, there has to be at least continuous function. Since the singularity is an algebraic singularity, the double integral is understood in the sense of improper integrals. The double integral of the form (1) appears in many problems of differential and integral calculus and in applied statistics. In many cases there are not analytic solutions of (1), therefore numerical approximation of double integration problems need to be carried out.

In this note cubature formulas are constructed to evaluate double integrals (1) by replacing the density function $f(x,y)$ by the $S_\Delta(P)$ [1] modified spline function interpolation of type (0,2). Convergence problem is investigated in the classes of function $f(x,y)\in C^{2,\alpha}(D)$ and the proof of the main theorem is presented.

2. Derivation of the cubature formula

Let us rewrite double integral (1) in the form

$$I(f)=\int_D \frac{f(\bar{x})}{|\bar{x}-\bar{x}_0|^{2-\alpha}}dA, \quad 0<\alpha\le 1, \tag{2}$$

where $D=[a,b]\times[c,d]$, $\bar{x}=(x,y)\in D$ and $\bar{x}_0=(x_0,y_0)$ is a fixed point in D. Assume that density function $f(\bar{x})=f(x,y)$ is continuous and smooth enough on rectangle region D. In the closed, finite rectangle domain D, let the function of two variables $f(x,y)$ be given and belong to the class of function $C^{2,\alpha}(D)$.

Here $f(\bar{x})\in C^{2,\alpha}(D)$ means that the functions $f(\bar{x})$, $f_x(\bar{x})$, $f_y(\bar{x})$, $f_{xx}(\bar{x})$, $f_{yy}(\bar{x})$ and $f_{xy}(\bar{x})=f_{yx}(\bar{x})$ in D are continuous and all second partial derivatives satisfy holder conditions.

Let region D be divided by sub rectangles $(x=x_i,\ y=x_j)$, where $x_i=x_{i-1}+h_i,\ x_0=a,\ i=1,2,...n,$ and $y_j=y_{j-1}+k_j,\ y_0=c,\ j=1,2,...,m$. Let D_{ij} denote the closed subdomain in D defined by

$$P\in D_{ij}=\{(x,y): x_i\le x\le x_{i+1},\ \ y_j\le y\le y_{j+1}\},$$

and for the domain D the system

$$\Delta=\begin{cases} a=x_0<x_1<...<x_i<x_{i+1}<...<x_{n-1}<x_n=b,\ n=2,3,... \\ c=y_0<y_1<...<y_j<y_{j+1}<...<y_{m-1}<y_m=d,\ m=2,3,... \end{cases}$$

is chosen and on Δ the interpolation nodes

$$\Delta: P_{i,j} = (x_i, y_j),\ i = 0,\dots,n,\ \ j = 0,\dots,m, \tag{3}$$

be given. By introducing notations

$$h_i = x_{i+1} - x_i,\ \max_i h_i = h,\ \min_i h_i = h^*,$$
$$k_j = y_{j+1} - y_j,\ \max_j k_j = k,\ \min_j k_j = k^*,$$

we define

$$c_1 = \frac{h}{k^*},\ c_2 = \frac{k}{h^*},\ \ c = \max\{c_1, c_2\}. \tag{4}$$

On the system of interpolation nodes (3) the modified spline function of type (0,2), $S_\Delta(P) = S_{i,j}(x, y)$ be given in the form [1]

$$S_\Delta(P) = \frac{1}{h_i \cdot k_j}\left[(x_{i+1} - x)(y_{j+1} - y)q_{i,j}(x, y) + (x - x_i)(y_{j+1} - y)q_{i+1,j}(x, y)\right.$$
$$\left. + (x_{i+1} - x)(y - y_j)q_{i,j+1}(x, y) + (x - x_i)(y - y_j)q_{i+1,j+1}(x, y)\right] \tag{5}$$

where

$$q_{i,j}(x, y) = f^{(i,j)} + \frac{1}{4}\left[(x_i - x)h_i\alpha^{(i,j)} + (y_j - y)k_j\beta^{(i,j)}\right],$$
$$q_{i+1,j}(x, y) = f^{(i+1,j)} + \frac{1}{4}\left[(x - x_{i+1})h_i\alpha^{(i+1,j)} + (y_j - y)k_j\beta^{(i+1,j)}\right],$$
$$q_{i,j+1}(x, y) = f^{(i,j+1)} + \frac{1}{4}\left[(x_i - x)h_i\alpha^{(i,j+1)} + (y - y_{j+1})k_j\beta^{(i,j+1)}\right],$$
$$q_{i+1,j+1}(x, y) = f^{(i+1,j+1)} + \frac{1}{4}\left[(x - x_{i+1})h_i\alpha^{(i+1,j+1)} + (y - y_{j+1})k_j\beta^{(i+1,j+1)}\right] \tag{6}$$

here

$$f^{(i,j)} = f(x_i, y_j) = f(P_{i,j}),\ \alpha^{(i,j)} = \frac{\partial^2 f(x_i, y_j)}{\partial x^2} \text{ and } \beta^{(i,j)} = \frac{\partial^2 f(x_i, y_j)}{\partial y^2}.$$

The modified spline function $S_\Delta(P)$ defined by (5) and (6) interpolates the given function $f(x, y)$ at the interpolation nodes (3), i.e. substituting $x = x_i\ (i = 0,\dots,n)$, $y = y_j\ (j = 0,\dots,m)$ into the expression of $S_\Delta(P)$, we obtain

$$S_\Delta(P_{i,j}) = S_{i,j}(x_i, y_j) = f^{(i,j)},$$
$$S_\Delta(P_{i+1,j}) = S_{i,j}(x_{i+1}, y_j) = f^{(i+1,j)},$$
$$S_\Delta(P_{i,j+1}) = S_{i,j}(x_i, y_{j+1}) = f^{(i,j+1)},$$
$$S_\Delta(P_{i+1,j+1}) = S_{i,j}(x_{i+1}, y_{j+1}) = f^{(i+1,j+1)}. \tag{7}$$

If $P \in D_{i,j}$, $i = 0,1,\dots,n-1$, $j = 0,1,\dots,m-1$ then differentiating the identity (5) twice with respect to x, taking into account (6) we obtain the equality

$$\frac{\partial^2 S_\Delta(P)}{\partial x^2} = \frac{1}{k_j}\left[(y_{j+1}-y)\frac{\alpha^{(i,j)}+\alpha^{(i+1,j)}}{2} + (y-y_j)\frac{\alpha^{(i,j+1)}+\alpha^{(i+1,j+1)}}{2}\right]. \tag{8}$$

In a similar way, by differentiating the identity (5) twice with respect to y, taking into account (6) we obtain the equality

$$\frac{\partial^2 S_\Delta(P)}{\partial y^2} = \frac{1}{h_i}\left[(x_{i+1}-x)\frac{\beta^{(i,j)}+\beta^{(i,j+1)}}{2} + (x-x_i)\frac{\alpha^{(i+1,j)}+\alpha^{(i+1,j+1)}}{2}\right] \tag{9}$$

From (8), for $y = y_j$ or $y = y_{j+1}$, $j = 0,1,...,m-1$, we obtain

$$\begin{aligned} \left.\frac{\partial^2 S_\Delta(P)}{\partial x^2}\right|_{y=y_j} &= \frac{1}{2}\left(\alpha^{(i,j)}+\alpha^{(i+1,j)}\right), \\ \left.\frac{\partial^2 S_\Delta(P)}{\partial x^2}\right|_{y=y_{j+1}} &= \frac{1}{2}\left(\alpha^{(i,j+1)}+\alpha^{(i+1,j+1)}\right). \end{aligned} \tag{10}$$

From (9), for $x = x_i$ or $x = x_{i+1}$, $i = 0,1,...,n-1$, we get

$$\begin{aligned} \left.\frac{\partial^2 S_\Delta(P)}{\partial y^2}\right|_{x=x_i} &= \frac{1}{2}\left(\beta^{(i,j)}+\beta^{(i,j+1)}\right), \\ \left.\frac{\partial^2 S_\Delta(P)}{\partial y^2}\right|_{x=x_{i+1}} &= \frac{1}{2}\left(\beta^{(i+1,j)}+\beta^{(i+1,j+1)}\right). \end{aligned} \tag{11}$$

To construct effective cubature formula for double integral (2) we cut out the singularity by the circle centred at $\bar{x}_0 = (x_0, y_0)$ with radius $\varepsilon < \delta$ in the following way

$$\begin{aligned} I(f) &= \int_D f(\bar{x})|\bar{x}-\bar{x}_0|^{\alpha-2}\,dA \\ &= \int_D (f(\bar{x}) - f(\bar{x}_0))|\bar{x}-\bar{x}_0|^{\alpha-2}\,dA + f(\bar{x}_0)\int_D |\bar{x}-\bar{x}_0|^{\alpha-2}\,dA \\ &= \int_D (S_\Delta(\bar{x}) - S_\Delta(\bar{x}_0))|\bar{x}-\bar{x}_0|^{\alpha-2}\,dA + f(\bar{x}_0)\int_D |\bar{x}-\bar{x}_0|^{\alpha-2}\,dA + R(f). \\ &= \sum_{i=0}^{n}\sum_{j=0}^{m} A_{ij}(\bar{x}_0) f(x_i, y_j) + R(f), \end{aligned} \tag{12}$$

where coefficients $A_{ij}(\vec{x}_0)$ are defined by the certain formula.

Cubature formula constructed in the form of (12) is accurate and error estimate is given in the following theorem.

Theorem 1: Let $f(x,y) \in C^{2,\alpha}(D)$ and $S_\Delta(P)$ be defined by (5) and D be a closed, finite rectangle domain in $\mathbf{R}^2$, then the error term $R(f)$ of (12) is

$$|R(f)| \le (8M + M^*)(3+c)\delta^{2+\alpha}, \tag{13}$$

where $0<\sqrt{h_i^2+k_j^2}=\delta\le\sqrt{h^2+k^2}$, and M, M^* are constants, c is defined by (4) and α is a Holder index.

Prof of the Theorem 1 is based on the following Lemmas.

Lemma 1: For the given system of interpolation nodes (3) and the arbitrary real numbers $\alpha^{(i,j)}$, $\beta^{(i,j)}$ and $f^{(i,j)}$, $S_\Delta(P)$ in (5) is a unique modified spline function of type (0,2) which satisfies (7), (10) and (11).

Lemma 2: Let D be closed, finite rectangle domain in $\mathbf{R}^2$, and $r(f,\bar{x})=f(\bar{x})-S_\Delta(\bar{x})$. Let $f(\bar{x})\in C^2(D)$ and denote $S_\Delta(P)=S_{i,j}(x,y)$ the spline function corresponding to the interpolation nodes (3) then for all $P\in D$ the following inequality holds:

$$|r(f)|=|f(P)-S_\Delta(P)|\le(7+6c)\omega(\delta)\delta^2,$$

where c is the constant at (4) and $0<\sqrt{h_i^2+k_j^2}=\delta\le\sqrt{h^2+k^2}$, with

$$\omega(\delta)=\max\{\omega_1(\delta),\omega_2(\delta),\omega_3(\delta)\},$$

the modulus of continuity of the functions f_{xx}, f_{yy}, f_{xy} i.e. if $|P_1P_2|\le h,\ P_1,P_2\in D$, then

$$\omega_1(h)=\max_{|P_1P_2|\le h}|f_{xx}(P_1)-f_{xx}(P_2)|,$$

$$\omega_2(h)=\max_{|P_1P_2|\le h}|f_{yy}(P_1)-f_{yy}(P_2)|,$$

$$\omega_3(h)=\max_{|P_1P_2|\le h}|f_{xy}(P_1)-f_{xy}(P_2)|.$$

Proofs of the Lemma 1 and 2 are given in [1].

As a consequence of Lemma 2, one can proof the following Lemma.

Lemma 3: Let D be closed, finite rectangle domain in $\mathbf{R}^2$, and $r(f,\bar{x})=f(\bar{x})-S_\Delta(\bar{x})$. If $f(\bar{x})\in C^{2,\alpha}(D)$ and $S_\Delta(P)=S_\Delta(x,y)$ is defined by (5) corresponding to the nodes (3) then for any $\bar{x}\in D$ the following inequalities holds:

(i) $\|r_N(f)\|_\infty=\|f(P)-S_\Delta(P)\|_\infty\le(3+c)\delta^{2+\alpha}$,

(ii) $\|r_N'(f)\|_\infty\le(2+c)\delta^{1+\alpha}$,

where norm here is the Chebyshev norm.

Proof of the Theorem: Let $D_\varepsilon=\{\bar{x}:|\bar{x}-\bar{x}_0|<\varepsilon\}$ be open domain in D with $\varepsilon<\delta$. Since $r(f,\bar{x})=f(\bar{x})-S_\Delta(\bar{x})$, then from (8) it follows that

$$\begin{aligned}R(f)&=\left|\int_D\left(r_N(f,\bar{x})-r_N(f,\bar{x}_0)\right)\left|\bar{x}-\bar{x}_0\right|^{\alpha-2}dA\right|\\&=\left|\left(\int_{D\setminus D_\varepsilon(x_0)}+\int_{D_\varepsilon(x_0)}\right)\left(r_N(f,\bar{x})-r_N(f,\bar{x}_0)\right)\left|\bar{x}-\bar{x}_0\right|^{\alpha-2}dA\right|\\&\le R_1(f)+R_2(f).\end{aligned}\qquad(14)$$

The geometric representation of (14) is given in Fig. 1

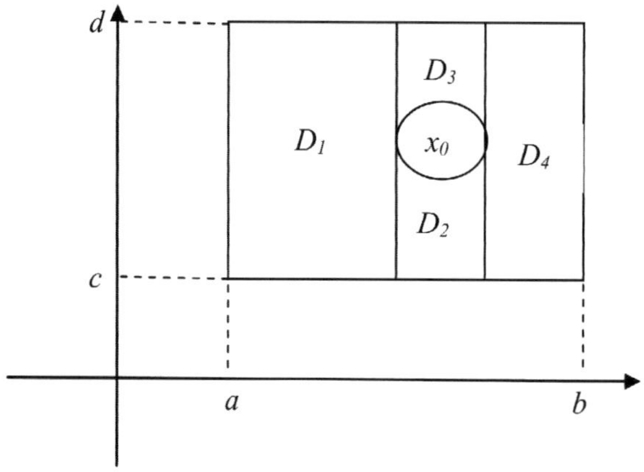

Figure 1.

Due to Lemma 3(i), $R_1(f)$ has upper bound in the form

$$R_1(f)=\left|\int_{D\setminus D_\varepsilon(x_0)}\left(r_N(f,\bar{x})-r_N(f,\bar{x}_0)\right)\left|\bar{x}-\bar{x}_0\right|^{\alpha-2}dA\right|\le 2\left\|r_N(f)\right\|_\infty\left|\int_{D\setminus D_\varepsilon(x_0)}\left|\bar{x}-\bar{x}_0\right|^{\alpha-2}dA\right|.\qquad(15)$$

Since integral in (15) has no singularity on the domain considered one can find upper bound i.e. there exists M independent of ε such that

$$\begin{aligned}\left|\int_{D\setminus D_\varepsilon(x_0)}\left|\bar{x}-\bar{x}_0\right|^{\alpha-2}dA\right|&\le\left|\int_{D_1}\left|\bar{x}-\bar{x}_0\right|^{\alpha-2}dA\right|+\left|\int_{D_2}\left|\bar{x}-\bar{x}_0\right|^{\alpha-2}dA\right|\\&\quad+\left|\int_{D_3}\left|\bar{x}-\bar{x}_0\right|^{\alpha-2}dA\right|+\left|\int_{D_4}\left|\bar{x}-\bar{x}_0\right|^{\alpha-2}dA\right|\\&\le M_1(\varepsilon)+M_2(\varepsilon)+M_3(\varepsilon)+M_4(\varepsilon)\le 4M\end{aligned}$$

Hence,

$$R_1(f)\le 4M(3+c)\delta^{2+\alpha}.\qquad(16)$$

Using mean value theorem and Lemma 3(ii), we get

$$R_2(f) \le \|r_N'(f)\|_\infty \left| \int_{D_\varepsilon(\bar{x}_0)} |\bar{x}-\bar{x}_0|^{\alpha-1} dA \right|. \quad (17)$$

Again estimating the integral in (17), we have

$$\left| \int_{D_\varepsilon(\bar{x}_0)} |\bar{x}-\bar{x}_0|^{\alpha-1} dA \right| \le \int_{D_\varepsilon(\bar{x}_0)} |\bar{x}-\bar{x}_0|^{\alpha-1} |dA| \le M * \delta ,$$

so that

$$R_2(f) \le M * (2+c)\delta^{2+\alpha} . \quad (18)$$

Substituting the estimates of $R_1(f)$ and $R_2(f)$ defined by (16) and (18) respectively into (14) completes the proof of Theorem 1.

Conclusion: Since the singularity of (2) is a week singularity from the Theorem we can conclude that rate of cubature formula (8) is the same of the rate of difference between function $f(x,y)$ and modified spline function $S_\Delta(P)$ defined by (5).

Acknowledgement

This work was supported by University Putra Malaysia under Research Universiti Grand Scheme (RUGS). Project code is 05-01-11-1277RU.

References

[1]. Rohman N.A.A. 1996 Annales Univ. Sci. Budapest., Sect. Comp. **16**, 303.

[2]. Lenard M., J. 1992 Approx. Theory. **68**(2) 113.

[3]. Domsta J. 1976 Studia Math. **58** 223.

[4]. Palamara A. O. 1990 J. Comput Appl Math **30** 191.

[5]. Gabdulkhaev B. G. 2007 Russian Mathematics (Iz. VUZ) **51**(7) 25.

[6]. Diethelm K. 1995 J. Comput Appl Math **65** 97.

[7]. Diethelm K. 1998 J. Comput Appl Math **89** 257.

[8]. Diethelim K. 2000 ANZIAM J. **42**(E) E1-E26.

[9]. Dragomir S. S., Cerone P., Barnett N.S., Roumeliotis J. 2000 Tamsui Oxford Journal of Mathematical Sciences **16**(1) 1-16.

[10]. Ujevic N. 2004 Period Math Hung **49**(1) 141.

[11]. Robert E. B., Gregory M. N. 1970 Mathematics of Computation, **24**(10) 301.

[12]. Yang Y. Atkinson K.E. 1995 SIAM J. Numer. Anal. **32**(3) 969.

Journal of Physics: Conference Series **435** (2013) 012010 doi:10.1088/1742-6596/435/1/012010

On the uniform convergence of the eigenfunction expansions.

Abdumalik Rakhimov

International Islamic University Malaysia, Jalan Gombak, 53100 Kuala Lumpur

abdumalik@iium.edu.my

Abstract. We study sufficient conditions for uniform convergence of eigenfunction expansions associated with Schrodinger's operator.

1. Preliminaries.

It is well known that a solution of the problem of heat conduction in a bounded rod with zero boundary conditions is a Fourier *sin* series and its coefficients evaluated from the given function describing initial distribution of heat in the rod. If this function is constant, then it cannot be uniformly approximated by a Fourier *sin* series near the boundary and the Gibbs phenomenon occurs at endpoints of the rod. However, this approximation is very precise in any compact subsets of the interval. This example of heat conduction in the rod and associated problems of uniformly approximation in closed domain have multidimensional analogies.

Thus, for representation of a function by eigenfunction expansions in the closed domain, it is expected that at least this function must satisfy a boundary conditions. Moreover, the boundary conditions are not sufficient for the uniform approximation and therefore it is necessary to require more smoothness near the boundary. In [1] V.A. Il'in proved that if a function in a certain Sobolev space and its repeated Laplacians satisfy the boundary conditions, then eigenfunction expansions associated with Laplace operator uniformly convergence in closed domain. In this an estimation of the behavior of the eigenfunctions near the boundary is very important. For the first boundary value problem an estimation of the eigenfunctions of the free Hamiltonian operator in closed domain obtained by E.I. Moiseev [3]. Recently in [6] this result was extended for the Schrodinger's operator with singular potential from the Sobolev spaces in two dimensional domain. In this a potential chosen such, that corresponding boundary value problem produces self-adjoint operator.

In quantum mechanics it is important for the Hamiltonian operator to be self-adjoin in the corresponding Hilbert space. Spectral expansions associated with self-adjoin operator always convergence in strong topology. However, spectral expansions of the given function and/or distribution from certain functional spaces may not converge at other topologies, for example in topology of uniformly convergence. Moreover, it is not guaranteed convergence in any good sense even in the domains where a function/distribution coincides with infinitely differentiable function. This circumstance can be explained with high impact of singularity points even if they are far from the domain where a problem is studying.

The most comprehensive study on the summability and convergence problems associated with the spectral expansions of differential operators done by Sh.A. Alimov and V.A. Il'in (see references in the book [1]). They developed methods of investigating of the spectral expansions associated with elliptic differential operators and established sharp conditions for convergence of spectral expansions in various functional spaces and topologies.

In [2] Sh.A. Alimov studied eigenfunction expansions in the functional spaces with negative smoothness. This work gave further impulse to deeper studies of the spectral expansions in the spaces of distributions. In its term the results of these studies proves applicability of the modern methods of mathematical physics in the spaces of singular functionals. The convergence and summability problems of spectral expansions of distributions associated with the development of mathematical tools for modern physics.

In the present paper we study localization and summability problems of eigenfunction expansions associated with Hamiltonian operator in generalized Sobolev spaces in close domain. In [3] it is studied uniformly convergence in a closed domain of the spectral expansions associated with the free Hamiltonian operator. In [4] the problem is studied for the operator with singular coefficients in the spaces of distributions and in the compact subsets of the bounded domain. The case of whole space (Fourier integrals) the study can be found in [5].

2. Convergence in closed domain.

Let Ω be a two dimensional bounded domain with smooth boundary $\partial\Omega$ and let $q(x)$ be a positive function from Sobolev's space $W_2^{\,1}(\Omega)$ with singularities at a point $x_0 \in \Omega$ (or in finite numbers of points) and enough smooth out of this point.

Suppose the Dirichlet problem associated with the time in depended Schrödinger equation $\Delta u + qu = f(x)$ produce self-adjoint operator in the Hilbert space $\boldsymbol{L_2(\Omega)}$. Let $\{u_n(x)\}$ be a complete system of orthonormal eigenfunctions associated with the eigenvalues λ_n.

Following equality defines the Riesz means of the partial sums of Fourier series of a function f

$$E_\lambda^{\,s} f(x) = \sum_{\lambda_n \le \lambda} \left(1 - \frac{\lambda_n}{\lambda}\right)^s f_n u_n(x), \qquad (1)$$

where f_n are Fourier coefficients of the function.

Let $\Sigma(\Omega)$ be a topological space of infinity differentiable functions on Ω with topology of uniformly convergence of the sequence of functions and their derivatives in compact subsets of this domain. Let $C_0^\infty(\Omega)$ be a subset of $\Sigma(\Omega)$ and any function from $C_0^\infty(\Omega)$ has a compact support in the domain Ω.

Denote $\Sigma'(\Omega)$ the space of all linear continuous functionals on $\Sigma(\Omega)$. Any functional from $\Sigma'(\Omega)$ has a compact support in the domain Ω.

Note, the eigenfunctions $u_n(x)$ of the above mentioned problem belong to the space $\Sigma(\Omega)$. Therefore, for any functional f from $\Sigma'(\Omega)$ its n-th Fourier coefficient f_n define as a value of this functional on an eigenfunction $u_n(x)$. Accordantly equality (1) defines the Riesz means of the partial sums of the Fourier series of the functional $f \in \Sigma'(\Omega)$.

Journal of Physics: Conference Series **435** (2013) 012010 doi:10.1088/1742-6596/435/1/012010

Definition. For any function $u \in L_2(\Omega)$ symbol $\omega(u,r)$ denotes module of continuity of second order in L_2 metrics:

$$\omega(u,r) \quad = -\sup_{|h|<r} \|u(x+h) - 2u(x) + u(x-h)\|_{L_2(\Omega_h)},$$

where $\Omega_h \;=\; \left\{ y \in \Omega:\ dist(y, \partial\Omega) < h \right\}.$

Let $\alpha > 0$, $\alpha = \nu + \zeta$, where ν - integer non negative number, $0 < \zeta \le 1$. Denote $H_{\ln}^{\alpha}(\Omega)$ a space of regular functions from $L_2(\Omega)$ with finite norm

$$\|\cdot\|_{\alpha,\chi} = \|\cdot\|_0 + \sum_{|\beta|=\nu} \sup_{0<\delta<1} \frac{\omega\left(D^{\beta}\cdot, \delta\right)}{\delta^{\varsigma} \cdot \ln\left(\frac{1}{\delta}\right)},$$

where $\|\ \|_0$ - denotes a norm in $L_2(\Omega)$ and denote $H_{\ln}^{-\alpha}(\Omega)$ a space of linear continuous functionals on $H_{\ln}^{\alpha}(\Omega)$.

Theorem 1. Let $f \in \Sigma'(\Omega) \bigcap H_{\ln}^{-\alpha}(\Omega)$, $\alpha > 0$ and let $s + \alpha \ge \frac{1}{2}$. Then the Riesz means of order s , of partial sums of the Fourier series of the distribution f convergence to zero uniformly in any compact set K from $\overline{\Omega} \setminus Suppf$.

Remark. From the embedding $W_2^{\alpha} \rightarrow H_{\chi}^{\alpha} \rightarrow W_2^{\alpha-\varepsilon}$, obtain, that a condition $s \ \ge \ \frac{1}{2} \ + \ \alpha$ is sharp. Moreover, from these embeddings and theorem 1 it also follows:

Theorem 2. Let $f \in \Sigma'(\Omega) \bigcap W_2^{-\alpha}(\Omega)$, $\alpha > 0$ and let $s + \alpha \ge \frac{1}{2}$. Then the Riesz means $E_{\lambda}^{s} f(x)$ convergence to zero in any compact K from $\overline{\Omega} \setminus Suppf$.

Let $\Phi(t) \;=\; \ln t \cdot t^{\frac{-\alpha}{2}}$ and let A be a self-adjoint extension of the Schrodinger's operator associated with system of eigenfunction $\{u_n(x)\}$:

$$Av(x) \;=\; \sum_{n=1}^{\infty} \lambda_n \; v_n \cdot u_n(x).$$

Let $\Omega^0 \subset\subset \Omega$. Then there exists a constant c such that for any function $g \in L_2(\Omega)$

$$\|\Phi(Ag)\|_{\alpha,\ln,0} \quad \le \quad c \cdot \|g\|_0 \, , \quad (2)$$

where symbol $\|\cdot\|_{\alpha,\ln,0}$ denotes a norm in the space $H_{\ln}^{\alpha}(\Omega^0)$.

Lemma 1. For any function $f \in C_0^{\infty}(\Omega)$

$$Av(x) = \sum_{n=1}^{\infty} \lambda_n^{-\alpha} \ln^2 \lambda_n \cdot f_n^2 \leq c \cdot \|f\|_{-\alpha,\ln},$$

where a constant c depends on α and the domain Ω.

Statement of the Lemma 1 follows from the inequality (2) and the Parsevval relation. From Lemma 1 we obtain:

Lemma 2. For any function $f \in C_0^{\infty}(\Omega)$ and number $\alpha > 0$ uniformly by x on any compact set $K \subset \bar{\Omega}$

$$| \sum_{n=1}^{\infty} \lambda_n^{-\frac{\alpha+\frac{1}{2}}{2}} \frac{1}{|\sqrt{\lambda_n} - \sqrt{\lambda}|+1} \cdot f_n \cdot u_n(x) | \leq c \cdot \|f\|_{-\alpha,\ln},$$

with constant depending from α and Ω.

Proof. Divide the sum as

$$\sum_{N=1}^{\infty} = \sum_{\sqrt{\lambda}\leq 1} + \sum_{\sqrt{\lambda}\geq 1} = S_1 + S_2. \quad (3)$$

It is obvious that

$$| S_1 | \leq c \cdot \|f\|_{-\alpha,\chi}. \quad (4)$$

Note that uniformly by x in closed domain $\bar{\Omega}$ [6]:

$$\sum_{k\leq\sqrt{\lambda_n}\leq k+1} u_n^2(x) = O(k^{N-1} \cdot \ln^2 k) \quad (5)$$

Thus,

$$\left| \sum_{k\leq\sqrt{\lambda_n}\leq k+1} \left(\sqrt{\lambda_n}\right)^{-\alpha-\frac{N-1}{2}} f_n \cdot u_n(x) \right| \leq c \cdot \left[\sum_{k\leq\sqrt{\lambda_n}\leq k+1} \lambda_n^{-\alpha} \cdot \ln^2 \lambda_n \cdot f_n^2 \right]^{\frac{1}{2}}. \quad (6)$$

Then by rearranging S_2 as

$$S_2 = \sum_{k=1}^{\infty} \sum_{k\leq\sqrt{\lambda_n}\leq k+1},$$

Obtain the inequality

$$| S_2 | \leq c \cdot \left(\sum_{n=1}^{\infty} \left[\frac{1}{1+|\sqrt{\lambda_n} - \sqrt{\lambda}|} \right] \right)^{\frac{1}{2}} \cdot \left(\sum_{k=1}^{\infty} \sum_{k\leq\sqrt{\lambda_n}\leq k+1} \lambda_n^{-\alpha} \cdot \ln^2 \lambda_n \cdot f_n^2 \right)^{\frac{1}{2}}. \quad (7)$$

Lemma 2 is proved.

From the statement of lemma 2 and method developed by V.A. Il'in and Sh. A. Alimov [1] obtain following estimation for the Riesz means (1):

$$E_{\lambda}^{s} f(x) = O(1) \cdot \|f\|_{-\alpha,\ln}, \quad (8)$$

uniformly with respect to any compact K from $\bar{\Omega} \setminus Suppf$

where $f \in C_0^\infty(\Omega)$ and $s = \dfrac{N-1}{2} + \alpha$.

The statements of the Theorem 1 immediately follow from inequality (8).

3. Acknowledgment.

This work done with the support of IIUM University Research Grant (Type B) EDW B12-397-0875

References

[1] Il'in V.A. 1991 *Spectral theory of differential operators. Moscow. «Nauka»,* 367 p.
[2] Alimov Sh. A.1993 *Doclades of Russian Acad Sciences,* **331:6** 661.
[3] Moiseev E.I. 1977 *Soviet Math. Dokl.* **18** 549.
[4] Abdumalik Rakhimov, Anvarjon Akhmedov, Hishamuddin Zainuddin. 2012 *Applied Mathematics Letters* **25:6** 921.
[5] Rakhimov A.A. 2011, *Abstract and Applied Analysis* doi:10.1155/2011/731249.
[6] Rakhimov A.A., Kamran Zakariya. 2010 *Mathematical Digest* **2** 1.

2012 iCAST: Contemporary Mathematics, Mathematical Physics and their Applications IOP Publishing
Journal of Physics: Conference Series **435** (2013) 012011 doi:10.1088/1742-6596/435/1/012011

On Invariant Measure of the Circle Maps

Habibulla Akhadkulov[1], **Mohd Salmi Md Noorani**[2], **Akhtam Dzhalilov**[3]

[1,2] School of Mathematical Sciences Faculty of Science and Technology University Kebangsaan Malaysia, 43600 UKM Bangi, Selangor Darul Ehsan, Malaysia
[3] Turin Polytechnic University, Kichik Halka yuli 17, Tashkent 100095, Uzbekistan.

E-mail: [1]`akhadkulov@yahoo.com`, [2]`msn@ukm.my`, [3]`a_dzhalilov@yahoo.com`

Abstract. Let f be piecewise smooth circle homeomorphisms with break points and the rotation number ρ is irrational. We provide a necessary condition for the absolute continuity of f-invariant measure with respect to Lebesgue measure.

1. Introduction and Statement of Results

Let f be an orientation preserving homeomorphism of the circle $S^1 \equiv \mathbb{R}/\mathbb{Z}$ with lift $F : \mathbb{R} \to \mathbb{R}$ continuous, strictly increasing and $F(x+1) = F(x) + 1, x \in \mathbb{R}$. The circle homeomorphism f is then defined by $f(x) = F(x) \mod 1$. The most important arithmetic characteristic of the homeomorphism f of the circle S^1 is the *rotation number*

$$\rho(f) = \lim_{i\to\infty} \frac{F^i(x)}{i} \mod 1.$$

Here and below, for a given map F, F^i denotes its i-th iteration. Denjoy in [2] proved that, if f is a circle diffeomorphism with irrational rotation number $\rho = \rho(f)$ and $\log Df$ is of bounded variation, then f is conjugate to the pure rotation $f_\rho : x \to x + \rho \mod 1$, that is, there exists an essentially unique homeomorphism φ of the circle with $\varphi \circ f = f_\rho \circ \varphi$. This classical result of Denjoy can be extended to circle homeomorphisms with break points. It is well known, that circle homeomorphisms f with irrational rotation number ρ admit a unique f- invariant probability measure μ. Since the conjugating map φ and the f-invariant measure μ are related by $\varphi(x) = \mu([0,x])$ (see [1]), regularity properties of the conjugating map φ imply corresponding properties of the density of the absolutely continuous invariant measure μ. This problem of smoothness of density function of μ for the smooth diffeomorphisms now very well understood by several authors [8], [9] and [10]. An important class of circle homeomorphisms are the homeomorphisms with break points so-called class **P-homeomorphisms**.

The class of **P-homeomorphisms** consists of orientation preserving circle homeomorphisms f whose lifts F are differentiable except in finite number points, so called *break points* of f, at which the left and right derivatives, denoted respectively by Df_- and Df_+ exist, and such that,

(i) there exist constants $0 < c < C < \infty$ with $c < Df(x) < C$ for all $S^1 \backslash BP(f), c < Df_-(x_b) < C$ and $c < Df_-(x_b) < C$ for all $x_b \in BP(f)$, with $BP(f)$ the set of break points of f on S^1;

(ii) $\log Df$ has bounded variation on S^1.

The ratio $\sigma(x_b) := Df(x_{b-})/Df(x_{b+})$ is called the *jump ratio* of f in x_b. The total variation of $\log Df$ we denote by v i.e. $v = Var_{S^1} \log Df$. f-invariant measures of P-homeomorphisms with one break point was first studied by A. Dzhalilov and K. Khanin in [3]. The main result of them is as follows:

Theorem 1.1. *Let $f \in C^{2+\alpha}(S^1 \setminus \{b\})$, $\alpha > 0$ be circle homeomorphism with one break point $b = b(f)$. Assume that the rotation number $\rho = \rho(f)$ of f is irrational. Then the f-invariant measure μ is mutually singular with Lebesgue measure ℓ.*

Piecewise-linear (PL) circle homeomorphisms with two break points are the simplest examples of homeomorphisms with two break points and feature in many areas of mathematics, for example, group theory and homotopy theory (see [12]). The f-invariant measures of such homeomorphisms were studied for the first time in [7]. There was shown that, the f-invariant measure of the (PL) circle homeomorphism f with two break points and irrational rotation number is absolutely continuous with respect to Lebesgue measure if and only if both break points lie on the same trajectory. General (non (PL)) class of P-homeomorphisms with two break points have been studied by A. Dzhalilov and I. Liousse in [4]. The main result of [4] is the following theorem.

Theorem 1.2. *Let $f \in C^2(S^1 \setminus \{a, b\})$, be circle homeomorphism with two break points. Assume that the rotation number $\rho = \rho(f)$ of f is irrational of bounded type i.e. the coefficients in the continued fraction expansion of ρ are bounded. Then the f-invariant measure is absolutely continuous with respect to Lebesgue measure if and only if both break points lie on the same trajectory and the product of their jump ratios is 1.*

The f-invariant measures of P-homeomorphisms with several break points for the case product of jump ratios not equal to one were studied by A. Dzhalilov D. Mayer and U. Safarov in [5]. The main result of [5] is the following statement.

Theorem 1.3. *Let $f \in C^2(S^1 \setminus \{b_i\})$, be circle homeomorphism with several break points $b_i = b_i(f)$. Assume that the rotation number $\rho = \rho(f)$ of f is irrational and product of jump ratios not equal to one. Then the f-invariant measure μ is singular with respect to Lebesgue measure ℓ.*

Note that the question of the regularity of the invariant measure of circle homeomorphisms with several break points that do not lie on the same trajectory and such that the product of the jumps is trivial, that is, product of jump ratios equal to one remains open. Let rotation number ρ be an irrational. We use the continued fraction $\rho = 1/(k_1 + 1/(k_2 + ...)) := [k_1, k_2, ..., k_n, ...)$ of the rotation number, which is understood as the limit of the sequence of convergents $p_n/q_n = [k_1, k_2, ..., k_n]$. The sequence of positive integer k_n with $n \geq 1$, which are called *incomplete multiples*, is uniquely determined for irrational ρ. The coprimes p_n and q_n satisfy the recursive relations $p_n = k_n p_{n-1} + p_{n-2}$ and $q_n = k_n q_{n-1} + q_{n-2}$ for $n \geq 1$, where we set for convenience, $p_{-1} = 0$, $q_{-1} = 1$, and $p_0 = 1$, $q_0 = k_1$. In this paper we will study P-homeomorphism f with irrational rotation number having several number break points. Our purpose in this paper is to give a necessary condition for the absolute continuity of the f-invariant measure μ w.r.t. Lebesgue measure ℓ. Our main result is the following:

Theorem 1.4. *Let f be a P-homeomorphism with irrational rotation number $\rho = \rho(f)$. If the f-invariant measure μ is absolutely continuous w.r.t. Lebesgue measure ℓ, then for all $\delta > 0$ and for all $0 \leq k \leq k_{n+1}$*

$$\lim_{n \to \infty} \ell(x : |\log Df^{kq_n + q_{n-1}}(x)| \geq \delta) = 0.$$

Notice that, Theorem 1.4 was proved in [4] for the case $k = 0$ and this fact played key role to prove main result (Theorem 1.2) of the paper [4]. To study the f-invariant measures, Theorem 1.4 provides an opportunity to analyze high iteration of $f^{q_{n-1}}$ i.e. we can analyze $f^{kq_n+q_{n-1}}$, for all $0 \leq k \leq k_{n+1}$.

2. Notations, terminology, background

The assertions listed below, which are valid for any orientation-preserving homeomorphism $f \in P$ with irrational rotation number ρ, constitute classical Denjoy's theory. Their elementary proofs can be found in [4] and [5].

I. Dynamical partition. We define the n-th *fundamental segment* $\Delta_0^n = \Delta^n(\xi)$ as the circle arc $[\xi, f^{q_n}(\xi)]$ if n is even and $[f^{q_n}(\xi), \xi]$ if n is odd. Let $\xi \in S^1$ and $\Delta_0^n(\xi)$ be n-th fundamental segment. We denote two sets of closed intervals of order n : q_n "lengthy" intervals: $\{\Delta_i^{n-1} = f^i(\Delta_0^{n-1}),\ 0 \leq i < q_n\}$ and q_{n-1} "short" intervals: $\left\{\Delta_j^n = f^j(\Delta_0^n),\ 0 \leq j < q_{n-1}\right\}$. The "lengthy" and "short" intervals are mutually disjoint except for the endpoints and cover the whole circle. The partition obtained by the above construction will be denoted by $\mathbf{P}_n$ and called the n-th *dynamical partition* of the point ξ. Obviously the partition $\mathbf{P}_{n+1}$ is a refinement of the partition $\mathbf{P}_n$. Indeed the "short" intervals are members of $\mathbf{P}_{n+1}$ and each "lengthy" interval $\Delta_i^{n-1} \in \mathbf{P}_n,\ 0 \leq i < q_n$, is partitioned into $k_{n+1}+1$ intervals belonging to $\mathbf{P}_{n+1}$ such that

$$\Delta_i^{n-1} = \Delta_i^{n+1} \cup \bigcup_{s=0}^{k_{n+1}-1} \Delta_{i+q_{n-1}+sq_n}^n. \tag{1}$$

II. Generalized Denjoy estimate. Let $\xi_0 \in S^1$ be a continuity point of Df^{q_n}, then the following inequality holds: $e^{-v} \leq Df^{q_n}(\xi_0) \leq e^v$, where $v = Var_{S^1} \log Df$.

III. Generalized Finzi estimate. Suppose $\xi \in S^1$, $\eta \in \Delta^{n-1}(\xi)$ and ξ, η are continuity points of Df^{q_n}. Then for any $0 \leq k < q_n$ the following inequality holds: $|\log Df^k(\xi) - \log Df^k(\eta)| \leq v$.

Let $(S^1, \mathfrak{G}, \mu)$ be a measure space and $T : S^1 \to S^1$ be a measurable map.

Definition 2.1. *The set $A \in \mathfrak{G}$ is said to be* ***invariant*** *with respect to the measurable T, if $A = T^{-1}A$.*

Definition 2.2. *A measurable map $T : S^1 \to S^1$ is said to be* ***ergodic*** *with respect Lebesgue measure ℓ if the Lebesgue measure $\ell(A)$ of any invariant set A equals 0 or 1.*

Let $\xi_0 \in S^1$, denote by $V_n = \Delta^n(\xi_0) \cup \Delta^{n-1}(\xi_0)$. Now we equip S^1 with the usual metric $|x - y| = \inf\{|\widetilde{x} - \widetilde{y}|$, where $\widetilde{x}, \widetilde{y}$ are ranges over the lifts of $x, y \in S^1$ respectively$\}$.

Lemma 2.3. *Let f be a P-homeomorphism with irrational rotation number ρ. Suppose $\xi \in V_n$ and ξ be a continuity point of Df^{q_n}. Then for any $0 \leq k < q_n$ the following inequality holds:*

$$e^{-v}\frac{|f^k(V_n)|}{|V_n|} \leq Df^k(\xi) \leq e^v\frac{|f^k(V_n)|}{|V_n|}. \tag{2}$$

Proof. Let the system of intervals $\mathcal{I} = \{I : I \subset V_n,$ and the map Df^{q_n} is continuous on $I\}$ be continuity intervals of Df^{q_n}. Let $\xi \in \Delta^{n-1}(\xi_0)$. Then, by the mean value theorem, for any $0 \leq k < q_n$, we have

$$\frac{|f^k(\Delta^{n-1}(\xi_0))|}{Df^k(\xi)|\Delta^{n-1}(\xi_0)|} = \frac{Df^k(z_1)|I_1| + Df^k(z_2)|I_2| + \ldots + Df^k(z_d)|I_d|}{Df^k(\xi)|\Delta^{n-1}(\xi_0)|}, \tag{3}$$

where $z_i \in I_i \subset \Delta^{n-1}(\xi_0)$ and $I_i \in \mathcal{I}$, $1 \leq i \leq d$. If $\xi \in \Delta^n(\xi_0)$ then we have

$$\frac{|f^k(\Delta^n(\xi_0))|}{Df^k(\xi)|\Delta^n(\xi_0)|} = \frac{Df^k(y_1)|J_1| + Df^k(y_2)|J_2| + ... + Df^k(y_t)|J_t|}{Df^k(\xi)|\Delta^n(\xi_0)|}, \tag{4}$$

where $y_i \in J_i \subset \Delta^n(\xi_0)$ and $J_i \in \mathcal{I}$, $1 \leq i \leq t$. Apply generalized Finzi estimate to the right-hand side of relations (12) and (13) we get

$$e^{-v} \leq \frac{|f^k(\Delta^{n-1}(\xi_0))|}{Df^k(\xi)|\Delta^{n-1}(\xi_0)|} \leq e^v \text{ and } e^{-v} \leq \frac{|f^k(\Delta^n(\xi_0))|}{Df^k(\xi)|\Delta^n(\xi_0)|} \leq e^v.$$

Finally, we get

$$e^{-v} \leq \frac{|f^k(V_n)|}{Df^k(\xi)|V_n|} \leq e^v.$$

□

Lemma 2.4. *Let f be a P-homeomorphism of the circle S^1 with irrational rotation number ρ, then f is ergodic with respect to Lebesgue measure ℓ.*

Proof. Suppose that there exists an invariant set A such that $0 < \ell(A) < 1$. Then by the Lebesgue Density Theorem, A has a density point z. We fix an arbitrary $\epsilon > 0$. By definition of density points, we can find a $\delta > 0$ such that for any interval $[a, b]$ satisfying the conditions $z \in [a, b]$, $[a, b] \subset (z - \delta, z + \delta)$, we have $\ell(A \cap [a, b]) \geq (1 - \epsilon)\ell([a, b])$, or, in other words, $\ell(A^c \cap [a, b]) < \epsilon\ell([a, b])$, where A^c denotes the complement of A. Now, we choose such n that $V_n = \Delta^n(z) \cup \Delta^{n-1}(z) \subset (z - \delta, z + \delta)$. We can check that $\bigcup_{k=0}^{q_n-1} T^k(V_n) = S^1$ and each point of the circle belongs to at most two intervals of this cover. Hence, the set A^c is invariant with respect to f, using above lemma we get:

$$\ell(A^c) = \sum_{k=0}^{q_n-1} \ell(A^c \cap f^k(V_n)) = \sum_{k=0}^{q_n-1} \int_{A^c \cap V_n} Df^k(x) d\ell \leq \frac{e^v \ell(A^c \cap V_n)}{\ell(V_n)} \sum_{k=0}^{q_n-1} \ell(f^k(V_n)) \leq 2e^v\epsilon.$$

Since ϵ was arbitrary, $\ell(A^c) = 0$. The theorem is proved. □

Lemma 2.5. *Let f be a P-homeomorphisms with irrational rotation number. Then conjugation map φ between f and f_ρ is either absolutely continuous or singular.*

Proof. Consider a P-homeomorphism f of the circle S^1 with irrational rotation number ρ. Let φ be a conjugating map between f and f_ρ, i.e.

$$\varphi \circ f = f_\rho \circ \varphi. \tag{5}$$

We know that conjugation function φ is strictly increasing function on S^1. Then $D\varphi$ exists almost everywhere on S^1. Denote by $A = \{x : x \in S^1, D\varphi(x) > 0\}$. It is clear that the set A is mod 0 invariant with respect to f. Since the class P-homeomorphism is ergodic with respect to the Lebesgue measure. Hence, the Lebesgue measure of A is either null or full. If Lebesque measure of A is null then φ is a singular function, if it is full then φ is an absolutely continuous function. □

Lemma 2.6. *Let f be a P-homeomorphism with irrational rotation number and φ be a conjugating map between f and f_ρ. Then the following properties are equivalent:*

(i) the map φ is singular;

(ii) the map φ^{-1} *is singular;*

(iii) the f*-invariant measure* μ *and the Lebesgue measure* ℓ *is mutually singular.*

Proof. Since the conjugating map φ and the invariant measure μ are related by $\varphi(x) = \mu([0, x])$ i.e. φ is a distribution function of measure μ. Considering this we get equivalence of relations (i) and (iii). We will prove that the properties (i) and (ii) are equivalent. Let the map φ be a singular function. Therefore, there exists a measurable subset $C \subset S^1$ such that $\ell(C) = 0$ and $\ell(\varphi(C)) > 0$. Denote by $B = \bigcup_{n\in\mathbb{Z}} f^n(C)$. It is clear that the subset $B \subset S^1$ is a invariant subset w.r.t. f and the Lebesgue measure of B is null, since f is P-homeomorphism. Using by equation (5) it is easy to see that the set $\varphi(B)$ is invariant w.r.t. f_ρ. Due to ergodicity of f_ρ w.r.t. Lebesgue measure, follows that $\ell(\varphi(B)) = 0$ or 1. But, since $\ell(\varphi(C)) > 0$, implies that $\ell(\varphi(B)) = 1$. The map φ^{-1} displays $S^1 \setminus \varphi(B)$ into $S^1 \setminus B$. Using above assertions we get $\ell(S^1 \setminus \varphi(B)) = 0$ and $\ell(S^1 \setminus B) = 1$. Therefore, φ^{-1} is singular. Similarly we can show that if φ^{-1} is singular, then φ is also singular. □

Using lemmas 2.5 and 2.6 we get the following remark.

Remark 2.7. *Let* f *be a* P*-homeomorphism with irrational rotation number and* φ *be a conjugating map between* f *and* f_ρ*. Then the following properties are equivalent:*

(i) the map φ *is absolutely continuous;*

(ii) the map φ^{-1} *is absolutely continuous;*

(iii) the f*-invariant measure* μ *and the Lebesgue measure* ℓ *is mutually absolutely continuous i.e. they are equivalent.*

3. Prove of Main Theorem

Proof. Since φ is conjugation map between f and f_ρ i.e. $\varphi \circ f = f_\rho \circ \varphi$. Then for all $0 \le k \le k_{n+1}$ we have

$$f^{kq_n+q_{n-1}} = \varphi^{-1} \circ f_\rho^{kq_n+q_{n-1}} \circ \varphi. \tag{6}$$

By assumption of Theorem 1.4 and Remark 2.7 the conjugation map φ is absolutely continuous. From remark 2.7 implies that the map φ^{-1} is absolutely continuous homeomorphism. Below we will use the following well known fact (see [6]). Let $[a, b], [c, d] \subset \mathbb{R}$ be two intervals, let $v \in AC([c, d])$ (i.e. absolutely continuous on $[c, d]$) and $u : [a, b] \to [c, d]$ be monotone and $u \in AC([a, b])$, then $v \circ u \in AC([a, b])$ and the chain rule equality holds i.e.

$$D(v(u(x)) = Dv(u(x))Du(x) \quad \text{a.e. (w.r.t. } \ell) \tag{7}$$

Since φ, $\varphi^{-1} \in AC(S^1)$ and $f_\rho^{kq_n+q_{n-1}}$ is linear, using above fact and (6) we get the following relation.

$$Df^{kq_n+q_{n-1}}(x) = D\varphi^{-1}(f_\rho^{kq_n+q_{n-1}}(\varphi(x)))D\varphi(x) \text{ a.e. (w.r.t. } \ell). \tag{8}$$

Hence, $\varphi^{-1}(\varphi(x)) = x$ and φ, $\varphi^{-1} \in AC(S^1)$ we have

$$D\varphi^{-1}(\varphi(x))D\varphi(x) = 1 \quad \text{a.e.} \quad (\text{w.r.t. } \ell). \tag{9}$$

Using equalities (8) and (9), we get

$$\log Df^{kq_n+q_{n-1}}(x) = \log D\varphi^{-1}(f_\rho^{kq_n+q_{n-1}}(\varphi(x))) - \log D\varphi^{-1}(\varphi(x)) \text{ a.e. (w.r.t. } \ell). \tag{10}$$

By assumption of Theorem 1.4 and Remark 2.7 the f-invariant measure μ and the Lebesgue measure ℓ are equivalent measures and therefore the relations (8), (9) and (10) hold with respect to f-invariant measure μ. Integrating last equality by measure $d\mu$ we have

$$\int_{S^1} |\log Df^{kq_n+q_{n-1}}(x)|d\mu = \int_{S^1} |\log D\varphi^{-1}(f_\rho^{kq_n+q_{n-1}}(\varphi(x))) - \log D\varphi^{-1}(\varphi(x))|d\mu. \tag{11}$$

Now we will show that the right site of (11) tending to zero when $n \to \infty$. Since, $\varphi_*\mu = \ell$ (i.e. $\mu(B) = \ell(\varphi(B))$ for any Borel set B) the equality $d\mu = D\varphi d\ell$ holds. Using this equality and changing of variable $x \to \varphi(x)$ we can write the right site of equality (11) as follows.

$$\int_{S^1} |\log D\varphi^{-1}(f_\rho^{kq_n+q_{n-1}}(\varphi(x))) - \log D\varphi^{-1}(\varphi(x))|d\mu =$$

$$= \int_{S^1} |\log D\varphi^{-1}(f_\rho^{kq_n+q_{n-1}}(x)) - \log D\varphi^{-1}(x)|d\ell. \tag{12}$$

Due to Remark 2.7 $\log D\varphi^{-1} \in L_1(S^1, d\ell)$. Well known fact that the class $C([a,b])$ of continuous functions on $[a,b]$ is dense (in $\|\cdot\|_{L_1}$) in $L_1([a,b], d\ell)$ (see [11]). From this fact implies that for any $\epsilon > 0$ there exists a continuous function ψ_ϵ and ℓ-integrable function ϕ_ϵ such that $\log D\varphi^{-1} = \psi_\epsilon + \phi_\epsilon$ and $\|\phi_\epsilon\|_{L_1} \leq \epsilon$. Using this fact, we have

$$\|\log D\varphi^{-1}(f_\rho^{kq_n+q_{n-1}}) - \log D\varphi^{-1}\|_{L_1} \leq$$

$$\leq \|\psi_\epsilon \circ f_\rho^{kq_n+q_{n-1}} - \psi_\epsilon\|_{L_1} + \|\phi_\epsilon \circ f_\rho^{kq_n+q_{n-1}} - \phi_\epsilon\|_{L_1}.$$

Using property of dynamical partition it is easy to see that $(x, f_\rho^{kq_n+q_{n-1}}(x)) \subset (x, f_\rho^{q_{n-1}}(x))$ for all $0 \leq k \leq k_{n+1}$ and $|f_\rho^{q_{n-1}}(x) - x| \leq q_n^{-1}$, for any $x \in S^1$. As ψ_ϵ is uniformly continuous on S^1 and by exponential refinement $f_\rho^{kq_n+q_{n-1}}(x)$ uniformly tends to x, there exists a positive integer $n_0 = n_0(\epsilon)$ such that for all $n \geq n_0$, $\|\psi_\epsilon \circ f_\rho^{kq_n+q_{n-1}} - \psi_\epsilon\|_{L_1} \leq \epsilon$. Since the function $f_\rho^{kq_n+q_{n-1}}(x)$ is linear function, by changing of variable $x \to f_\rho^{kq_n+q_{n-1}}(x)$ we have $\|\phi_\epsilon \circ f_\rho^{kq_n+q_{n-1}}\|_{L_1} = \|\phi_\epsilon\|_{L_1}$. Therefore,

$$\|\log D\varphi^{-1}(f_\rho^{kq_n+q_{n-1}}) - \log D\varphi^{-1}\|_{L_1} \leq 3\epsilon. \tag{13}$$

Since $\epsilon > 0$ was arbitrary and sufficiently small and using relations (11)-(13) we get

$$\lim_{n\to\infty} \int_{S^1} |\log Df^{kq_n+q_{n-1}}(x)|d\mu = 0. \tag{14}$$

Denote by $B(\delta, k, n) = \{x : |\log Df^{kq_n+q_{n-1}}(x)| \geq \delta\}$. It is clear that

$$\int_{S^1} |\log Df^{kq_n+q_{n-1}}(x)|d\mu \geq \delta\mu(B(\delta, k, n)).$$

Using this inequality together with relation (14) we get $\lim_{n\to\infty} \mu(B(\delta, k, n)) = 0$. Therefore, the f-invariant measure μ and the Lebesgue measure ℓ are equivalent measures we get

$$\lim_{n\to\infty} \ell(B(\delta, k, n)) = 0.$$

□

4. Acknowledgements
The first and second authors are grateful to Universiti Kebangsaan Malaysia for providing financial support via the grants UKM-MI-OUP-2011 and UKM-DIP-2012-31. The third author A. Dzhalilov visited Stony Brook university by Fulbright Grant (for 2011-2012 year). A. Dzhalilov very grateful Fulbright program for financial support and Stony Brook university for hospitality.

References
[1] I.P. Cornfeld, S.V. Fomin and Ya.G. Sinai, Ergodic Theory, Springer Verlag, Berlin (1982).
[2] A. Denjoy, Sur les courbes définies par les équations differentielles à la surface du tore. J. Math. Pures Appl., **11**, 333-375 (1932).
[3] A.A. Dzhalilov, K. M. Khanin, On invariant measure for homeomorphisms of a circle with a break point. Funct. Anal. Appl. **32**, 153-161 (1998).
[4] A.A. Dzhalilov and I. Liousse, Circle homeomorphisms with two break points. Nonlinearity 19:8, 1951-1968, (2006).
[5] A.A. Dzhalilov D. Mayer and U.A. Safarov, Pieswise-smooth circle homeomorphisms with several break points. Izvestiya: Mathematics 76:1 95-113. Trans: Izvestiya RAN: Ser. Mat. 76:1 101-120, (2012).
[6] Giovanni Leoni, A fitst course in Sobolev spaces. Amer. Math. Soc. Providence, Rhode Island, (1967).
[7] M. Herman, Sur la conjugaison différentiable des difféomorphismes du cercle à des rotations. Inst. Hautes Etudes Sci. Publ. Math., **49**, 5-234 (1979).
[8] Y. Katznelson and D. Ornstein, The differentiability of the conjugation of certain diffeomorphisms of the circle. Ergod. Theor. Dyn. Syst., **9**, 643-680 (1989).
[9] K.M. Khanin and Ya.G. Sinai, Smoothness of conjugacies of diffeomorphisms of the circle with rotations. Russ. Math. Surv., **44**, 69-99 (1989), translation of Usp. Mat. Nauk, **44**, 57-82 (1989).
[10] K.M. Khanin and A.Yu. Teplinskii, Herman's theory revisited. Invent. math., **178**, 333-344 (2009).
[11] M. Reed, B. Simon, Methods of modern mathematical physics. San Diego. (1980).
[12] M. Stein, Groups of piecewise linear homeomorphisms, Trans. Amer. Math. Soc. 332:2 (1992), 477-514.

Journal of Physics: Conference Series **435** (2013) 012012 doi:10.1088/1742-6596/435/1/012012

On a p-adic Cubic Generalized Logistic Dynamical System

Farrukh Mukhamedov[1], **Wan Nur Fairuz Alwani Wan Rozali**[2]

[1,2] Computational And Theoretical Sciences Department, Faculty of Science, International Islamic University Malaysia, Bandar Indera Mahkota, 25200, Kuantan, Malaysia

E-mail: [1]farrukh_m@iium.edu.my, [2]fairuzalwani@gmail.com

Abstract. Applications of p-adic numbers mathematical physics, quantum mechanics stimulated increasing interest in the study of p-adic dynamical system. One of the interesting investigations is p-adic logistics map. In this paper, we consider a new generalization, namely we study a dynamical system of the form $f_a(x) = ax(1-x^2)$. The paper is devoted to the investigation of a trajectory of the given system. We investigate the generalized logistic dynamical system with respect to parameter a and we restrict ourselves for the investigation of the case $|a|_p < 1$. We study the existence of the fixed points and their behavior. Moreover, we describe their size of attractors and Siegel discs since the structure of the orbits of the system is related to the geometry of the p-adic Siegel discs.

1. Introduction

Over the last century, p-adic numbers and p-adic analysis have come to play major role in the number theory. P-adic numbers were first introduced by K. Hensel. During a century after their discovery they were considered mainly objects of pure mathematics. Starting from 1980's various models described in the language of p-adic analysis have been actively studied. P-adic numbers have been widely used in many applications mostly in p-adic mathematical physics, quantum mechanics and others which roses the interest in the study of p-adic dynamical systems (see for example, [6, 18, 42, 44, 45]).

On the other hand, the study of p-adic dynamical systems arises in Diophantine geometry in the constructions of canonical heights, used for counting rational points on algebraic vertices over a number field, as in [12]. In [24] the p-adic field have arisen in physics in the theory of superstrings, promoting questions about their dynamics. Also some applications of p-adic dynamical systems to some biological, physical systems were proposed in [8, 3, 4, 15, 25]. In [9],[27] dynamical systems (not only monomial) over finite field extensions of the p-adic numbers were considered. Other studies of non-Archimedean dynamics in the neighborhood of a periodic and of the counting of periodic points over global fields using local fields appeared in [26, 20, 29, 28, 36]. Certain rational p-adic dynamical systems were investigated in [22],[31],[33], which appear from problems of p-adic Gibbs measures [23, 32, 34, 35]. Note that in [37, 10, 11] a general theory of p-adic rational dynamical systems over complex p-adic filed $\mathbb{C}_p$ has been developed.

The most studied discrete p-adic dynamical systems (iterations of maps) are so-called monomial systems. In [5],[25] the behavior of a p-adic dynamical system $f(x) = x^n$ in the fields of p-adic numbers $\mathbb{Q}_p$ and $\mathbb{C}_p$ was investigated. In [25] perturbated monomial dynamical systems defined by functions $f_q(x) = x^n + q(x)$, where the perturbation $q(x)$ is a polynomial whose coefficients have small p-adic absolute value, have been studied. It was investigated the connection between monomial and perturbated monomial systems. These investigations show that the study of perturbated dynamical systems is important (see [16]). Note that for a quadratic function $f(x) = x^2 + c$, $c \in \mathbb{Q}_p$ its chaotic behavior is complicated (see [41, 4, 40, 46]). In [40, 14] the Fatou and Julia sets of logistic p-adic dynamical system, i.e. $f(x) = rx(1-x)$, have been studied. Note that the logistic map and generalized logistic maps are well known in the literature and it is of great importance in the study of dynamical systems (see [7, 13, 21]). Certain ergodic and mixing properties of monomial and perturbated dynamical systems have been considered in [1, 17, 19].

In [30] the asymptotic behavior of a nonlinear p-adic dynamical system, especially a generalized p-adic logistic map $G(x) = (ax)^2(x+1)$ has been investigated. On the other hand, such a cubic dynamical system is also a perturbated cubic dynamical system, since it can be reduced to the form $f(x) = x^3 + ax^2$. In the present paper we are going to consider an other kind of cubic logistic mapping, i.e. $f_a(x) = ax(1-x^2)$. We will show that the defined two cubic dynamical systems are not almost conjugate. In this paper, we restrict ourselves to the case $|a|_p < 1$. Other values will be studied elsewhere. So, for such a mapping we will investigate the basin of attraction of such a dynamical system. Note that globally attracting sets play an important role in dynamics, restricting the asymptotic behavior to certain regions of the phase space. However, descriptions of the global attractor can be difficult as it may contain complicated chaotic dynamics. Moreover, we also describe the Siegel discs of the system, since the structure of the orbits of the system is related to the geometry of the p-adic Siegel discs (see [2]).

2. Preliminaries

2.1. p-adic numbers

Let $\mathbb{Q}$ be the field of rational numbers. Throughout the paper p will be a fixed prime number. Every rational number $x \neq 0$ can be represented in the form $x = p^r \frac{n}{m}$, where $r, n \in \mathbb{Z}$, m is a positive integer and p, n, m are relatively prime. The p-adic norm of x is given by $|x|_p = p^{-r}$ and $|0|_p = 0$. This norm satisfies so called the strong triangle inequality

$$|x+y|_p \leq \max\{|x|_p, |y|_p\}.$$

From this inequality one can infer that

$$\text{if } |x|_p \neq |y|_p, \text{ then } |x-y|_p = \max\{|x|_p, |y|_p\} \quad (1)$$

$$\text{if } |x|_p = |y|_p, \text{ then } |x-y|_p \leq |2x|_p. \quad (2)$$

This is a ultrametricity of the norm. The completion of $\mathbb{Q}$ with respect to the p-adic norm defines the p-adic field which is denoted by $\mathbb{Q}_p$. Note that any p-adic number $x \neq 0$ can be uniquely represented in the canonical series:

$$x = p^{\gamma(x)}(x_0 + x_1 p + x_2 p^2 + ...), \quad (3)$$

where $\gamma = \gamma(x) \in \mathbb{Z}$ and x_j are integers, $0 \leq x_j \leq p-1$, $x_0 > 0$, $j = 0, 1, 2, ...$ (see more detail [?, 38, 39]). Observe that in this case $|x|_p = p^{-\gamma(x)}$.

Denote

$$\mathbb{Z}_p = \{x \in \mathbb{Q}_p : |x|_p \leq 1\}, \quad \mathbb{Z}_p^* = \{x \in \mathbb{Q}_p : |x|_p = 1\}.$$

We recall that an integer $a \in \mathbb{Z}$ is called *a quadratic residue modulo* p if the equation $x^2 \equiv a (\text{mod } p)$ has a solution $x \in \mathbb{Z}$.

Definition 2.1 *Let p be an odd prime i.e. $p > 2$ and a be an integer not divisible by p; i.e. $(a, p) = 1$. The Legendre Symbol*

$$\left(\frac{a}{b}\right) = \begin{cases} 1, & \textit{If } a \textit{ is a quadratic residue of } p; \\ -1, & \textit{If } a \textit{ is not a quadratic residue of } p. \end{cases}$$

One has the following useful

Theorem 2.2 *(Euler's criterion) Let $p > 2$ be prime number and a be an integer not divisible by p. Then*

$$\left(\frac{a}{p}\right) \equiv a^{\frac{p-1}{2}} (\mod p)$$

Now using the last theorem we infer the following

Lemma 2.3 *[38] Let $a \in \mathbb{Q}_p$, $a \neq 0$ and $p \geq 3$ such that*

$$a = \frac{a^*}{|a|_p}, a^* \in \mathbb{Z}_p^*$$

Then, $\sqrt{a}$ exists if and only if it satisfies these two conditions:

(i) $\log_p |a|_p$ *is even*

(ii) $(a^*)^{\frac{p-1}{2}} \equiv 1(\mod p)$

However, at $p = 2$, $\sqrt{a}$ exists if and only if it satisfies these two conditions:

(i) $\log_2 |a|_2$ *is even*

(ii) $(a^*) \equiv 1(\mod 2^3)$

In particular case, let $a^* \in \mathbb{Z}_p^*$ $(p > 2)$, then $\sqrt{a}$ exists if and only if

$$(a^*)^{\frac{p-1}{2}} \equiv 1(\mod p)$$

Theorem 2.4 *Let p be prime number and $p > 2$. Then we have*

(i) The p-adic number -1 is a square in $\mathbb{Q}_p$ if and only if $p \equiv 1(\mod 4)$ and the p-adic number -1 is not a square in $\mathbb{Q}_p$ if and only if $p \equiv 3 \mod (4)$

(ii) The p-adic number 2 is a square in $\mathbb{Q}_p$ if and only if $p \equiv \pm 1(\mod 8)$ and the p-adic number 2 is not a square in $\mathbb{Q}_p$ if and only if $p \equiv \pm 3 \mod (8)$

(iii) The p-adic number -2 is a square in $\mathbb{Q}_p$ if and only if $p \equiv 1$ or $3(\mod 8)$ and the p-adic number -2 is not a square in $\mathbb{Q}_p$ if and only if $p \equiv -1$ or $-3 \mod (8)$

(iv) Let $p > 3$ then the p-adic number 3 is a square in $\mathbb{Q}_p$ if and only if $p \equiv \pm 1(\mod 12)$ and the p-adic number 3 is not a square in $\mathbb{Q}_p$ if and only if $p \equiv \pm 5 \mod (12)$

(v) Let $p > 3$ then the p-adic number -3 is a square in $\mathbb{Q}_p$ if and only if $p \equiv 1$ or $-5(\mod 12)$ and the p-adic number -3 is not a square in $\mathbb{Q}_p$ if and only if $p \equiv -1$ or $5 \mod (12)$

2.2. Dynamical Systems in $\mathbb{Q}_p$

For any $a \in \mathbb{Q}_p$ and $r > 0$ denote

$$\bar{B}_r(a) = \{x \in \mathbb{Q}_p : |x - a|_p \leq r\}, \ B_r(a) = \{x \in \mathbb{Q}_p : |x - a|_p < r\},$$

$$S_r(a) = \{x \in \mathbb{Q}_p : |x - a|_p = r\}.$$

A function $f : B_r(a) \to \mathbb{Q}_p$ is said to be *analytic* if it can be represented by

$$f(x) = \sum_{n=0}^{\infty} f_n(x-a)^n, \quad f_n \in \mathbb{Q}_p,$$

which converges uniformly on the ball $B_r(a)$.

Note the basics of p-adic analysis, p-adic mathematical physics are explained in [38, 39, 43]

Consider a dynamical system (f, B) in $\mathbb{Q}_p$, where $f : x \in B \to f(x) \in B$ is an analytic function and $B = B_r(a)$ or $\mathbb{Q}_p$. Denote $x^{(n)} = f^n(x^{(0)})$, where $x^0 \in B$ and $f^n(x) = \underbrace{f \circ \cdots \circ f(x)}_{n}$. If $f(x^{(0)}) = x^{(0)}$ then $x^{(0)}$ is called a *fixed point*. A fixed point $x^{(0)}$ is called an *attractor* if there exists a neighborhood $U(x^{(0)})(\subset B)$ of $x^{(0)}$ such that for all points $y \in U(x^{(0)})$ it holds $\lim_{n\to\infty} y^{(n)} = x^{(0)}$, where $y^{(n)} = f^n(y)$. If $x^{(0)}$ is an attractor then its *basin of attraction* is

$$A(x^{(0)}) = \{y \in \mathbb{Q}_p : \ y^{(n)} \to x^{(0)}, \ n \to \infty\}.$$

A fixed point $x^{(0)}$ is called *repeller* if there exists a neighborhood $U(x^{(0)})$ of $x^{(0)}$ such that $|f(x) - x^{(0)}|_p > |x - x^{(0)}|_p$ for $x \in U(x^{(0)})$, $x \neq x^{(0)}$. For a fixed point $x^{(0)}$ of a function $f(x)$ a ball $B_r(x^{(0)})$ (contained in B) is said to be a *Siegel disc* if each sphere $S_\rho(x^{(0)})$, $\rho < r$ is an invariant sphere of $f(x)$, i.e. if $x \in S_\rho(x^{(0)})$ then all iterated points $x^{(n)} \in S_\rho(x^{(0)})$ for all $n = 1, 2 \ldots$. The union of all Siegel discs with the center at $x^{(0)}$ is said to *a maximum Siegel disc* and is denoted by $SI(x^{(0)})$.

Remark 2.5 *In non-Archimedean geometry, a center of a disc is nothing but a point which belongs to the disc, therefore, in principle, different fixed points may have the same Siegel disc.*

Let $x^{(0)}$ be a fixed point of an analytic function $f(x)$. Set

$$\lambda = \frac{d}{dx} f(x^{(0)}).$$

The point $x^{(0)}$ is called *attractive* if $0 \leq |\lambda|_p < 1$, *indifferent* if $|\lambda|_p = 1$, and *repelling* if $|\lambda|_p > 1$.

Theorem 2.6 *[25] Let a be a fixed point of the analytic function $f : B \to K$. Then,*

i. If $r > 0$ satisfies the inequality

$$\max_{1\leq n\leq\infty} \left| \frac{1}{n!} \cdot \frac{d^n f}{dx^n}(a) \right|_k r^{n-1} < 1 \tag{4}$$

and $B_r(a) \subset B$, then $B_r(a) \subset A(a)$.

ii. If a is a neutral point of f then it is the center of a Siegel disc. If $r > 0$ satisfies the inequality

$$S = \max_{2\leq n<\infty} \left| \frac{1}{n!} \cdot \frac{d^n f}{dx^n}(a) \right|_k r^{n-1} < |f'(a)|_k$$

and $B_r(a) \subset B$, then $B_r(a) \subset SI(a)$

iii. If a is a repelling point of f, then a is a repeller of the dynamical system.

3. A generalized cubic p-adic logistic map

In this section we are going to study the dynamical system given by

$$f_a(x) = ax(1 - x^2), x \in \mathbb{Q}_p \tag{5}$$

where $a \in\mathbb{Q}_p$. Such a mapping is call a *generalized cubic logistic mapping*, and the rest of the paper is devoted to the investigation of such a dynamical system. As mentioned before, another kind of generalize logistic map given by

$$G_b(x) = x^3 + bx^2, x \in \mathbb{Q}_p \tag{6}$$

where $b \in\mathbb{Q}_p$, has been studied in [30]. We will show for a few certain values of parameter a and b, they are conjugate but for the rest they are not. Here by conjugacy we mean a one-to-one mapping $\tau : \mathbb{Q}_p \to \mathbb{Q}_p$ such that $\tau \circ f = g \circ \tau$.

Proposition 3.1 *The mappings f_a and G_b are conjugate via $\tau(x) = \alpha x + \beta$ iff one has*

$$\alpha = \sqrt{-\frac{3}{2}}, \ \beta = \sqrt{-\frac{1}{2}}, \ a = \frac{3}{2}, \ b = -3\sqrt{-\frac{1}{2}}.$$

Now using Lemma 2.3 one can establish the following

Proposition 3.2 *The numbers $-\frac{3}{2}$ and $-\frac{1}{2}$ are the square of some numbers in $\mathbb{Q}_p$ (the same time) if and only if*

$$\left(\frac{p-3}{2}\right)^{\frac{p-1}{2}} \equiv 1(\mod p)$$

and

$$\left(\frac{p-1}{2}\right)^{\frac{p-1}{2}} \equiv 1(\mod p)$$

Remark 3.3 *From the Propositions 3.1, 3.2, we conclude that if f_a and G_b are conjugate, then $|a|_p = |b|_p = 1$. Moreover, such a case in [30] is not fully investigated. For the other values of a, this two mappings are not conjugate. Therefore, it is reasonable to investigate behavior of the dynamical system $f_a(x)$.*

3.1. Description and existence of the fixed points

In this section we will investigate the dynamics of generalized cubic logistic function by finding their fixed points, and their existence for the case $|a|_p < 1$.

Let us find its fixed points which means we need to solve the equation $f_a(x) = x$. One can find three formal fixed points:

$$x_1 = 0 \text{ and } x_{2,3} = \pm\sqrt{\frac{a-1}{a}}. \tag{7}$$

We can denote the fixed points as

$$Fix = \left\{x_1 = 0, x_2 = \sqrt{\frac{a-1}{a}}, x_3 = -\sqrt{\frac{a-1}{a}}\right\} \tag{8}$$

We have done finding the formal solutions for the fixed points. Note that not all these fixed points do exist. Therefore, let us move on to the existence of the fixed points. Note that Lemma 2.3 is useful in most parts of our investigation in establishing the existence of the fixed points.

Take into account that the fixed point $x_1 = 0$ is a trivial case and it always exist for any prime number. But the fixed points $x_{2,3} = \pm\sqrt{\frac{a-1}{a}}$ are not always exist for any prime number. Let us check the existence of the fixed points $x_{2/3}$ .Here, we restrict ourselves for the case $|a|_p < 1$.

As before, we have $|a|_p < 1$, then from the canonical representation of a, one finds

$$a = p^k(a_0 + a_1 p + \cdots), \ k \geq 1, \ a_0 \neq 0. \tag{9}$$

Proposition 3.4 *Let $|a|_p < 1$, then the following assertions hold true:*

(i) Let $p \geq 3$, then $\sqrt{\frac{a-1}{a}}$ exists if and only if $\log_p |a|_p$ is even and $(-a_0)^{\frac{p-1}{2}} \equiv 1(\mod p)$ has solution.

(ii) Let $p = 2$, then $\sqrt{\frac{a-1}{a}}$ exists if and only if $\log_p |a|_p$ is even number which is less than -3 and $a_0 \equiv 1(\mod 2^3)$.

Proof. From (9) we have

$$1 - a = 1 - p^k(a_0 + a_1p + \cdots) \tag{10}$$

If $p \geq 3$, then according to Lemma 2.3, the equation

$$x^2 \equiv 1(mod p)$$

has solution in $\mathbb{Q}$ (i.e. $x = p - 1$). Therefore, $\sqrt{1-a}$ exists. If $p = 2$, then from (10) and again Lemma 2.3 implies that $\sqrt{1-a}$ exists if and only if $k \geq 3$. Hence from $\sqrt{\frac{a-1}{a}} = \sqrt{\frac{1-a}{-a}}$, we conclude that $\sqrt{\frac{a-1}{a}}$ exists if and only if $\sqrt{-a}$ exists. Now using Lemma 2.3 we get the desired assertion. This completes the proof.

3.2. Description Of the Behavior

In this section, we will describe the behavior of the function (5) with respect to the parameter a ($|a|_p < 1$) whether the fixed points x_1, x_2, x_3 is attracting, neutral or repelling. Form (5), we have

$$f'(x) = a - 3ax^2 \tag{11}$$

We already know those fixed points exist for some values of prime number. Therefore in what follows, we assume the existence ones. Now,let us check their behavior and study the value of

$$|f'(x^{(0)})|_p = |a - 3a(x^{(0)})^2|_p, \tag{12}$$

where $x^{(0)}$ is a fixed point. We substitute the fixed points into (12), one gets

$$|f'(x_1)|_p = |a|_p \text{ and } |f'(x_{2/3})|_p = |a - 3a\left(\frac{a-1}{a}\right)|_p = |3 - 2a|_p \tag{13}$$

We have used the equality $x_{2/3}^2 = \frac{a-1}{a}$.

Let us start with the first fixed point $x_1 = 0$. From (13) immediately find that x_1 is an attracting fixed point.

Next, let us turn to fixed point $x_{2/3}$. If we consider $p = 3$, then from (13) we find

$$|3 - 2a|_3 < 1 \tag{14}$$

This means $x_{2/3}$ are also attracting fixed points. However, if we let $p \neq 3$, from the strong triangle inequality we obtain

$$|f'(x_{2/3})|_p = |3 - 2a| = 1$$

Hence, the fixed points $x_{2/3}$ are neutral or indifferent fixed point. So taking into account Proposition 3.4 we can formulate the following:

Proposition 3.5 *Let* $0 < |a|_p < 1$, *and* $a = \frac{a^*}{|a|_p}$ *with* $a^* = a_0 + a_1p + a_2p^2 + \cdots$, *where* $a^* \in \mathbb{Z}_p^*$. *Then the following statements hold true:*

(I) Let $p > 3$,

(a) If $\log_p |a|_p$ *is even and* $(-a_0)^{\frac{p-1}{2}} \equiv 1(\mod p)$, *then* $x_1 = 0$ *is an attracting and* $x_{2/3}$ *are indifferent fixed points.*

(b) In the rest cases, the dynamical system has unique attracting fixed point $x_1 = 0$.

(II) Let $p = 3$,

(a) *If* $\log_3 |a|_3$ *is even and* $a_0 = 2$, *then all three fixed points are attracting;*

(b) *In the rest cases, the dynamical system has unique attracting fixed point* $x_1 = 0$.

(III) *Let* $p = 2$,

(a) *If* $\log_2 |a|_2 \leq 4$ *and is even with* $a^* \equiv -1(\mod 8)$. *Then the fixed point* $x_1 = 0$ *is an attracting and* $x_{2/3}$ *are indifferent fixed points.*

(b) *In the rest cases, the dynamical system has unique attracting fixed point* $x_1 = 0$.

4. Attractors and Siegel Discs of Fixed Points

In the previous section, we have established the behavior of the fixed points of the dynamical system. Using those results, in this section we are going to describe the size of attractors and Siegel discs of the system. Before going to details, let us formulate ceratain auxiliary facts. Let us assume that $x^{(0)}$ is a fixed point of f. Now taking into account Theorem 2.6 one gets

Proposition 4.1 *Let* $x^{(0)}$ *be a fixed point of* f *given by* $f_a(x) = ax(1-x^2)$ *then the following assertions hold true:*

(i) *If* $x^{(0)}$ *is an attractive point of* f, *then it is an attractor of the dynamical system. If* $r > 0$ *satisfies the inequality*

$$\max\{|3ax^{(0)}|_p \cdot r; |a|_p \cdot r^2\} < 1 \tag{15}$$

then $\overline{B_r(x^{(0)})} \subset A(x^{(0)})$.

(ii) *If* $x^{(0)}$ *is a neutral point of* f, *then it is the center of a Siegel disc. If* $r > 0$ *satisfies the inequality* (15) *then*

$$\overline{B_r(x^{(0)})} \subset SI(x^{(0)})$$

Let us study the fixed point x_1, which is attractive. One can see that

$$\max\{|3ax^{(0)}|_p \cdot r; |a|_p \cdot r^2\} < 1 \text{ if and only if } r < \frac{1}{\sqrt{|a|_p}}.$$

Therefore due to Proposition 4.1 we have

$$\overline{B_r(x_1)} \subset A(x_1), \quad r < \frac{1}{\sqrt{|a|_p}}.$$

Now assume that $|x|_p \geq \frac{1}{\sqrt{|a|_p}}$. From $|a|_p < 1$ one gets $|x|_p > 1$. By means of the strong triangle inequality we get

$$|f_a(x)|_p = |a|_p |x|_p |1 - x^2|_p = |a|_p |x|_p^3 \geq |a|_p \cdot \frac{1}{|a|_p \sqrt{|a|_p}} = \frac{1}{\sqrt{|a|_p}} > 1$$

Therefore

$$A(x_1) \subset B_{\frac{1}{\sqrt{|a|_p}}}(x_1) \text{ and } \overline{B_r(x_1)} \subset A(x_1) \text{ for any } r < \frac{1}{\sqrt{|a|_p}}$$

Therefore, one may formulate the following:

Proposition 4.2 *Let* $|a|_p < 1$, *then* $A(x_1) = B_{\frac{1}{\sqrt{|a|_p}}}(x_1)$.

Now, let us turn to the fixed points $x_{2/3} = \sqrt{\frac{a-1}{a}}$. Due to Proposition 3.5, we should consider two cases (i) $p \neq 3$ and (ii) $p = 3$.

Proposition 4.3 *Let $|a|_p < 1$ and $p = 3$, then $x_{2/3}$ is attractive, and one has*

$$A(x_{2/3}) = B_{\frac{1}{\sqrt{|a|_p}}}(x_{2/3}).$$

Proof. Due to Proposition 4.1 we have $r < \frac{1}{\sqrt{|a|_p}}$, so $\overline{B_r(x_{2/3})} \subset A(x_{2/3})$.

Let $|x - x_{2/3}|_p \geq \frac{1}{\sqrt{|a|_p}}$, then one has

$$|f_a(x) - x_2|_p = |a|_p|x - x_2|_p|x_2^2 + x \cdot x_2 + x^2 - 1|_p$$

By letting $x - x_2 = \gamma$, where $|\gamma|_p \geq \frac{1}{\sqrt{|a|_p}}$, one gets

$$|f_a(x) - x_2|_p = |\gamma|_p|a\gamma^2 + 3a\gamma x_2 + 2a - 3|_p. \tag{16}$$

By putting $P(\gamma) = a\gamma^2 + 3a\gamma x_2 + 2a - 3$, we find

$$|P(\gamma)|_p = |a\gamma^2 + 3a\gamma x^2 + 2a - 3|_p = |a\gamma(\gamma + 3x_2) + 2a - 3|_p. \tag{17}$$

Now taking into account $|2a - 3|_p < 1$, $|\gamma + 3x_2|_p = |\gamma|_p$, $|\gamma|_p \geq \frac{1}{\sqrt{|a|_p}}$ and $|3x_2|_p = \frac{1}{\sqrt[3]{|a|_p}}$ with the strong triangle inequality we obtain $|P(\gamma)|_p = |a\gamma^2|_p$. This with (16) yields

$$|f_a(x) - x_2|_p = |a\gamma^3|_p \geq \frac{1}{\sqrt{|a|_p}}.$$

Therefore,

$$B_{\frac{1}{\sqrt{|a|_p}}}(x_{2/3}) \supset A(x_{2/3})$$

which gives us that

$$\overline{B_r}(x_{2/3}) \subset A(x_{2/3}) \text{ for any } r < \frac{1}{\sqrt{|a|_p}}$$

this completes the proof.

Proposition 4.4 *Let $|a|_p < 1$, $p \neq 3$ and $\sqrt{-a}$ exists. Then the following statements hold true:*

i. If $|P(\gamma)|_p = 1$ for all γ with $|\gamma| = \frac{1}{\sqrt{|a|_p}}$, then

$$SI(x_{2/3}) = \bar{B}_{\frac{1}{\sqrt{|a|_p}}}(x_{2/3}).$$

ii. If $|P(\gamma_0)|_p < 1$ for some γ_0 with $|\gamma_0|_p = \frac{1}{\sqrt{|a|_p}}$, then

$$SI(x_{2/3}) = B_{\frac{1}{|a|_p}}(x_{2/3})$$

(iii) If $p \neq 2$, then $SI(x_2) \cap SI(x_3) = \varnothing$. If $p = 2$, then $SI(x_2) = SI(x_3)$.

Proof. The items (i) and (ii) can be proven by the same argument as above. To establish (iii) it is enough to find a relation between two the balls $B_{\frac{1}{\sqrt{|a|_p}}}(x_2)$ and $B_{\frac{1}{\sqrt{|a|_p}}}(x_3)$ whether they are coincide to each other or not. First, let us consider

$$|x_2 - x_3|_p = \left|\sqrt{\frac{a-1}{a}} + \sqrt{\frac{a-1}{a}}\right|_p = \left|2\sqrt{\frac{a-1}{a}}\right|$$

Now, suppose $p \neq 2$, then we obtain

$$|x_2 - x_3|_p = \frac{1}{\sqrt{|a|_p}}$$

Hence, at $p \neq 2$ we have

$$x_2 \notin B_{\frac{1}{\sqrt{|a|_p}}}(x_3),$$

this means

$$B_{\frac{1}{\sqrt{|a|_p}}}(x_2) \cap B_{\frac{1}{\sqrt{|a|_p}}}(x_3) = \varnothing.$$

Next, let us consider the case at $p = 2$. Again by distance we have

$$|x_2 - x_3|_p = \frac{1}{2} \cdot \frac{1}{\sqrt{|a|_p}} < \frac{1}{\sqrt{|a|_p}}$$

This means

$$x_3 \in B_{\frac{1}{\sqrt{|a|_p}}}(x_2) \Rightarrow B_{\frac{1}{\sqrt{|a|_p}}}(x_2) = B_{\frac{1}{\sqrt{|a|_p}}}(x_3)$$

This completes the proof.

Now we want to find some sufficient conditions for the equality $SI(x_{2/3}) = B_{\frac{1}{|a|_p}}(x_{2/3})$.

Corollary 4.5 *Let $|a|_p < 1$, $p \neq 2, 3$ and $\sqrt{-a}$ exist. If $\sqrt{-3}$ exists, then*

$$SI(x_{2/3}) = B_{\frac{1}{\sqrt{|a|_p}}}(x_{2/3})$$

Proof. To establish $SI(x_{2/3}) = B_{\frac{1}{|a|_p}}(x_{2/3})$ it is enough to solve the equation $P(\gamma) = 0$. One can see the discriminant of the last equation is

$$D = a^2 + 3a.$$

Let $a = p^k \cdot \varepsilon$, then one gets

$$3a + a^2 = 3p^k \cdot \varepsilon + (p^{2k} \cdot \varepsilon^2) = p^k(3\varepsilon + p^k \cdot \varepsilon^2),$$

therefore $\sqrt{a^2 + 3a}$ exists if and only if k is even and

$$x^2 \equiv 3\varepsilon (\mod p)$$

has a solution at $p \neq 3$ (see Lemma 2.3).But in the last conclusion equivalent to the existence of $\sqrt{3a}$. We know that $\sqrt{-a}$ exists, therefore, $\sqrt{3a}$ exists iff $\sqrt{-3}$ exists. Hence, we conclude $\sqrt{a^2 + 3a}$ exists if and only if $\sqrt{-3}$ exists. Now assume that $\sqrt{a^2 + 3a}$ exists. Then one can see that the solutions of (17) are the following:

$$\gamma_{1/2} = \frac{-3ax_2 \pm \sqrt{D}}{2a}. \tag{18}$$

From $p \neq 2$ and Vietta's theorem, we can directly find that

$$|\gamma_1 + \gamma_2|_p| = | - 3x_2|_p = \frac{1}{\sqrt{|a|_p}} \tag{19}$$

$$|\gamma_1 \cdot \gamma_2| = \frac{|2a - 3|_p}{|a|_p} = \frac{1}{|a|_p}. \tag{20}$$

Now, we are interested in finding the explicit value of $|\gamma_1|_p$. Assume $|\gamma_1|_p \neq |\gamma_2|_p$ or more explicitly $|\gamma_1|_p > |\gamma_2|_p$. Then, from the strong triangle inequality, we have

$$|\gamma_1 + \gamma_2|_p = |\gamma_1| = \frac{1}{\sqrt{|a|_p}}$$

On the other hand,

$$|\gamma_1 \cdot \gamma_2|_p = |\gamma_1|_p \cdot |\gamma_2|_p = \frac{1}{\sqrt{|a|_p}} \cdot |\gamma_2|_p = \frac{1}{|a|_p}.$$

Correspondingly, $|\gamma|_2 = \frac{1}{\sqrt{|a|_p}}$, but this contradicts to our assumption. Hence, $|\gamma_1|_p = |\gamma_2|_p$. So from (19), we got

$$|\gamma_1|_p = |\gamma_2|_p = \frac{1}{\sqrt{|a|_p}}$$

This with (16) implies

$$|f_a(\widetilde{x_0}) - x_2|_p = 0;$$

where $\widetilde{x_0} = x_2 + \gamma_1$, which means

$$S_{\frac{1}{\sqrt{|a|_p}}} \neq SI(x_2).$$

So, we have

$$SI(x_2) = B_{\frac{1}{\sqrt{|a|_p}}}(x_2)$$

This completes the proof.

Acknowledgments
The first named author acknowledges the Junior Associate scheme of the Abdus Salam International Centre for Theoretical Physics, Trieste, Italy.

References

[1] Anashin V 2006 In book: *p-adic Mathematical Physics, AIP Conference Proc.* **826**, 3
[2] Arrowsmith D K and Vivaldi F 1994 *Physica D.* **71** 222
[3] Albeverio S, Khrennikov A and Kloeden P E 1999 *BioSys.* **49** 105
[4] Albeverio S, Kloeden P E and Khrennikov A 1998 *Theor.Math. Phys.* **117** 1414
[5] Albeverio S, Khrennikov A, Tirozzi B, and De Smedt S, 1998 *Theor.Math. Phys.* **114** 276
[6] Aref'eva I Ya, Dragovich B, Frampton P H , and Volovich I V 1991 *Internat. J. Modern Phys. A.* **6** 4341
[7] Aulbach B and Kininger B 2004 *J. Diff. Eqs. Appl.* **10** 1243
[8] Avetisov V A, Bikulov A H, Kozyrev S V, and Osipov V A, 2002 *J. Phys. A: Math.Gen.* **35** 177
[9] Batra A, and Morton P, 1994 *Rocky Mountain J. of Math.* **24** 453; 905
[10] Benedetto R 2003 *Amer. J. Math.* **125** 581
[11] Bézivin J-P 2004 *Acta Arith.* **112** 151
[12] Call G and Silverman J 1993 *Composito Math.* **89** 163
[13] Devaney R L 1989 *An Introduction to Chaotic Dynamical Systems*, (Addison- Wesley: New York)
[14] Dremov V, Shabat G, and Vymova P 2006 In book: *p-adic Mathematical Physics, AIP Conference Proc.* **826** 43
[15] Dubischer D, Gundlach VM, Khrennikov A and Steinkamp O 1999 *Physica D.* **130** 1
[16] Fan A, Liao L, Wang Y F, and Zhao D 2007 *C.R. Acad. Sci.Paris* **344** 219
[17] Fan A., Li M.T., and Zhao D., 2007 Srict ergodicity of affine p-adic dynamical system on $\mathbb{Z}_p$, *Adv. Math.* **214**, 666-700.
[18] Freund P G O, and Witten E, 1987 *Phys. Lett. B* **199** 191
[19] Gundlach V M, Khrennikov A, and Lindahl K O, 2001 *Infin. Dimen. Anal. Quantum Probab. Relat. Top.* **4** 569
[20] Hsia L 2000 *J. London Math. Soc.* **62** 685
[21] Jaganathan R, and Sinha S 2005 *Phys. Lett. A.* **338** 277
[22] Khamraev M, and Mukhamedov FM, 2006 *Jour. Math. Anal. Appl.* **315**, 76
[23] Khamraev M, Mukhamedov F M, and Rozikov U A , 2004 *Lett.Math.Phys.* **70** 17

[24] Khrennikov A Yu 1997 *Non-Archimedean analysis: quantum paradoxes, dynamical systems and biological models*, (Kluwer: Dordreht) .
[25] Khrennikov A Yu and Nilsson M 2004 *p-adic deterministic and random dynamical systems* (Kluwer:Dordreht)
[26] Herman M., and Yoccoz J.-C., 1983 *Lec. Notes in Math.* **1007** (Springer:Berlin), 408
[27] Lindahl K O 2004 *Nonlinearity* **17** 745
[28] Ledrappier F and Pollicott M 2005 *Bull. Braz. Math. Soc. (N.S.)* **36** 143
[29] Lubin J 1994 *Composito Math.* **94** 321
[30] Mukhamedov F M and Mendes J F F 2007 *J. Diff. Eqs.* 59
[31] Mukhamedov F M 2007 *Appl. Math. Lett.* **20** 88
[32] Mukhamedov F M 2012 *Rep. Math. Phys.* **70** 385
[33] Mukhamedov F M, and Rozikov U A 2004 *Methods Funct. Anal. Topology*,**10** (2) 21
[34] Mukhamedov F M, and Rozikov U A 2004 *Indag. Math. N.S.* **15** 85
[35] Mukhamedov F M, and Rozikov U A 2005 *Infin. Dimens. Anal. Quantum Probab. Relat. Top.* **8** 277
[36] Pezda T 1994 *Acta Arith.* **66** 11
[37] Rivera-Letelier J 2003 *Astérisque* **287** 147
[38] Robert A M 2000 *A course of p-adic analysis* (Springer: New York)
[39] Schikhof W H 1984 *Ultrametric Calculus* (Cambridge Univ. Press: Cambridge)
[40] Shabat B 2004 *Proc. Steklov Math. Inst.* **245** 257
[41] Thiran E, Verstegen D, and Weters J 1989 *J.Stat. Phys.* **54** 893
[42] van Rooji A 1978 *Non-Archimedean functional analysis* (Marcel Dekker: New York)
[43] Vladimirov V S, Volovich I V , and Zelenov E I, 1994 *p-adic Analysis and Mathematical Physics* (World Scientific: Singapore)
[44] Volovich I V 2010 *p-Adic Numbers, Ultrametric Analysis Appl.* **2** 77
[45] Volovich I V 1987 *Class. Quantum Grav.* **4** L83
[46] Woodcock C F 1998 *Experimental Math.* **7** 333

On quadratic stochastic processes and related differential equations

Farrukh Mukhamedov[1], **Nurul Akma Supar** [2], **Pah Chin Hee**[3]

[1,2,3] Computational And Theoretical Sciences Department, Faculty of Science, International Islamic University Malaysia, Bandar Indera Mahkota, 25200, Kuantan, Malaysia

E-mail: [1]farrukh_m@iium.edu.my, [2]akmasupar@yahoo.com.my, [3]pahchinhee@gmail.com

Abstract. It is known that the theory of Markov process is a rapidly developing field with numerous applications to many branches of mathematics and physics, biology and so on. But there are some physical models which cannot be described by such processes. One of such models is related to population genetics. These processes are called quadratic stochastic processes (q.s.p.). In this theory it is important to construct nontrivial examples of such processes. In the present paper we are going to provide a construction of q.s.p. by means of two given processes. We should stress that such a construction allows us to produce lots of nontrivial examples of q.s.o. We also associate to given q.s.p. two kind of processes. Note that one of such processes is Markov. It is proved that such kind of processes uniquely define q.s.p. Moreover, we also derive some differential equations for q.s.p.

1. Introduction

It is known that Markov processes are well-developed field of mathematics which have various applications in physics, biology and so on. But there are some physical models which cannot be described by such processes. One of such models is a model related to population genetics. Namely, this model is described by quadratic stochastic processes (see [7, 1] for review). Let us recall the definition of the said process.

Let $E = \{1, 2, \dots, n\}$ be a finite set. Denote

$$S^{n-1} = \{(x_1, \dots, x_n) \in \mathbb{R}^n : x_n \geq 0, \sum_{k=1}^{n} x_k = 1\}.$$

Consider a family of functions $\{P_{ij,k}^{[s,t]} : i, j, k \in E, s, t \in \mathbb{R}_+, t - s \geq 1$ with an initial state $\mathbf{x}^{(0)} = (x_1^{(0)}, \dots, x_n^{(0)}) \in S^{n-1}$. This family is said to be *quadratic stochastic process (q.s.p.)* if for fixed $s, t \in \mathbb{R}_+$, it satisfies the following conditions :

i. $P_{ij,k}^{[s,t]} = P_{ji,k}^{[s,t]}$ for any $i, j, k \in E$;

ii. $P_{ij,k}^{[s,t]} \geq 0$ and $\sum_{k=1}^{n} P_{ij,k}^{[s,t]} = 1$ for any $i, j, k \in E$;

iii. An analogue of Kolmogorov-Chapman equation: there are two variants: for the initial point $\mathbf{x}^{(0)}$ and $s < r < t$ such that $t - r \geq 1, r - s \geq 1$ one has

(iii$_A$)

$$P^{[s,t]}_{ij,k} = \sum_{m,l,k=1}^{n} P^{[s,r]}_{ij,m} P^{[r,t]}_{ml,k} x^{(r)}_l ,$$

where $x^{(r)}_k$ is defined as follows :

$$x^{(r)}_k = \sum_{i,j=1}^{n} P^{[0,r]}_{ij,k} x^{[0]}_i x^{(0)}_j ;$$

(iii$_B$)

$$P^{[s,t]}_{ij,k} = \sum_{m,l,g,h=1}^{n} P^{[s,r]}_{im,l} P^{[s,r]}_{jg,h} P^{[r,t]}_{lh,k} x^{(s)}_m x^{(s)}_g .$$

In what follows by, $(E, P^{[s,t]}_{ij,k}, \mathbf{x}^{(0)})$ we denote the definite process. We say that the q.s.p is of type (A) or (B) if it satisfies the fundamental equations either (iii_A) or (iii_B), respectively.

In this definition the functions $P^{[s,t]}_{ij,k}$ denote the probability that under the interaction of the elements i and j at time s the element k comes into effect at time t. Since for physical, chemical and biological phenomena, a certain time is necessary for the realization of an interaction, we shall take the greatest such time to be equal to 1 (see the Boltzmann model [4] or the biological model [7]). Thus the probability $P^{[s,t]}_{ij,k}$ is defined for $t - s \geq 1$.

A set of quadratic stochastic processes one can decompose into following three classes:

(i) *homogeneous*, that is a transition functions $P^{[s,t]}_{ij,k}$ depends only on $t-s$ for arbitrary $x, y \in E,\ s, t \in \mathbb{R}_+$ such that $t - s \geq 1$

(ii) *homogeneous per unit time*, that is transition functions satisfy $P^{[t,t+1]}_{ij,k} = P^{[0,1]}_{ij,k}$ for all $t \geq 1$ and $i, j, k \in E$, but don't belong to the first class;

(iii) *nonhomogeneous*, if the process doesn't belong to the second class.

It should be noted that the quadratic stochastic processes are related to quadratic transformations (see [4, 6]) in the same way as Markov processes are related to linear transformations. The equations (iii$_A$) and (iii$_B$) can be interpreted as different laws of behavior of the off spring. Some examples of q.s.p. were given in [2],[3]. We note that quadratic processes of type (A) were considered in [2],[9]. In this theory it is important to construct nontrivial examples of such processes. In the present paper we are going to provide a construction of q.s.p. by means of two given processes. We should stress that such a construction allows us to produce lots of nontrivial examples of q.s.o. We also associate to given q.s.p. two kind of processes. Note that one of such processes is Markov. It is proved that such kind of processes uniquely define q.s.p. Moreover, we also derive some differential equations for q.s.p. Note that similar kind of equations were obtained in [9] for homogeneous per unit time q.s.o.

2. Marginal Markov processes related to q.s.p.

In this section we are going to prove that every q.s.p. can be defined uniquely by two kind of Markov processes.

Recall that a matrix $(Q_{ij})_{i,j \in E}$ is called *stochastic* if

$$Q_{ij} \geq 0, \ \sum_{j=1}^{n} Q_{ij} = 1.$$

A family of stochastic matrices $\{(Q^{[s,t]}_{ij})_{i,j \in E} :\ s, t \in \mathbb{R}_+, t - s \geq 1\}$ is called *Markov process* if the following condition holds: for every $s < r < t$ one has

$$Q^{[s,t]}_{ij} = \sum_{k=1}^{n} Q^{[s,r]}_{ik} Q^{[r,t]}_{kj} . \tag{1}$$

This equation is known as the Kolmogorov-Chapman equation .

Let $(E, P_{ij,k}^{[s,t]}, \mathbf{x}^{(0)})$ be a q.s.p. Let us define

$$H_{ij}^{[s,t]} = \sum_{\ell=1}^{n} P_{i\ell,j}^{[s,t]} x_l^{(s)}, \tag{2}$$

$$Q_{(ij)(uv)}^{[s,t]} = P_{ij,u}^{[s,t]} x_v^{(t)}, \tag{3}$$

where $i, j, u, v \in E$.

It is clear that $H_{ij}^{[s,t]}$ and $Q_{(ij)(uv)}^{[s,t]}$ are stochastic matrices.

Theorem 2.1 *Let $(E, P_{ij,k}^{[s,t]}, x^{(0)})$ be a q.s.p. type (A). Then defined processes $(H_{ij}^{[s,t]})_{i,j\in E}$ and $(Q_{(ij)(uv)}^{[s,t]})_{(ij),(uv)\in E\times E}$ are Markov processes.*

Proof. We have

$$\begin{aligned}
\sum_{k=1}^{n} H_{ik}^{[s,r]} H_{kj}^{[r,t]} &= \sum_{k=1}^{n}\Big(\sum_{l=1}^{n} P_{il,k}^{[s,r]} x_l^{(s)}\Big)\Big(\sum_{m=1}^{n} P_{km,j}^{[r,t]} x_m^{(r)}\Big) \\
&= \sum_{l=1}^{n}\Big(\sum_{k,m=1}^{n} P_{il,k}^{[s,r]} P_{km,j}^{[r,t]} x_m^{(r)}\Big) x_l^{(s)} \\
&= \sum_{l=1}^{n}\Big(\sum_{k=1}^{n} P_{il,k}^{[s,r]} P_{kj}^{[r,t]} x_l^{(s)}\Big) \\
&= \sum_{l=1}^{n} P_{il,j}^{[s,t]} x_l^{(s)} \\
&= H_{ij}^{[s,t]},
\end{aligned}$$

hence $(H_{ij}^{[s,t]})_{i,j\in E}$ is a Markov process.

Similarly, one gets

$$\begin{aligned}
Q_{(ij)(uv)}^{[s,t]} &= \Big(\sum_{m,l=1}^{n} P_{ij,m}^{[s,r]} P_{ml,u}^{[r,t]} x_l^{(t)}\Big) x_v^{(t)} \\
&= \sum_{m,l=1}^{n} P_{ij,m}^{[s,r]} x_l^{(t)} P_{ml,u}^{[r,t]} x_v^{(t)} \\
&= \sum_{m,l=1}^{n} Q_{(ij)(ml)}^{[s,r]} Q_{(ml)(uv)}^{[r,t]}
\end{aligned}$$

So, $Q_{(ij)(uv)}^{[s,t]}$ is also a Markov processes. This completes the proof.

Theorem 2.2 *Let $(E, P_{ij,k}^{[s,t]}, \mathbf{x}^{(0)})$ be a q.s.p type (B). Then the process $(H_{ij}^{[s,t]})$ defined by (2) is Markov. Moreover, the process $Q_{(ij)(uv)}^{[s,t]}$ defined by (3) satisfies the following equation*

$$Q_{(ij)(uv)}^{[s,t]} = \sum_{l,k=1}^{n} H_{il}^{[s,\tau]} H_{jk}^{[s,\tau]} Q_{(lk)(uv)}^{[\tau,t]}$$

Proof. First we want to show that

$$x_k^{(t)} = \sum_{i,j=1}^{n} P_{i,j,k}^{[s,t]} x_i^{(s)} x_j^{(s)},$$

Indeed, from

$$x_k^{(s)} = \sum_{i,j=1}^{n} P_{i,j,k}^{[0,s]} x_i^{(0)} x_j^{(0)}$$

and

$$P_{i,j,k}^{[0,t]} = \sum_{m,l,g,u=1}^{n} P_{i,m,l}^{[0,s]} P_{j,g,u}^{[0,s]} P_{l,u,k}^{[s,t]} x_m^{(0)} x_g^{(0)},$$

one finds

$$\begin{aligned} x_k^{(t)} &= \sum_{i,j=1}^{n} \Big(\sum_{m,l,g,u=1}^{n} P_{im,l}^{[0,s]} P_{jg,u}^{[0,s]} P_{lu,k}^{[s,t]} x_m^{(0)} x_g^{(0)} \Big) x_i^{(0)} x_j^{(0)} \\ &= \sum_{l,u=1}^{n} P_{lu,k}^{[s,t]} \Big(\sum_{i,m=1}^{n} P_{im,l}^{[0,s]} x_i^{0} x_m^{(0)} \Big) \Big(\sum_{j,g} P_{jg,u}^{[0,s]} x_j^{(0)} x_g^{(0)} \Big) \\ &= \sum_{l,u=1}^{n} P_{lu,k}^{[s,t]} x_l^{(s)} x_u^{s}. \end{aligned}$$

Now using the last equality, we obtain

$$\begin{aligned} H_{ij}^{[s,t]} &= \sum_{l=1}^{n} P_{il,j}^{[s,t]} x_l^{(s)} \\ &= \sum_{l=1}^{n} \Big(\sum_{m,u,g,v=1}^{n} P_{i,m,u}^{[s,\tau]} P_{l,g,v}^{[s,\tau]} P_{u,v,j}^{[\tau,t]} x_m^{(s)} x_g^{(s)} \Big) x_l^{(s)} \\ &= \sum_{u,v=1}^{n} \Big(\sum_{m=1}^{n} P_{i,m,u}^{[s,\tau]} x_m^{(s)} \Big) \Big(\sum_{l,g=1}^{n} P_{l,g,v}^{[s,\tau]} x_g^{(s)} x_l^{(s)} \Big) P_{u,v,j}^{[\tau,t]} \\ &= \sum_{u,v=1}^{n} H_{i,u}^{[s,\tau]} P_{u,v,j}^{[\tau,t]} x_v^{(\tau)} \\ &= \sum_{u=1}^{n} H_{i,u}^{[s,\tau]} H_{u,j}^{[\tau,t]} \end{aligned}$$

This shows $H_{ik}^{[s,t]}$ is a Markov process.

From (iii$_B$) we immediately find

$$\begin{aligned} Q_{(ij)(uv)}^{[s,t]} = P_{i,j,u}^{[s,t]} x_v^{(t)} &= \Big(\sum_{m,l,g,k=1}^{n} P_{i,m,l}^{[s,\tau]} P_{j,g,k}^{[s,\tau]} P_{lk,u}^{[\tau,t]} x_m^{(s)} x_g^{(s)} \Big) x_v^{(t)} \\ &= \sum_{l,k=1}^{n} \Big(\sum_{m=1}^{n} P_{i,m,l}^{[s,\tau]} x_m^{(s)} \Big) \Big(\sum_{g=1}^{n} P_{j,g,k}^{[s,\tau]} x_g^{(s)} \Big) P_{l,k,u}^{[\tau,t]} x_v^{(t)} \\ &= \sum_{l,k=1}^{n} H_{il}^{[s,\tau]} H_{jk}^{[s,\tau]} Q_{(lk)(uv)}^{[\tau,t]}. \end{aligned}$$

This completes the proof.

Now we have two processes associated with q.s.p. Note that they relate with each other by the following formula

$$\sum_{j=1}^{n} Q^{[s,t]}_{(ij)(uv)} x^{(s)}_j = H^{[s,t]}_{iu} x^{(t)}_v$$

We are interested in the reverse problem. Namely assume that $E = \{1, 2, dots, n\}$ be a finite set, and we are given two non-homogeneous processes $(\mathcal{Q}^{[s,t]}_{(ij)(uv)})_{i,j,u,v\in\ E}$ and $(\mathcal{H}^{[s,t]}_{ij})_{i,j\in\ E}$. Under what conditions these two processes uniquely determine some q.s.p? To answer this question, we first fix an initial state $\mathbf{x}^{(0)} = (x^0_i)^n_{i=1} \in S^{n-1}$. Let us define

$$z^{(t)}_k = \sum_i \mathcal{H}^{[0,t]}_{ik} x^{(0)}_i, \quad y^{(t)}_k = \sum_{i,j,l=1}^{n} \mathcal{Q}^{[0,t]}_{(ij)(kl)} x^{(0)}_i x^{(0)}_j, \quad k \in E.$$

Theorem 2.3 *Let $E = \{1, 2, \dots, n\}$ and $(\mathcal{Q}^{[s,t]}_{(ij)(uv)})$, $(\mathcal{H}^{[s,t]}_{ij})$ be two stochastic processes on $E\times\ E$ and E, respectively. Assume that*

(a) $\mathcal{Q}^{[s,t]}_{(ij)(uv)} = \mathcal{Q}^{[s,t]}_{(ji)(uv)}$ *for any $i, j, u, v \in\ E$;*

(b) $\sum\limits_{j=1}^{n} \mathcal{Q}^{[s,t]}_{(ij)(uv)} y^{(s)}_j = \mathcal{H}^{[s,t]}_{iu} z^{(t)}_v$ *for any $i, u, v \in\ E$;*

(c) $\mathcal{Q}^{[s,t]}_{(ij)(uv)} = \sum\limits_{l=1}^{n} \mathcal{Q}^{[s,t]}_{(ij)(ul)} y^{(t)}_v$ *for any $i, j, u, v \in\ E$.*

Let

$$P^{[s,t]}_{ij,k} = \sum_{l=1}^{n} \mathcal{Q}^{[s,t]}_{(ij)(kl)}. \tag{4}$$

Then the following assertions hold true:

(i) $z^{(t)}_k = y^{(t)}_k$ *for any $k \in\ E$;*

(ii) If $(\mathcal{Q}^{[s,t]}_{(ij)(uv)})$ and $(\mathcal{H}^{[s,t]}_{ij})$ are Markov processes, then $(E, P^{[s,t]}_{ij,k}, \mathbf{x}^{(0)})$ is a q.s.p. of type (A).

(iii) If $(\mathcal{H}^{[s,t]}_{ij})$ is a Markov process and $(\mathcal{Q}^{[s,t]}_{(ij)(uv)})$ satisfies

$$\mathcal{Q}^{[s,t]}_{(ij)(uv)} = \sum_{m,l=1}^{n} \mathcal{H}^{[s,\tau]}_{im} \mathcal{H}^{[s,\tau]}_{jl} \mathcal{Q}^{[\tau,t]}_{(ml)(uv)} \tag{5}$$

then $(E, P^{[s,t]}_{ij,k}, \mathbf{x}^{(0)})$ is a q.s.p. of type (B).

Moreover, one has $x^{(t)}_k = z^{(t)}_k$ ($k \in E$) and

$$\mathcal{H}^{[s,t]}_{ik} = \sum_{j=1}^{n} P^{[s,t]}_{ij,k} z^{(s)}_j \tag{6}$$

Proof. (i) Let us show that $z_k^{(t)} = y_k^{(t)}, (k \in E)$. Indeed from condition (c) and the stochasticity of $\mathcal{Q}^{[s,t]}_{(ij)(uv)}$, one finds

$$\sum_{m=1}^{n} \mathcal{Q}^{[s,t]}_{(ij)(mk)} = \sum_{m,l=1} \mathcal{Q}^{[s,t]}_{(ij)(ml)} y_k^{(t)} = y_k^{(t)},$$

for any $k \in E$. Hence, the last equality with (b) yields

$$y_k^{(t)} = \sum_{m,j=1}^{n} \mathcal{Q}^{[s,t]}_{(ij)(mk)} y_j^{(s)} = \sum_{m=1}^{n}\left(\sum_{j=1}^{n} \mathcal{Q}^{[s,t]}_{(ij)(mk)} y_j^{(s)}\right) = \sum_{m=1}^{n} \mathcal{H}^{[s,t]}_{im} z_k^{(t)} = z_k^{(t)},$$

which is the required assertion.

(ii) From (4) one can see that $(P^{[s,t]}_{ij,k})$ satisfies the conditions (i), (ii) of definition q.s.p. We need to check the equality (iii_A). Denote

$$x_k^{(t)} = \sum_{i,j=1}^{n} P^{[0.t]}_{ij} x_i^{(0)} x_j^{(0)}.$$

Then we find

$$x_k^{(t)} = \sum_{i,j=1}^{n}\left(\sum_{l=1}^{n} \mathcal{Q}^{[s,t]}_{(ij)(kl)}\right) x_i^{(0)} x_j^{(0)} = y_k^{(t)}$$

Hence, due to (i) one has

$$x_k^{(t)} = y_k^{(t)} = z_k^{(t)},$$

for any $k \in E$. Let us directly check the fundamental equation (iii_A). Indeed, for $s, \tau, t \in \mathbb{R}_+$ with $\tau - s \geq 1, t - \tau \geq 1$ and due to Markovianity of $(\mathcal{Q}^{[s,t]}_{(ij)(uv)})$, we have

$$\begin{aligned}
P^{[s,t]}_{ij,k} &= \sum_{l=1}^{n} \mathcal{Q}^{[s,t]}_{(ij)(kl)} = \\
&= \sum_{l=1}^{n}\left(\sum_{m,h=1}^{n} \mathcal{Q}^{[s,\tau]}_{(ij)(mh)} \mathcal{Q}^{[\tau,t]}_{(mh)(kl)}\right) = \\
&= \sum_{l=1}^{n}\left(\sum_{m,h=1}^{n}\left(\sum_{g=1}^{n} \mathcal{Q}^{[s,\tau]}_{(ij)(mg)} y_h^{(\tau)}\right) \mathcal{Q}^{[\tau,t]}_{(mh)(kl)}\right) = \\
&= \sum_{m,h=1}^{n} P^{[s,\tau]}_{ij,m} P^{[\tau,t]}_{mh,k} x_h^{(\tau)}.
\end{aligned}$$

(iii) From (b) one gets

$$\sum_{j,l=1}^{n} \mathcal{Q}^{[s,t]}_{(ij)(kl)} y_j^{(s)} = \mathcal{H}^{[s,t]}_{ik}$$

for any $i, k \in E$. Then using the last equality and (5) we obtain

$$\sum_{m,l,h,g=1}^{n} P^{[s,\tau]}_{im,l} P^{[s,\tau]}_{jh,g} P^{[\tau,t]}_{lg,k} x_m^{(s)} x_h^{(s)} =$$

$$
\begin{aligned}
&= \sum_{m,l,h,g=1}^{n} \left(\sum_{u=1}^{n} \mathcal{Q}^{[s,\tau]}_{(im)(lu)}\right)\left(\sum_{v=1}^{n} \mathcal{Q}^{[s,\tau]}_{(jh)(gv)}\right)\left(\sum_{a=1}^{n} \mathcal{Q}^{[\tau,t]}_{(lg)(ka)}\right) x^{(s)}_m x^{(s)}_h = \\
&= \sum_{a,l,g=1}^{n} \mathcal{H}^{[s,\tau]}_{il}\mathcal{H}^{[s,\tau]}_{jg}\mathcal{Q}^{[\tau,t]}_{(lg)(ka)} = \\
&= \sum_{a=1}^{n} \mathcal{Q}^{[s,t]}_{(ij)(ka)} = P^{[s,t]}_{ij,k},
\end{aligned}
$$

which means the q.s.p. satisfies the equation (iii_B).

Note that from (b) we immediately find (6). This completes the proof.

From the proved Theorems 2.1,2.2 and 2.3 we conclude that any q.s.p. can be uniquely defined by two kind of Markov processes. Such processes are called *marginal Markov processes* associated with q.s.p.

3. Construction of q.s.p.

In this section we propose a construction of q.s.p. by means of two other such kind of processes. Using the proposed construction we provide certain concrete examples of q.s.p.

Theorem 3.1 *Let* $(E, P^{[s,t]}_{ij,k}, \mathbf{x}^{(0)})$ *and* $(F, G^{[s,t]}_{ij,k}, \mathbf{y}^{(0)})$ *be two q.s.o. of type (A). Then the process* $(E\times F, (P\otimes G)^{[s,t]}_{\bar{i}\bar{j},\bar{k}}, \mathbf{x}^{(0)}\times\mathbf{y}^{(0)})$ *defined by*

$$
(P\otimes G)^{[s,t]}_{\bar{i}\bar{j},\bar{k}} = P^{[s,t]}_{i_1j_1k_1}G^{[s,t]}_{i_2j_2k_2}, \ \textit{where}\ \bar{i}=(i_1,i_2), \bar{j}=(j_1,j_2),\ \bar{k}=(k_1,k_2),
$$

$$
\mathbf{x}^{(0)}\times\mathbf{y}^{(0)} = (x^{(0)}_i y^{(0)}_j)_{i\in E, j\in F}
$$

is a q.s.o. of type (A).

Proof. It is clear that $(P\otimes G)^{[s,t]}_{\bar{i}\bar{j},\bar{k}} \geq 0$ and

$$
\sum_{\bar{k}=(k_1,k_2)}^{n,m} (P\otimes G)^{[s,t]}_{\bar{i}\bar{j},\bar{k}} = \sum_{k_1=1,k_2=1}^{n,m} P^{[s,t]}_{i_1,j_1,k_1}G^{[s,t]}_{i_2,j_2,k_2} = \sum_{k_1=1}^{n} P^{[s,t]}_{i_1,j_1,k_1}\sum_{k_2=1}^{m} G^{[s,t]}_{i_2,j_2,k_2} = 1.
$$

From

$$
P^{[s,t]}_{i_1j_1,k_1}P^{[s,t]}_{i_2j_2,k_2} = P^{[s,t]}_{j_1i_1,k_1}G^{[s,t]}_{j_2i_2,k_2},
$$

we immediately get $(P\otimes G)^{[s,t]}_{\bar{i}\bar{j},\bar{k}} = (P\otimes G)^{[s,t]}_{\bar{j}\bar{i},\bar{k}}$. Let us denote that $X^{(0)}_{\bar{i}} = x^{(0)}_{i_1}y^{(0)}_{i_2}$. Then, one can see that that $\mathbf{x}^{(0)}\times\mathbf{y}^{(0)} = (X^{(0)}_{\bar{i}})$. Let us find $X^{(s)}_{\bar{i}}$, which is given by

$$
\begin{aligned}
X^{(s)}_{\bar{k}} &= \sum_{\bar{i},\bar{j}} (P\otimes G)^{[0,s]}_{\bar{i}\bar{j},\bar{k}}\bar{x}^{(0)}_{\bar{i}}\bar{x}^{(0)}_{\bar{j}} \\
&= \sum_{(i_1i_2)(j_1j_2)} P^{[0,s]}_{i_1,j_1,k_1}G^{[0,s]}_{i_2,j_2,k_2}x^{(0)}_{i_1}y^{(0)}_{i_2}x^{(0)}_{j_1}y^{(0)}_{j_2} \\
&= \left(\sum_{(i_1i_2)} P^{[0,s]}_{i_1,j_1,k_1}x^{(0)}_{i_1}x^{(0)}_{j_1}\right)\left(\sum_{(j_1j_2)} G^{[0,s]}_{i_2,j_2,k_2}y^{(0)}_{i_2}y^{(0)}_{j_2}\right) \\
&= x^{(s)}_{k_1}y^{(s)}_{k_2}
\end{aligned} \tag{7}
$$

Now let us check the equation (iii$_A$).

Taking into account that the given processes $P^{[s,t]}_{ij,k}$ and $G^{[s,t]}_{ij,k}$ have type (A), then we have

$$\begin{aligned}\sum_{\bar{u},\bar{l}}(P\otimes G)^{[s,\tau]}_{\bar{i}\bar{j},\bar{u}}(P\otimes G)^{[\tau,t]}_{\bar{u}\bar{l},\bar{k}}X^{(s)}_{\bar{l}} &= \sum_{(u_1u_2)(l_1l_2)} P^{[s,\tau]}_{i_1,j_1u_1}G^{[s,\tau]}_{i_2j_2u_2}P^{[\tau,t]}_{u_1,l_1k_1}G^{[\tau,t]}_{u_2l_2k_2}x^{(s)}_{l_1}y^{(s)}_{l_2}\\ &= \left(\sum_{(u_1l_1)} P^{[s,\tau]}_{i_1,j_1u_1}P^{[\tau,t]}_{u_1,l_1k_1}x^{(s)}_{l_1}\right)\left(\sum_{(u_2l_2)} G^{[s,\tau]}_{i_2j_2u_2}G^{[\tau,t]}_{u_2l_2k_2}y^{(s)}_{l_2}\right)\\ &= P^{[s,t]}_{i_1,j_1k_1}G^{[s,t]}_{i_2j_2k_2}\\ &= (P\otimes G)^{[s,t]}_{\bar{i}\bar{j},\bar{k}}.\end{aligned}$$

This completes the proof.

Theorem 3.2 *Let* $(E,P^{[s,t]}_{ij,k},\mathbf{x}^{(0)})$ *and* $(F,G^{[s,t]}_{ij,k},\mathbf{y}^{(0)})$ *be two q.s.o. of type (B). Then the process* $(E\times F,(P\otimes G)^{[s,t]}_{\bar{i}\bar{j},\bar{k}},\mathbf{x}^{(0)}\times\mathbf{y}^{(0)})$ *defined by*

$$(P\otimes G)^{[s,t]}_{\bar{i}\bar{j},\bar{k}} = P^{[s,t]}_{i_1j_1k_1}G^{[s,t]}_{i_2j_2k_2},\ \text{where}\ \bar{i}=(i_1,i_2),\bar{j}=(j_1,j_2),\ \bar{k}=(k_1,k_2),$$

$$\mathbf{x}^{(0)}\times\mathbf{y}^{(0)} = (x^{(0)}_i y^{(0)}_j)_{i\in E,j\in F}$$

is a q.s.o. of type (B).

Proof. To prove the theorem it is enough to check the equation $((iii_B))$. Noting that $X^{(s)}_{\bar{m}} = x^{(s)}_{m_1}y^{(s)}_{m_2}$, where $\bar{m}=(m_1,m_2)$, we get

$$\begin{aligned}&\sum_{\bar{m},\bar{l},\bar{j},\bar{u}}(P\otimes G)^{[s,\tau]}_{\bar{i},\bar{m},\bar{l}}(P\otimes G)^{[s,\tau]}_{\bar{j},\bar{g},\bar{u}}(P\otimes G)^{[\tau,t]}_{\bar{l},\bar{u},k}X^{(s)}_{\bar{m}}X^{(s)}_{\bar{g}}\\ &= \sum_{(m_1m_2)(l_1l_2)(j_1j_2)(u_1u_2)} P^{[s,\tau]}_{i_1,m_1,l_1}G^{[s,\tau]}_{j_2,m_2,l_2}P^{[s,\tau]}_{j_1,g_1,u_1}G^{[s,\tau]}_{j_2,g_2,u_2}\\ &\quad P^{[\tau,t]}_{l_1,u_1,k_1}G^{[\tau,t]}_{l_2,u_2,k_2}x^{(s)}_{m_1}y^{(s)}_{m_2}x^{(s)}_{g_1}y^{(s)}_{g_2}\\ &= \left(\sum_{(m_1l_1j_1u_1)} P^{[s,\tau]}_{i_1,m_1,l_1}P^{[s,\tau]}_{j_1,g_1,u_1}P^{[\tau,t]}_{l_1,u_1,k_1}x^{(s)}_{m_1}x^{(s)}_{g_1}\right)\\ &\quad\left(\sum_{(m_2l_2j_2u_2)} G^{[s,\tau]}_{j_2,m_2,l_2}G^{[s,\tau]}_{j_2,g_2,u_2}G^{[\tau,t]}_{l_2,u_2,k_2}y^{(s)}_{m_2}y^{(s)}_{g_2}\right)\\ &= P^{[s,t]}_{i_1,j_1k_1}G^{[s,t]}_{i_2j_2k_2}\\ &= (P\otimes G)^{[s,t]}_{\bar{i}\bar{j},\bar{k}}\end{aligned}$$

This is the desired assertion.

From these theorems and Theorem 2.3 we immediately get the following

Theorem 3.3 *Let* $(E,P^{[s,t]}_{ij,k},\mathbf{x}^{(0)})$ *and* $(F,G^{[s,t]}_{ij,k},\mathbf{y}^{(0)})$ *be two q.s.o. with marginal processes* $(Q^{[s,t]}_{P,(ij)(uv)}),(H^{[s,t]}_{P,ij})$ *and* $(Q^{[s,t]}_{G,(ij)(uv)}),(H^{[s,t]}_{G,ij})$*, respectively. Then the marginal processes of* $(E\times F,(P\otimes G)^{[s,t]}_{\bar{i}\bar{j},\bar{k}},\mathbf{x}^{(0)}\times\mathbf{y}^{(0)})$ *are given by*

$$\mathbf{Q}^{[s,t]}_{(\bar{i}\bar{j}),(\bar{u},\bar{v})} = Q^{[s,t]}_{P,(i_1j_1)(u_1v_1)}Q^{[s,t]}_{G,(i_2j_2)(u_2v_2)},\quad \mathbf{H}^{[s,t]}_{\bar{i}\bar{j}} = H^{[s,t]}_{P,i_1j_1}H^{[s,t]}_{G,i_2j_2},$$

where $\bar{i} = (i_1, i_2), \bar{j} = (j_1, j_2),\ \bar{u} = (u_1, u_2), \bar{v} = (v_1, v_2)$.

3.1. Example

In this paragraph, using the proved theorem, we are going to construct new examples of q.s.p. Let us consider the following process:

Let $E = 1, 2$ and $(x, 1 - x)$ be an initial distribution on E, where $x \in [0, 1]$. Consider the following system of transition probabilities:

$$P^{[s,t]}_{11,1} = \frac{(1-2\epsilon)^{t-s}}{2^{t-s-1}}[(2^{t-s-1}-1)(1-2\epsilon)^s x + 1];$$

$$P^{[s,t]}_{12,1} = P^{[s,t]}_{21,1} = \frac{(1-2\epsilon)^{t-s}}{2^{t-s-1}}[(2^{t-s-1}-1)(1-2)^s x + \frac{1}{2}];$$

$$P^{[s,t]}_{22,1} = \frac{(1-2\epsilon)^{t-s}}{2^{t-s-1}}(2^{t-s-1}-1)x;$$

$$P^{[s,t]}_{ij,2} = 1 - P^{[s,t]}_{ij,1}, i, j = 1, 2.$$

It is known [9] that for $\epsilon \in [0, \frac{1}{2}]$ these transition probabilities generate a q.s.o. which is of type (A) and type (B) simultaneously. In this case we have

$$x^{(t)}_1 = (1-2\epsilon)^t x$$

$$x^{(t)}_2 = 1 - (1-2\epsilon)^t x$$

For $\epsilon = 0$ this q.s.p is homogeneous; for $\epsilon \neq 0$ it is from class (II). Now, we multiply the same process to itself according to the theorem.

Let

$$F = E \times E = (1, 1), (1, 2), (2, 1), (2, 2)$$

and

$$X^0 = \big(xy, x(1-y), (1-x)y, (1-x)(1-y)\big)$$

where $x, y \in [0, 1]$ and define

$$\mathcal{P}^{[s,t]}_{(i_1 i_2)(j_1 j_2)(k_1 k_2)} = P^{[s,t]}_{i_1,j_1,k_1} P^{[s,t]}_{i_2,j_2 k_2}$$

According to the theorem, such a process is q.s.p type of A and B. Let us describe the process more precisely,

$$\mathcal{P}_{11,11,11} = \frac{1}{2^{2(t-s-1)}}[x^2(2^{t-s-1}-1)^2 + 2x(2^{t-s-1}-1) + 1]$$

$$\mathcal{P}_{11,11,12} = \frac{1}{2^{2(t-s-1)}}[-x^2(2^{t-s-1}-1)^2 + x(2^{t-s-1}-1)(2^{t-s-1}-2) + 2^{t-s-1} - 1]$$

$$\mathcal{P}_{11,11,22} = \frac{1}{2^{2(t-s-1)}}[x^2(2^{t-s-1}-1)^2 - 2x(2^{t-s-1}-1)^2 + 2^2 t - s - 1 - 2^{t-s-1} + 1]$$

$$\mathcal{P}_{11,12,11} = \frac{1}{2^{2(t-s-1)}}[x^2(2^{t-s-1}-1)^2 + \frac{3}{2}x(2^{t-s-1}-1) + \frac{1}{2}]$$

$$\mathcal{P}_{11,12,12} = \frac{1}{2^{2(t-s-1)}}[x^2(2^{t-s-1}-1)^2 - x(2^{t-s-1} - \frac{3}{2}2^{t-s-1} - \frac{3}{2}) + 2^{t-s-1} + \frac{1}{2}]$$

$$\mathcal{P}_{11,12,21} = \frac{1}{2^{2(t-s-1)}}[-x^2(2^{t-s-1}-1)^2 + x(2^{t-s-1} - \frac{3}{2}2^{t-s-1} + \frac{3}{2}) + \frac{1}{4}2^{t-s} - \frac{1}{2}]$$

$$\begin{aligned}\mathcal{P}_{11,12,22} &= \frac{1}{2^{2(t-s-1)}}[x^2(2^{t-s-1}+1)^2 - x(2(2^{2(t-s-1)}) - \frac{7}{2}2^{t-s-1} - \frac{3}{2}) \\ &\quad +(2^{2(t-s-1)} + 2^{t-s-1} + \frac{1}{2})]\end{aligned}$$

$$\mathcal{P}_{11,22,11} = \frac{1}{2^{2(t-s-1)}}[x^2(2^{t-s-1}-1)^2 + x(2^{t-s-1}-1) + \frac{1}{4}]$$

$$\mathcal{P}_{11,22,12} = \frac{1}{2^{2(t-s-1)}}[x^2(1-2^{t-s-1})(1+2^{t-s-1}) + x(2^{2(t-s-1)} - 2^{t-s-1} - 1) + \frac{1}{4}2^{t-s} + \frac{1}{4}]$$

$$\mathcal{P}_{11,22,22} = \frac{1}{2^{2(t-s-1)}}[x^2(2^{t-s-1}+1)^2 - x(2(2^{t-s-1})+1)(2^{t-s-1}+1) + (2^{t-s-1}+\frac{1}{2})^2]$$

$$\mathcal{P}_{22,12,11} = \frac{1}{2^{2(t-s-1)}}[x^2(2^{t-s-1}-1)^2 + \frac{1}{4}2^{t-s}x - \frac{1}{2}x]$$

$$\mathcal{P}_{22,12,12} = \frac{1}{2^{2(t-s-1)}}[-x^2(2^{t-s-1} - 2^{t-s-1} - 1) + x(2^{2(t-s-1)} - \frac{1}{2}2^{t-s-1} + \frac{1}{2}) + \frac{1}{4}2^{t-s}]$$

$$\mathcal{P}_{22,12,21} = \frac{1}{2^{2(t-s-1)}}[-x^2(2^{t-s-1}-1)^2 + x((2^{t-s-1}-1)+1)(2(2^{t-s-1})-1)]$$

$$\mathcal{P}_{22,12,22} = \frac{1}{2^{2(t-s-1)}}[x^2(2^{t-s-1}-1)^2 - x(4(2^{t-s-1}))(2^{t-s-1}-1) + 2^{t-s-1}(2^{t-s-1}-\frac{1}{2})]$$

$$\mathcal{P}_{12,12,11} = \frac{1}{2^{2(t-s-1)}}[x^2(2^{2(t-s-1)} - 2^{t-s-1} + 1) - x]$$

$$\mathcal{P}_{12,12,12} = \frac{1}{2^{2(t-s-1)}}[-x^2(2^{2(t-s-1)} - 2^{t-s-1} + 1) + x(2^{2(t-s-1)} - 2^{t-s-1} + 1) + 2^{t-s-1}]$$

$$\mathcal{P}_{12,12,21} = \frac{1}{2^{2(t-s-1)}}[-x^2(2^{t-s-1}-1)^2 - x(1-2^{t-s-1})(1+2^{t-s-1})]$$

$$\mathcal{P}_{12,12,22} = \frac{1}{2^{2(t-s-1)}}[x^2(2^{t-s-1}-1)^2 - x(2(2^{t-s-1})-1)(2^{t-s-1}-1) + 2^{t-s-1}(2^{t-s-1}-1)]$$

$$\mathcal{P}_{12,22,11} = \frac{1}{2^{2(t-s-1)}}[x^2(2^{t-s-1}-1)^2 + \frac{1}{2}x(2^{t-s-1}-1)]$$

$$\mathcal{P}_{12,22,12} = \frac{1}{2^{2(t-s-1)}}[-x^2(2^{t-s-1}-1)^2 + x(2^{2(t-s-1)} - \frac{1}{4}2^{t-s} + \frac{1}{2}) + \frac{1}{4}2^{t-s}]$$

$$\mathcal{P}_{12,22,21} = \frac{1}{2^{2(t-s-1)}}[-x^2(2^{t-s-1}-1)^2 + x(4(2^{t-s-1}-2)(2^{t-s-1}-1)]$$

$$\mathcal{P}_{12,22,22} = \frac{1}{2^{2(t-s-1)}}[x^2(2^{t-s-1}-1)^2 - x(4(2^{t-s-1}-1)(2^{t-s-1}-1) + 2^{t-s-1}(2^{t-s-1}-\frac{1}{2})]$$

$$\mathcal{P}_{22,22,11} = \frac{1}{2^{t-s-1}}[x^2 - \frac{x^2}{2^{t-s-1}}]$$

$$\mathcal{P}_{22,22,12} = \frac{x}{2^{t-s-1}}[2^{t-s-1} - 1 - x + \frac{x}{2^{t-s-1}}]$$

$$\mathcal{P}_{22,22,22} = \frac{1}{2^{2(t-s-1)}}[x^2(2^{t-s-1}+1)^2 - 2(2^{t-s-1}x)(2^{t-s-1}+1) + 2^{2(t-s-1)}]$$

4. Analytical theory for quadratic stochastic processes

In this section we are going to derive differential equations related to q.s.p. In what follows, we always assume that the functions $P_{ij,k}^{[s,t]}$ are continuous with respect to variables s and t as well they are differentiable with respect to s and t with $t > s + 1$. We will consider two separate subcases with respect to types of q.s.p.

4.1. Differential equations for q.s.p. type A

In this subsection we consider quadratic stochastic processes of type (A). Assume that $(E, P_{ij,k}^{[s,t]}, \mathbf{x}^{(0)})$ be a q.s.p. of type (A). Then using (iii$_A$) for $t > s + 1$ one gets

$$\frac{P_{ij,k}^{[s,t+h]} - P_{ij,k}^{[s,t]}}{h} = \frac{\sum\limits_{m,l=1}^{n} P_{ij,m}^{[s,t-1]} P_{ml,k}^{[t-1,t+h]} x_l^{(t-1)} - \sum\limits_{m,l=1}^{n} P_{ij,m}^{[s,t-1]} P_{ml,k}^{[t-1,t]} x_l^{(t-1)}}{h}$$
$$= \sum_{m,l=1}^{m} P_{ij,m}^{[s,t-1]} \left(\frac{P_{ml,k}^{[t-1,t+h]} - P_{ml,k}^{[t-1,t]}}{h} \right) x_l^{(t-1)}. \tag{8}$$

Now assuming

$$\alpha_{ml,k}(t) = \lim_{h \to 0+} \frac{P_{ml,k}^{[t-1,t+h]} - P_{ml,k}^{[t-1,t]}}{h}, \tag{9}$$

provided that limit exists, and passing the limit for $h \to 0$, we have first system of differential equations

$$\frac{\partial P_{ij,k}^{[s,t]}}{\partial t} = \sum_{m,l=1}^{m} \alpha_{ml,k}(t) x_l^{(t-1)} P_{ij,m}^{[s,t-1]}, \quad i,j,k = 1, \dots, n \tag{10}$$

Due to

$$x_k^{(t)} = \sum_{i,j=1}^{n} P_{ij,k}^{[0,t]} x_i^{(0)} x_j^{(0)} \tag{11}$$

we rewrite the equation (10) as follows

$$\frac{\partial P_{ij,k}^{[s,t]}}{\partial t} = \sum_{m,l,r,q=1}^{n} \alpha_{ml,k}(t) x_r^{(0)} x_q^{(0)} P_{rq,l}^{[0,t-1]} P_{ij,m}^{[s,t-1]} \tag{12}$$

Similarly for $t > s + 2$ we have

$$P_{ij,k}^{[s,t]} - P_{ij,k}^{[s+h,t]} = \sum_{m,l} P_{ij,m}^{[s,s+1+h]} P_{ml,k}^{[s+1+h,t]} x_l^{(s+1+h)}$$
$$- \sum_{m,l} P_{ij,m}^{[s+h,s+1+h]} P_{ml,k}^{[s+1+h]} x_l^{(s+1+h)}$$
$$= \sum_{m,l} \left(P_{ij,m}^{[s,s+1+h]} - P_{ij,m}^{[s,s+1]} \right) x_l^{(s+1+h)} P_{ml,k}^{[s+1+h,t]}. \tag{13}$$

Dividing both sides of this equality to h and passing to limit for $h \to 0$, one gets the second system of differential equations

$$\frac{\partial P_{ij,k}^{[s,t]}}{\partial s} = - \sum_{m,l=1}^{n} \alpha_{ij,m}(s+1) x_l^{(s+1)} P_{ml,k}^{[s+1,t]}, \quad i,j,k = 1, \dots, n \tag{14}$$

Again due to equation (11), we can rewrite (14) as follows

$$\frac{\partial P_{ij,k}^{[s,t]}}{\partial s} = -\sum_{m,l,r,q=1}^{n} \alpha_{ij,m}(s+1)x_r^{(0)}x_q^{(0)}P_{rq,l}^{[0,s+1]}P_{ml,k}^{[s+1,t]} \tag{15}$$

4.2. Differential equations for q.s.p. type B

Assume that $(E, P_{ij,k}^{[s,t]}, \mathbf{x}^{(0)})$ be a q.s.p. of type (B). Let us denote

$$\tilde{\alpha}_{ml,k}(t) = \lim_{h\to 0+} \frac{P_{ij,k}^{[t-1,t+h]} - P_{ij,k}^{[t-1,t]}}{h}, \tag{16}$$

and assume this limit exists.

For $t > s+2$, due to (iii$_B$) we have

$$\begin{aligned} P_{ij,k}^{[s,t+h]} - P_{ij,k}^{[s,t]} &= \sum_{m,l,r,q} P_{im,l}^{[s,t-1]}P_{jr,q}^{[s,t-1]}P_{lq,k}^{[t-1,t+h]}x_m^{(s)}x_r^{(s)} - \\ &\quad - \sum_{m,l,r,q} P_{im,l}^{[s,t-1]}P_{jr,q}^{[s,t-1]}P_{lq,k}^{[t-1,t]}x_m^{(s)}x_r^{(s)} \\ &= \sum_{m,l,r,q} P_{im,l}^{[s,t-1]}P_{jr,q}^{[s,t-1]}\left(P_{lq,k}^{[t-1,t+h]} - P_{lq,k}^{[t-1,t]}\right)x_m^{(s)}x_r^{(s)}. \end{aligned} \tag{17}$$

Now, dividing both sides the last equality by h and passing the limit $h \to 0$, we derive the following system of differential equations.

$$\frac{\partial P_{ij,k}^{[s,t]}}{\partial t} = \sum_{m,l,r,q} \tilde{\alpha}_{lq,k}(t)x_m^{(s)}x_r^{(s)}P_{im,l}^{[s,t-1]}P_{jr,q}^{[s,t-1]}, \quad i,j,k = 1,\dots,n. \tag{18}$$

Now let us consider the following expressions

$$\begin{aligned} P_{ij,k}^{[s,t]} - P_{ij,k}^{[s+h,t]} &= \sum_{m,l,r,q} \Big(P_{im,l}^{[s,s+1+h]}P_{jr,q}^{[s,s+1+h]}x_m^{(s)}x_r^{(s)} - \\ &\quad P_{im,l}^{[s+h,s+1+h]}P_{jr,q}^{[s+h,s+1+h]}x_m^{(s+h)}x_r^{(s+h)}\Big)P_{lq,r}^{[s+1+h,t]} \\ &= \sum_{m,l,r,q} \Big(P_{im,l}^{[s,s+1+h]}P_{jr,q}^{[s,s+1+h]}x_m^{(s)}x_r^{(s)} - \\ &\quad P_{im,l}^{[s,s+1+h]}P_{jr,q}^{[s+h,s+1+h]}x_m^{(s+h)}x_r^{(s+h)} + \\ &\quad P_{im,l}^{[s,s+1+h]}P_{jr,q}^{[s+h,s+1+h]}x_m^{(s+h)}x_r^{(s)})P_{lq,r}^{[s+1+h,t]} \\ &= \sum_{m,l,r,q} \underbrace{[P_{jr,q}^{[s,s+1+h]}x_m^{(s)} - P_{jr,q}^{[s+h,s+1+h]}x_m^{(s+h)}x_r^{(s+h)}]}_{I} \\ &\quad \times P_{im,l}^{[s,s+1+h]}x_r^{(s)}P_{lq,r}^{[s+1+h,t]} + \\ &\quad \sum_{m,l,r,q} \underbrace{[P_{im,l}^{[s,s+1+h]}x_r^{(s)} - P_{im,l}^{[s+h,s+1+h]}x_r^{(s+h)}]}_{II} \\ &\quad \times P_{jr,q}^{[s+h,s+1+h]}x_m^{(s+h)}P_{lq,r}^{[s+1+h,t]}. \end{aligned} \tag{19}$$

Now let us separately compute I and II, respectively. One gets

$$\begin{aligned} I &= P^{[s,s+1+h]}_{jr,q} x^{(s)}_m - P^{[s+h,s+1+h]}_{jr,q} x^{(s+h)}_m x^{(s+h)}_r \\ &= P^{[s,s+1+h]}_{jr,q} x^{(s)}_m - P^{[s+h,s+1+h]}_{jr,q} x^{(s)}_m + \\ &\quad P^{[s+h,s+1+h]}_{jr,q} x^{(s)}_m - P^{[s+h,s+1+h]}_{jr,q} x^{(s+h)}_m \\ &= x^{(s)}_m (P^{[s,s+1+h]}_{jr,q} - P^{[s+h,s+1+h]}_{jr,q}) + \\ &\quad P^{[s+h,s+1+h]}_{jr,q} (x^{(s}_m - x^{(s+h)}_m) \end{aligned} \tag{20}$$

Similarly, one finds

$$\begin{aligned} II &= P^{[s,s+1+h]}_{im,l} x^{(s)}_r - P^{[s+h,s+1+h]}_{im,l} x^{(s+h)}_r \\ &= P^{[s,s+1+h]}_{im,l} x^{(s)}_r - P^{[s,s+1+h]}_{im,l} x^{(s+h)}_r + \\ &\quad P^{[s,s+1+h]}_{im,l} x^{(s+h)}_r - P^{[s+h,s+1+h]}_{im,l} x^{(s+h)}_r \\ &= x^{(s+h)}_r (P^{[s,s+1+h]}_{im,l} - P^{[s+h,s+1+h]}_{im,l}) + \\ &\quad P^{[s,s+1+h]}_{im,l} (x^{(s}_r - x^{(s+h)}_r) \end{aligned} \tag{21}$$

Let us denote,

$$\beta_{ml,q}(s) = \lim_{\Delta \to 0_+} \frac{P^{[s,s+1+\Delta]}_{ml,q} - P^{[s+\Delta,s+1+\Delta]}_{ml,q}}{\Delta}, \tag{22}$$

Then dividing (19) by h and passing to the limit (keeping in mind (20) and (21)) we obtain

$$\begin{aligned} \frac{\partial P^{[s,t]}_{ij,k}}{\partial s} &= \sum_{m,l,r,q} P^{[s,s+1]}_{im,l} x^{(s)}_r (x^{(s)}_m \beta_{jr,q}(s) - P^{[s,s+1]}_{jr,q} (x^{(s)}_m)') P^{[s+1,t]}_{lr,q} \\ &\quad + \sum_{m,l,r,q} P^{[s,s+1]}_{jr,q} x^{(s)}_m (-P^{[s,s+1]}_{im,l} (x^{(s)}_r)' + x^{(s)}_r \alpha_{im,l}(s)) P^{[s+1,t]}_{lq,r} \\ &= \sum_{m,l,r,q} (P^{[s,s+1]}_{im,l} x^{(s)}_r (x^{(s)}_m \beta_{jr,q}(s) - P^{[s,s+1]}_{im,l} x^{(s)}_r P^{[s,s+1]}_{jr,q} (x^{(s)}_m)' \\ &\quad - P^{[s,s+1]}_{jr,q} x^{(s)}_m P^{[s,s+1]}_{im,l} (x^{(s)}_r)' + P^{[s,s+1]}_{jr,q} x^{(s)}_m x^{(s)}_r \alpha_{im,l}(s)) P^{[s+1,t]}_{lq,r} \\ &= \sum_{m,l,r,q} (-P^{[s,s+1]}_{im,l} P^{[s,s+1]}_{jr,q} (x^{(s)}_r x^{(s)}_m)' + (P^{[s,s+1]}_{im,l} \beta_{jr,q}(s) \\ &\quad + P^{[s,s+1]}_{jr,q} \beta_{im,l}(s))) P^{[s+1,t]}_{lq,r} \end{aligned} \tag{23}$$

Hence, we get the second system of differential equations

$$\begin{aligned} \frac{\partial P^{[s,t]}_{ij,k}}{\partial s} &= \sum_{m,l,r,q} (-\widetilde{P}^{[s,s+1]}_{im,l} P^{[s,s+1]}_{jr,q} (x^{(s)}_r x^{(s)}_m)' + \\ &\quad (P^{[s,s+1]}_{im,l} \alpha_{jr,q}(s) + P^{[s,s+1]}_{jr,q} \alpha_{im,l}(s))) P^{[s+1,t]}_{lq,r} \end{aligned} \tag{24}$$

for $i, j, k = 1, \ldots, n$.

Note that similar kind of differential equations have been obtained in [9] for homogeneous per unit time q.s.p.

Acknowledgments
The authors acknowledge the MOHE grant FRGS11-022-0170 for the financial support.

References

[1] Ganikhodzhaev R, Mukhamedov F and Rozikov U 2011 *Inf. Dim. Anal. Quantum Probab. and Related Topics* **14** 279-335
[2] Ganikhodjaev N N 1991 *J. Theor. Probab.* **4** 639–653
[3] Ganikhodjaev N N and Mukhamedov F M 2000 *Izvestiya Math.* **64** 873–890
[4] Jenks R D 1969 *J. Differ. Equations* **5** 497–514
[5] Kesten H 1970 *Adv. Appl. Probab.* **2** 1–82; 179–228
[6] Kolmogorov A N 1938 *Uspekhi Math. Nauk* **5** 5–51; English transl., selected works of A. N. Kolmogorov, vol. II, Kluwer, Dortrecht 1992, article 9.
[7] Lyubich Yu I 1992 *Mathematical structures in population genetics* (Springer-Verlag: Berlin).
[8] Mukhamedov F M 2002 *Russian Math. Sur.* **57** 1236–1237
[9] Sarymsakov T A and Ganikhodjaev N N 1990 *J. Theor. Probab.* **3** 51–70
[10] Sarymsakov T A and Ganikhodjaev N N 1991 *Soviet Math. Dokl.* **43** 279–283

Uniformly Convergence Of The Spectral Expansions Of The Schrödinger Operator On A Closed Domain

Anvarjon A. Ahmedov[1,2] **Nur Amalina Binti Jamaludin** [2,4]
Abdumalik Rakhimov [2,3]

[1]Department of Process and Food Engineering, Faculty of Engineering, University Putra Malaysia
[2]Institute for Mathematical Research, Universiti Putra Malaysia
[3]Department of Science in Engineering, International Islamic University Malaysia
[4]Department of Mathematic, National Defense University of Malaysia

E-mail: anvar@eng.upm.edu.my, nuramalina@upnm.edu.my , abdumalik@iium.edu.my

Abstract. In this work uniformly convergent problems of the eigenfunction expansions of the Schrödinger operator $-\Delta+q(y_1,y_2)$ with singular potential from $W_2^1(\Omega)$ are investigated. Using the estimation of the spectral function of the Schrödinger operator on closed domain and mean value formula for the eigenfunction the uniformly convergent of the eigenfunction expansions of the functions continuous on the closed domain is proved.

1. Introduction

Solving the boundary value problems of equations of mathematical physics in bounded domains leads to the investigation of convergence and summability problems, related to the eigenfunction expansion on a closed domain. This particular problem requires estimating eigenfunctions near boundary of the domain. Methodology of estimation eigenfunctions in compact subsets of the domain is well developed and known (see in [1]) . But in estimation of eigenfunctions in closed domains some difficulties occurs near the boundary. These difficulties can be avoided if we consider boundary conditions that help to estimate eigenfunctions near the boundary with more accurate values. E.I. Moiseev [2] proved an estimation of eigenfunctions of the first boundary value problem for the Laplace operator. Some latest results on the uniformly convergent of the spectral expansions of the distributions are obtained by Rakhimov (see [6]).

Let $\Omega \subset R^2$ be a domain with smooth boundary $\partial\Omega$ and $q(y_1,y_2) \geq 0$ be a potential functions from Sobolev's space $W_2^1(\Omega)$.

We investigate the solutions of the following Schrödinger equation:

$$-\Delta\psi(y_1,y_2) + q(y_1,y_2)\psi(y_1,y_2) + \lambda\psi(y_1,y_2) = 0, (y_1,y_2) \in \Omega, \tag{1}$$

with Dirichlet boundary condition:

$$\psi(y_1,y_2) = 0, (y_1,y_2) \in \partial\Omega. \tag{2}$$

It is well known that the (1) and (2) have the countable set of solutions: $\{\psi_n(x_1,x_2)\}_{n=1}^{\infty}$; $\{\lambda_n\}$, which are called eigenfunctions and eigenvalues of Schrödinger operator, respectively.

The main aim of the paper is to investigate conditions for uniformly convergence of eigenfunctions expansions

$$\sum_{n=1}^{\infty} f_n \psi_n(x_1, x_2)$$

by Riesz means, which can be defined as follows:

$$E_\lambda^\alpha f(x_1,x_2) = \sum_{\lambda_n<\lambda} \left(1-\frac{\lambda_n}{\lambda}\right)^\alpha f_n\psi_n(x_1,x_2), \alpha>0, \tag{3}$$

where f_n denotes the Fourier coefficients of the function $f(x_1, x_2)$:

$$f_n = \int\int_\Omega f(x_1,x_2)\psi_n(x_1,x_2)dx_1dx_2, \qquad n=1,2,\cdots.$$

2. Main results and formulation

Theorem 2.1. For the eigenfunctions of the Schrödinger operator on $\Omega \subset R^2$ with Dirichlet boundary conditions one has

$$\sum_{|\sqrt{\lambda_n}-\lambda|\le 1} \psi_n^2(x_1,x_2) \le C\lambda\ln^2\lambda, \qquad \lambda\to+\infty$$

uniformly on $(x_1, x_2) \in \overline{\Omega}$.
Theorem 2.1 is proved in [5] using the techniques of the work [2].

Using the estimation in Theorem 2.1 we prove uniformly convergence of spectral expansions on closed domain as follows:
Theorem 2.2. Let $f \in C(\overline{\Omega})$. If $\alpha > \frac{1}{2}$, then for the Riesz mean $E_\lambda^\alpha f(x_1, x_2)$ of spectral expansions of Schrödinger operator corresponding to Dirichlet boundary conditions we have

$$\lim_{\lambda\to\infty} E_\lambda^\alpha f(x_1,x_2) = f(x_1,x_2),$$

uniformly on $(x_1, x_2) \in \overline{\Omega}$.

In the proof of the estimation for eigenfunctions and in establishing the facts on convergent the following Lemma plays main role.
Lemma 2.3 (Mean Value Formula): For the eigenfunctions of the Schrödinger operator one has

$$\int_0^{2\pi} \psi_n(x_1+r\cos\theta, x_2+r\sin\theta)d\theta$$
$$= 2\pi\psi_n(x_1,x_2)J_o(r\sqrt{\lambda_n}) + \frac{\pi}{2}\int_0^r \{J_o(t\sqrt{\lambda_n})Y_o(r\sqrt{\lambda_n}) - Y_o(t\sqrt{\lambda_n})J_o(r\sqrt{\lambda_n})\}\rho_n(t)tdt, r\ge 0, \tag{4}$$

where $J_\nu(x)$ and $Y_\nu(x)$ are Bessel Functions of the first and second kind, respectively and we use notation:

$$\rho_n(t) = \int_0^{2\pi} q(x_1+t\cos\theta, x_2+t\sin\theta)\psi_n(x_1+t\cos\theta, x_2+t\sin\theta)d\theta. \tag{5}$$

Lemma 2.3 is proved in Titchmarsh [10].

We fix $(x_1, x_2) \in \overline{\Omega}$ and consider

$$g(r) = \begin{cases} r^{-1-\alpha} J_{1+\alpha}(r\sqrt{\lambda}), & r \leq R \\ 0, & r > R \end{cases}$$

where

$$r = \sqrt{(x_1 - y_1)^2 + (x_2 - y_2)^2}.$$

For the Fourier coefficients of $g(r)$ we obtain

$$g_n = \int\int_{\Omega} g(y_1, y_2)\psi_n(y_1, y_2)dy_1dy_2 = \int\int_{r\leq R} r^{-1-\alpha} J_{1+\alpha}(r\sqrt{\lambda})\psi_n(y_1, y_2)dy_1dy_2.$$

After transforming to polar coordinates we obtain by using Mean Value Formula

$$\begin{aligned} g_n &= \int_0^R r^{-\alpha} J_{1+\alpha}(r\sqrt{\lambda}) \int_0^{2\pi} \psi_n(x_1 + r\cos\theta, x_2 + r\sin\theta)d\theta dr \\ &= \int_0^R r^{-\alpha} J_{1+\alpha}(r\sqrt{\lambda})[2\pi\psi_n(x_1, x_2)J_o(r\sqrt{\lambda_n}) \\ &+ \frac{\pi}{2}\int_0^r \{J_o(t\sqrt{\lambda_n})Y_o(r\sqrt{\lambda_n}) - Y_o(t\sqrt{\lambda_n})J_o(r\sqrt{\lambda_n})\}\rho_n(t)tdt]dr \\ &= \int_0^R r^{-\alpha} J_{1+\alpha}(r\sqrt{\lambda})[2\pi\psi_n(x_1, x_2)J_o(r\sqrt{\lambda_n}) + S_q^n(r)]dr, \end{aligned} \tag{6}$$

where it is denoted

$$S_q^n(r) = \frac{\pi}{2}\int_0^r [J_o(t\sqrt{\lambda_n})Y_o(r\sqrt{\lambda_n}) - Y_o(t\sqrt{\lambda_n})J_o(r\sqrt{\lambda_n})]\rho_n(t)tdt.$$

The integral on the right side of the latter (6) can be simplified as follows:

$$g_n = 2\pi\psi_n(x_1, x_2)\{\frac{2^{-\alpha}}{\lambda^{\frac{1-\alpha}{2}}\Gamma(\alpha+1)}(1 - \frac{\lambda_n}{\lambda})^{\alpha}\} - 2\pi\psi_n(x_1, x_2)I_n(R) + T_n(R). \tag{7}$$

Here we used the notations:

$$\begin{aligned} I_n(R) &= \int_R^{\infty} r^{-\alpha} J_{1+\alpha}(r\sqrt{\lambda})J_o(r\sqrt{\lambda_n})dr \\ T_n(R) &= \int_0^R r^{-\alpha} J_{1+\alpha}(r\sqrt{\lambda})S_q^n(r)dr. \end{aligned}$$

From well known Integral formula for Bessel functions (see [10]):

$$\int_0^{\infty} r^{-\alpha} J_{1+\alpha}(r\sqrt{\lambda})J_o(r\sqrt{\lambda_n})dr = \begin{cases} \dfrac{2^{-\alpha}}{\left(\lambda^{\frac{1-\alpha}{2}}\right)\Gamma(\alpha+1)}(1 - \dfrac{\lambda_n}{\lambda})^{\alpha}, & \lambda_n \leq \lambda \\ 0, & \lambda_n > \lambda, \end{cases}$$

we derive

$$g_n = 2\pi\psi_n(x_1, x_2)\{\frac{2^{-\alpha}}{\lambda^{\frac{1-\alpha}{2}}\Gamma(\alpha+1)}(1 - \frac{\lambda_n}{\lambda})^{\alpha}\delta_{\lambda_n}^{\lambda}\} - 2\pi\psi_n(x_1, x_2)I_n(R) + T_n(R), \tag{8}$$

where $\delta^{\lambda}_{\lambda_n}$ is a Kronecker number:

$$\delta^{\lambda}_{\lambda_n} = \begin{cases} 1, & \lambda_n \le \lambda \\ 0, & \lambda_n > \lambda. \end{cases}$$

Substitution of (8) into Parseval's formula

$$\int\int_{\Omega} f(x_1,x_2)g(x_1,x_2)dx_1dx_2 = \sum_{n=1}^{\infty} f_n g_n$$

gives

$$\int\int_{r\le R} f(y_1,y_2)r^{-1-\alpha}J_{1+\alpha}(r\sqrt{\lambda})dy_1dy_2$$
$$= \sum_{n=1}^{\infty} f_n\{2\pi\psi_n(x_1,x_2)[\frac{2^{-\alpha}}{\lambda^{\frac{1-\alpha}{2}}\Gamma(\alpha+1)}(1-\frac{\lambda_n}{\lambda})^{\alpha}\delta^{\lambda}_{\lambda_n}] - 2\pi\psi_n(x_1,x_2)I_n(R) + T_n(R)\}$$
$$= 2\pi\sum_{n=1}^{\infty} f_n\psi_n(x_1,x_2)[\frac{2^{-\alpha}}{\lambda^{\frac{1-\alpha}{2}}\Gamma(\alpha+1)}(1-\frac{\lambda_n}{\lambda})^{\alpha}\delta^{\lambda}_{\lambda_n}] - 2\pi\sum_{n=1}^{\infty} f_n\psi_n(x_1,x_2)I_n(R) + \sum_{n=1}^{\infty} f_nT_n(R).$$

From the definition of Riesz Means we have a representation:

$$E^{\alpha}_{\lambda}f(x_1,x_2) = \sum_{\lambda_n<\lambda}(1-\frac{\lambda_n}{\lambda})^{\alpha} f_n\psi_n(x_1,x_2), \tag{9}$$

where

$$E^{\alpha}_{\lambda}f(x_1,x_2) = \frac{1}{2\pi}(2^{\alpha}\lambda^{\frac{1-\alpha}{2}}\Gamma(\alpha+1))\int\int_{r\le R} f(y_1,y_2)r^{-1-\alpha}J_{1+\alpha}(r\sqrt{\lambda})dy_1dy_2$$
$$+ (2^{\alpha}\lambda^{\frac{1-\alpha}{2}}\Gamma(\alpha+1))\sum_{n=1}^{\infty} f_n\psi_n(x_1,x_2)I_n(R) - \frac{1}{2\pi}(2^{\alpha}\lambda^{\frac{1-\alpha}{2}}\Gamma(\alpha+1))\sum_{n=1}^{\infty} f_nT_n(R)$$
$$= \frac{1}{2\pi}(2^{\alpha}\lambda^{\frac{1-\alpha}{2}}\Gamma(\alpha+1))\int\int_{r\le R} f(y_1,y_2)r^{-1-\alpha}J_{1+\alpha}(r\sqrt{\lambda})dy_1dy_2$$
$$+ (2^{\alpha}\lambda^{\frac{1-\alpha}{2}}\Gamma(\alpha+1))\sum_{n=1}^{\infty} f_n\psi_n(x_1,x_2)\int_R^{\infty} r^{-\alpha}J_{1+\alpha}(r\sqrt{\lambda})J_o(r\sqrt{\lambda_n})dr$$
$$- \frac{1}{2\pi}(2^{\alpha}\lambda^{\frac{1-\alpha}{2}}\Gamma(\alpha+1))\sum_{n=1}^{\infty} f_n\int_0^R r^{-\alpha}J_{1+\alpha}(r\sqrt{\lambda})S^n_q(r)dr.$$

The last expression for $E^{\alpha}_{\lambda}f(x_1,x_2)$ can be simplified as follows

$$E^{\alpha}_{\lambda}f(x_1,x_2) = \frac{1}{2\pi}(2^{\alpha}\lambda^{\frac{1-\alpha}{2}}\Gamma(\alpha+1))\int\int_{r\le R} f(y_1,y_2)r^{-1-\alpha}J_{1+\alpha}(r\sqrt{\lambda})dy_1dy_2 + \sigma^{\alpha}_{\lambda}f(x_1,x_2) - \tau^{\alpha}_{\lambda}f(x_1,x_2),$$

where

$$\sigma^{\alpha}_{\lambda}f(x_1,x_2) = (2^{\alpha}\lambda^{\frac{1-\alpha}{2}}\Gamma(\alpha+1))\sum_{n=1}^{\infty} f_n\psi_n(x_1,x_2)\int_R^{\infty} r^{-\alpha}J_{1+\alpha}(r\sqrt{\lambda})J_o(r\sqrt{\lambda_n})dr,$$
$$\tau^{\alpha}_{\lambda}f(x_1,x_2) = -\frac{1}{2\pi}(2^{\alpha}\lambda^{\frac{1-\alpha}{2}}\Gamma(\alpha+1))\sum_{n=1}^{\infty} f_n\int_0^R r^{-\alpha}J_{1+\alpha}(r\sqrt{\lambda})S^n_q(r)dr.$$

To prove the Theorem 2.2 it is sufficient to establish the following results:

Lemma 2.4: Let $f(x_1, x_2)$ be any continuous function with compact support in Ω. For any $\alpha > 0$, one has uniformly for all $(x_1, x_2) \in \overline{\Omega}$:

$$| \frac{1}{2\pi}(2^\alpha \lambda^{\frac{1-\alpha}{2}} \Gamma(\alpha+1)) \int\int_{r \le R} f(x_1, x_2) r^{-1-\alpha} J_{1+\alpha}(r\sqrt{\lambda}) dy_1 dy_2 | \le C \parallel f \parallel_\infty, R > 0, \lambda \to +\infty$$

For the proof see we refer to [6].

To estimate $\sigma_\lambda^\alpha f(x_1, x_2)$ we have
Lemma 2.5: Let $f(x_1, x_2)$ be any continuous function with compact support in Ω. If $\alpha > \frac{1}{2}$, then we have

$$| \sigma_\lambda^\alpha f(x_1, x_2) | \le C \parallel f \parallel_\infty, \lambda \to +\infty$$

uniformly for all $(x_1, x_2) \in \overline{\Omega}$.
For the proof see [6].

Lemma 2.6: Let $f(x_1, x_2)$ be any continuous function in Ω with compact support . If $\alpha > \frac{1}{2}$,we have

$$| \tau_\lambda^\alpha f(x_1, x_2) | \le C(\sqrt{\lambda})^{\frac{1}{2}-\alpha} \parallel f \parallel_{L_2(\Omega)} \parallel q \parallel_{W_2^1(\Omega)}$$

uniformly on $(x_1, x_2) \in \overline{\Omega}$.

Proof of the Lemma 2.6:

$$\tau_\lambda^\alpha f(x_1, x_2) = -\frac{1}{2\pi}(2^\alpha \lambda^{\frac{1-\alpha}{2}} \Gamma(\alpha+1)) \sum_{n=1}^\infty f_n \int_0^R r^{-\alpha} J_{1+\alpha}(r\sqrt{\lambda}) S_q^n(r) dr.$$

Substitute equation (8) into $\tau_\lambda^\alpha f(x_1, x_2)$ then we have

$$\tau_\lambda^\alpha f(x_1, x_2) = -\frac{1}{2\pi}(2^\alpha \lambda^{\frac{1-\alpha}{2}} \Gamma(\alpha+1)) \sum_{n=1}^\infty f_n \int_0^R r^{-\alpha} J_{1+\alpha}(r\sqrt{\lambda})$$
$$\frac{\pi}{2} \int_0^r [J_o(t\sqrt{\lambda_n}) Y_o(r\sqrt{\lambda_n}) - Y_o(t\sqrt{\lambda_n}) J_o(r\sqrt{\lambda_n})] \rho_n(t) t dt dr.$$

We expand $\tau_\lambda^\alpha f(x_1, x_2)$ into two main parts as follows:

$$\tau_\lambda^\alpha f(x_1, x_2) = \frac{1}{4}(2^\alpha \lambda^{\frac{1-\alpha}{2}} \Gamma(\alpha+1)) \sum_{n=1}^\infty f_n \int_0^R r^{-\alpha} J_{1+\alpha}(r\sqrt{\lambda}) \int_0^r Y_o(t\sqrt{\lambda_n}) J_o(r\sqrt{\lambda_n}) \rho_n(t) t dt dr$$
$$-\frac{1}{4}(2^\alpha \lambda^{\frac{1-\alpha}{2}} \Gamma(\alpha+1)) \sum_{n=1}^\infty f_n \int_0^R r^{-\alpha} J_{1+\alpha}(r\sqrt{\lambda}) \int_0^r J_o(t\sqrt{\lambda_n}) Y_o(r\sqrt{\lambda_n}) \rho_n(t) t dt dr.$$

And denoted

$$\tau_\lambda^\alpha f(x_1, x_2) = (\tau_\lambda^\alpha f(x_1, x_2))_1 - (\tau_\lambda^\alpha f(x_1, x_2))_2 = (\tau_\lambda^\alpha f)_1 - (\tau_\lambda^\alpha f)_2,$$

where we defined

$$(\tau_\lambda^\alpha f)_1 = \frac{1}{4}(2^\alpha \lambda^{\frac{1-\alpha}{2}} \Gamma(\alpha+1)) \sum_{n=1}^\infty f_n \int_0^R r^{-\alpha} J_{1+\alpha}(r\sqrt{\lambda}) \int_0^r Y_o(t\sqrt{\lambda_n}) J_o(r\sqrt{\lambda_n}) \rho_n(t) t dt dr$$
$$(\tau_\lambda^\alpha f)_2 = -\frac{1}{4}(2^\alpha \lambda^{\frac{1-\alpha}{2}} \Gamma(\alpha+1)) \sum_{n=1}^\infty f_n \int_0^R r^{-\alpha} J_{1+\alpha}(r\sqrt{\lambda}) \int_0^r J_o(t\sqrt{\lambda_n}) Y_o(r\sqrt{\lambda_n}) \rho_n(t) t dt dr.$$

We substitute $\rho_n(t)$ in equation (5) into $(\tau_\lambda^\alpha f)_1$ and $(\tau_\lambda^\alpha f)_2$ and we obtain;

$$(\tau_\lambda^\alpha f)_1 = \frac{1}{4}(2^\alpha \lambda^{\frac{1-\alpha}{2}}\Gamma(\alpha+1))$$
$$\sum_{n=1}^{\infty} f_n \int_0^R r^{-\alpha} J_{1+\alpha}(r\sqrt{\lambda})\{\int_0^r Y_o(t\sqrt{\lambda_n})J_o(r\sqrt{\lambda_n})[\int_{\gamma\le r}\psi_n(y_1,y_2)q(y_1,y_2)d\theta]tdt\}dr$$
$$(\tau_\lambda^\alpha f)_2 = -\frac{1}{4}(2^\alpha \lambda^{\frac{1-\alpha}{2}}\Gamma(\alpha+1))$$
$$\sum_{n=1}^{\infty} f_n \int_0^R r^{-\alpha} J_{1+\alpha}(r\sqrt{\lambda})\{\int_0^r J_o(t\sqrt{\lambda_n})Y_o(r\sqrt{\lambda_n})[\int_{\gamma\le r}\psi_n(y_1,y_2)q(y_1,y_2)d\theta]tdt\}dr.$$

Now we estimate $(\tau_\lambda^\alpha f)_1$ and $(\tau_\lambda^\alpha f)_2$;

$$(\tau_\lambda^\alpha f)_1 = \frac{1}{4}(2^\alpha \lambda^{\frac{1-\alpha}{2}}\Gamma(\alpha+1))$$
$$\sum_{n=1}^{\infty} f_n \int_0^R r^{-\alpha} J_{1+\alpha}(r\sqrt{\lambda})J_o(r\sqrt{\lambda_n})\{\int_0^r Y_o(t\sqrt{\lambda_n})[\int_{\gamma\le r}\psi_n(y_1,y_2)q(y_1,y_2)d\theta]tdt\}dr$$
$$(\tau_\lambda^\alpha f)_2 = -\frac{1}{4}(2^\alpha \lambda^{\frac{1-\alpha}{2}}\Gamma(\alpha+1))$$
$$\sum_{n=1}^{\infty} f_n \int_0^R r^{-\alpha} J_{1+\alpha}(r\sqrt{\lambda})Y_o(r\sqrt{\lambda_n})\{\int_0^r J_o(t\sqrt{\lambda_n})[\int_{\gamma\le r}\psi_n(y_1,y_2)q(y_1,y_2)d\theta]tdt\}dr.$$

Estimations for $(\tau_\lambda^\alpha f)_1$ and $(\tau_\lambda^\alpha f)_2$ are similar.
We estimate $(\tau_\lambda^\alpha f)_2$ by using the recurrence relation of Bessel functions as follows:

$$\int r^{-\nu} J_{\nu+1}(r\sqrt{\lambda})dr = -r^{-\nu}(\sqrt{\lambda})^{-1}J_\nu(r\sqrt{\lambda})$$

$$-r^{-\nu} J_{\nu+1}(r\sqrt{\lambda}) = \frac{d}{dr}\{r^{-\nu}(\sqrt{\lambda})^{-1}J_\nu(r\sqrt{\lambda})\}$$
$$-r^{-\nu}\sqrt{\lambda} J_{\nu+1}(r\sqrt{\lambda}) = \frac{d}{dr}\{r^{-\nu}J_\nu(r\sqrt{\lambda})\}$$

and using integration by parts and we obtain

$$(\tau_\lambda^\alpha f)_2 = -\frac{1}{4}(2^\alpha \lambda^{\frac{1-\alpha}{2}}\Gamma(\alpha+1))$$
$$\sum_{n=1}^{\infty} f_n Y_o(r\sqrt{\lambda_n})[J_o(t\sqrt{\lambda_n})\int_{r\le t}\psi_n(y_1,y_2)q(y_1,y_2)d\theta|_0^r - \int_0^r \frac{d}{dt}(J_o(t\sqrt{\lambda_n}))\int_{r\le t}\psi_n(y_1,y_2)q(y_1,y_2)d\theta dt]$$
$$= -\frac{1}{4}(2^\alpha \lambda^{\frac{1-\alpha}{2}}\Gamma(\alpha+1))\sum_{n=1}^{\infty} f_n Y_o(r\sqrt{\lambda_n})$$
$$[J_o(r\sqrt{\lambda_n})\int_{\beta\le r}\psi_n(y_1,y_2)q(y_1,y_2)d\theta + \int_0^r \sqrt{\lambda_n}(J_o(t\sqrt{\lambda_n}))\int_{\beta\le t}\psi_n(y_1,y_2)q(y_1,y_2)d\theta dt]$$
$$= -\frac{1}{4}(2^\alpha \lambda^{\frac{1-\alpha}{2}}\Gamma(\alpha+1))\sum_{n=1}^{\infty} f_n \int_0^R r^{-\alpha} J_{1+\alpha}(r\sqrt{\lambda_n})Y_o(r\sqrt{\lambda_n})J_o(r\sqrt{\lambda_n})dr\int_{r\le R} q\psi_n d\theta.$$

Finally we have

$$|(\tau_\lambda^\alpha f)_2|^2 \le C\lambda^{\frac{1-\alpha}{2}}(\sum_{n=1}^{\infty} f_n^2)(\int_0^R r^{-2\alpha} J_{1+\alpha}^2(r\sqrt{\lambda_n})Y_o^2(r\sqrt{\lambda_n})J_o^2(r\sqrt{\lambda_n})dr)(\int_{r\le R}\psi_n^2 d\theta)\|q\|_2 \quad (10)$$

The following estimations can be obtained from Theorem 2.1. Applying them we have

$$\sum_{\lambda_n>\lambda} \psi_n^2 \lambda_n^{(-\varepsilon-1)} = O(\lambda^{-\varepsilon}) ln^2\lambda, \lambda > 0$$

$$\sum_{\lambda_n<\lambda} \psi_n^2 \lambda_n^{(\varepsilon-1)} = O(\lambda^{\varepsilon}) ln^2\lambda.$$

After we have proved this estimation, the Theorem 2.2 can be proved as follows,

$$|(\tau_\lambda^\alpha f)_2| \leq C\lambda^{(\frac{1}{4}-\frac{\alpha}{2})} \|f\|_{L_2(\Omega)} \|q\|_{W_2^1(\Omega)}.$$

For any function from the space $W_2^1(\Omega)$ convergence of Riesz Means $E_\lambda^\alpha f(x)$, $\alpha > 0$ will be uniformly convergent on closed domain $\overline{\Omega}$. From compactness of $W_2^1(\Omega)$ in the space of continuous function on Ω,we conclude that the function f which satisfies conditions of the theorem can be approximated by functions from $W_2^1(\Omega)$,which has supports in Ω.

The proof of Theorem 2.2 is completed.

Acknowledgement
This paper has been supported by University Putra Malaysia under Research Grant (RUGS), project number 05-01-11-1273RU.

References

[1] Il'in V A 1990 *Spectral theory of differential operators* Nauka Moscow
[2] Moiseev E I 1977 *Soviet. Math. Dokl.* **18** 549-533
[3] Alimov S A and Rakhimov A A 1986 *Doclades of Uzbec Acad. Science* **10** 5-7
[4] Alimov S A and Rakhimov A A 1997 *J. Diff. eqn.* **33** 1
[5] Rakhimov A A 2003 *Modern problems of math.* **1** 167-172
[6] Rakhimov A A and Zakaria K 2008 *MathDigest Research Bulletin of Institute for Mathematical Research* **2** 1
[7] Rakhimov A A, Ahmedov A and Zainuddin H 2012 *Appl. Math. Lett.* **25** 921-924
[8] Ravshan A and Ahmedov A 2012 *Malaysian J. Math. Sci.* **5(2)** 185-196
[9] Ahmedov A and Zainuddin H 2008 *MathDigest Research Bulletin of Institute for Mathematical Research* **2** 1
[10] Titchmarsh E 1961 *Eigenfunctions expansion connected with differential equations of second order* **2** Moscow IL

The Localization Principle of Spectral Expansions of Distributions on Closed Domain

Anvarjon A. Ahmedov[1,2], **Siti Nor Aini binti Mohd Aslam**[2]
[1]Department of Process and Food Engineering, Faculty of Engineering, Universiti Putra Malaysia
[2]Institute for Mathematical Research, Universiti Putra Malaysia
E-mail: anvar@eng.upm.edu.my,aini_lily@yahoo.com

Abstract. The aim of this work is to investigate the problems of localization of spectral expansions of the distributions on closed domain, related to biharmonic operator. Using the mean value formula for the eigenfunctions of the biharmonic operator the Riesz means of the spectral function is estimated. Moreover, isomorphism of the generalized Holder spaces is applied.

1. Introduction

The first use of biharmonic operator Δ^2 for the modeling of an elastic plate is attributed to a correction of Lagrange of a manuscript by Sophie Germain from 1811. For a more elaborate history of the biharmonic problem and the relation with elasticity from an engineering point of view one may consult a survey of Meleshko [8]. Mathematically interesting questions came up around 1900 when Almansi [2], [3], Boggio [5] and Hadamard [7] addressed existence and positivity questions.

In order to have physically meaningful and mathematically well-posed problems the plate equation $\Delta^2 u = f$ has to be complemented with prescribing a suitable set of boundary data. The most commonly studied boundary value problems for second order elliptic equations are named Dirichlet, Neumann and Robin. These three types appear since they have a physical meaning. For fourth order differential equations such as the plate equation, the variety of possible boundary conditions is much larger.

We assume that the plate with the vertical projection of which is the planar region $\Omega \subset R^2$, is free to move horizontally at the boundary. Then a simple model for the elastic energy is

$$J(u) = \int\limits_{\Omega} \left(\frac{1}{2}(\Delta u)^2 - fu\right) dxdy,$$

where $f(x, y)$ is the external vertical load and $u(x, y)$ is the deflection of the plate in vertical direction and, as above for the beam, first order derivatives are left out which indicates that the plate is free to move horizontally.

Minimizing the energy functional leads to the weak Euler-Lagrange equation $(dJ(u), v) = 0$. In the case of pure hinged plate the latter is equivalent to the following boundary value problem

$$\Delta^2 u(x, y) = f(x, y), \quad (x, y) \in \Omega$$

$$u(x, y) = \Delta u(x, y) = 0, \quad (x, y) \in \partial\Omega.$$

This boundary value problem is in general referred to as the one with homogeneous Navier boundary conditions. The differential equation $\Delta^2 u = f$ is called the Kirchhoff-Love model for the vertical deflection of a thin elastic plate. In the current research we consider the problems of the representing any function from Sobolev's space by the eigenfunctions of the biharmonic operator.

2. The spectral expansions of the biharmonic operator

Let $\Omega \subset R^2$ be a domain with smooth boundary $\partial\Omega$. We consider the biharmonic operator $\Delta^2 = \left(\frac{\partial^2}{\partial x_1^2} + \frac{\partial^2}{\partial x_2^2}\right)^2$ with domain $C_0^\infty(\Omega)$.

In the present work we consider spectral expansions connected with the biharmonic operator on closed domain, and extend the localization properties of the spectral expansions to the distributions.

We define the space $C^l(\Omega)$ as a set of functions in Ω all of whose derivatives of order at most $l \in Z_+$ are continuous. The symbol $C^\infty(\Omega)$ denotes the space of all infinitely differentiable functions on Ω. The subset of $C^\infty(\Omega)$ which contains the functions from $C^\infty(\Omega)$ with compact support on Ω is denoted by $C_0^\infty(R^N)$.

The dual spaces (i.e., the spaces of continuous linear functionals on the sets of test functions) are denoted by

$$(C_0^\infty(\Omega))' = D'(\Omega)$$

and

$$(C^\infty(\Omega))' = \varepsilon'(\Omega).$$

The dual spaces are nested as follows:

$$\varepsilon' \subset D'.$$

Elements of the space D' are called distributions. Elements of the space ε' are distributions with compact support [6].

For $\ell > 0$ we define a functional $\|\cdot\|_\ell$ as follows

$$\|u\|_\ell = \left(\sum_{0\le|\alpha|\le\ell} \|D^\alpha u\|_{L_2}{}^2\right)^{1/2} \tag{1}$$

for any function u for which the right side makes sense.

We use notation $W_2^\ell(\Omega)$ to denote the Sobolev's space:

$$W_2^\ell(\Omega) = \{u \in L_2(\Omega) : D^\alpha u \in L_2(\Omega), 0 \le |\alpha| \le \ell\}$$

Equipped with the norm (1) the space $W_2^\ell(\Omega)$ is called the Sobolev's space. Next we define the dual space as follows

$$(W_2^l(\Omega))' = W_2^{-l}(\Omega).$$

Let $\hat{A}$ be a self-adjoint extension of the biharmonic operator in $L_2(\Omega)$. If E_λ is the corresponding spectral resolution unity, then the spectral expansion of any element $f \in L_2(\Omega)$ has the form

$$E_\lambda f(x) = \sum_{\lambda_n<\lambda} f_n u_n(x), \tag{2}$$

where f_n is the Fourier coefficients of the function f, with respect to the eigenfunctions $\{u_n(x)\}, n = 1, 2, ...$

It is well known that the operators E_λ are integral operators whose kernels $\Theta(x, y, \lambda)$ belong to the $C^\infty(\Omega)$ with respect to the both variables $x = (x_1, x_2)$ and $y = (y_1, y_2)$ and for any $\lambda > 0$. The spectral decomposition of an arbitrary function $g \in L_2(\Omega)$ is defined by the following

$$E_\lambda g(x) = \int\limits_\Omega g(y)\Theta(x, y, \lambda)dy.$$

Since $\Theta(x, y, \lambda) \in C^\infty(\Omega \times \Omega)$, it follows that for any $f \in \varepsilon'(\Omega)$ we can define the spectral decomposition of f by the formula

$$E_\lambda f(x) = \langle f, \Theta(x, y, \lambda)\rangle,$$

where the functional f acts on $\Theta(x, y, \lambda)$ with respect to the second argument. While the Riesz means of the spectral decomposition of f can be defined as follows

$$E_\lambda^s f(x) = \langle f, \Theta^s(x, y, \lambda)\rangle, \; s \geq 0$$

where

$$\Theta^s(x, y, \lambda) = \int\limits_0^\lambda \left(1 - \frac{t}{\lambda}\right)^s d_t\Theta(x, y, t).$$

Note that for any $f \in \varepsilon'(\Omega)$ its spectral expansions $E_\lambda^s f(x)$ belongs to the space $C^\infty(\Omega)$ for all $\lambda > 0$.

The problems of convergent of the eigenfunction expansions of the differential operators have a lot of practical applications in physics: in gas dynamics, in hydrodynamics, in elasticity theory and in other areas of mechanics, an important role is played by the theory of discontinuous solutions of the differential equations. In this work we investigate the principles of the localization of spectral decompositions and their Riesz means: a dependence of the convergence of spectral decompositions or their Riesz means on the behavior of the function in a small neighborhood of the given point.

Studies of the spectral expansions of distributions in the spaces with negative smoothness were carried out first in the work of Alimov (see [1]). He obtained the precise conditions for localization of the spectral expansions of distributions in Hilbert spaces, related to the Laplace operator. This work of Alimov gave impetus to conducting in-depth research into the spectral expansions of distributions, related to differential operators, in the development of new methods of investigation of the spectral expansions of distributions and in obtaining the final conditions for decomposability in different functional spaces with negative smoothness. Spectral decompositions of distributions associated with elliptic differential operators of order 2m were considered in the works [10], [11]. It is proved that the localization principle holds for the Riesz means of the spectral expansions of the distribution from the Sobolev space $W_p^-\ell(\Omega)$, $\Omega \subset R^N$ under the following conditions: $1 < p \leq 2$, $\ell > 0, s > (N-1)/2 + \ell$. In the present work we consider spectral expansions connected with the biharmonic operator on the closed domain, and extend the localization properties of the spectral expansions related to distributions.

Let us formulate the main results of the work

Theorem 2.1 *Let $f(x) \in \varepsilon'(\Omega) \cap W_2^{-\ell}(\Omega)$, $\ell > 0$ and $s \geq \ell + \frac{1}{2}$. Then the Riesz means $E_\lambda^s f(x)$ uniformly converge to 0 on any compact $K \subset \overline{\Omega} \setminus supp(f)$.*

The following is essential in proving the Theorem 2.1.

Theorem 2.2 *For $u_n(x)$ and λ_n, respectively, eigenfunctions and eigenvalues of the biharmonic operator Δ^2 on $\Omega \subset R^2$, corresponding to the boundary conditions $u|_{\partial\Omega} = \Delta\ u|_{\partial\Omega} = 0$ one has*

$$\sum_{\mu < \sqrt[2]{\lambda_n} < \mu+1} u_n^2(x) \leq C(\mu+1)\ln^2(\mu+1), \tag{3}$$

for all $x \in \overline{\Omega} = \Omega \cup \partial\Omega$ and $\mu \to +\infty$.

3. Mean value formula for the eigenfunctions of the biharmonic operator

Let $\Omega \subset R^2$ be a domain with smooth boundary $\partial\Omega$. We denote the eigenfunctions and eigenvalues of the biharmonic operator Δ^2 by $u_n(x)$ and λ_n respectively where $\Delta = \partial_{x_1}^2 + \partial_{x_2}^2$ is the Laplace operator. Then we have

$$\Delta^2 u_n(x) - \lambda_n u_n(x) = 0,$$

$$u_n|_{\partial\Omega} = 0 = \Delta u_n|_{\partial\Omega}.$$

Let $u_n(x) = \varphi(\sqrt[4]{\lambda_n}x)$.

$$\Delta^2 u_n(x) - \lambda_n u_n(x) = \lambda_n \Delta^2 \varphi - \lambda_n \varphi = \lambda_n(\Delta^2 \varphi_n - \varphi_n) = 0.$$

We use notations

$$\psi_n(x) = (\Delta + 1)\varphi_n \tag{4}$$

and

$$\phi_n(x) = (\Delta - 1)\varphi_n \tag{5}$$

Note that,

$$\Delta\psi_n(x) = \Delta^2\varphi_n + \Delta\varphi_n = \varphi_n + \Delta\varphi_n = \psi_n(x),$$

$$\Delta\phi_n(x) = \Delta^2\varphi_n - \Delta\varphi_n = \varphi_n - \Delta\varphi_n = -\phi_n(x).$$

Therefore,

$$\Delta\psi_n(x) - \psi_n(x) = 0,$$

and

$$\Delta\phi_n(x) + \phi_n(x) = 0.$$

For the eigenfunctions $v_n(x)$ of the Laplace operator

$$\Delta v_n + \lambda_n v_n = 0,$$

one has the following mean value formula

$$\int_0^{2\pi} v_n(x_1 + rcos\theta, x_2 + rsin\theta)d\theta = 2\pi J_0(r\sqrt{\lambda_n})v_n(x_1, x_2). \tag{6}$$

by using the latter, we obtain

for $\Delta\phi_n(x) + \phi_n(x) = 0$,

$$\int_0^{2\pi} \phi_n(x + r\theta)d\theta = 2\pi J_0(r)\phi_n(x_1, x_2),$$

for $\Delta\psi_n(x) - \psi_n(x) = 0$,

$$\int_0^{2\pi} \psi_n(x + r\theta)d\theta = 2\pi J_0(r\sqrt{-1})\psi_n(x_1, x_2).$$

From (4) and (5),

$$\varphi_n(x) = \frac{1}{2}(-\phi_n(x) + \psi_n(x)),$$

so

$$\int_0^{2\pi} \varphi_n(x + r\theta)d\theta = \frac{1}{2}(-2\pi J_0(r)\phi_n(x_1, x_2) + 2\pi J_0(ri)\psi_n(x_1, x_2)).$$

by using $u_n(x) = \varphi(\sqrt[4]{\lambda_n}x)$ then from (4), $\phi(\sqrt[4]{\lambda_n}x) = \sqrt{\lambda_n}\Delta - u_n$ and from (5), $\psi(\sqrt[4]{\lambda_n}x) = \sqrt{\lambda_n}\Delta + u_n$.

For the eigenfunctions of the biharmonic operator, we get

$$\int_0^{2\pi} u_n(x + r\theta)d\theta = \frac{1}{2}(-2\pi J_0(r\sqrt[4]{\lambda_n})(\sqrt{\lambda_n}\Delta - u_n) + 2\pi J_0(ri\sqrt[4]{\lambda_n})(\sqrt{\lambda_n}\Delta + u_n)).$$

Finally,

$$\int_0^{2\pi} u_n(x + r\theta)d\theta = \pi J_0(r\sqrt[4]{\lambda_n})u_n(x) - \pi J_0(r\sqrt[4]{\lambda_n})\sqrt{\lambda_n}\Delta u_n + \pi J_0(ri\sqrt[4]{\lambda_n})(\sqrt{\lambda_n}\Delta u_n + u_n).$$

We denote

$$S_r(\lambda_n) = \pi[J_0(ri\sqrt[4]{\lambda_n}) - (r\sqrt[4]{\lambda_n})]\sqrt{\lambda_n}\Delta + u_n + \pi J_0(ri\sqrt[4]{\lambda_n})u_n.$$

We note here that

$$\int_0^{2\pi} u_n(x_1 + rcos\theta, x_2 + rsin\theta)d\theta = \pi J_0(r\sqrt[4]{\lambda_n})u_n(x_1, x_2) + O(e^{-(r_o - r)\sqrt[4]{\lambda_n}}),$$

where $0 < r_0 < r$.

Proof of Theorem 2.1

Let $\{u_n\}$ be the set of eigenfunctions of Biharmonic operator corresponding to eigenvalues $\{\lambda_n\}$ as the following

$$\Delta^2 u_n(x) - \lambda_n u_n(x) = 0.$$

For eigenfunctions $u_n(x)$ of the biharmonic operator, we have

$$\int_0^{2\pi} u_n(x_1 + rcos\theta, x_2 + rsin\theta)d\theta = 2\pi J_0(|x-y|\sqrt[4]{\lambda_n})u_n(x) + O(e^{-\beta_o(r_o - r)\sqrt[4]{\lambda_n}}),$$

where $0 < r < r_0 < dist(x, \partial\Omega)$.

Let fix $\delta > 0$ and $R > \delta$. Let $x \in \Omega_\delta$, the radial function

$$V(r) = \begin{cases} \Gamma(s+1)2^s(2\pi)^{-1}\lambda^{\frac{1-s}{2}} \times \frac{J_{s+1}(r\sqrt{\lambda})}{r^{s+1}} & ,r \leq R \\ 0 & ,r > R \end{cases}$$

We use Parseval's formula to establish representation for Riesz means

$$\sum_{n=1}^{\infty} f_n d_n = \int_\Omega f(x)g(x)dx \tag{7}$$

where

$$f_n = \int_\Omega f_n u_n(x)dx,$$

$$d_n = \int_\Omega g(x)u_n(x)dx.$$

Substituting $g(x)$ with the radial function $V(r)$, one has

$$d_n = \int_\Omega V(r)u_n(y)dy$$

$$= \iint_\Omega V(r)u_n(y_1, y_2)dy_1dy_2$$

$$= \int \int_{|x-y| \leq R} V(r)u_n(y_1, y_2)dy_1dy_2$$

Journal of Physics: Conference Series **435** (2013) 012015 doi:10.1088/1742-6596/435/1/012015

$$= \int_0^R V(r) \int_0^{2\pi} u_n(x_1 + rcos\theta, x_2 + rsin\theta) r dr d\theta$$

$$= \int_0^R V(r) r (J_0(r\sqrt[4]{\lambda_n}) u_n(x_1, x_2) + S_r) d\theta dr.$$

Applying mean value formula, we get

$$d_n = u_n(x) \int_0^R \Gamma(s+1) 2^s (2\pi)^{-1} \lambda^{\frac{1-s}{2}} \times \frac{J_{1+s}(r\sqrt{\lambda})}{r^{1+s}} r J_0(r\sqrt[4]{\lambda_n}) dr +$$

$$u_n(x) \int_0^R \Gamma(s+1) 2^s (2\pi)^{-1} \lambda^{\frac{1-s}{2}} \times \frac{J_{1+s}(r\sqrt{\lambda})}{r^{1+s}} r S_r dr.$$

Now, we will consider the integration part from the first term of the following

$$d_n = \Gamma(s+1) 2^s (2\pi)^{-1} \lambda^{\frac{1-s}{2}} u_n(x) \int_0^R J_{1+s}(r\sqrt{\lambda}) J_0(r\sqrt[4]{\lambda}) r^{-s} dr + u_n(x) B_n^\lambda(R),$$

where

$$B_n^\lambda(R) = \int_0^R \Gamma(s+1) 2^s (2\pi)^{-1} \lambda^{\frac{1-s}{2}} \times \frac{J_{1+s}(r\sqrt{\lambda})}{r^{1+s}} r S_r dr.$$

Note that,(see[12])

$$\int_0^\infty J_{1+s}(r\sqrt{\lambda}) J_0(r\sqrt[4]{\lambda}) r^{-s} dr = \begin{cases} \frac{2^{-s}}{\lambda^{\frac{1-s}{2}} \Gamma(s+1)} \times (1 - \frac{\sqrt{\lambda_n}}{\lambda})^s & , \sqrt{\lambda_n} \le \lambda \\ 0 & , \sqrt{\lambda_n} > \lambda \end{cases} \tag{8}$$

Separating the integral into two parts i.e. $\int_0^R = \int_0^\infty - \int_R^\infty$, we obtain

$$d_n = \Gamma(s+1) 2^s (2\pi)^{-1} \lambda^{\frac{1-s}{2}} u_n(x) \int_0^\infty J_{1+s}(r\sqrt{\lambda}) J_0(r\sqrt[4]{\lambda}) r^{-s} dr$$

$$-\Gamma(s+1) 2^s (2\pi)^{-1} \lambda^{\frac{1-s}{2}} u_n(x) \int_R^\infty J_{1+s}(r\sqrt{\lambda}) J_0(r\sqrt[4]{\lambda}) r^{-s} dr + u_n B_n^\lambda(R).$$

Substituting the value of integral (8), we have

$$d_n = u_n(x)(1 - \frac{\sqrt[2]{\lambda_n}}{\lambda})^s \delta_n^\lambda + \Gamma(s+1)2^s(2\pi)^{-1}\lambda^{\frac{1-s}{2}} u_n(x) \int_R^\infty J_{1+s}(r\sqrt{\lambda})J_0(r\sqrt[4]{\lambda})r^{-s}dr + u_n B_n^\lambda(R),$$

where

$$\delta_n^\lambda = \begin{cases} 1 & ,if \quad \sqrt{\lambda_n} \le \lambda \\ 0 & ,if \quad \sqrt{\lambda_n} > \lambda \end{cases}$$

by using the Parseval's formula, we obtain

$$\Gamma(s+1)2^s(2\pi)^{-1} \int_\Omega f(x)\lambda^{\frac{1-s}{2}} \times \frac{J_{1+s}(r\sqrt{\lambda})}{r^{1+s}} dx$$

$$= \sum_{n=1}^\infty f_n(u_n(x)(1 - \frac{\sqrt{\lambda_n}}{\lambda})^s) - \sum_{n=1}^\infty f_n \Gamma(s+1)2^s(2\pi)^{-1}\lambda^{\frac{1-s}{2}} u_n(x)$$

$$\times \int_R^\infty J_{1+s}(r\sqrt{\lambda})J_0(r\sqrt[4]{\lambda})r^{-s}dr + \sum_{n=1}^\infty f_n u_n B_n^\lambda(R).$$

The first term is equal to Riesz means, so we have

$$\Gamma(s+1)2^s(2\pi)^{-1} \int_\Omega f(x)\lambda^{\frac{1-s}{2}} \times \frac{J_{1+s}(r\sqrt{\lambda})}{r^{1+s}} dx = E_\lambda^s f(x) - \sum_{n=1}^\infty f_n u_n(x) I_n(\lambda, \lambda_n) + \sum_{n=1}^\infty f_n u_n B_n^\lambda,$$

where

$$I_n(\lambda, \lambda_n) = \Gamma(s+1)2^s(2\pi)^{-1}\lambda^{\frac{1-s}{2}} \times \int_R^\infty J_{1+s}(r\sqrt{\lambda})J_0(r\sqrt[4]{\lambda})r^{-s}dr.$$

Thus, for Riesz means we obtain

$$E_\lambda^s f(x) = \Gamma(s+1)2^s(2\pi)^{-1} \int_{r\le\delta} f(x)\lambda^{\frac{1-s}{2}} \times \eta(x)\frac{J_{1+s}(r\sqrt{\lambda})}{r^{1+s}} dx$$

$$+ \sum_{n=1}^\infty f_n u_n(x) I_n(\eta, \lambda, \lambda_n) - \sum_{n=1}^\infty f_n u_n B_n^\lambda(R).$$

The first part is equal to zero, so

$$E_\lambda^s f(x) = \sum_{n=1}^\infty f_n u_n(x) I_n(\eta, \lambda, \lambda_n) - \sum_{n=1}^\infty f_n u_n B_n^\lambda(R).$$

Lemma 3.1 *For any compact $K \subset \Omega$ there exists $C = C(k)$, such that uniformly for all $X \in K$ one has*

$$\sum_{n=1}^{\infty} \frac{u_n^2(x)}{1 + (\sqrt[2]{\lambda_n} - \lambda)^2} \times (\sqrt{\lambda_n})^{-1} \leq C(k.)$$

For the proof we refer the readers to [13].

Let denote A by a self-adjoint extension of Biharmonic operator in $L_2(\Omega)$. Then we have

$$E_\lambda^s f(x) =< A^{-\frac{\ell}{2}} f, A^{\frac{\ell}{2}} \theta^s(x, y, \lambda) >$$

We prove that

$$||A^{\frac{\ell}{2}} \theta^s(x, y, \lambda)||_{L_2} \leq C(\sqrt{\lambda})^{\ell - s + \frac{1}{2}}.$$

where

$$|E_\lambda^s f(x)| \leq |\sum f_n (1 + \lambda_n)^{-\frac{\ell}{2}} (1 + \lambda_n)^{\frac{\ell}{2}} u_n(x) I_{n(\lambda, \lambda_n)}|$$

$$\leq (\sum_{n=1}^{\infty} f_n^2 (1 + \lambda_n)^{-\ell})^{\frac{1}{2}} (\sum_{n=1}^{\infty} (1 + \lambda_n)^{\ell} u_n^2(x) I_n(\lambda, \lambda_n))^{\frac{1}{2}}$$

$$\leq C||f||_{-\ell} (\sqrt[2]{\lambda})^{\ell - s + \frac{1}{2}}.$$

So we have

$$|E_\lambda^s f(x)| \leq C(\sqrt{\lambda})^{\ell - s + \frac{1}{2}} ||f||_{W_2^{-\ell}}$$

For any function from $C_0^\infty(\Omega)$, the Riesz means, $E_\lambda^s f(x)$ uniformly convergence to f on closed domain $\overline{\Omega}$. From density of $C_0^\infty(\Omega)$ in $W_2^{-\ell}(\Omega)$ and using techniques from the work [13], we obtain uniformly convergence of $E_\lambda^s f(x)$ to $f(x)$ on closed domain.

4. Acknowledgement

This work has been supported by Universiti Putra Malaysia under the Research University Grant (RUGS), project number 05-01-11-1273RU.

References

[1] Alimov Sh A 1993 *Russian: Doclady of Russian Academy Sciences* **331 (6)** 661
[2] Almansi E 1899 *Rom. Acc. L. Rend* **(5) 81**104
[3] Almansi E 1899 *Annali di Mat.* (3) 2:1
[4] Anvarjon Ahmedov, Hishamuddin Zainuddin 2009 *Malaysia: MathDigest, Bulletin of the Institute for Mathematical Research, Universiti Putra Malaysia* **2 (1)** pp. 6
[5] Boggio T 1905 *Rend. Circ. Mat. Palermo* **20** 97
[6] Egorov U V 1984 *Linear Differential Equations of Principle Type*(Moscow: Nauka)
[7] Hadamard J 1968 *Me moire Sur le Proble'me Danalyse Relatif a' Le quilibre des Plaques E lastiques Encastre es. In: uvres de Jacques Hadamard* (Paris: Tome II CNRS) pages 515641
[8] Meleshko V V 2003 *Applied Mechanics Reviews* **56** 33
[9] Moiseev E I 1977 *Dokl. Akad. Nauk* **233(6)** 1042
[10] Alimov Sh, Rakhimov A 1996 *Differential Equations* **32(6)** 798
[11] Rakhimov A A 1996 *Mathematical Notes* **56 (3)** 422

[12] Titchmarsh E C 1958 *Eigenfunction Expansions Associated with Second Order Differential Equations II* (Oxford: Clarendon Press)

[13] Alimov Sh, Rakhimov A 1997 *Differential Equations* **33 (1)** 80

[14] Triebel H 1978 *Interpolation Theory Function Spaces, Differential Operators* (North-Holland, Amsterdam, New York, Oxford)

Journal of Physics: Conference Series **435** (2013) 012016 doi:10.1088/1742-6596/435/1/012016

On The Sufficient Conditions Of The Localization Of The Fourier-Laplace Series Of Distributions From Liouville Classes

Anvarjon A. Ahmedov[1,2], **Ahmad Fadly Nurullah bin Rasedee**[2], **Abdumalik Rakhimov**[2,3]

[1]Department of Process and Food Engineering, Faculty of Engineering, Universiti Putra Malaysia

[2]Institute for Mathematical Research, Universiti Putra Malaysia

[3]Kulliyyah of Engineering, International Islamic University Malaysia

E-mail: [1]anvar@eng.upm.edu.my, [2]ahmadfadlynurullah@yahoo.com, [3]abdumalik@iium.edu.my

Abstract. In this work we investigate the localization principle of the Fourier-Laplace series of the distribution. Here we prove the sufficient conditions of the localization of the Riesz means of the spectral expansions of the Laplace-Beltrami operator on the unit sphere.

1. Introduction

Let S^N be the N-dimensional unit sphere. We investigate the spectral expansions connected with Laplace-Beltrami operator on sphere which is related to the Laplace operator on R^{N+1} by the following

$$\Delta f = \frac{\partial^2 f}{\partial r^2} + \frac{N}{r}\frac{\partial f}{\partial r} + \frac{1}{r^2}\Delta_S f$$

where Δ_S is the Laplace-Beltrami operator on the N- dimensional sphere, which is symmetric and nonegative elliptic operator.

The Laplace-Beltrami operator Δ_s has a complete system of eigenfunctions

$$\left\{Y_1^{(k)}(x), Y_2^{(k)}(x), ..., Y_{a_k}^{(k)}(x)\right\}, \quad k = 0, 1, 2, ...,$$

corresponding to the eigenvalues $\{\lambda_k = k(k+N-1)\}, k = 0, 1, 2..$ where

$$a_k = \begin{cases} 1, & if \quad k = 0, \\ N, & if \quad k = 1, \\ \frac{(N+k)!}{N!k!} - \frac{(N+k-2)!}{N!(k-2)!}, & if \quad k \geq 2 \end{cases}$$

is the multiplicity of the eigenvalues λ_k.

The Riesz means of the order $\alpha \geq 0$ of the spectral expansions of the Laplace-Beltrami of the f is defined by

$$E_n^\alpha f(x) = \sum_{k=0}^{n} \left(1 - \frac{\lambda_k}{\lambda_n}\right)^\alpha Y_k(f, x)$$

where,

$$Y_k(f, x) = \sum_{j=1}^{a_k} Y_j^{(k)}(x) \int_{S^N} f(y) Y_j^{(k)}(y) d\sigma(y).$$

In this work the localization principle of the Spectral expansions of the distibutions on unit sphere is investigated. In the case of Spectral expansions of the integrable functions on unit sphere the same problems are considered in the papers [1], [2], [3], [6], [8], [12] and [15] . For more reference related to the problems of the convergence of the Fourier-Laplace series on unit sphere we refer the readers to [20]. Investigating the problems of the localization of the Fourier-Laplace series of the distribution on unit sphere was started by Rakhimov (see [16]). He proved the sufficient conditions for the localization of the Cesaro means of the Fourier-Laplace series of the distirbution on unit sphere. In current work we extend the latter result for the case of the Riesz means of the spectral expansions of the Laplace-Beltrami operator on unit sphere. Here we use the estimation for the Riesz means of the spectral function of the Laplace-Beltrami operator on sphere (see [2]).

2. Distributions on The Sphere

By $D(S^N)$ we denote the space of infinite times differentiable functions on sphere. Let $D'\left(S^N\right)$ denotes the conjugate space of $D\left(S^N\right)$ i.e the space of the linear functional defined on $D\left(S^N\right)$. We define the Riesz means of the spectral expansions of the is distirbutions by

$$E_n^\alpha f(x) = \langle f, \Theta^\alpha(x, y, n) \rangle = \sum_{k=0}^{n} \left(1 - \frac{\lambda_k}{\lambda_n}\right)^\alpha \sum_{j=1}^{a_k} Y_j^k(x) \left\langle f, Y_j^k(y) \right\rangle,$$

where $\left\langle f, Y_j^k(y) \right\rangle$ is the value of the functional f on the eigenfunction $Y_j^k(y)$ of the Laplace-Beltrami operator. The following results can be obtained by usng properties of E_n^α in $D\left(S^N\right)$ and $D'\left(S^N\right)$.

Theorem 2.1 *Let $\psi \in D\left(S^N\right)$, then the Riesz means $E_n^\alpha, \alpha \geq 0$, of the Spectral expansions of the φ converges in the topology of the space $D\left(S^N\right)$,*

$$\lim_{n\to\infty} E_n^\alpha \varphi(x) = \varphi(x).$$

Theorem 2.2 *If $f \in D'\left(S^N\right)$, then the Riesz means, $E_n^\alpha, \alpha \geq 0$, of the Spectral expansions of the f converges in the topology of the space $D'\left(S^N\right)$,*

$$\langle E_n^\alpha f(x), g(x) \rangle \to \langle f(x), g(x) \rangle, \forall g \in D'\left(S^N\right).$$

Definition 2.3 *For any $l \in R$ we define the class of Liouville $L_p^l\left(S^N\right)$, which consist of all distributions $f \in D'\left(S^N\right)$, such that*

$$\left\|\sum_{k=0}^{\infty}(1+\lambda_k)^l Y_k(f,x)\right\|_{L_p} < \infty.$$

The norm of f in $L_p^l\left(S^N\right)$ is defined by

$$\|f\|_{L_p^l} = \left\|\sum_{k=0}^{\infty}(1+\lambda_k)^l Y_k(f,x)\right\|_{L_p}.$$

For any open set $V \subset S^N$ we define by V^t the following symmetric image of V by $V^t = \left\{x \in S^N : -x \in V\right\}$. For more properties of $L_p^l\left(S^N\right)$ we refer the reader to to [19]. Let us proceed with the formulation of the main results of this work:

Theorem 2.4 *Let $f \in L_p^{-l}\left(S^N\right)$, $l > 0$, $p > 1$. If $f = 0$ on $V \cup V^t \subset S^N$ and $\alpha \geq l + \frac{N-1}{2}$, then uniformly on any compact K, $K \subset V \cup V^t$ one has*

$$\lim_{n\to\infty} E_n^{\alpha} f(x) = 0.$$

The case when the distribution will coincide with continous function in some domain we have the following result:

Theorem 2.5 *If $f \in L_p^{-l}\left(S^N\right)$, $l > 0$, $p > 1$ coincides with continous function $g(x)$ on $V \cup V^t$ and $\alpha \geq l + \frac{N-1}{2}$, then uniformly on any compact, $K \subset V \cup V^t$ one has*

$$\lim_{n\to\infty} E_n^{\alpha} f(x) = g(x).$$

Proof of Theorem 2.4. Let consider fractional order $\beta \in R$ of $1 - \Delta_s$:

$$\left(\overline{1-\Delta_s}\right)^{\beta}\phi = \sum_{k=0}^{\infty}(1+\lambda_k)^{\beta} Y_k(\phi,x), \quad \phi \in C^{\infty}\left(S^N\right).$$

It is well known that

$$\left(\overline{1-\Delta_s}\right)^{\frac{l}{2}}\Theta^{\alpha}(x,y,n,) = \sum_{k=0}^{\infty}(1+\lambda_k)^{\frac{l}{2}}\left(1-\frac{\lambda_k}{\lambda_n}\right)^{\alpha}\sum_{j=1}^{a_k} Y_j^{(k)}(x)Y_j^{(k)}(y).$$

For the distibution (functionals with compact support) $f = 0$, in V and V^t we have

$$E_n^{\alpha} f(x) = \langle f(y), \Theta^{\alpha}(x,y,n)\rangle$$

$$= \left\langle \left(\overline{1-\Delta_s}\right)^{-\frac{l}{2}} f, \left(\overline{1-\Delta_s}\right)^{\frac{l}{2}}\Theta^{\alpha}(x,y,n)\right\rangle.$$

Taking into account the definition of norm in L_p^{-l}, $p > 1$ we obtain

$$\|f\|_{-l} = \left\|\left(\overline{1-\Delta_s}\right)^{-\frac{l}{2}} f\right\|_{L_p} = \left\|\sum_{k=0}^{\infty}(1+\lambda_k)^{-\frac{l}{2}}\sum_{j=1}^{a_k}\left\langle f, Y_j^{(k)}\right\rangle Y_j^{(k)}\right\|_{L_p} =$$

$$= \left\| \sum_{k=0}^{\infty} (1+\lambda_k)^{-\frac{l}{2}} Y_k(f,x) \right\|_{L_p}, \qquad l \in R.$$

The following Lemma will play an essential role in the proof of the Theorem 2.4

Lemma 2.6 *Let $V \subset S^N$ be an open domain and $K \subset V$ any compact in V. Then one has*

$$\|\Theta^\alpha(x,y,n)\|_{L_q(S^N \setminus (V \cup V^t))} \le C n^{\frac{N-1}{2}-\alpha}, \qquad \frac{1}{p}+\frac{1}{q}=1,$$

uniformly $x \in K$

For the Riesz means of the spectral function

$$\Theta^\alpha(x,y,n) = \sum_{k=0}^{\infty} \left(1-\frac{\lambda_k}{\lambda_n}\right)^\alpha \sum_{j=1}^{a_k} Y_j^{(k)}(x) Y_j^{(k)}(y),$$

we have

Theorem 2.7 *Let $\Theta^\alpha(x,y,n)$ is the kernel of Riesz means of the spectral expansions. If*
1) $\left|\frac{\pi}{2}-\gamma(x,y)\right| < \frac{\pi n}{2(n+1)}, n \to \infty$ *then*

$$|\Theta^\alpha(x,y,n)| \le \frac{C_1 n^{\frac{N-1}{2}-\alpha}}{(\sin\gamma)^{\frac{N-1}{2}} \left(\sin\frac{\gamma}{2}\right)^{1+\gamma}} + \frac{C_2 n^{\frac{N-3}{2}-\alpha}}{(\sin\gamma)^{\frac{N+1}{2}} \left(\sin\frac{\gamma}{2}\right)^{1+\gamma}} + \frac{C_3 . n^{-1}}{\left(\sin\frac{\gamma}{2}\right)^{N+1}}$$

2) $0 < \gamma_0 \le \gamma \le \pi$

$$|\Theta^\alpha(x,y,n)| \le C_4 n^{N-1-\alpha}$$

3) $0 \le \gamma \le \pi$

$$|\Theta^\alpha(x,y,n)| \le C_5 n^N$$

For the proof we refer the reader to [2].

Proof of Lemma 2.6:
Let K be any compact in $V \cup V^T$. For any $x \in K$ and $y \in S^N \setminus (V \cup V^T)$, we have $\gamma(x,\ y) > \delta > 0$. We denote by δ the distance between K and $S^N \setminus (V \cup V^T)$. We estimate

$$\|\Theta^\alpha(x,y,n)\|_{L_q(S^N \setminus (V \cup V^t))} \le \|I_1+I_2+I_3\|_{L_q}, \quad q>1, \tag{1}$$

where

$$I_1 = \frac{C_1 n^{\frac{N-1}{2}-\alpha}}{(\sin\gamma)^{\frac{N-1}{2}} \left(\sin\frac{\gamma}{2}\right)^{1+\gamma}}, \quad I_2 = \frac{C_2 n^{\frac{N-3}{2}-\alpha}}{(\sin\gamma)^{\frac{N+1}{2}} \left(\sin\frac{\gamma}{2}\right)^{1+\gamma}}, \quad I_3 = \frac{C_3 n^{-1}}{\left(\sin\frac{\gamma}{2}\right)^{N+1}}.$$

Journal of Physics: Conference Series **435** (2013) 012016 doi:10.1088/1742-6596/435/1/012016

Applying Holder's inequality (1) becomes

$$\leq \|I_1\|_{L_q} + \|I_2\|_{L_q} + \|I_3\|_{L_q}.$$

Estimation of I_1

$$I_1 = C_1 n^{\frac{N-1}{2}-\alpha}\left(\int (\sin\gamma)^{-q\frac{N-1}{2}}(\sin\frac{\gamma}{2})^{-q(1+\alpha)}d\sigma(x)\right)^{\frac{1}{q}}$$

$\mathrm{dist}\big(K, S^N \setminus (V \cup V^t)\big) = \delta$

$$I_1 = C_1 n^{\frac{N-1}{2}-\alpha}\left(\int_{\delta}^{\pi-\delta} (\sin\gamma)^{-q\frac{N-1}{2}}(\sin\frac{\gamma}{2})^{-q(1+\alpha)}\left(\int_{(x,y)=\cos\gamma} df(y)\right)d\sigma(x)\right)^{\frac{1}{q}} \tag{2}$$

where $\int\limits_{(x,y)=\cos\gamma} df(y) = |S^{N-1}|\sin^{N-1}\gamma = \frac{2\pi^{\frac{N}{2}}}{\Gamma(\frac{N}{2})}\sin^{n-1}\gamma$. Hence $|S_\rho^N|$ as

$$|S^N| = \frac{2\pi^{\frac{N+1}{2}}}{\Gamma\left(\frac{N+1}{2}\right)}.$$

and then substituting

$$\sin\gamma \geq \sin\delta, \quad \sin\frac{\gamma}{2} \leq \sin\frac{\delta}{2}, \quad \sin\gamma = 2\cos\frac{\gamma}{2}\sin\frac{\gamma}{2}, \quad \sin\gamma \leq \gamma$$

into (2) yields the following inequality,

$$I_1 \leq C_N n^{\frac{N-1}{2}-\alpha}\left(\int_{\delta}^{\pi-\delta} (\gamma)^{-p\frac{N-1}{2}+N-1}(\frac{\gamma}{2})^{-p(1+\alpha)}d\gamma\right)^{\frac{1}{p}}$$

$$= C_N^{\alpha} n^{\frac{N-1}{2}-\alpha}\frac{\gamma^{-q\frac{N-1}{2}+N-1-q(1+\alpha)+1}}{-q\frac{N-1}{2}+N-1-q(1+\alpha)+1}\Bigg|_{\delta}^{\pi-\delta}$$

where $\delta > 0$

$$q\left(-\frac{N-1}{2}+\frac{N}{q}-1-\alpha\right) \leq q\left(-\frac{N-1}{2}+\frac{N}{q}-1-\frac{N-1}{2}\right)$$

and $\alpha \geq \frac{N-1}{2}$, $\quad -\alpha \leq -\frac{N-1}{2}$ giving,

$$q\left(-\frac{N-1}{2}+\frac{N}{q}-1-\frac{N-1}{2}\right) = q\left(-N+1+\frac{N}{q}-1\right) = -qN+N \leq 0.$$

Estimation of I_2

$$I_2 = C_2 n^{\frac{N-3}{2}-\alpha}\left(\int (\sin\gamma)^{-q\frac{N+1}{2}}(\sin\frac{\gamma}{2})^{-q(1+\alpha)}d\sigma(x)\right)^{\frac{1}{q}}$$

By similiar substitutions used for derving I_1, I_2 yields the inequality,

$$I_2 \leq C_2 n^{\frac{N-3}{2}-\alpha} \left(\int_{\delta}^{\pi-\delta} (\gamma)^{-q\frac{N+1}{2}+N-1} (\frac{\gamma}{2})^{-q(1+\alpha)} d\gamma \right)^{\frac{1}{q}}$$

$$= C_2^{\alpha} n^{\frac{N-3}{2}-\alpha} \frac{\gamma^{-q\frac{N+1}{2}+N-1-q(1+\alpha)+1}}{-q\frac{N+1}{2}+N-1-q(1+\alpha)+1} \Bigg|_{\delta}^{\pi-\delta}$$

where $\delta > 0$

$$q\left(-\frac{N+1}{2}+\frac{N}{q}-1-\alpha\right) \leq q\left(-\frac{N+1}{2}+\frac{N}{q}-1-\frac{N-3}{2}\right)$$

and $\alpha \geq \frac{N-3}{2}$, $\quad -\alpha \leq -\frac{N-3}{2}$ giving,

$$q\left(-\frac{N+1}{2}+\frac{N}{q}-1-\frac{N-3}{2}\right) = q\left(-N+1+\frac{N}{q}-1\right) = -qN+N \leq 0.$$

Estimation of I_3

$$I_3 = C_3 n^{-1} \left(\int_{\delta}^{\pi-\delta} \sin\left(\frac{\gamma}{2}\right)^{-q(1+N)} d\sigma(x) \right)^{\frac{1}{q}}.$$

Similiar techniques used in I_1 and I_2 gives us

$$\leq C_3 n^{-1} \left(\int_{\delta}^{\pi-\delta} \left(\frac{\gamma}{2}\right)^{-q(1+N)} d\gamma \right)^{\frac{1}{q}}$$

$$\leq C_3^{\alpha} n^{-1} \frac{\gamma^{-q(1+N)+1}}{-q(1+N)+1} \Bigg|_{\delta}^{\pi-\delta} \leq \frac{C}{n}.$$

This complete the proof for Lemma 2.6.

We are ready to complete prove for Theorem 2.4.

Let $f(y)$ be a distribution vanishing on the function ϕ: $\phi = 0$; $\phi \in L_p^l$

$$E_n^{\alpha} f(x) = \int_{S^N \setminus (V \cup V^t)} f(y) \Theta^{\alpha}(x, y, n)\, d(y).$$

Using Holder Inequality we have

$$|E_n^{\alpha} f(x)| \leq \left\| \left(1-\Delta_s\right)^{-\frac{l}{2}} f \right\|_{L_p} \cdot \left\| \Theta_{\frac{l}{2}}^{\alpha}(x, y, n) \right\|_{L_q}$$

where $\frac{1}{p}+\frac{1}{q}=1$, $\Theta_{\frac{l}{2}}^{\alpha}(x, y, n) = \left(1-\Delta_s\right)^{\frac{l}{2}} \Theta^{\alpha}(x, y, n)$.

We denote

$$\Theta^{\alpha}_{\frac{l}{2}}\left(x,y,n\right) = \left(\overline{1-\Delta_s}\right)^{\frac{l}{2}}\Theta^{\alpha}\left(x,y,n\right)$$

$$= \left(\overline{1-\Delta_s}\right)^{\frac{l}{2}}\sum_{k=0}^{n}\left(1-\frac{\lambda_k}{\lambda_n}\right)^{\alpha}\sum_{j=1}^{a_k}Y_j^k(x)Y_j^k(y).$$

Using the summation by parts we have

$$\Theta^{\alpha}_{\frac{l}{2}}(x,y,n) = \sum_{k=0}^{n}\left(1-\frac{\lambda_k}{\lambda_n}\right)^{\alpha}(1+\lambda_k)^{\frac{l}{2}}\sum_{j=1}^{a_k}Y_j^k(x)Y_j^k(y)$$

$$= \sum_{k=0}^{n-1}\left[(1+\lambda_k)^{\frac{l}{2}}-(1+\lambda_{k+1})^{\frac{l}{2}}\right]\Theta^{\alpha}\left(x,y,k\right)+(1+\lambda_n)^{\frac{l}{2}}\,\Theta^{\alpha}\left(x,y,n\right).$$

In the norm of L_q we obtain,

$$\|\Theta^{\alpha}_{\frac{l}{2}}\left(x,y,n\right)\|_{L_q} \leq \sum_{k=0}^{n-1}\left|(1+\lambda_k)^{\frac{l}{2}}-(1+\lambda_{k+1})^{\frac{l}{2}}\right|\|\Theta^{\alpha}(x,y,k)\|_{L_q}+(1+\lambda_n)^{\frac{l}{2}}\|\Theta^{\alpha}(x,y,n)\|_{L_q}.$$

For the second term we estimate as follows.

$$(1+\lambda_n)^{\frac{l}{2}}\|\Theta^{\alpha}(x,y,n)\|_{L_q(S^N\setminus(V\cup V^t))} \leq$$

$$\leq (1+\lambda_n)^{\frac{l}{2}}\,Cn^{\frac{N-1}{2}-\alpha} \leq$$

$$\leq C(1+n)^{\frac{N-1}{2}-\alpha+l}.$$

To estimate the first term we use Lagrange's Formula. Its not diffficult to see that

$$\left|(1+\lambda_k)^{\frac{l}{2}}-(1+\lambda_{k+1})^{\frac{l}{2}}\right| = \left|\frac{1}{2}\left(1+\eta_k\right)^{\frac{l}{2}-1}\left(\lambda_k-\lambda_{k+1}\right)\right|, \quad l>0,$$

where $\lambda_k < \eta_k < \lambda_{k+1}, \quad k=0,1,2,3,\cdots,n-1.$

If $l \geq 2$

$$\left|(1+\lambda_k)^{\frac{l}{2}}-(1+\lambda_{k+1})^{\frac{l}{2}}\right| \leq \frac{l}{2}\left(1-\lambda_{k+1}\right)^{\frac{l}{2}-1}|k(k+N-1)-(k+1)(k+N)| =$$

$$= \frac{l}{2}\left(1+k^2+kN\right)^{\frac{l}{2}-1}\left|k^2+kN-k-k^2-KN-k-N\right| \leq Ck^{l-1}.$$

If $0 < l < 2$

$$\left|(1+\lambda_k)^{\frac{l}{2}}-(1+\lambda_{k+1})^{\frac{l}{2}}\right| \leq \frac{l}{2}\left(1+\lambda_k\right)^{\frac{l}{2}-1}\left(\lambda_{k+1}-\lambda_k\right) =$$

$$= \frac{l}{2}\left(1 + k^2 + kN - k\right)^{\frac{l}{2}-1}\left((k+1)(k+N) - k(k+N-1)\right) \leq Ck^{l-1}.$$

Finally we have

$$\left|(1+\lambda_k)^{\frac{l}{2}} - (1+\lambda_{k+1})^{\frac{l}{2}}\right| \leq C(1+k)^{l-1} \;\; l > 0, \;\; k = 1, 2, 3, \cdots, n-1.$$

Finally for thr first term we obtain:

$$\sum_{k=0}^{n-1}\left|(1+\lambda_k)^{\frac{l}{2}} - (1+\lambda_{k+1})^{\frac{l}{2}}\right| \|\Theta^{\alpha}(x,y,k)\|_{L_p^l(S^N\setminus(V\cup V^t))}$$

$$\leq \sum_{k=1}^{n-1} C_1 k^{l-1} C_2 (1+k)^{\frac{N-1}{2}-\alpha} \leq C(1+n)^{\frac{N-1}{2}-\alpha+l}.$$

So we have proven

$$\|\Theta_{\frac{l}{2}}^{\alpha}\left(x,y,n\right)\|_{L_q(S^N\setminus(V\cup V^t))} \leq Cn^{\frac{N-1}{2}-\alpha+l}.$$

So for the Riesz means of distributions f we have

$$|E_n^{\alpha} f\left(x\right)| \leq \|f\|_{L_p^{-l}} \|\Theta^{\alpha}\left(x,y,n\right)\|_{L_q^l(S^N\setminus(V\cup V^t))}$$

$$\leq C(1+n)^{\frac{N-1}{2}-\alpha+l}\|f\|_{L_p^{-l}}$$

Using the ideas of the works of [16] we prove that $E_n^{\alpha} f(x) \to 0$, as $n \to \infty$ when $\alpha \geq \frac{N-1}{2} + l$. The proof is completed.

Acknowledgments
This work has been supported by Universiti Putra Malaysia under the Research University Grant (RUGS), project number 05-01-12-1630RU.

References

[1] Ahmedov A A 2009 *Journal of Mathematical Analysis and Applications* **56** 310.
[2] Ahmedov A A 1997 *Uzbek Mathematical Journal,* **4** 20.
[3] Ahmedov A A 2011 *Applied Mathematics Letter* **24(5)** 615.
[4] Ahmedov A A, Rakhimov A A and Zainuddin H 2011 *Australian Journal of Basic and Applied Sciences* **5(5)** 1.
[5] Ashurov R R 1983 *Mathematical Notes of the Academy of Sciences of the USSR* **34 (6)** 913.
[6] Bastis A I 1983 *Mathematical Notes of the Academy of Sciences of the USSR* **33(6)** 440.
[7] Bochner S 1936 *Trans. Amer. Math. Soc.* **40** 175.
[8] Bonami A and Clerc J L 1973 *Trans. Amer. Math. Soc.* **183** 223.
[9] Carleson L 1966 *Acta Mathematica* **116** 135.
[10] Hermander L 1965-1966 *On the Riesz means of spectral functions and eigenfunction expansions for elliptic differential operators, Recent advances in the Basic Sciences,* (Proc. Annual Sci. Conf. Belfer Grad. School Sci. **2**) (Yeshiva Univ. New York) 155.
[11] Kogbetliantz E 1924 *J. Math. Pures Appl.* **9(3)** 107.

[12] Meaney C 1984 *Monatsh Math* **98** 65.
[13] Mitchell J 1951 *Trans. Amer. Math. Soc.* **71** 136.
[14] Peetre J 1964 *Math. Scand.* **15** 83.
[15] Rakhimov A A 1983 *On the Almost Everywhere Localization of Fourier-Laplace Series* Manuscript of Diploma Thesis, Preprint Tashkent State University.
[16] Rakhimov A A 2003 *J Izvestiya Vuzov, Phys. Math. Sci.* **3-4** 47.
[17] Sjolin P 1971 *Ark. Mat.* **9(1)** 65.
[18] Stein E M and Weiss G 1971 *Introduction to Fourier Analysis on Euclidean Spaces* (Princeton, New Jersey: Princeton University Press).
[19] Triebrl Hans 1986 *Arkiv fü matematik* **24(1)** 299.
[20] Zhizhiashvili L V and Topuriya S B 1979 *Journal of Soviet Mathematics* **12(6)** 682.

Journal of Physics: Conference Series **435** (2013) 012017 doi:10.1088/1742-6596/435/1/012017

Fixed Duration Pursuit-Evasion Differential Game with Integral Constraints

Ibragimov G.I.[1], **and Kuchkarov A.Sh.**[2]

[1]INSPEM & Department of Mathematics, FS, UPM, 43400, Serdang, Selangor, Malaysia

[2] Institute of Mathematics, Do'rman yuli str., 29,100125, Tashkent, Uzbekistan

E-mail: `gafur@science.upm.edu.my`, `kuchkarov1@yandex.ru`

Abstract. We investigate a pursuit-evasion differential game of countably many pursuers and one evader. Integral constraints are imposed on control functions of the players. Duration of the game is fixed and the payoff of the game is infimum of the distances between the evader and pursuers when the game is completed. Purpose of the pursuers is to minimize the payoff and that of the evader is to maximize it. Optimal strategies of the players are constructed, and the value of the game is found. It should be noted that energy resource of any pursuer may be less than that of the evader.

1. Statement of the problem.

The study of two person zero-sum differential games was initiated by Isaacs [16]. Berkovitz [3], Fleming [5], Friedman [6], Hajek [8], Krasovskii, [19], Petrosyan [22], Pontryagin [23], Subbotin [26] and others developed mathematical foundations for the theory of differential games.

Many investigations were devoted to study the differential games with integral constraints; e.g., [1], [2], [4], [7], [9]-[15], [18], [20], [21], [25], [27].

Constructing the optimal strategies, and finding the value of the game are of interest in differential games, e.g., see [10], [11], [17], [24], [26], [27]. Such problems in the case of many pursuers were studied, for example, in [10] and [11]. In [10], a differential game of optimal approach of countably many pursuers to one evader was studied in Hilbert space with geometric constraints on controls of players. In [11], such differential game was studied for inertial players with integral constraints under the assumption that the control resource of the evader less than that of each pursuer.

In the present paper, we also discuss an optimal pursuit problem with countably many pursuers and one evader in Hilbert space l_2, and control resource of the evader σ can be greater than that of any pursuer. In the space l_2 with elements

$$\alpha = (\alpha_1, \alpha_2, ..., \alpha_k, ...), \sum_{k=1}^{1} \alpha_k^2 < \infty,$$

and inner product and norm

$$(\alpha, \beta) = \sum_{k=1}^{1} \alpha_k \beta_k, \quad ||\alpha|| = \left(\sum_{k=1}^{1} \alpha_k^2 \right)^{1/2},$$

motions of the pursuers P_i and the evader E are described by the equations

$$\begin{array}{ll} P_i: & \dot{x}_i = u_i, \quad x_i(0) = x_{i0}, \\ E: & \dot{y} = v, \quad\ \ y(0) = y_0, \end{array} \tag{1}$$

where $x_i,\ x_0,\ u_i,\ y,\ y_0,\ v \in l_2$, $u_i = (u_{i1}, u_{i2}, ..., u_{ik}, ...)$ is control parameter of the pursuer P_i, and $v = (v_1, v_2, ..., v_k, ...)$ is that of the evader E; throughout, $i = 1, 2, ..., m, ...$ Let ϑ be a fixed time, $I = \{1, 2, ..., m, ...\}$, and

$$H(x_0, r) = \{x \in l_2 : ||x - x_0|| \leq r\} \quad S(x_0, r) = \{x \in l_2 : ||x - x_0|| = r\}.$$

Definition 1. A function $u_i = u_i(t),\ 0 \leq t \leq \vartheta$, with the Borel measurable coordinates $u_{ik} : [0, \vartheta] \to R^1$, $k = 1, 2, ...$, subjected to the condition

$$\left(\int_0^{\vartheta} ||u_i(t)||^2 dt \right)^{1/2} \leq \rho_i,$$

is called an admissible control of the pursuer P_i, where ρ_i are given positive numbers.

Definition 2. A function $v = v(t),\ 0 \leq t \leq \vartheta$, with the Borel measurable coordinates $v_k : [0, \vartheta] \to R^1$, $k = 1, 2, ...$, subjected to the condition

$$\left(\int_0^{\vartheta} ||v(t)||^2 dt \right)^{1/2} \leq \sigma,$$

is called an admissible control of the evader E, where σ is given positive number.

If it has been chosen admissible controls $u_i(\cdot),\ v(\cdot)$ of players, then corresponding to them motions $x_i(\cdot), y(\cdot)$ are defined by formulas

$$x_i(t) = (x_{i1}(t), x_{i2}(t), ..., x_{ik}(t), ...), \quad y = (y_1(t), y_2(t), ..., y_k(t), ...),$$

$$x_{ik}(t) = x_{i0} + \int_0^t u_{ik}(s) ds, \quad y_k(t) = y_0 + \int_0^t v_k(s) ds.$$

It is not difficult to verify that $x_i(\cdot), y(\cdot) \in C(0, \vartheta; l_2)$, where $C(0, \vartheta; l_2)$ is the space of continuous functions $f(t) = (f_1(t), f_2(t), ..., f_k(t), ...) \in l_2$, $0 \leq t \leq \vartheta$, with absolutely continuous components $f_k(t),\ 0 \leq t \leq \vartheta$.

Definition 3. A function

$$U_i(\xi_i, v),\ U_i : [0, \rho_i^2] \times l_2 \to l_2,$$

is called a strategy of the pursuer P_i if for any admissible control $v = v(t),\ 0 \leq t \leq \vartheta$, of evader E, the system of equations

$$\begin{array}{ll} \dot{x}_i = U_i(\xi_i, v), & x_i(0) = x_{i0}, \\ \dot{\xi}_i = -||U_i(\xi_i, v)||^2, & \xi_i(0) = \rho_i^2, \\ \dot{y} = v, & y(0) = y_0, \end{array}$$

has a unique solution $(x_i(\cdot), \xi_i(\cdot), y(\cdot))$, where $x_i(\cdot), y(\cdot) \in C(0, \vartheta; l_2)$, and $\xi_i(\cdot)$ is absolutely continuous scalar function on $[0, \vartheta]$. A strategy U_i is said to be admissible, if every control generated by U_i is admissible.

Definition 4. Strategies U_{i0} of the pursuers P_i are referred to as the optimal strategies if

$$\inf_{U_1,\dots,U_m,\dots} \Gamma_1(U_1,\dots,U_m,\dots) = \Gamma_1(U_{10},\dots,U_{m0},\dots),$$

where

$$\Gamma_1(U_1,\dots,U_m,\dots) = \sup_{v(\cdot)} \inf_{i\in I} ||x_i(\vartheta) - y(\vartheta)||,$$

U_i are admissible strategies of the pursuers P_i, and $v(\cdot)$ is admissible control of the evader E.

Definition 5. A function

$$V(x_1,\dots,x_m,\dots,y),\ V: l_2 \times \dots \times l_2 \times \dots \times l_2 \to l_2,$$

is called a strategy of the evader E, if for any admissible controls $u_i = u_i(t),\ 0 \le t \le \vartheta$, of the pursuers P_i, the system of equations

$$\begin{aligned} \dot{x}_k &= u_k, & x_k(0) &= x_{k0},\ k = 1,2,\dots,m,\dots \\ \dot{y} &= V(x_1,\dots,x_m,\dots,y), & y(0) &= y_0, \end{aligned}$$

has a unique solution $(x_1(\cdot),\dots,x_m(\cdot),\dots,y(\cdot))$, $x_i(\cdot), y(\cdot) \in C(0,\vartheta,l_2)$. A strategy V is said to be admissible, if every control generated by V is admissible.

Definition 6. Strategy V_0 of the evader E is said to be optimal, if

$$\sup_V \Gamma_2(V) = \Gamma_2(V_0),$$

where

$$\Gamma_2(V) = \inf_{u_1(\cdot),\dots,u_m(\cdot),\dots} \inf_{i\in I} ||x_i(\vartheta) - y(\vartheta)||,$$

$u_i(\cdot)$ are admissible controls of pursuers P_i, V is admissible control of evader E.

If $\Gamma_1(U_{10},\dots,U_{m0},\dots) = \Gamma_2(V_0) = \gamma$, then we say that the game has the value γ [26].

The problem is to find the optimal strategies U_{i0}, V_0 of the players P_i and E, respectively, and the value of the game.

2. An auxiliary game.

The attainability set of the pursuer P_i from the initial position x_{i0} up to the time ϑ, i.e., the set of all points $x(\vartheta) = x_{i0} + \int_0^{\vartheta} u_i(s)ds$, where $u_i(\cdot)$ is an admissible control of the ith pursuer, is the ball $H(x_{i0}, \rho_i\sqrt{\vartheta})$.

Indeed, by the Cauchy-Schwartz inequality we have

$$||x_i(\vartheta) - x_{i0}|| = ||\int_0^{\vartheta} u_i(s)ds|| \le \int_0^{\vartheta} ||u_i(s)||ds$$

$$\le \left(\int_0^{\vartheta} 1ds \cdot \int_0^{\vartheta} ||u_i(s)||^2 ds\right)^{1/2} \le \rho_i\sqrt{\vartheta}.$$

On the other hand, if $\bar{x} \in H(x_{i0}, \rho_i\sqrt{\vartheta})$, then for the control

$$u_i(s) = (\bar{x} - x_{i0})/\vartheta,\ 0 \le s \le \vartheta,$$

of the pursuer we get $x_i(\vartheta) = \bar{x}$. Admissibility of this control follows from the relations

$$\int_0^\vartheta ||u_i(s)||^2 ds = \int_0^\vartheta \frac{1}{\vartheta^2}||\bar{x} - x_{i0}||^2 ds \le \frac{1}{\vartheta} \cdot \rho_i^2 \vartheta = \rho_i^2.$$

In a similar fashion we can show that the attainability set of the evader E from the initial position y_0 up to the time ϑ is the ball $H(y_0, \sigma\sqrt{\vartheta})$.

For the simplicity in this section we will drop index i, i.e., we use designations $\rho_i = \rho,\ x_{i0} = x_0,\ x_i = x$. Let $x_0 \ne y_0$ and

$$X = \{z \in l_2 : 2(y_0 - x_0, z) \le (\rho^2 - \sigma^2)\vartheta + ||y_0||^2 - ||x_0||^2\},\ e = \frac{y_0 - x_0}{||y_0 - x_0||},$$

Consider the following game with one pursuer

$$P : \dot{x} = u,\ \ x(0) = x_0, \quad E : \dot{y} = v,\ y(0) = y_0. \tag{2}$$

The goal of the pursuer P is to realize the equality $x(\tau) = y(\tau)$ at some τ, $0 \le \tau \le \vartheta$, and that of the evader E is opposite. Construct the strategy of the pursuer as follows:

$$u(t) = \frac{1}{\vartheta}(y_0 - x_0) + v(t),\ 0 \le t \le \vartheta. \tag{3}$$

Lemma 1. If $y(\vartheta) \in X$, then the strategy (3) of the pursuer P is admissible and ensures the equality $x(\vartheta) = y(\vartheta)$ in the game (2).

We show that if $y(\vartheta) \in X$, then the strategy (3) is admissible. As

$$y(\vartheta) = y_0 + \int_0^\vartheta v(s)ds,$$

then from the inequality

$$2(y_0 - x_0, y(\vartheta)) \le (\rho^2 - \sigma^2)\vartheta + ||y_0||^2 - ||x_0||^2$$

we have

$$2\int_0^\vartheta (y_0 - x_0, v(s))ds \le (\rho^2 - \sigma^2)\vartheta - ||x_0 - y_0||^2.$$

Hence from (3) we have

$$\int_0^\vartheta ||u(s)||^2 ds = \frac{1}{\vartheta}||y_0 - x_0||^2 + \frac{2}{\vartheta}\int_0^\vartheta (y_0 - x_0, v(s))ds + \int_0^\vartheta ||v(s)||^2 ds$$

$$\le \frac{1}{\vartheta}||y_0 - x_0||^2 + \frac{1}{\vartheta}\left((\rho^2 - \sigma^2)\vartheta - ||x_0 - y_0||^2\right) + \sigma^2 = \rho^2,$$

and so strategy (3) is admissible. Then

$$x(\vartheta) = x_0 + \int_0^{\vartheta} u(s)ds = x_0 + y_0 - x_0 + \int_0^{\vartheta} v(s)ds = y(\vartheta).$$

This proves lemma.

3. Main Result.

Now we consider the game (1). Define

$$\gamma = \inf\{l \geq 0 : \; H(y_0, \sigma\sqrt{\vartheta}) \subset \bigcup_{i=1}^{\infty} H(x_{i0}, \rho_i\sqrt{\vartheta} + l)\}. \tag{4}$$

Theorem. If there exists a non-zero vector $p_0 \in l_2$ such that $(y_0 - x_{i0}, p_0) \geq 0$ for all $i \in I$, then the number γ defined by the formula (4) is the value of the game (1).

Proof. To prove this theorem we need the following statement (see, [10], Assertions 4 and 5 in Appendix)

Lemma 2. Suppose there exists a non-zero vector $p_0 \in l_2$ such that $(y_0 - x_{i0}, p_0) \geq 0$ for all $i \in I$.

(a) If $H(y_0, r) \subset \bigcup_{i=1}^{\infty} H(x_{i0}, R_i)$ then

$$H(y_0, r) \subset \bigcup_{i=1}^{\infty} \{y : 2(y_0 - x_{i0}, y) \leq R_i^2 - r^2 + ||y_0||^2 - ||x_{i0}||^2\}.$$

(b) If for any $\varepsilon > 0$

$$H(y_0, r) \not\subset \bigcup_{i \in I} H(x_{i0}, (R_i - \varepsilon)_+) \quad (a_+ = \max\{0, a\}),$$

then there exists a point $\bar{y} \in S(y_0, r)$ such that $||\bar{y} - x_{i0}|| \geq R_i$ for all $i \in I$.

1^0. *Construction the strategies of the pursuers.* We introduce fictitious pursuers (FPs) z_i, whose motions are described by the equations

$$\dot{z}_i = w_i^{\varepsilon}, \; z_i(0) = x_{i0}, \; \left(\int_0^{\vartheta} ||w_i^{\varepsilon}(s)||^2 ds \right)^{1/2} \leq \bar{\rho}_i(\varepsilon) = \rho_i + \frac{\gamma}{\sqrt{\vartheta}} + \frac{\varepsilon}{k_i\sqrt{\vartheta}}, \tag{5}$$

where ε is an arbitrary positive number, $k_i = \max\{1, \rho_i\}$. It can be shown easily that the attainability set of the FP z_i from the initial position x_{i0} up to the time ϑ is the ball

$$H(x_{i0}, \bar{\rho}_i(\varepsilon)\sqrt{\vartheta}) = H(x_{i0}, \rho_i\sqrt{\vartheta} + \gamma + \varepsilon/k_i).$$

We define strategies for FPs z_i on the time interval $[0, \vartheta]$ as follows.

$$w_i^{\varepsilon}(t) = \begin{cases} \frac{1}{\vartheta}(y_0 - x_{i0}) + v(t), & 0 \leq t \leq \tau_i^{\varepsilon}, \\ 0 & \tau_i^{\varepsilon} < t \leq \vartheta, \end{cases} \tag{6}$$

where τ_i^{ε}, $0 \leq \tau_i^{\varepsilon} \leq \vartheta$, is the time for which

$$\int_0^{\tau_i^\varepsilon} ||w_i^\varepsilon(s)||^2 ds = \bar{\rho}_i^2(\varepsilon),$$

if such a time exists. We define strategies of the pursuers x_i by the strategies of the FPs as follows.

$$u_i(t) = \frac{\rho_i}{\bar{\rho}_i} w_i(t), \ 0 \le t \le \vartheta, \tag{7}$$

where $\bar{\rho}_i = \bar{\rho}_i(0) = \rho_i + \frac{\gamma}{\sqrt{\vartheta}}$,

$$w_i(t) = \begin{cases} \frac{1}{\vartheta}(y_0 - x_{i0}) + v(t), & 0 \le t \le \tau_i, \\ 0 & \tau_i < t \le \vartheta, \end{cases} \tag{8}$$

where τ_i, $0 \le \tau_i \le \vartheta$, is the time for which

$$\int_0^{\tau_i} ||w_i(s)||^2 ds = \bar{\rho}_i^2,$$

that is, $w_i(t)$ is obtained from (6) at $\varepsilon = 0$.

As $\bar{\rho}_i(\varepsilon) > \bar{\rho}_i$, then

$$\int_0^{\tau_i^\varepsilon} ||w_i^\varepsilon(s)||^2 ds = \bar{\rho}_i^2(\varepsilon) > \bar{\rho}_i^2 = \int_0^{\tau_i} ||w_i(s)||^2 ds,$$

that is,

$$\int_0^{\tau_i^\varepsilon} \left|\left| \frac{y_0 - x_{i0}}{\vartheta} + v(t) \right|\right|^2 dt > \int_0^{\tau_i} \left|\left| \frac{y_0 - x_{i0}}{\vartheta} + v(t) \right|\right|^2 ds.$$

Hence $\tau_i^\varepsilon > \tau_i$.

2^0. *Guaranteed result for the pursuers.* We shall show that strategies (7) of pursuers guarantee that

$$\sup_{v(\cdot)} \inf_{i \in I} ||y(\vartheta) - x_i(\vartheta)|| \le \gamma. \tag{9}$$

In accordance with the definition of the number γ we have

$$H(y_0, \sigma\sqrt{\vartheta}) \subset \bigcup_{i=1}^{\infty} H(x_{i0}, \rho_i\sqrt{\vartheta} + \gamma + \varepsilon/k_i). \tag{10}$$

Denote

$$X_i = \{z: \ 2(y_0 - x_{i0}, z) \le (\rho_i\sqrt{\vartheta} + \gamma + \varepsilon/k_i)^2 - (\sigma\sqrt{\vartheta})^2 + ||y_0||^2 - ||x_{i0}||^2\}.$$

According to the condition of the theorem $(y_0 - x_{i0}, p_0) \geq 0$ for all $i \in I$. Then by Lemma 2 it follows from (10) that

$$H(y_0, \sigma\sqrt{\vartheta}) \subset \bigcup_{i=1}^{\infty} X_i.$$

Consequently for the point $y(\vartheta) \in H(y_0, \sigma\sqrt{\vartheta})$ at some $s \in I$ we have

$$2(y_0 - x_{s0}, y(\vartheta)) \leq (\rho_s\sqrt{\vartheta} + \gamma + \varepsilon/k_s)^2 - (\sigma\sqrt{\vartheta})^2 + ||y_0||^2 - ||x_{s0}||^2. \tag{11}$$

If $x_{s0} = y_0$, then according to (6) the strategy of sth FP takes the form

$$w_s^\varepsilon(t) = v(t), \;\; 0 \leq t \leq \vartheta$$

and from (11) we get $\rho_s\sqrt{\vartheta} + \gamma + \varepsilon/k_s \geq \sigma\sqrt{\vartheta}$, and so

$$\int_0^\vartheta ||w_s^\varepsilon(t)||^2 dt = \int_0^\vartheta ||v(t)||^2 dt \leq \sigma^2 \leq \left(\rho_s + \frac{\gamma}{\sqrt{\vartheta}} + \frac{\varepsilon}{k_s\sqrt{\vartheta}}\right)^2,$$

that is, $w_s^\varepsilon(\cdot)$ is admissible and, moreover, $z_s(\vartheta) = y(\vartheta)$.

If $x_{i0} \neq y_0$ then by Lemma 1 for the strategies (6) of FPs we get $z_s(\vartheta) = y(\vartheta)$. Then taking into account (5) and (7), we get

$$||y(\vartheta) - x_s(\vartheta)|| = ||z_s(\vartheta) - x_s(\vartheta)|| = \left|\left|\int_0^\vartheta \left(w_s^\varepsilon(t) - \frac{\rho_s}{\bar{\rho}_s} w_s(t)\right) dt\right|\right|$$

$$\leq \int_0^\vartheta ||w_s^\varepsilon(t) - w_s(t)||dt + \int_0^\vartheta ||w_s(t) - \frac{\rho_s}{\bar{\rho}_s} w_s(t)||dt. \tag{12}$$

We now estimate right-hand side of the inequality (12). Firstly, we show that

$$\int_0^\vartheta ||w_i^\varepsilon(t) - w_i(t)||dt \leq K\sqrt{\varepsilon} \tag{13}$$

for all $i \in I$ and some constant K. Indeed, as we noted above that $\tau_i^\varepsilon > \tau_i$ and according to (6) and (8) $w_i^\varepsilon(t) = w_i(t)$ for $0 \leq t \leq \tau_i$; $w_i(t) = 0$ for $t > \tau_i$, $w_i^\varepsilon(t) = 0$ for $t > \tau_i^\varepsilon$, then we have

$$\int_0^\vartheta ||w_i^\varepsilon(t) - w_i(t)||dt = \int_0^{\tau_i} ||w_i^\varepsilon(t) - w_i(t)||dt + \int_{\tau_i}^{\tau_i^\varepsilon} ||w_i^\varepsilon(t) - w_i(t)||dt$$

$$+ \int_{\tau_i^\varepsilon}^\vartheta ||w_i^\varepsilon(t) - w_i(t)||dt = \int_{\tau_i}^{\tau_i^\varepsilon} ||w_i^\varepsilon(t)||dt \leq \sqrt{\tau_i^\varepsilon - \tau_i}\left(\int_{\tau_i}^{\tau_i^\varepsilon} ||w_i^\varepsilon(t)||^2 dt\right)^{1/2}$$

$$\leq \sqrt{\vartheta}\left(\int_0^{\tau_i^\varepsilon} ||w_i^\varepsilon(t)||^2 dt - \int_0^{\tau_i} ||w_i^\varepsilon(t)||^2 dt\right)^{1/2} = \sqrt{\vartheta}\left(\bar{\rho}_i^2(\varepsilon) - \bar{\rho}_i^2\right)^{1/2}$$

$$= \sqrt{\vartheta}\left(\frac{\varepsilon}{k_i\sqrt{\vartheta}}\left(2\rho_i + \frac{2\gamma}{\sqrt{\vartheta}} + \frac{\varepsilon}{k_i\sqrt{\vartheta}}\right)\right)^{1/2} \leq K\sqrt{\varepsilon},$$

where K does not depend on i. We note that the sequence $\rho_1, ..., \rho_m, ...$ may be unbounded or their infimum may equal to 0. Thus, we have shown the inequality (13). For the second integral of (12) we have

$$\left|\left|\int_0^{\vartheta}\left(1 - \frac{\rho_s}{\bar{\rho}_s}\right) w_s(t)\right|\right| \leq \left(1 - \frac{\rho_s}{\bar{\rho}_s}\right)\int_0^{\vartheta} ||w_s(t)|| dt \leq \left(1 - \frac{\rho_s}{\bar{\rho}_s}\right)\sqrt{\vartheta}\bar{\rho}_s = \gamma.$$

It follows then from (12) that

$$||y(\vartheta) - x_s(\vartheta)|| \leq \gamma + K\sqrt{\varepsilon}.$$

Thus, if pursuers use strategies (7), then inequality (9) holds. So the result γ is guaranteed for the pursuers.

3^0. *Guaranteed result for the evader.* We construct a strategy for the evader, which provides the inequality

$$\inf_{u_1(\cdot),...,u_m(\cdot),...} \inf_{i\in I} ||y(\vartheta) - x_i(\vartheta)|| \geq \gamma, \tag{14}$$

where $u_1(\cdot), ..., u_m(\cdot), ...$ are arbitrary admissible controls of pursuers.

If $\gamma = 0$, then for any admissible control of the evader validity of (14) is clear. Let $\gamma > 0$. Then by definition of the number γ, for any $\varepsilon > 0$, the set $\bigcup_{i=1}^{\infty} H(x_{i0}, \rho_i\sqrt{\vartheta} + \gamma - \varepsilon)$ does not contain the ball $H(y_0, \sigma\sqrt{\vartheta})$. Consequently, by Lemma 2 there exists a point $\bar{y} \in S(y_0, \sigma\sqrt{\vartheta})$ such that $||\bar{y} - x_{i0}|| \geq \rho_i\sqrt{\vartheta} + \gamma$. Hence

$$||\bar{y} - x_i(\vartheta)|| \geq ||\bar{y} - x_{i0}|| - ||x_i(\vartheta) - x_{i0}|| \geq \rho_i\sqrt{\vartheta} + \gamma - \rho_i\sqrt{\vartheta} = \gamma.$$

The control

$$v(t) = \frac{1}{\vartheta}(\bar{y} - y_0), \ 0 \leq t \leq \vartheta,$$

supplies validity of the inequality (14), since for this control we have

$$y(\vartheta) = y_0 + \int_0^{\vartheta} v(s) ds = \bar{y}.$$

Now, we give an illustrative example.

Example. Let $\rho_i = 1$, $\vartheta = 9$, $\sigma = 4$ in the game (1). We consider the following initial positions

$$x_{i0} = (0, ..., 0, 5, 0, ...), \quad y_0 = (0, 0, ...),$$

of the players, where number 5 is ith coordinate of the point x_{i0}. Note that $\rho_i\sqrt{\vartheta} = 3$, $\sigma\sqrt{\vartheta} = 12$. We show that the value of the game is $\gamma = 10$. It is sufficient to show that

1) for any $\varepsilon > 0$ the inclusion

$$H(O, 12) \subset \bigcup_{i=1}^{\infty} H(x_{i0}, 13 + \varepsilon),$$

holds, where O is the origin;

2) the ball $H(O, 12)$ is not contained in the set $\bigcup_{i=1}^{\infty} H(x_{i0}, 13)$.

Indeed, let $z = (z_1, z_2, ...)$ be an arbitrary point of the ball $H(O, 12)$. So $\sum\limits_{i=1}^{\infty} z_i^2 \le 144$. Then either the vector z has a nonnegative coordinate or all the coordinates of the vector z are negative. In the former case, z has a non negative coordinate z_k. Then

$$||z - x_{k0}|| = \left(z_1^2 + ... + z_{k-1}^2 + (5 - z_k)^2 + z_{k+1}^2 + ...\right)^{1/2}$$

$$= \left(\sum_{i=1}^{\infty} z_i^2 + 25 - 10z_k\right)^{1/2} \le (169 - 10z_k)^{1/2} \le 13 < 13 + \varepsilon.$$

Hence, $z \in H(x_{k0}, 13 + \varepsilon)$.

In the latter case, since $\sum\limits_{i=1}^{\infty} z_i^2$ is convergent then $z_k \to 0$ as $k \to \infty$ and therefore

$$||z - x_{k0}|| = \left(\sum_{i=1}^{\infty} z_i^2 + 25 - 10z_k\right)^{1/2} \le (169 - 10z_k)^{1/2} < 13 + \varepsilon$$

for an index k.

On the other hand, any point $z \in S(O, 12)$ with negative coordinates does not belong to the set

$$\bigcup_{i=1}^{\infty} H(x_{i0}, 13),$$

since for any number i

$$||z - x_{i0}|| = (169 - 10z_i)^{1/2} > 13.$$

Therefore by the theorem the number

$$\gamma = \inf\{l \ge 0 : \; H(y_0, \sigma\sqrt{\vartheta}) \subset \bigcup_{i=1}^{\infty} H(x_{i0}, \rho_i\sqrt{\vartheta} + l)\}$$

$$= \inf\{l \ge 0 : \; H(O, 12) \subset \bigcup_{i=1}^{\infty} H(x_{i0}, 3 + l)\} = 10$$

is the value of the game.

4. Conclusion

We have studied a simple motion pursuit-evasion differential game of fixed duration with countably many pursuers. Control functions of all players are subjected to integral constraints. The value of the game has been found, and the optimal strategies of players have been constructed. The main assumption in the game is existence of a non-zero vector $p_0 \in l_2$ such that $(y_0 - x_{i0}, p_0) \geq 0$ for all $i \in I$. If the initial positions of the players do not satisfy this condition then the stated problem is open. The advantage of the pursuers' strategies in this work with respect to those of the paper [11] is that σ can be greater than ρ_i for any $i \in I$.

Acknowledgements. This research was partially supported by the Research Grant (RUGS) of the Universiti Putra Malaysia, No. 05-02-12-1868RU.

References

[1] Azamov A.A., Samatov B. 2000. П-*strategy. An elementary introduction to the theory of differential games.* (Tashkent, NUU press).
[2] Azimov A.Ya. 1974. A linear evasion differential game with integral constraints on the controls. *USSR Computational Mathematics and Mathematical Physics* **14**, (2), 56–65. http://dx.doi.org/10.1016/0041-5553(74)90169-4.
[3] Berkovitz L. D. 1967. A Survey of Differential Games, Mathematical Theory of Control. (New York, Academic Press) 373–385.
[4] Chikrii A.A., Belousov A.A. 2009. On linear differential games with integral constraints. *Memoirs of Institute of Mathematics and Mechanics, Ural Division of RAS.* **15**, No.4, 290–301. (Russian)
[5] Fleming W.H. 1961. The convergence problem for differential games. *Journal of Mathematical Analysis and Applications.***3**, 102–116.
[6] Friedman A. 1971 *Differential Games* (New York, John Wiley & Sons).
[7] Gusiatnikov P.B., Mohon'ko E.Z. 1980. On $l\infty$-escape in a linear many-person differential game with integral constraints. *Journal of Applied Mathematics and Mechanics.* **44**, No 4, 436–440.
[8] Hajek O. 1975. *Pursuit games.* (New York, Academic Press).
[9] Ibragimov G.I. 2004 Collective pursuit with integral constraints on the controls of players. *Siberian Advances in Mathematics.* **14**, No 2, 14–26.
[10] Ibragimov G.I. 2005. Optimal Pursuit with Countable Many Pursuers and One Evader. *Differential Equations* **41**, No. 5, 627–635.
[11] Ibragimov G.I, Mehdi Salimi.2009. Pursuit-Evasion differential game with many inertial players. *Mathematical Problems in Engineering*, **2009**, doi:10.1155/2009/653723.
[12] Ibragimov G.I, Risman Mat Hasim. 2010. An Evasion Differential Game in Hilbert Space. *International Game Theory Review (IGTR).* **12**, No.3, 239–251.
[13] Ibragimov G.I, Azamov A., Khakestari M. 2011 Solution of a Linear Pursuit-Evasion Game with Integral Constraints. *ANZIAM J.* **52(E)** 59–75.
[14] Ibragimov G.I., Mehdi Salimi, Massoud Amini. 2012. Evasion from Many Pursuers in Simple Motion Differential Game with Integral Constraints. *European Journal of Operational Research.* **218(2012)**, No 2, 505–511. doi:10.1016/j.ejor.2011.11.026.
[15] Ibragimov G.I., Satimov N.Yu. 2012. A Multi Player Pursuit Differential Game on Closed Convex Set with Integral Constraints. *Abstract and Applied Analysis.* **2012**, doi:10.1155/2012/460171.
[16] Isaacs R. 1967 *Differential games. A Mathematical theory with Applications to Warfare and Pursuit, Control and Optimization.* (New York, John Wiley and Sons)
[17] Ivanov R.P., Ledyayev Yu.S. 1981 Optimality of pursuit time in differential game of many objects with simple motion. *Trudi MIAN USSR.* **158**. 87–97. (Russian)
[18] Krasovskii N.N. 1968 *The Theory of Motion Control.* (Moscow, Nauka.)
[19] Krasovskii N.N., Subbotin A.I. 1988. *Game-theoretical control problems.* (N.Y., Springer).1988.
[20] Mezentsev A.V. 1974 A sufficient condition for linear evasion games with integral constraints. *DAN USSR*, **218**, No 5, 1021–1023.
[21] Nikolskii M.S. 1969 The direct method in linear differential games with integral constraints. *Controlled systems, IM, IK, SO AN SSSR.* **2**, 49–59.
[22] Petrosyan L.A. 1993. *Differential games of pursuit.* (London, World Scientific).
[23] Pontryagin L.S. 1988. *Collected works.* (Moscow, Nauka). (Russian)
[24] Pshenichnii B.N., and Onopchuk Yu.N. 1968 Linear differential games with integral constraints. *Izvestige Akademii Nauk SSSR, Tekhnicheskaya Kibernetika* **1**, 13–22.
[25] Satimov N.Yu., Rikhsiev B.B. 2000 *Methods of Solving of Evasion Problems in Mathematical Control Theory.* (Tashkent, Fan). (Russian)

[26] Subbotin A.I., Chentsov A.G. 1981. *Optimization of the guarantee in control problems.* (Moscow, Nauka) (Russian)

[27] Ushakov V.N. 1972 Extremal strategies in differential games with integral restrictions. *J. Appl. Math.Mech.*, **36**, No 1, 12–19.

Journal of Physics: Conference Series **435** (2013) 012018 doi:10.1088/1742-6596/435/1/012018

On bistochastic Kadison-Schwarz operators on $M_2(\mathbb{C})$

Farrukh Mukhamedov[1], **Abduaziz Abduganiev**[2]

[1,2] Department of Computational & Theoretical Sciences, Faculty of Science, International Islamic University Malaysia, Kuantan, Pahang, Malaysia

E-mail: [1]`farrukh_m@iium.edu.my`, [2]`azizi85@yandex.ru`

Abstract. In this paper we describe bistochastic Kadison-Schawrz operators acting on $M_2(\mathbb{C})$. Such a description allows us to find positive, but not Kadison-Schwarz operators. Moreover, by means of that characterization we are able to construct Kadison-Schawrz operators, which are not completely positive.

1. Introduction

It is known that the theory of quantum dynamical systems provides a convenient mathematical description of irreversible dynamics of an open quantum system (see [3]) investigation of various properties of such dynamical systems have had a considerable growth. In a quantum setting, the matter is more complicated than in the classical case. Some differences between classical and quantum situations are pointed out in [13]. This motivates an interest to study dynamics of quantum systems (see[13]). One of the main objects of this theory is mapping (or channel) defined on matrix algebras. One of the main constraints to such a mapping is positivity and complete positivity. There are many papers devoted to this problem (see for example [4, 9, 17, 18]). In the literature the most tractable maps, the completely positive ones, have proved to be of great importance in the structure theory of C*-algebras. However, general positive (order-preserving) linear maps are very intractable[9, 10, 11]. It is therefore of interest to study conditions stronger than positivity, but weaker than complete positivity. Such a condition is called *Kadison-Schwarz (KS) property*, i.e a map ϕ satisfies the KS property if $\phi(a)^*\phi(a) \leq \phi(a^*a)$ holds for every a. Note that every unital completely positive map satisfies this inequality, and a famous result of Kadison states that any positive unital map satisfies the inequality for self-adjoint elements a. But KS-operators no need to be completely positive. In [16] relations between n-positivity of a map ϕ and the KS property of certain map is established (see also [2]). Some ergodic properties of the Kadison-Schwarz maps were investigated in [8, 5, 15].

In this paper we are going to describe KS-operators which are unital, trace preserving linear mappings (i.e. bistochastic operators) defined on the algebra of 2 by 2 matrices $M_2(\mathbb{C})$. In Section 3 we provide a characterization of KS-operators form $M_2(\mathbb{C})$ to $M_2(\mathbb{C})$. In section 4, we give a sufficient condition for a class of bistochastic mappings from $M_2(\mathbb{C})$ to $M_2(\mathbb{C}) \otimes M_2(\mathbb{C})$ to be KS-operator. Such a description allows us to find positive, but not Kadison-Schwarz operators. Moreover, by means of that conditions we are able to construct KS-operators, which are not completely positive. Note that trace-preserving maps arise naturally in quantum information theory [6, 7, 13, 14] and other situations in which one wishes to restrict attention to a quantum system that should properly be considered a subsystem of a larger system with which it interacts.

2. Preliminaries

Let $M_n(\mathbb{C})$ be the algebra of $n \times n$ matrices over the complex field $\mathbb{C}$. Recall that a linear mapping $\Phi : M_n(\mathbb{C}) \to M_m(\mathbb{C})$ is called

(i) *positive* if $\Phi(x) \geq 0$ whenever $x \geq 0$;
(ii) *unital* if $\Phi(\mathbb{1}) = \mathbb{1}$;
(iii) *trace preserving* if $\tau(\Phi(x)) = \tau(x)$, where τ is the normalized trace;
(iv) *bistochastic* if Φ is unital and trace preserving;
(v) *n-positive* if the mapping $\Phi_n : M_n(A) \to M_n(B)$ defined by $\Phi_n(a_{ij}) = (\Phi(a_{ij}))$ is positive. Here $M_n(A)$ denotes the algebra of $n \times n$ matrices with A-valued entries;
(vi) *completely positive* if it is n-positive for all $n \in \mathbb{N}$;
(vii) *Kadison-Schwarz operator (KS-operator)*, if one has

$$\Phi(x)^*\Phi(x) \leq \Phi(x^*x) \quad \text{for all} \quad x \in A. \tag{1}$$

It is clear that any KS-operator is positive. Note that every unital 2- positive map is KS-operator, and a famous result of Kadison states that any positive unital map satisfies the inequality (1) for all self-adjoint elements $x \in A$.

By $\mathcal{KS}(M_n, M_m)$ we denote the set of all KS-operators mapping from $M_n(\mathbb{C})$ to $M_m(\mathbb{C})$.

Theorem 2.1 ([12]) *The following assertions hold true:*

(i) Let $\Phi, \Psi \in \mathcal{KS}(M_n, M_m)$, then for any $\lambda \in [0,1]$ the mapping $\Gamma = \lambda\Phi + (1-\lambda)\Psi$ belongs to $\mathcal{KS}(M_n, M_m)$. This means $\mathcal{KS}(M_n, M_m)$ is convex;
(ii) Let U, V be unitaries in $M_n(\mathbb{C})$ and $M_m(\mathbb{C})$, respectively, then for any $\Phi \in \mathcal{KS}(M_n, M_m)$ the mapping $\Psi_{U,V}(x) = U\Phi(VxV^*)U^*$ belongs to $\mathcal{KS}(M_n, M_m)$.

3. Kadison-Schwarz operators from $M_2(\mathbb{C})$ to $M_2(\mathbb{C})$

In this section we are going to provide description of bistochastic mappings from $M_2(\mathbb{C})$ to $M_2(\mathbb{C})$.

It is known (see [3, 6]) that the identity and the Pauli matrices $\{\mathbb{1}, \sigma_1, \sigma_2, \sigma_3\}$ form a basis for $M_2(\mathbb{C})$, where

$$\sigma_1 = \begin{pmatrix} 0 & 1 \\ 1 & 0 \end{pmatrix} \; \sigma_2 = \begin{pmatrix} 0 & -i \\ i & 0 \end{pmatrix} \; \sigma_3 = \begin{pmatrix} 1 & 0 \\ 0 & -1 \end{pmatrix}.$$

Every matrix $a \in M_2(\mathbb{C})$ can be written in this basis as $a = w_0\mathbb{1} + \mathbf{w} \cdot \sigma$ with $w_0 \in \mathbb{C}, \mathbf{w} = (w_1, w_2, w_3) \in \mathbb{C}^3$, here by $\mathbf{w} \cdot \sigma$ we mean the following

$$\mathbf{w} \cdot \sigma = w_1\sigma_1 + w_2\sigma_2 + w_3\sigma_3.$$

The following facts holds (see [14]):

(a) a matrix $a \in M_2(\mathbb{C})$ is self-adjoint if and only if w_0 and $\mathbf{w}$ are real;
(b) a matrix $a \in M_2(\mathbb{C})$ is positive if and only if $\|\mathbf{w}\| \leq w_0$, where

$$\|\mathbf{w}\| = \sqrt{|w_1|^2 + |w_2|^2 + |w_3|^2} \; ;$$

(c) a matrix $a \in M_2(\mathbb{C})$ is normal if and only if $[\mathbf{w}, \overline{\mathbf{w}}] = [\overline{\mathbf{w}}, \mathbf{w}]$ for every $\mathbf{w} \in \mathbb{C}^3$, where $[\cdot, \cdot]$ stands for the cross product of vectors in $\mathbb{C}^3$.

Every $\Phi : M_2(\mathbb{C}) \to M_2(\mathbb{C})$ linear mapping can also be represented in this basis by a unique 4×4 matrix $\mathbf{F}$. It is trace preserving if and only if $\mathbf{F} = \begin{pmatrix} 1 & 0 \\ t & T \end{pmatrix}$ where T is a 3×3 matrix and 0 and t are row and column vectors respectively so that

$$\Phi(w_0\mathbf{1} + \mathbf{w} \cdot \sigma) = w_0\mathbf{1} + (w_0 t + T\mathbf{w}) \cdot \sigma. \tag{2}$$

When Φ is also positive then it maps the subspace of self-adjoint matrices of $M_2(\mathbb{C})$ into itself, which implies that T is real. A linear mapping Φ is unital if and only if $t = 0$. So, in this case we have

$$\Phi(w_0\mathbf{1} + \mathbf{w} \cdot \sigma) = w_0\mathbf{1} + (T\mathbf{w}) \cdot \sigma. \tag{3}$$

Hence, any bistochastic mapping $\Phi : M_2(\mathbb{C}) \to M_2(\mathbb{C})$ has a form (3). Now we are going to give a characterization bistochastic KS-operators.

Theorem 3.1 **([12])** *Any bistochastic mapping $\Phi : M_2(\mathbb{C}) \to M_2(\mathbb{C})$ is KS-operator if and only if one has*

$$\|T\mathbf{w}\| \leq \|\mathbf{w}\|, \quad T\overline{\mathbf{w}} = \overline{T\mathbf{w}} \tag{4}$$

$$\left\| T[\mathbf{w}, \overline{\mathbf{w}}] - \left[T\mathbf{w}, \overline{T\mathbf{w}}\right] \right\| \leq \|\mathbf{w}\|^2 - \|T\mathbf{w}\|^2 \tag{5}$$

for all $\mathbf{w} \in \mathbb{C}^3$.

Let Φ be a bistochastic KS-operator on $M_2(\mathbb{C})$, then it can be represented by (3). Following [6] let us decompose the matrix T as follows $T = RS$, here R is a rotation and S is a self-adjoint matrix (see [6]). Define a mapping Φ_S as follows

$$\Phi_S(w_0\mathbf{1} + \mathbf{w} \cdot \sigma) = w_0\mathbf{1} + (S\mathbf{w}) \cdot \sigma. \tag{6}$$

Every rotation is implemented by a unitary matrix in $M_2(\mathbb{C})$, therefore there is a unitary $U \in M_2(\mathbb{C})$ such that

$$\Phi(x) = U\Phi_S(x)U^*, \quad x \in M_2(\mathbb{C}). \tag{7}$$

On the other hand, every self-adjoint operator S can be diagonalized by some unitary operator, i.e. there is a unitary $V \in M_2(\mathbb{C})$ such that $S = VD_{\lambda_1,\lambda_2,\lambda_3}V^*$, where

$$D_{\lambda_1,\lambda_2,\lambda_3} = \begin{pmatrix} \lambda_1 & 0 & 0 \\ 0 & \lambda_2 & 0 \\ 0 & 0 & \lambda_3 \end{pmatrix}, \tag{8}$$

where $\lambda_1, \lambda_2, \lambda_3 \in \mathbb{R}$.

Consequently, the mapping Φ can be represented by

$$\Phi(x) = \tilde{U}\Phi_{D_{\lambda_1,\lambda_2,\lambda_3}}(x)\tilde{U}^*, \quad x \in M_2(\mathbb{C}) \tag{9}$$

for some unitary $\tilde{U}$. Due to Theorem 2.1 the mapping $\Phi_{D_{\lambda_1,\lambda_2,\lambda_3}}$ is also KS-operator. Hence, all bistochastic KS-operators can be characterized by $\Phi_{D_{\lambda_1,\lambda_2,\lambda_3}}$ and unitaries. In what follows, for the sake of shortness by $\Phi_{(\lambda_1,\lambda_2,\lambda_3)}$ we denote the mapping $\Phi_{D_{\lambda_1,\lambda_2,\lambda_3}}$. It is clear to observe from (4) that $|\lambda_k| \leq 1, k = 1, 2, 3$.

Let us characterize KS operators of the form $\Phi_{(\lambda_1,\lambda_2,\lambda_3)}$.

Theorem 3.2 ([12]) *If*

$$(1+\lambda_1^2)(3+\lambda_2^2+\lambda_3^2-\lambda_1^2) \le 4(1+\lambda_1\lambda_2\lambda_3); \tag{10}$$
$$(1+\lambda_2^2)(3+\lambda_1^2+\lambda_3^2-\lambda_2^2) \le 4(1+\lambda_1\lambda_2\lambda_3); \tag{11}$$
$$(1+\lambda_3^2)(3+\lambda_1^2+\lambda_2^2-\lambda_3^2) \le 4(1+\lambda_1\lambda_2\lambda_3); \tag{12}$$
$$\lambda_1^2+\lambda_2^2+\lambda_3^2 \le 1+2\lambda_1\lambda_2\lambda_3. \tag{13}$$

are satisfied, then a map $\Phi_{(\lambda_1,\lambda_2,\lambda_3)}$ *is a KS-operator.*

The last theorem allows us to construct lots of KS-operators, which are not completely positive.

4. A class of Kadison-Schwarz operators from $M_2(\mathbb{C})$ to $M_2(\mathbb{C})\otimes M_2(\mathbb{C})$

In this section we are going to provide description of a class of bistochastic mappings from $M_2(\mathbb{C})$ to $M_2(\mathbb{C})\otimes M_2(\mathbb{C})$.

First we need the following auxiliary

Lemma 4.1 *Let* $x = w_0\mathbb{1}\otimes\mathbb{1}+\mathbf{w}\cdot\sigma\otimes\mathbb{1}+\mathbb{1}\otimes\mathbf{r}\cdot\sigma$. *Then the following statements hold true:*

(i) x *is self-adjoint if and only if* $w_0\in\mathbb{R}$ *and* $\mathbf{w},\mathbf{r}\in\mathbb{R}^3$;

(ii) x *is positive if and only if* $w_0>0$ *and* $\|\mathbf{w}\|+\|\mathbf{r}\|\le w_0$.

Now consider the following operator $T: M_2(\mathbb{C})\to M_2(\mathbb{C})\otimes M_2(\mathbb{C})$ given by

$$T(w_0\mathbb{1}+\mathbf{w}\cdot\sigma) = w_0\mathbb{1}\otimes\mathbb{1}+\mathbf{A}\mathbf{w}\cdot\sigma\otimes\mathbb{1}+\mathbb{1}\otimes\mathbf{C}\mathbf{w}\cdot\sigma \tag{14}$$

where $\mathbf{A},\mathbf{C}$ are linear operators on $\mathbb{C}^3$.

Theorem 4.2 *The mapping* T *given by* (14) *is positive if and only if*

$$\|\mathbf{A}\mathbf{w}\|+\|\mathbf{C}\mathbf{w}\|\le 1,$$

for all $\mathbf{w}\in\mathbb{R}^3$ *with* $\|\mathbf{w}\|=1$.

The proof immediately follows from Lemma 4.1.

Corollary 4.3 *Let* $\mathbf{A}=\mathbf{C}$ *then* T *is positive if and only if* $\|\mathbf{A}\|\le\frac{1}{2}$.

Let us define the following mappings

$$\Phi(x) = w_0\mathbb{1}+2\mathbf{A}\mathbf{w}\cdot\sigma, \quad \Psi(x) = w_0\mathbb{1}+2\mathbf{C}\mathbf{w}\cdot\sigma \tag{15}$$

Then one finds

$$T(x) = \frac{1}{2}\Big(\Phi(x)\otimes\mathbb{1}+\mathbb{1}\otimes\Psi(x)\Big). \tag{16}$$

Theorem 4.4 *Let* T *be a mapping given by* (16). *If one has*

$$\|\mathbf{w}\|^2-2\|\mathbf{A}\mathbf{w}\|^2-2\|\mathbf{C}\mathbf{w}\|^2\ge 0 \tag{17}$$

$$\|\mathbf{A}[\mathbf{w},\overline{\mathbf{w}}]-2[\mathbf{A}\mathbf{w},\mathbf{A}\overline{\mathbf{w}}]\|+\|\mathbf{C}[\mathbf{w},\overline{\mathbf{w}}]-2[\mathbf{C}\mathbf{w},\mathbf{C}\overline{\mathbf{w}}]\|\le\|\mathbf{w}\|^2-2\|\mathbf{A}\mathbf{w}\|^2-2\|\mathbf{C}\mathbf{w}\|^2 \tag{18}$$

Then T *is a Kadison-Schwarz operator.*

Proof. From (16) one finds that

$$\begin{aligned} T(x^*x) - T(x)^*T(x) &= \frac{1}{2}\Big(\big(\Phi(x^*x) - \Phi(x)^*\Phi(x)\big)\otimes \mathbb{1} \\ &+ \mathbb{1}\otimes\big(\Psi(x^*x) - \Psi(x)^*\Psi(x)\big)\Big) \\ &+ \frac{1}{4}\Big(\mathbb{1}\otimes\Psi(x) - \Phi(x)\otimes\mathbb{1}\Big)^*\Big(\mathbb{1}\otimes\Psi(x) - \Phi(x)\otimes\mathbb{1}\Big). \end{aligned} \tag{19}$$

Now taking into account the following formula

$$x^*x = \big(|w_0|^2 + \|\mathbf{w}\|^2\big)\mathbb{1} + \big(w_0\overline{\mathbf{w}} + \overline{w_0}\mathbf{w} - i[\mathbf{w},\overline{\mathbf{w}}]\big)\cdot\sigma$$

from (15) we have

$$\begin{aligned} \Phi(x^*x) - \Phi(x)^*\Phi(x) &= \big(\|\mathbf{w}\|^2 - \|2\mathbf{A}\mathbf{w}\|^2\big)\mathbb{1} - 2i\big(\mathbf{A}[\mathbf{w},\overline{\mathbf{w}}] - 2[\mathbf{A}\mathbf{w},\mathbf{A}\overline{\mathbf{w}}]\big)\sigma, \\ \Psi(x^*x) - \Psi(x)^*\Psi(x) &= \big(\|\mathbf{w}\|^2 - \|2\mathbf{C}\mathbf{w}\|^2\big)\mathbb{1} - 2i\big(\mathbf{C}[\mathbf{w},\overline{\mathbf{w}}] - 2[\mathbf{C}\mathbf{w},\mathbf{C}\overline{\mathbf{w}}]\big)\sigma. \end{aligned}$$

Therefore, one gets

$$\begin{aligned} &\Big(\Phi(x^*x) - \Phi(x)^*\Phi(x)\Big)\otimes\mathbb{1} + \mathbb{1}\otimes\Big(\Psi(x^*x) - \Psi(x)^*\Psi(x)\Big) \\ = &\Big(\big(\|\mathbf{w}\|^2 - 4\|\mathbf{A}\mathbf{w}\|^2\big)\mathbb{1} - 2i\big(\mathbf{A}[\mathbf{w},\overline{\mathbf{w}}] - 2[\mathbf{A}\mathbf{w},\mathbf{A}\overline{\mathbf{w}}]\big)\sigma\Big)\otimes\mathbb{1} \\ + &\mathbb{1}\otimes\Big(\big(\|\mathbf{w}\|^2 - 4\|\mathbf{C}\mathbf{w}\|^2\big)\mathbb{1} - 2i\big(\mathbf{C}[\mathbf{w},\overline{\mathbf{w}}] - 2[\mathbf{C}\mathbf{w},\mathbf{C}\overline{\mathbf{w}}]\big)\sigma\Big) \\ = &\big(2\|\mathbf{w}\|^2 - 4\|\mathbf{A}\mathbf{w}\|^2 - 4\|\mathbf{C}\mathbf{w}\|^2\big)\mathbb{1}\otimes\mathbb{1} \\ - &2i\big(\mathbf{A}[\mathbf{w},\overline{\mathbf{w}}] - 2[\mathbf{A}\mathbf{w},\mathbf{A}\overline{\mathbf{w}}]\big)\sigma\otimes\mathbb{1} - \mathbb{1}\otimes 2i\big(\mathbf{C}[\mathbf{w},\overline{\mathbf{w}}] - 2[\mathbf{C}\mathbf{w},\mathbf{C}\overline{\mathbf{w}}]\big)\sigma \end{aligned}$$

According to Lemma 4.1 we conclude that the last expression is positive if and only if (17) and (18) are satisfied. Consequently, from (19) we infer that under the last conditions the mapping T is a KS operator. This completes the proof.

Remark 4.5 *We should stress that the conditions* (17), (18) *are sufficient to be KS-operator.*

Corollary 4.6 *If the mappings Φ and Ψ are KS operators, then T is also KS operator.*

The proof immediately follows from (19).

Remark 4.7 *We have to stress that if T is KS operator, then the mappings Φ and Ψ no need to be KS.*

Let us consider a more concrete case, namely, we assume that $\mathbf{A}\mathbf{w} = \lambda\mathbf{w}$ and $\mathbf{C}\mathbf{w} = \mu\mathbf{w}$. Then by $T_{\lambda,\mu}$ we denote the corresponding operator (see (14)). Then one can see that $\Phi_{(2\lambda,2\lambda,2\lambda)}$ and $\Psi_{(2\mu,2\mu,2\mu)}$ are the corresponding mappings. Due to Theorem 3.1 one can find that $\Phi_{(2\lambda,2\lambda,2\lambda)}$ is a KS-operator if and only if

$$2|\lambda||1 - 2\lambda|\|[\mathbf{w},\overline{\mathbf{w}}]\| \leq \big(1 - 4\lambda^2\big)\|\mathbf{w}\|^2.$$

From $\|[\mathbf{w},\overline{\mathbf{w}}]\| \leq \|\mathbf{w}\|^2$ (if we choose $\mathbf{w} = (0, 1, i)$, one gets $\|[\mathbf{w},\overline{\mathbf{w}}]\| = \|\mathbf{w}\|^2$) one finds

$$2|\lambda|(1 - 2\lambda) \leq 1 - 4\lambda^2.$$

The solution of the last inequality is $\lambda \in \left[-\frac{1}{4}; \frac{1}{2}\right]$.

Similarly, one finds that $\Psi_{(2\mu,2\mu,2\mu)}$ is a KS-operator if and only if $\mu \in \left[-\frac{1}{4}; \frac{1}{2}\right]$.

From Corollary 4.6 we immediately conclude that if $\lambda, \mu \in \left[-\frac{1}{4}; \frac{1}{2}\right]$ then $T_{\lambda,\mu}$ is a KS-operator.

Next we want to provide for other values of λ and μ for which $T_{\lambda,\mu}$ is Kadison-Schwarz.

Theorem 4.8 *Let* $T_{\lambda,\mu} : M_2(\mathbb{C}) \to M_2(\mathbb{C}) \otimes M_2(\mathbb{C})$ *be given by* (14). *If*

$$|\lambda||1-2\lambda| + |\mu||1-2\mu| \leq 1 - 2\lambda^2 - 2\mu^2$$

is satisfied, then the map $T_{\lambda,\mu}$ *is KS-operator.*

Acknowledgments
The author acknowledges the MOHE Grant FRGS11-022-0170. He also thanks the Junior Associate scheme of the Abdus Salam International Centre for Theoretical Physics, Trieste, Italy.

References

[1] Amosov G G, Holevo A S and Werner R F 2000 *Probl. Inf. Transm.* **36** 305
[2] Bhatt S J 1998 *Proc. Indian Acad. Sci. (Math. Sci.)* **108** 283
[3] Bratteli O and Robertson D W 1979 *Operator algebras and quantum statistical mechanics* (Berlin:Springer)
[4] Choi M D 1975 *Lin. Alg. Appl.* **10** 285
[5] Groh U 1984 *J. Operator Theory* **11** 395
[6] King C and Ruskai M B 2001 *IEEE Trans. Info. Theory* **47** 192
[7] Kossakowski A 2003 *Open Sys. & Information Dyn.* **10** 213
[8] Labuschagne L E, Majewski W A and Marciniak M 2006 *Expo. Math.* **24** 103
[9] Majewski W A and Marciniak M 2001 *J. Phys. A: Math. Gen.* **34** 5863
[10] Majewski W A 2007 *J. Phys. A: Math. Gen.* **40** 11539
[11] Majewski W A 2007 *Rep. Math. Phys.* **59** 289
[12] Mukhamedov F and Abduganiev A 2010 *Open Sys. & Information Dyn.* **17**, 245
[13] Nielsen M A and Chuang I L 2000 *Quantum Computation and Quantum Information* (Cambridge University Press)
[14] Ruskai M B, Szarek S and Werner E 2002 *Lin. Alg. Appl.* **347** 159
[15] Robertson A G 1977 *Math. Z.* **156** 205
[16] Robertson A G 1983 *Math. Proc. Camb. Philos. Soc.* **94** 291
[17] Stormer E 1963 *Acta Math.* **110** 233
[18] Woronowicz S L 1976 *Rep. Math. Phys.* **10** 165

Results in coupled fixed point in non-linear contractive conditions

Mona Khandaqji*, **Sharifa Al-Sharif**† **and Mohammad Al-Khaleel**‡

*Department of Mathematics, Hashemite University, Jordan

E-mail: * `mkhan@hu.edu.jo`

†,‡ Department of Mathematics, Yarmouk University, Jordan

E-mail: †,‡`sharifa@yu.edu.jo,khaleel@yu.edu.jo`

Abstract. In this paper we have established some coupled coincidence and coupled common fixed point theorems on (ψ, ϕ)-weakly contractive condition for mapping having the g-mixed monotone property in partially ordered generalized metric spaces which generalize some recent fixed point theorems given in the literature.

1. Introduction

The Banach contraction principle have played a major role in solving problems in many branches of pure and applied mathematics as Boyd and Wong [1] have extended the Banach contraction principle to the case of nonlinear contraction mappings. Moreover Ran and Reurings [2], Nieto and Rodriguez-López [3], [4], Bhaskar and Lakshmikantham [5], and Agarwal et. al. [6] have presented some new results for contractions in partially ordered metric spaces. Bhashkar and Lakshmikantham [5] have introduced the concept of a coupled fixed point. Furthermore, they established some coupled fixed point theorems in partially ordered complete metric spaces.

Later on, Lakshmikantham and Ćirić [7] have introduced some results of coupled coincidence and coupled common fixed point results for nonlinear mappings that satisfied certain contractive conditions in partially ordered complete metric spaces. On the other hand, Mustafa and Sims [8] recently, have generalized the concept of a metric space (G-metric space), where many authors have obtained many different results in these spaces, e.g., [9],[10], [11].

Aydi et. al. [12] has proved some coupled coincidence and coupled common fixed point theorems for a mixed g-monotone mapping satisfying nonlinear contractions in partially ordered generalized metric spaces. Saadati et. al. [13] has established some fixed point results in generalized ordered metric space. Recently, Choudhury and Maity [14], and Cho et. al. [15] initiated the study of coupled fixed point in generalized ordered metric spaces.

The purpose of this work is to establish the coupled coincidence and coupled common fixed point theorems for nonlinear contraction condition related to a pair of altering distance functions for mappings with mixed monotone property in partially ordered G-metric spaces.

Consistent with Mustafa and Sims [8], the following definitions and results will be needed in the sequel.

Definition 1 *Let X be a non-empty set and $G : X \times X \times X \to R^+$ satisfy the following properties:*

$(G1)$ $G(x, y, z) = 0$ *if* $x = y = z = 0$*(coincidence),*

(G2) $0 < G(x,x,y)$, *for all* $x,y \in X$, *where* $x \neq y$,
(G3) $G(x,x,y) \leq G(x,y,z)$, *for all* $x,y,z \in X$, *with* $z \neq y$,
(G4) $G(x,y,z) = G(p\{x,y,z\})$, *where p is a permutation of* x,y,z *(symmetry)*,
(G5) $G(x,y,z) \leq G(x,a,a) + G(a,y,z)$, *for all* $x,y,z,a \in X$ *(rectangle inequality)*.
Then the function G is called the G-metric on X, and the pair (X,G) *is called the G-metric space.*

Definition 2 *Let* (X,G) *be a G-metric space, let* (x_n) *be a sequence of points in* X,
(i) a point $x \in X$ *is said to be the limit of the sequence*(x_n) *if* $G(x_n,x_m,x) \to 0$ *(as* $n,m \to \infty$*), and we say that the sequence* (x_n) *is G-convergent to x (we say* $x_n \overset{(G)}{\to} x$*).*
(ii) A sequence (x_n)*is said to be G-Cauchy if* $G(x_n,x_m,x_l) \to 0$ *(as* $n,m,l \to \infty$*).*
(iii) (X,G)*is called a complete G-metric space if for every G-Cauchy sequence in X is G-converge in X.*

Proposition 3 *Let* (X,G) *be a G-metric space, then the following are equivalent*
(1) $x_n \overset{(G)}{\to} x$.
(2) $G(x_n,x_n,x) \to 0$ *(as* $n \to \infty$*)*.
(3) $G(x_n,x,x) \to 0$ *(as* $n \to \infty$*)*.
(4) $G(x_n,x_m,x) \to 0$ *(as* $n \to \infty$*)*.

Proposition 4 *Let* (X,G) *be a G-metric space, then the following are equivalent*
(1) (x_n) *is be G-Cauchy in X.*
(2) $G(x_n,x_m,x_m) \to 0$ *(as* $n,m \to \infty$*)*.

Proposition 5 *Let* (X,G) *be a G-metric space. A mapping* $f : X \to X$ *is G-continuous at* $x \in X$ *if and only if it is G-sequentially continuous at x, that is, whenever* (x_n) *is G-convergent to x,* $(f(x_n))$ *is G-convergent to* $f(x)$.

Proposition 6 *Let* (X,G) *be a G-metric space. Then, the function* $G(x,y,z)$ *is jointly continuous in all three of its variables.*

Proposition 7 *Let* (X,G) *be a G-metric space. A mapping* $F : X \times X \to X$ *is said to be continuous at* $(x,y) \in X \times X$, *if for any two G-convergent sequences* (x_n) *and* (y_n) *convergent to x and y respectively,* $(F(x_n,y_n))$ *is G- convergnet to* $F(x,y)$.

The concept of a mixed monotone property has been introduced by Bhaskar and Lakshmikantham in [5].

Definition 8 *Let* $(X,\leq)$ *be apartially ordered set. A mapping* $F : X \times X \to X$ *is said to have mixed monotone property if* $F(x,y)$ *is monotone non-decreasing in x and is monotone non-increasing in y, that is, for any* $x,y \in X$,
$x_1,x_2 \in X$, $x_1 \leq x_2$ *implies* $F(x_1,y) \leq F(x_2,y)$,
$y_1,y_2 \in X$, $y_1 \leq y_2$ *implies* $F(x,y_2) \leq F(x,y_1)$.

Lakshmikantham and Ćirić in [7] introduced the concept of a g-mixed monotone mapping.

Definition 9 *Let* $(X,\leq)$ *be a partially ordered set. Consider the mappings* $F : X \times X \to X$ *and* $g : X \to X$. *The map F is said to have mixed g-monotone property if* $F(x,y)$ *is monotone g-non-decreasing in x and is monotone g-non-increasing in y; that is, for any* $x,y \in X$,
$x_1,x_2 \in X$, $gx_1 \leq gx_2$ *implies* $F(x_1,y) \leq F(x_2,y)$,
$y_1,y_2 \in X$, $gy_1 \leq gy_2$ *implies* $F(x,y_2) \leq F(x,y_1)$.

Journal of Physics: Conference Series **435** (2013) 012019 doi:10.1088/1742-6596/435/1/012019

Definition 10 ([7]) *An element $(x, y) \in X \times X$ is called*

(1) a coupled fixed point of mapping $F : X \times X \to X$ if $x = F(x, y)$ and $y = F(y, x)$.

(2) a coupled coincidence point of mappings $F : X \times X \to X$ and $g : X \to X$ if $g(x) = F(x, y)$ and $g(y) = F(y, x)$, and in this case $(gx,\ gy)$ is called coupled point of coincidence.

(3) a common coupled fixed point of mappings $F : X \times X \to X$ and $g : X \to X$ if $x = g(x) = F(x, y)$ and $y = g(y) = F(y, x)$.

Definition 11 ([7]). *We say that mappings $F : X \times X \to X$ and $g : X \to X$ are commutative if $g(F(x, y)) = F(gx, gy)$, for all $x, y \in X$.*

2. Main results

We begin with the definition of altering distance function.

Definition 12 ([16]) *The function ψ: $[0, \infty) \to [0, \infty)$ is called altering distance if the following properties are satisfied:*

(1) *ψ is continuous and non-decreasing,*

(2) *$\psi(t) = 0$ if and only if $t = 0$.*

Now, we introduce our main result.

Theorem 13 *Let $(X, \leq)$ be a partially ordered set and (X, G) a complete G-metric space. Let $F : X \times X \to X$ and $g : X \to X$ be two continuous mappings such that F has the mixed g-monotone property on X, g commutes with F. Assume that there exist ψ, ϕ altering distance functions and non-negative real numbers α, β with $\alpha + \beta < 1$ such that*

$$\begin{aligned} \psi\left(G(F(x, y), F(u, v), F(w, z))\right) \ \leq \ & \psi\left(\alpha G(gx, gu, gw) + \beta G(gy, gv, gz)\right) \\ & - \phi\left(\alpha G(gx, gu, gw) + \beta G(gy, gv, gz)\right), \end{aligned} \tag{1}$$

for all $x, y, u, v, w, z \in X$ with $gw \leq gu \leq gx$ and $gy \leq gv \leq gz$. Also, suppose that $F(X \times X) \subseteq g(X)$. If there exist two elements $x_0, y_0 \in X$ with $g(x_0) \leq F(x_0, y_0)$ and $g(y_0) \geq F(y_0, x_0)$, then F and g have a coupled coincidence point.

Proof. Let $x_0, y_0 \in X$ such that $x_0 \leq F(x_0, y_0)$ and $y_0 \geq F(y_0, x_0)$. Given that $F(X \times X) \subseteq g(X)$, then we can choose $x_1 = F(x_0, y_0)$ and $y_1 = F(y_0, x_0)$. Also $x_2 = F(x_1, y_1)$ and $y_2 = F(y_1, x_1)$. Since F has the mixed g-monotone property, then $gx_0 \leq gx_1 \leq gx_2$ and $gy_2 \leq gy_1 \leq gy_0$. Continuing this process we can construct two sequences (x_n) and (y_n) in X such that

$$gx_n = F(x_{n-1}, y_{n-1}) \leq gx_{n+1} = F(x_n, y_n),$$

and

$$gy_{n+1} = F(y_n, x_n) \leq gy_n = F(y_{n-1}, x_{n-1}).$$

We can assume that $(gx_{n+1}, gy_{n+1}) \neq (gx_n, gy_n)$, for all $n \in N$. Otherwise if $(gx_{n+1}, gy_{n+1}) = (gx_n, gy_n)$, then $F(x_n, y_n) = gx_n$ and $F(y_n, x_n) = gy_n$, that is F and g have a coincidence point. Thus we can assume that either $gx_{n+1} = F(x_n, y_n) \neq gx_n$ or $gy_{n+1} = F(y_n, x_n) \neq gy_n$.

To complete the proof we need the following steps.

Step 1: We show that $\lim\limits_{n \to \infty} [G(gx_n, gx_n, gx_{n-1}) + G(gy_n, gy_n, gy_{n-1})] = 0$.

For each $n \in N$, using inequality (1), we have

$$\begin{aligned} \psi\left(G(gx_{n+1}, gx_{n+1}, gx_n)\right) &= \psi\left(G\left(F(x_n, y_n), F(x_n, y_n), F(x_{n-1}, y_{n-1})\right)\right) \\ &\leq \psi\left(\alpha G(gx_n, gx_n, gx_{n-1}) + \beta G(gy_n, gy_n, gy_{n-1})\right) \\ &\quad -\phi\left(\alpha G(gx_n, gx_n, gx_{n-1}) + \beta G(gy_n, gy_n, gy_{n-1})\right) \\ &\leq \psi\left(\alpha G(gx_n, gx_n, gx_{n-1}) + \beta G(gy_n, gy_n, gy_{n-1})\right). \end{aligned} \tag{2}$$

Since ψ is a non-decreasing function, we get from (2)

$$G(gx_{n+1}, gx_{n+1}, gx_n) \leq \alpha G(gx_n, gx_n, gx_{n-1}) + \beta G(gy_n, gy_n, gy_{n-1}). \tag{3}$$

On the other hand, we have

$$\begin{aligned} \psi\left(G(gy_{n+1}, gy_{n+1}, gy_n)\right) &= \psi\left(G\left(F(y_n, x_n), F(y_n, x_n), F(y_{n-1}, x_{n-1})\right)\right) \\ &\leq \psi\left(\alpha G(gy_n, gy_n, gy_{n-1}) + \beta G(gx_n, gx_n, gx_{n-1})\right) \\ &\quad -\phi\left(\alpha G(gy_n, gy_n, gy_{n-1}) + \beta G(gx_n, gx_n, gx_{n-1})\right) \\ &\leq \psi\left(\alpha G(gy_n, gy_n, gy_{n-1}) + \beta G(gx_n, gx_n, gx_{n-1})\right). \end{aligned} \tag{4}$$

Since ψ is a non-decreasing function, we get from (4)

$$G(gy_{n+1}, gy_{n+1}, gy_n) \leq \alpha G(gy_n, gy_n, gy_{n-1}) + \beta G(gx_n, gx_n, gx_{n-1}). \tag{5}$$

Adding (3) and (5), we have

$$\begin{aligned} & G(gx_{n+1}, gx_{n+1}, gx_n) + G(gy_{n+1}, gy_{n+1}, gy_n) \\ \leq\ & (\alpha + \beta)\left[G(gx_n, gx_n, gx_{n-1}) + G(gy_n, gy_n, gy_{n-1})\right]. \end{aligned}$$

Set $z_n = G(gx_{n+1}, gx_{n+1}, gx_n) + G\left(gy_{n+1}, gy_{n+1}, gy_n\right)$ and $\delta = \alpha + \beta$, then the sequence (z_n) is a non-increasing and bounded below sequence as

$$0 \leq z_n \leq \delta z_{n-1} \leq \delta^2 z_{n-2} \leq \cdots \leq \delta^n z_0,$$

which implies that

$$\lim_{n\to\infty} z_n = \lim_{n\to\infty} \left[G(gx_n, gx_n, gx_{n-1}) + G(gy_n, gy_n, gy_{n-1})\right] = 0. \tag{6}$$

Step 2: We show that (gx_n) and (gy_n) are G-Cauchy sequences in X.
From (6), it follows that

$$\lim_{n\to\infty} G(gx_n, gx_n, gx_{n-1}) = 0 \quad \text{and} \quad \lim_{n\to\infty} G(gy_n, gy_n, gy_{n-1}) = 0.$$

For each $m \geqslant n$, using rectangle inequality, we have

$$\begin{aligned} G(gx_m, gx_m, gx_n) &\leq G(gx_m, gx_m, gx_{m-1}) + G(gx_{m-1}, gx_{m-1}, gx_{m-2}) \\ &\quad + G(gx_{m-2}, gx_{m-2}, gx_{m-3}) + G(gx_{m-3}, gx_{m-3}, gx_{m-4}) \\ &\quad + \cdots + G(gx_{n-2}, gx_{n-2}, gx_{n-1}) + G(gx_{n-1}, gx_{n-1}, gx_n), \end{aligned} \tag{7}$$

and

$$\begin{aligned} G(gy_m, gy_m, gy_n) &\leq G(gy_m, gy_m, gy_{m-1}) + G(gy_{m-1}, gy_{m-1}, gy_{m-2}) \\ &\quad + \cdots + G(gx_{n-1}, gx_{n-1}, gx_n). \end{aligned} \tag{8}$$

Adding (7) and (8), we get

$$\begin{aligned} G(gx_m, gx_m, gx_n) + G(gy_m, gy_m, gy_n) &\leq z_{m-1} + z_{m-2} + \cdots\cdots + z_0 \\ &\leq \left(\delta^{m-1} + \delta^{m-1} + \cdots + \delta^n\right) z_0 \\ &\leq \left(\frac{\delta^n}{1-\delta}\right) z_0. \end{aligned}$$

Since $0 < \delta < 1$, we have

$$\lim_{n,m\to\infty} G(gx_m, gx_m, gx_n) + G(gy_m, gy_m, gy_n) = 0.$$

This implies that (gx_n) and (gy_n) are G-Cauchy sequences in X.

Step 3: We show the existence of a coupled coincidence sequence.

Since (gx_n) and (gy_n) are G-Cauchy sequences in a complete G-metric space (X, G), then there exist $x, y \in X$ such that (gx_n) and (gy_n) are G-convergent to points x and y, respectively, that is

$$\lim_{n\to\infty} G(gx_n, gx_n, x) = \lim_{n\to\infty} G(gx_n, x, x) = 0, \tag{9}$$

and

$$\lim_{n\to\infty} G(gy_n, gy_n, y) = \lim_{n\to\infty} G(gy_n, y, y) = 0. \tag{10}$$

Then, by the continuity of g and the continuity of G on its variables, we have

$$\lim_{n\to\infty} G(g(gx_n), g(gx_n), gx) = \lim_{n\to\infty} G(g(gx_n), gx, gx) = 0, \tag{11}$$

and

$$\lim_{n\to\infty} G(g(gy_n), g(gy_n), gy) = \lim_{n\to\infty} G(g(gy_n), gy, gy) = 0, \tag{12}$$

Therefore, $(g(gx_n))$ is G-convergent to $g(x)$ and $(g(gy_n))$ is G-convergent to $g(y)$. Since F and g commute, we get

$$g(gx_{n+1}) = g(F(x_n, y_n)) = F(gx_n, gy_n), \tag{13}$$

and

$$g(gy_{n+1}) = g(F(y_n, x_n)) = F(gy_n, gx_n). \tag{14}$$

Using the continuity of F on $X \times X$ and the fact that the sequences (gx_n) and (gy_n) are G-convergent to points x and y respectively, we get from Definition 7 that the sequence $(F(gx_n, gy_n))$ is G-convergent to $F(x,y)$ and the sequence $(F(gy_n, gx_n))$ is G-convergent to $F(y,x)$. Therefore, from (13) and (14), $g(gx_{n+1})$ is G-convergent to $F(x,y)$ and $g(gy_{n+1})$ is G-convergent to $F(y,x)$. By the uniqueness of the limit and (11), we have $gx = F(x,y)$. Similarly we can show that $gy = F(y,x)$, which implies that (x,y) is a coupled coincidence point of F and g. This ends the proof. ∎

Taking $g = I_X$ (the identity mapping) in Theorem 13, we obtain the following coupled fixed point result.

Corollary 14 *Let $(X, \leq)$ be a partially ordered set and (X, G) a complete G-metric space. Let $F : X \times X \to X$ be a continuous mapping satisfying the mixed monotone property on X. Suppose that there exist ψ, ϕ altering distance functions and non-negative real numbers α, β with $\alpha + \beta < 1$ such that*

$$\begin{aligned} \psi(G(F(x,y), F(u,v), F(w,z))) \;\leq\; & \psi(\alpha G(x,u,w) + \beta G(y,v,z)) \\ & -\phi(\alpha G(x,u,w) + \beta G(y,v,z)), \end{aligned} \tag{15}$$

for all $x, y, u, v, w, z \in X$ with $w \leq u \leq x$ and $y \leq v \leq z$. If there exist two elements $x_0, y_0 \in X$ such that $x_0 \leq F(x_0, y_0)$ and $y_0 \geq F(y_0, x_0)$, then F has a coupled fixed point.

Taking $\psi(x) = x$, $\phi(x) = 0$ in Theorem 13, we obtain the following coupled coincidence point.

Corollary 15 *Let $(X,\leq)$ be a partially ordered set and (X,G) a complete G-metric space. Let $F: X\times X\to X$ and $g: X\to X$ be two continuous mappings such that F has the mixed g-monotone property on X, g commutes with F. Assume that there exist two non-negative real numbers α,β with $\alpha+\beta<1$ such that*

$$G(F(x,y),F(u,v),F(w,z))\leq\alpha G(gx,gu,gw)+\beta G(gy,gv,gz), \tag{16}$$

for all $x,y,u,v,w,z\in X$ with $gw\leq gu\leq gx$ and $gy\leq gv\leq gz$. Also, suppose that $F(X\times X)\subseteq g(X)$. If there exist two elements $x_0,y_0\in X$ with $g(x_0)\leq F(x_0,y_0)$ and $g(y_0)\geq F(y_0,x_0)$, then F and g have a coupled coincidence point.

Taking $g=I_X$ (the identity mapping) and $\alpha=\beta=\frac{k}{2}$ in Corollary 15, we obtain the following coupled fixed point which is Theorem (3.1) in [14].

Corollary 16 *Let $(X,\leq)$ be a partially ordered set and (X,G) a complete G-metric space. Let $F: X\times X\to X$ a continuous mapping having the mixed monotone property on X. Assume that there exist a $k\in[0,1)$ such that*

$$G(F(x,y),F(u,v),F(w,z))\leq\frac{k}{2}\left[G(x,u,w)+G(y,v,z)\right], \tag{17}$$

for all $x,y,u,v,w,z\in X$ with $w\leq u\leq x$ and $y\leq v\leq z$. If there exist two elements $x_0,y_0\in X$ with $x_0\leq F(x_0,y_0)$ and $y_0\geq F(y_0,x_0)$, then F has a coupled fixed point in X, that is, there exist $x,y\in X$ such that $x=F(x,y)$ and $y=F(y,x)$.

Corollary 17 *Let $(X,\leq)$ be a partially ordered set and (X,G) a complete G-metric space. Let $F: X\times X\to X$ a continuous mapping having the mixed monotone property on X. Assume that there exist a $k\in[0,1)$ such that*

$$G(F(x,y),F(u,v),F(w,z))\leq\frac{k}{2}\max\left\{G(x,u,w),G(y,v,z)\right\},$$

or all $x,y,u,v,w,z\in X$ with $w\leq u\leq x$ and $y\leq v\leq z$. If there exist two elements $x_0,y_0\in X$ with $x_0\leq F(x_0,y_0)$ and $y_0\geq F(y_0,x_0)$, then F has a coupled fixed point in X.

Proof. We have

$$\begin{aligned} G(F(x,y),F(u,v),F(w,z)) &\leq \frac{k}{2}\max\left\{G(x,u,gw),G(y,v,z)\right\} \\ &\leq \frac{k}{2}\left[G(x,u,w)+G(y,v,z)\right], \end{aligned}$$

Then the result follows from Corollary 16 with $\alpha=\beta=\frac{k}{2}$. ■

The following example was introduced by Choudhury and Maity in [14]:

Example 18 *Let $X=0,1,2,\ldots$ and $G: X\times X\times X\to R^{+}$ be defined as follows:*

$$G(x,y,z)=\begin{cases} x+y+z & \text{if } x,y,z \text{ are all distinct and different from zero,}\\ x+z & \text{if } x=y\neq z \text{ and all different from zero,}\\ y+z+1, & \text{if } x=0,\ y\neq z \text{ and } y,\ z \text{ are different from zero,}\\ y+2 & \text{if } x=0, z=y\neq 0,\\ 1+z, & \text{if } x=0,\ y=0,\ z\neq 0,\\ 0 & \text{if } x=y=z. \end{cases}$$

Then, (X, G) is a complete G-metric space. Let a partial order $\preceq$ on X be defined as follows: For $x, y \in X$, $x \preceq y$ holds if $x > y$ and 3 divides $(x - y)$ and $3 \preceq 1$ and $0 \preceq 1$ hold. Let $F : X \times X \to X$ be defined as follows:

$$F(x,y) = \begin{cases} 1, & \text{if } x \prec y \\ 0. & \text{if otherwise.} \end{cases}$$

Let $w \preceq u \preceq x \preceq y \preceq v \preceq z$ hold, then equivalently, we have $w \geq u \geq x \geq y \geq v \geq z$. Then

$$F(x,y) = F(u,v) = F(w,z) = 1$$

Let $\psi(t) = t$, $\phi(t) = \left(1 - \frac{k}{2}\right)t$. for $t \geq 0$, $k \in [0,1)$ and $g(x) = x$, $x \in X$. Thus the left-hand side of (1) is $G(1,1,1) = 0$ and hence (1) is satisfied. Then, with $x_0 = 81$ and $y_0 = 0$ Corollary (16) is applicable to this example. It may be observed that in this example the coupled fixed point is not unique, $(0,0)$ and $(1,0)$ are two coupled fixed point of F.

Definition 19 *Let $(X, \leq)$ be a partially ordered set and (X, G) a G-metric space . If X has the following properties:*
(1) if a non-decreasing sequence $x_n \to x$, then $x_n \leq x$ for all n,
(2) if a non-increasing sequence $x_n \to x$, then $x \leq x_n$ for all n.
Then $(X, G, \leq)$ is called a regular space.

In the following Theroem, we obtain coupled coincidence point results similar to the result in Theorem 13 without the continuity condition of the mappings F and g and the commutativity hypotheses of F and g. But we give additional conditions on the partially ordered set $(X, \leq)$.

Theorem 20 *Let $(X, \leq)$ be a partially ordered set and $(X, G, \leq)$ a regular space. Assume that there exist ψ, ϕ altering distance functions, the mappings $F : X \times X \to X$ and $g : X \to X$ such that F has the mixed g-monotone property, and $F(X \times X) \subseteq g(X)$. Also, assume that there exist non-negative real numbers α, β with $\alpha + \beta < 1$ such that*

$$\begin{aligned} \psi\left(G(F(x,y), F(u,v), F(w,z))\right) &\leq \psi\left(\alpha G(gx, gu, gw) + \beta G(gy, gv, gz)\right) \\ &\quad - \phi\left(\alpha G(gx, gu, gw) + \beta G(gy, gv, gz)\right), \end{aligned} \tag{18}$$

for all $x, y, u, v, w, z \in X$ with $gw \leq gu \leq gx$ and $gy \leq gv \leq gz$. Also, suppose that $(g(X), G)$ is a complete G-metric space. If there exist two elements $x_0, y_0 \in X$ with $g(x_0) \leq F(x_0, y_0)$ and $g(y_0) \geq F(y_0, x_0)$, then F and g have a coupled coincidence point.

Proof. Following step 1 and step 2 in Theorem 13, we get that (gx_n) and (gy_n) are G-Cauchy sequences in $g(X)$ with $gx_n \leq gx_{n+1}$ and $gy_n \geq gy_{n+1}$, for all $n \in N$. Since $(g(X), G)$ is a complete G-metric space, then there exist $x, y \in X$ such that (gx_n) and (gy_n) are G-convergent to points $g(x)$ and $g(y)$ in $g(X)$, respetively, that is

$$\lim_{n\to\infty} G(gx_n, gx_n, gx) = 0,$$

and

$$\lim_{n\to\infty} G(gy_n, gy_n, gy) = 0.$$

Using the fact that the sequence (gx_n) is non-decreasing and the sequence (gy_n) is non-increasing, and the space $(X, G, \leq)$ is a regular space, we have

$$gx_n \leq gx \text{ and } gy \leq gy_n, \text{ for all } n \in N.$$

If there exist $n \in N$ such that $gx_n = gx$ and $gy = gy_n$, then $gx = gx_n \leq gx_{n+1} \leq gx$ and $gy \leq gy_{n+1} \leq gy_n = gy$. This implies that $gx_n = gx_{n+1} = F(x_n, y_n)$ and $gy_n = gy_{n+1} = F(y_n, x_n)$, therefore, (x_n, y_n) is a coupled coincidence point of F and g. Then we can assume that $(gx_n, gy_n) \neq (gx, gy)$ for all $n \in N$. Using inequality (18), we have

$$\begin{aligned} \psi(G(gx_{n+1}, gx_{n+1}, F(x,y))) &= \psi(G(F(x_n, y_n), F(x_n, y_n), F(x,y))) \\ &\leq \psi(\alpha G(gx_n, gx_n, gx) + \beta G(gy_n, gy_n, gy)) \\ &\quad -\phi(\alpha G(gx_n, gx_n, gx) + \beta G(gy_n, gy_n, gy))) . \end{aligned} \tag{19}$$

Letting $n \to \infty$ in (19) and using the continuity of ψ and ϕ, we obtain that $\psi(G(gx, gx, F(x,y))) = 0$, which implies that $G(gx, gx, F(x,y)) = 0$. Therefore, $gx = F(x,y)$. Similarly, one can show that $gy = F(y,x)$. Therefore (x,y) is a coupled coincidence point of F and g. ■

Taking $g = I_X$ (the identity mapping) in Theorem 20, we obtain the following result.

Corollary 21 *Let $(X, \leq)$ be a partially ordered set, $(X, G, \leq)$ a regular space and (X, G) is a complete G-metric space. Assume that there exist ψ, ϕ altering distance functions and the mappings $F : X \times X \to X$ has the mixed monotone property. Also assume that there exist non-negative real numbers α, β with $\alpha + \beta < 1$ such that*

$$\begin{aligned} \psi(G(F(x,y), F(u,v), F(w,z))) &\leq \psi(\alpha G(x,u,w) + \beta G(y,v,z)) \\ &\quad - \phi(\alpha G(x,u,w) + \beta G(y,v,z)), \end{aligned}$$

for all $x, y, u, v, w, z \in X$ with $w \leq u \leq x$ and $y \leq v \leq z$. If there exist two elements $x_0, y_0 \in X$ with $x_0 \leq F(x_0, y_0)$ and $y_0 \geq F(y_0, x_0)$, then F has a coupled fixed point.

Taking $\psi(x) = x$, $\phi(x) = 0$ in Corollary 21, we obtain the following result.

Corollary 22 *Let $(X, \leq)$ be a partially ordered set, $(X, G, \leq)$ a regular space, and (X, G) is a complete G-metric space. Assume that the mappings $F : X \times X \to X$ has the mixed monotone property and there exist non-negative real numbers α, β with $\alpha + \beta < 1$ such that*

$$\psi(G(F(x,y), F(u,v), F(w,z))) \leq \alpha G(x,u,w) + \beta G(y,v,z),$$

for all $x, y, u, v, w, z \in X$ with $w \leq u \leq x$ and $y \leq v \leq z$. If there exist two elements $x_0, y_0 \in X$ with $x_0 \leq F(x_0, y_0)$ and $y_0 \geq F(y_0, x_0)$, then F has a coupled fixed point.

Now we shall prove the uniqueness of a coupled common fixed point. If $(X, \leq)$ is a partially ordered set, then we endow the product $X \times X$ with the following partial order for $(x,y), (u,v) \in X \times X$

$$(x,y) \leq (u,v) \iff x \leq u,\ v \leq y.$$

Theorem 23 *In addition to hypotheses of Theorem 13, suppose that for every (x,y), $(x^*, y^*) \in X \times X$, there exists a (u,v) in $X \times X$ such that $(F(u,v), F(v,u))$ is comparable with $(F(x,y), F(y,x))$ and $(F(x^*, y^*), F(y^*, x^*))$. Then F and g have a unique coupled common fixed point, that is there exists a unique $(x,y) \in X \times X$ such that $x = gx = F(x,y)$, $y = gy = F(y,x)$.*

Proof. From Theorem 13, the set of coupled coincidence points of F is non-empty. Suppose (x,y) and (x^*, y^*) are coupled coincidence points of F and g, that is, $gx = F(x,y)$, $gy = F(y,x)$, $gx^* = F(x^*, y^*)$, and $gy^* = F(y^*, x^*)$. We shall show that

$$gx = gx^* \text{ and } gy = gy^*. \tag{20}$$

By assumption, there exists $(u,v) \in X \times X$ such that $(F(u,v),F(v,u))$ is comparable with $(F(x,y),F(y,x))$ and $(F(x^*,y^*),F(y^*,x^*))$. One can assume that $(F(x,y),F(y,x)) \leq (F(u,v),F(v,u))$ and $(F(x^*,y^*),F(y^*,x^*)) \leq (F(u,v),F(v,u))$. Put $u_0 = u, v_0 = v$ and define inductively sequences $(gu_n),(gv_n)$ in X as follows

$$gu_{n+1} = F(u_n,v_n) \text{ and } gv_{n+1} = F(v_n,u_n) \text{ for all } n \in N.$$

Set $x_0 = x, y_0 = y,\ x_0^* = x^*, y_0^* = y^*$ and, on the same way, define sequences (gx_n), (gy_n), (gx_n^*), (gy_n^*). Since

$$(F(x,y),F(y,x)) = (gx_1,gy_1) = (gx,gy),$$

and

$$(F(u,v),F(v,u)) = (gu_1,gv_1),$$

are comparable, then $gx \leq gu_1$ and $gv_1 \leq gy$. One can show by induction that

$$gx \leq gu_n \text{ and } gv_n \leq gy, \text{ for all } n \in N.$$

From inequality (1), we have

$$\begin{aligned}\psi\left(G(gx,gx,gu_{n+1})\right) &= \psi\left(G(F(x,y),F(x,y),F(u_n,v_n))\right)\\ &\leq \psi\left(\alpha G(gx,gx,gu_n)+\beta G(gy,gy,gv_n)\right)\\ &\quad -\phi\left(\alpha G(gx,gx,gu_n)+\beta G(gy,gy,gv_n)\right)\\ &\leq \psi\left(\alpha G(gx,gx,gu_n)+\beta G(gy,gy,gv_n)\right),\end{aligned}$$

and

$$\begin{aligned}\psi\left(G(gy,gy,gv_{n+1})\right) &= \psi\left(G(F(y,x),F(y,x),F(v_n,u_n))\right)\\ &\leq \psi\left(\alpha G(gy,gy,gv_n)+\beta G(gx,gx,gu_n)\right)\\ &\quad -\phi\left(\alpha G(gy,gy,gv_n)+\beta G(gx,gx,gu_n)\right)\\ &\leq \psi\left(\alpha G(gy,gy,gv_n)+\beta G(gx,gx,gu_n)\right).\end{aligned}$$

Since ψ is non-decreasing, it follows that

$$G(gx,gx,gu_{n+1}) \leq \alpha G(gx,gx,gu_n) + \beta G(gy,gy,gv_n), \tag{21}$$

and

$$G(gy,gy,gv_{n+1}) \leq \alpha G(gy,gy,gv_n) + \beta G(gx,gx,gu_n). \tag{22}$$

Adding (21) and (22), we get

$$G(gx,gx,gu_{n+1}) + G(gy,gy,gv_{n+1}) \leq (\alpha+\beta)\left[G(gx,gx,gu_n) + G(gy,gy,gv_n)\right].$$

Therefore,

$$\begin{aligned}G(gx,gx,gu_{n+1}) + G(gy,gy,gv_{n+1}) &\leq (\alpha+\beta)\left[G(gx,gx,gu_n)+G(gy,gy,gv_n)\right] \qquad (23)\\ &\leq (\alpha+\beta)^2\left[G(gx,gx,gu_{n-1})+G(gy,gy,gv_{n-1})\right]\\ &\vdots\\ &\leq (\alpha+\beta)^{n+1}\left[G(gx,gx,gu_1)+G(gy,gy,gv_1)\right].\end{aligned}$$

Taking the limit in (23) as $n \to \infty$, we get

$$\lim_{n\to\infty} G(gx,gx,gu_{n+1}) + G(gy,gy,gv_{n+1}) = 0.$$

This implies that

$$\lim_{n\to\infty} G(gx, gx, gu_{n+1}) = 0 \text{ and } \lim_{n\to\infty} G(gy, gy, gv_{n+1}) = 0. \tag{24}$$

Similarly, we show that

$$\lim_{n\to\infty} G(gx^*, gx^*, gu_{n+1}) = 0 \text{ and } \lim_{n\to\infty} G(gy^*, gy^*, gv_{n+1}) = 0. \tag{25}$$

Thus by $(G5)$, (24) and (25) , we have as $n \to \infty$,

$$G(gx, gx, gx^*) \leq G(gx, gx, gu_{n+1}) + G(gu_{n+1}, , gu_{n+1}, gx^*) \to 0,$$

and

$$G(gy, gy, gy^*) \leq G(gy, gy, gv_{n+1}) + G(gv_{n+1}, , gv_{n+1}, gy^*) \to 0.$$

Hence, $gx = gx^*$ and $gy = gy^*$. Thus, we have (20) .

On the other hand, since $gx = F(x, y)$ and $gy = F(y, x)$, by commutativity of F and g, we have

$$g(gx) = g(F(x, y)) = F(gx, gy), \; g(gy) = g(F(y, x)) = F(gy, gx). \tag{26}$$

Letting $z = gx$ and $w = gy$. Then from (26), it follows that

$$gz = F(z, w) \text{ and } gw = F(w, z). \tag{27}$$

Thus, (z, w) is a coupled coincindence point. Then, from (20) with $x^* = z$ and $y^* = w$ it follows that, $gz = gx$ and $gw = gy$. That is,

$$gz = z \text{ and } gw = w.$$

Thus, from (26) and (27), we have

$$z = gz = F(z, w) \text{ and } w = gw = F(w, z).$$

Therefore, (z, w) is a coupled common fixed point of F and g. To show the uniqueness of the point (z, w), assume that (z^*, w^*) is another coupled common fixed point of F and g. Then, we have

$$z^* = gz^* = F(z^*, w^*) \text{ and } w^* = gw^* = F(w^*, z^*).$$

Since the pair (z^*, w^*) is a coupled coincident point of F and g, we have $gz^* = gx = z$ and $gw^* = gy = w$. From (20), we have

$$z = gz^* = gz = z \text{ and } w^* = gw^* = gw = w.$$

This complete of the uniqueness of the coupled common fixed point. ■

We have the following result.

Theorem 24 *Let $(X, \preceq)$ be a partially ordered set, (X, G) a complete G-metric space and, $(X, G, \leq)$ is regular. Suppose that $F : X \times X \to X$ having the mixed monotone property such that, there exist two elements $x_0, y_0 \in X$ with $x_0 \leq F(x_0, y_0)$ and $y_0 \geq F(y_0, x_0)$. Suppose that there exist ψ, ϕ altering distance functions and non-negative real numbers $\alpha + \beta < 1$ such that*

$$\begin{aligned} \psi\left(G(F(x, y), F(u, v), F(w, z))\right) \;\leq\; & \psi\left(\alpha G(x, u, w) + \beta G(y, v, z)\right) \\ & -\phi\left(\alpha G(x, u, w) + \beta G(y, v, z)\right), \end{aligned} \tag{28}$$

for all $x, y, u, v, w, z \in X$ with $w \leq u \leq x$ and $y \leq v \leq z$. If there exist $x_0, y_0 \in X$ such that $x_0 \leq F(x_0, y_0)$ and $F(y_0, x_0) \leq y_0$, then F has a coupled fixed point. Furthermore, if $y_0 \leq x_0$, then $x = y$, that is, $x = F(x, x)$.

Proof. Taking $g = I_X$ in Theorem 20 and following the same proof, there exist $x, y \in X$ such that $x = F(x, y)$ and $y = F(y, x)$. If $y_0 \leq x_0$, then we need to show that $x = y$ (i.e., $x = F(x, x)$). Since $y_0 \leq x_0$, we get

$$y \leq y_n \leq \cdots y_1 \leq y_0 \leq x_0 \leq x_1 \leq \cdots \leq x_n \leq x.$$

Thus, we have $y \leq x$. Suppose that $x \neq y$. Then (28) gives the following

$$\begin{aligned} \psi(G(x,x,y)) &= \psi(G(F(x,y),F(x,y),F(y,x))) \\ &\leq \psi(\alpha G(x,x,y) + \beta G(y,y,x)) \\ &\quad -\phi(\alpha G(x,x,y) + \beta G(y,y,x) \\ &\leq \psi(\alpha G(x,x,y) + \beta G(y,y,x)), \end{aligned} \tag{29}$$

and

$$\begin{aligned} \psi(G(x,y,y)) &= \psi(G(F(x,y),F(y,x),F(y,x))) \\ &\leq \psi(\alpha G(x,y,y) + \beta G(y,x,x)) \\ &\quad -\phi(\alpha G(x,y,y) + \beta G(y,x,x)) \\ &\leq \psi(\alpha G(x,y,y) + \beta G(y,x,x)). \end{aligned} \tag{30}$$

Since ψ is non-decreasing, (29) and (30) give the following

$$G(x,x,y) \leq \alpha G(x,x,y) + \beta G(y,y,x), \tag{31}$$

and

$$G(x,y,y) \leq \alpha G(x,y,y) + \beta G(y,x,x). \tag{32}$$

Adding (31) and (32), we have

$$G(x,x,y) + G(x,y,y) \leq (\alpha + \beta)\left[G(x,x,y) + G(y,y,x)\right].$$

This is a contradiction since $\alpha + \beta < 1$. Therefore, $x = y = F(x, x)$. ■

References

[1] Boyd D W and Wong J S 1969 *Proc. Amer. Math. Soc. 20* On nonlinear contractions P 458
[2] Ran ACM and Reurings MCB *Proc Am. Math. Soc. 132* A fixed point theorem in partially ordered sets and some applications to matrix equations P 1435
[3] Nieto J J and López R R 2005 *Order* Contractive mapping theorems in partially ordered sets and applications to ordinary differential equations. **22**, 223–239
[4] Nieto J J and López R R 2007 *Acta Math Sinica Engl Ser. 23* Existence and uniqueness of fixed point in partially ordered sets and applications to ordinary differential equations **12** 2205–2212
[5] Bhaskar T G and Lakshmikantham V 2006 *Nonlinear Anal.TMA* Fixed point theorems in partially ordered metric spaces and applications **65** 1379–1393
[6] Agarwal R P, El-Gebeily M A and O'Regan D 2008 *Appl Anal.* Generalized contractions in partially ordered metric spaces **87** 1–8
[7] Lakshmikantham V and Cirić L J 2009 *Nonlinear Anal.* Coupled fixed point theorems for nonlinear contractions in partially ordered metric spaces. **70**, 4341–4349
[8] Mustafa Z and Sims B 2006 *J. Nonlinear Convex Anal.* A new approach to generalized metric spaces **7(2)** 289–297
[9] Mustafa Z and Sims B 2009 *Fixed Point Theory Appl.* Fixed point theorems for contractive mappings in complete G-metric space **2009** 10 pages Article ID 917175
[10] Abbas M, Khan A R and Nazir T 2011 *Appl Math Comput.* Coupled common fixed point results in two generalized metric spaces **217** 6328-6336

[11] Shatanawi W 2011 *Hacet J Math Stat.* Coupled fixed point theorems in generalized metric spaces **40(3)** 441–447

[12] Aydi H Damjanović B, Samet B and Shatanawi W 2011 *Math.Comput. Modelling* Coupled fixed point theorems for nonlinear contractions in partially ordered G-metric spaces **54** 2443–2450

[13] Saadati R, Vaezpour S M, Vetro P and Rhoades B E 2010 *Math Comput Model* Fixed point theorems in generalized partially ordered G-metric spaces **52** 797–801

[14] Choudhury B S and Maity P 2011 *Math Comput Model* Coupled fixed point results in generalized metric spaces **54** 73–79

[15] Cho Y J, Rhoades B E, Saadati R, Samet B and Shatanawi W 2012 *Fixed Point Theory and Appl.* Nonlinear coupled fixed point theorems in ordered generalized metric spaces with integral type **2012** 8 pages doi:10.1186/1687-1812-2012-8

[16] Khan M Swaleh and Sessa S 1984 *Bull. Aust. Math. Soc.* Fixed points theorems by altering distances between the points **30** 1–9

Journal of Physics: Conference Series **435** (2013) 012020 doi:10.1088/1742-6596/435/1/012020

Noncommutative of space-time and the Relativistic Hydrogen Atom

Slimane Zaim and Yazid Delenda

Département des Sciences de la Matière, Faculté des Sciences, Université Hadj Lakhdar – Batna, Algeria.

E-mail: zaim69slimane@yahoo.fr, yazid.delenda@yahoo.com

Abstract. We study the Klein-Gordon equation in a non-commutative space-time as applied to the Hydrogen atom to extract the energy levels, by considering the second-order corrections in the non-commutativity parameter. By comparing to the 2S - 1S transition energies we obtain an upper bound on the non-commutativity parameter. Phenomenologically we show that non-commutativity is the source of lamb shift corrections and spin of the electron.

1. Introduction

The non-commutativity of spatial rotations in three and more dimensions is an idea which is deeply ingrained in our theories. The non-commutativity has received a wide appreciation as an alternative approach to understanding many physical phenomenon such as the ultraviolet and infrared divergences [1], unitarity violation [2], causality [3], and new physics at very short distances of the Planck-length order [4, 5, 6].

The non-commutative field theory is motivated by the natural extension of the usual quantum mechanical commutation relations between position and momentum, by imposing further commutation relations between position coordinates themselves. As in usual quantum mechanics, the non-commutativity of position coordinates immediately implies a set of uncertainty relations between position coordinates analogous to the Heisenberg uncertainty relations between position and momentum; namely:

$$[\hat{x}^{\mu}, \hat{x}^{\nu}]_* = i\theta^{\mu\nu}, \tag{1}$$

where $\hat{x}^{\mu}$ are the coordinate operators and $\theta^{\mu\nu}$ are the non-commutativity parameters of dimension of area that signify the smallest area in space that can be probed in principle.We use the symbol $*$ in equation (1) to denote the product of the non-commutative structure. This idea is similar to the physical meaning of the Plank constant in the relation $[\hat{x}_i, \hat{p}_j] = i\hbar\delta_{ij}$, which as is known is the smallest phase-space in quantum mechanics. The space-time non-commutativity has gained considerable interest in current literature and a search for any possible upper bound on the space-time non-commutativity parameter using recent experimental feedback is very much desirable. The issue of space-time non-commutativity is worth pursuing in its own right because of its deep connection with such fundamental notions as unitarity and causality. It was argued that introduction of space-time non-commutativity spoils unitarity or even causality. Much attention has been devoted in recent times to circumvent these difficulties in formulating theories with $\theta^{0\nu} \neq 0$ [7, 8, 9, 10]. We do not consider momentum space non-commutative effects as have been done by [11]. Our motivation is to study the effect of non-commutativity on the level of quantum mechanics when time-space non-commutativity is accounted for. One can study the physical consequences of this theory by making detailed analytical estimates for measurable physical quantities and compare the results with experimental data to find an upper bound on the θ parameter. The most obvious natural phenomena to use in hunting for non-commutative effects are simple quantum mechanics systems, such as the hydrogen atom [12, 13, 14]. In the non-commutative space-time one expects the degeneracy of the initial spectral line to be lifted, thus one may say that non-commutativity plays the role of spin.

In this work we present an important contribution to the non-commutative approach to the hydrogen atom. Our goal is to solve the Klein-Gorden equation for the Coulomb potential in a non-commutative space-time up to second-order of the non-commutativity parameter using the Seiberg-Witten maps and the Moyal product. We thus find the non-commutative modification of the energy levels of the hydrogen atom and we show that the non-commutativity is the source of spin as a result of rotation of space-time dimensions.

This paper is organized as follows. In section 2 we derive the corresponding Seiberg-Witten maps up to the second order of θ for the various dynamical fields, and we propose an invariant action of the non-commutative charged scalar field in the presence of an electric field. In section 3, using the generalised Euler-Lagrange field equation, we derive the deformed Klein-Gordon (KG) equation. Applying these results to the hydrogen atom, we solve the deformed KG equation and obtain the non-commutative modification of the energy levels. The last section is devoted to a discussion.

2. Seiberg-Witten maps

Here we look for a mapping $\phi^A \to \hat{\phi}^A$ and $\lambda \to \hat{\lambda}(\lambda, A_\mu)$, where $\phi^A = (A_\mu, \varphi)$ is a generic field, A_μ and φ are the gauge and charged scalar fields respectively (the Greek and Latin indices denote curved and tangent space-time respectively), and λ is the U(1) gauge Lie-valued infinitesimal transformation parameter, such that:

$$\hat{\phi}^A(A) + \hat{\delta}_{\hat{\lambda}}\hat{\phi}^A(A) = \hat{\phi}^A(A + \delta_\lambda A), \tag{2}$$

where δ_λ is the ordinary gauge transformation and $\hat{\delta}_{\hat{\lambda}}$ is a noncommutative gauge transformation which are defined by:

$$\hat{\delta}_{\hat{\lambda}}\hat{\varphi} = i\hat{\lambda} * \hat{\varphi}, \qquad \delta_\lambda\varphi = i\lambda\varphi, \tag{3}$$

$$\hat{\delta}_{\hat{\lambda}}\hat{A}_\mu = \partial_\mu\hat{\lambda} + i\left[\hat{\lambda}, \hat{A}_\mu\right]_*, \qquad \delta_\lambda A_\mu = \partial_\mu\lambda. \tag{4}$$

In accordance with the general method of gauge theories, in the non-commutative space-time, using these transformations one can get at second order in the non-commutative parameter $\theta^{\mu\nu}$ (or equivalently θ) the following Seiberg–Witten maps [15]:

$$\hat{\varphi} = \varphi + \theta\varphi^1 + \theta^2\varphi^2 + \mathcal{O}(\theta^3), \tag{5}$$

$$\hat{\lambda} = \lambda + \theta\lambda^1(\lambda, A_\mu) + \theta^2\lambda^2(\lambda, A_\mu) + \mathcal{O}(\theta^3), \tag{6}$$

$$\hat{A}_\xi = A_\xi + \theta A_\xi^1(A_\xi) + \theta^2 A_\xi^2(A_\xi) + \mathcal{O}(\theta^3), \tag{7}$$

$$\hat{F}_{\mu\xi} = F_{\mu\xi}(A_\xi) + \theta F_{\mu\xi}^1(A_\xi) + \theta^2 F_{\mu\xi}^2(A_\xi) + \mathcal{O}(\theta^3), \tag{8}$$

where

$$F_{\mu\nu} = \partial_\mu A_\nu - \partial_\nu A_\mu. \tag{9}$$

To begin, we consider a non-commutative field theory with a charged scalar particle in the presence of an electrodynamic gauge field in a Minkowski space-time. We can write the action as:

$$\mathcal{S}=\int d^4x\left(\eta^{\mu\nu}\left(\hat{D}_\mu\hat{\varphi}\right)^\dagger*\hat{D}_\nu\hat{\varphi}+m^2\hat{\varphi}^\dagger*\hat{\varphi}-\frac{1}{4}\hat{F}_{\mu\nu}*\hat{F}^{\mu\nu}\right), \tag{10}$$

where the gauge covariant derivative is defined as: $\hat{D}_\mu\hat{\varphi}=\left(\partial_\mu+ie\hat{A}_\mu\right)*\hat{\varphi}$.

Next we use the generic-field infinitesimal transformations (3) and (4) and the star-product tensor relations to prove that the action in eq. (10) is invariant. By varying the scalar density under the gauge transformation and from the generalised field equation and the Noether theorem we obtain [16]:

$$\frac{\partial\mathcal{L}}{\partial\hat{\varphi}}-\partial_\mu\frac{\partial\mathcal{L}}{\partial\left(\partial_\mu\hat{\varphi}\right)}+\partial_\mu\partial_\nu\frac{\partial\mathcal{L}}{\partial\left(\partial_\mu\partial_\nu\hat{\varphi}\right)}-\partial_\mu\partial_\nu\partial_\sigma\frac{\partial\mathcal{L}}{\partial\left(\partial_\mu\partial_\nu\partial_\sigma\hat{\varphi}\right)}+\mathcal{O}\left(\theta^3\right)=0. \tag{11}$$

3. Non-commutative Klein-Gordon equation

In this section we study the Klein-Gordon equation for a Coulomb interaction $(-e/r)$ in the free non-commutative space-time. This means that we will deal with solutions of the U(1) gauge-free non-commutative field equations [17]. For this we use the modified field equations in eq. (11) and the generic field $\hat{A}_\mu$ so that:

$$\delta\hat{A}_\mu=\partial_\mu\hat{\lambda}-ie\hat{A}_\mu*\hat{\lambda}+ie\hat{\lambda}*\hat{A}_\mu, \tag{12}$$

and the free non-commutative field equations:

$$\partial^\mu\hat{F}_{\mu\nu}-ie\left[\hat{A}^\mu,\hat{F}_{\mu\nu}\right]_*=0. \tag{13}$$

Using the Seiberg-Witten maps (7)–(8) and the choice (13), we can obtain the following deformed Coulomb potential [17]:

$$\hat{a}_0=-\frac{e}{r}-\frac{e^3}{r^4}\theta^{0j}x_j+\frac{e^5}{2\,r^5}\left[\left(\theta^{0j}\right)^2-5\left(\frac{\theta^{0j}x_j}{r}\right)^2\right]+\mathcal{O}\left(\theta^3\right). \tag{14}$$

Using the modified field equation (11) and the generic field $\hat{\varphi}$ so that:

$$\delta_{\hat{\lambda}}\hat{\varphi}=i\hat{\lambda}*\hat{\varphi}, \tag{15}$$

the Klein-Gordon equation in a non-commutative space-time in the presence of the vector potential $\hat{A}_\mu$ can be cast into:

$$\left(\eta^{\mu\nu}\partial_\mu\partial_\nu-m^2\right)\hat{\varphi}+\left(ie\eta^{\mu\nu}\partial_\mu\hat{A}_\nu-e^2\eta^{\mu\nu}\hat{A}_\mu*\hat{A}_\nu+2ie\eta^{\mu\nu}\hat{A}_\mu\partial_\nu\right)\hat{\varphi}=0. \tag{16}$$

Now using the fact that:

$$\eta^{\mu\nu}\partial_\mu\partial_\nu=-\partial_0^2+\Delta, \tag{17}$$

and

$$2ie\eta^{\mu\nu}\hat{A}_\mu\partial_\nu=i\frac{2e^2}{r}\partial_0+2i\frac{e^4}{r^4}\theta^{0j}x_j\partial_0-i\frac{e^6}{r^5}\left[\left(\theta^{0j}\right)^2-5\left(\frac{\theta^{0j}x_j}{r}\right)^2\right]\partial_0, \tag{18}$$

and

$$-e^2\eta^{\mu\nu}\hat{A}_\mu*\hat{A}_\nu=\frac{e^4}{r^2}+2\frac{e^6}{r^5}\theta^{0j}x_j+\frac{e^8}{r^6}\left[\left(\theta^{0j}\right)^2-4\left(\frac{\theta^{0j}x_j}{r}\right)^2\right], \tag{19}$$

then the Klein-Gordon equation (16) up to $\mathcal{O}\left(\theta^3\right)$ takes the form:

$$\left[-\partial_0^2+\Delta-m_e^2+\frac{e^4}{r^2}+i\frac{2e^2}{r}\partial_0+2i\frac{e^4}{r^4}\theta^{0j}x_j\partial_0-i\frac{e^6}{r^5}\left[\left(\theta^{0j}\right)^2-5\left(\frac{\theta^{0j}x_j}{r}\right)^2\right]\partial_0 \right.$$
$$\left. +2\frac{e^6}{r^5}\theta^{0j}x_j+\frac{e^8}{r^6}\left[\left(\theta^{0j}\right)^2-4\left(\frac{\theta^{0j}x_j}{r}\right)^2\right]\right]\hat{\varphi}=0. \quad (20)$$

The solution to eq. (20) in spherical polar coordinates (r,θ,ϕ) takes the separable form [18]:

$$\hat{\varphi}(r,\theta,\phi,t)=\frac{1}{r}\hat{R}(r)\hat{Y}\left(\theta,\phi\right)\exp(-iEt). \quad (21)$$

Then eq. (20) reduces to the radial equation:

$$\left[\frac{d^2}{dr^2}-\frac{l(l+1)-e^4}{r^2}+\frac{2Ee^2}{r}+E^2-m_e^2+2E\frac{e^4}{r^4}\theta^{0j}x_j+2\frac{e^6}{r^5}\theta^{0j}x_j \right.$$
$$\left. -E\frac{e^6}{r^5}\left[\left(\theta^{0j}\right)^2-5\left(\frac{\theta^{0j}x_j}{r}\right)^2\right]+\frac{e^8}{r^6}\left[\left(\theta^{0j}\right)^2-4\left(\frac{\theta^{0j}x_j}{r}\right)^2\right]\right]\hat{R}(r)=0. \quad (22)$$

In eq. (22) the coulomb potential in non-commutative space-time appears within the perturbation terms:

$$H^{\theta}_{\text{pert}}=2E\frac{e^4}{r^4}\theta^{0j}x_j+2\frac{e^6}{r^5}\theta^{0j}x_j+5E\frac{e^6}{r^5}\left(\frac{\theta^{0j}x_j}{r}\right)^2-E\frac{e^6}{r^5}\left(\theta^{0j}\right)^2-4\frac{e^8}{r^6}\left(\frac{\theta^{0j}x_j}{r}\right)^2+\frac{e^8}{r^6}\left(\theta^{0j}\right)^2, \quad (23)$$

where the first term is the electric dipole–dipole interaction created by the non-commutativity, the second is the electric dipole–quadruple interactions, the third and fourth terms are the electric quadruple–quadruple interaction, and the last two terms are similar to the Van-der-Waals potential energy forces between two atoms. These interactions show us that the effect of space-time non-commutativity on the interaction of the electron and the proton is equivalent to an extension of two nucleus interactions at a considerable distance. This idea effectively confirms the presence of gravity at this level. To investigate the modification of the energy levels by eq. (23), we use the first-order perturbation theory. The spectrum of H_0 and the corresponding wave functions are well known and given by:

$$R_{nl}(r)=\sqrt{\frac{a}{n+\nu+1}}\left(\frac{n!}{\Gamma\left(n+2\nu+2\right)}\right)^{1/2}x^{\nu+1}e^{-x/2}L_n^{2\nu+1}(x), \quad (24)$$

where the relativistic energy levels are given by:

$$E=E_{n,l}=\frac{m_e\left(n+\frac{1}{2}+\sqrt{\left(l+\frac{1}{2}\right)^2-\alpha^2}\right)}{\left[\left(n+\frac{1}{2}\right)^2+\left(l+\frac{1}{2}\right)^2+2\left(n+\frac{1}{2}\right)\sqrt{\left(l+\frac{1}{2}\right)^2-\alpha^2}\right]^{\frac{1}{2}}}, \quad (25)$$

and $L_n^{2\nu+1}$ are the associated Laguerre polynomials [19], with the following notations:

$$\nu=-\frac{1}{2}+\sqrt{\left(l+\frac{1}{2}\right)^2-\alpha^2}, \qquad \alpha=e^2, \qquad a=\sqrt{m_e^2-E^2}. \quad (26)$$

4. Noncommutative corrections of the energy

Now to obtain the modification to the energy levels as a result of the terms (23) due to the non-commutativity of space-time we use perturbation theory. For simplicity, first of all, we take $\theta_i = \delta_{i3}\theta$ $(\theta^{0j}x_j = \theta r\cos\vartheta)$ and assume that the other components are all zero and also the fact that in the first-order perturbation theory the expectation value of $\cos\vartheta/r^3$, $\cos\vartheta/r^4$, $\cos\vartheta^2/r^5$, $\cos\vartheta^2/r^6$, $1/r^5$ and $1/r^6$ are as follows:

$$\langle\psi^0_{nlm} + \psi^1 \left| H^{\theta(1)}_{\text{pert}} + H^{\theta(2)}_{\text{pert}} \right| \psi^0_{nlm} + \psi^1\rangle, \tag{27}$$

where

$$H^{\theta(1)}_{\text{pert}} = \theta 2e^4 \left(E\frac{\cos\vartheta}{r^3} + e^2\frac{\cos\vartheta}{r^4} \right), \tag{28}$$

$$H^{\theta(2)}_{\text{pert}} = \theta^2 e^6 \left(5E\frac{\cos\vartheta^2}{r^5} - 4e^2\frac{\cos\vartheta^2}{r^6} - E\frac{1}{r^5} + e^2\frac{1}{r^6} \right), \tag{29}$$

$$\psi^0_{nlm} = R_{nl}(r)Y^m_l(\vartheta,\phi), \tag{30}$$

and

$$\psi^1 = \sum \frac{\langle\psi^0_{nlm} \left| H^{\theta(1)}_{\text{pert}} \right| \psi^0_{nkm}\rangle}{E_{nl} - E_{nk}} \psi^0_{nkm}. \tag{31}$$

By taking into account the fact that

$$\langle\psi^0_{nlm} \mid \cos\vartheta \mid \psi^0_{nl'm'}\rangle = B^{m'}_{l'+1}\delta_{l,l'+1}\delta_{m,m'} + B^{m'}_{l'}\delta_{l,l'-1}\delta_{m,m'}, \tag{32}$$

$$\begin{aligned} \langle\psi^0_{nlm} \mid \cos\vartheta^2 \mid \psi^0_{nl'm'}\rangle &= B^{m'}_{l'+1}B^{m'}_{l'+2}\delta_{l,l'+2}\delta_{m,m'} + B^{m'}_{l'}B^{m'}_{l'-1}\delta_{l'-2}\delta_{m,m'} + \\ &+ \left[B^{m'}_{l'+1}B^{m'}_{l'+1} + B^{m'}_{l'}B^{m'}_{l'} \right] \delta_{l'l}\delta_{mm'}, \end{aligned} \tag{33}$$

where

$$B^m_l = \sqrt{\frac{(l+m)(l-m)}{(2l+1)(2l-1)}}, \tag{34}$$

we have

$$\psi^1 = \theta^2 \frac{e^4 B^m_l}{E_{nl} - E_{nl-1}} \left[E_{nl}f(3) + e^2 f(4) \right] \psi^0_{nl-1m} + \theta^2 \frac{e^4 B^m_{l+1}}{E_{nl} - E_{nl+1}} \left[E_{nl}f(3) + e^2 f(4) \right] \psi^0_{nl+1m}. \tag{35}$$

where $f(k) = \langle r^{-k} \rangle$ and is completely defined by

$$\begin{aligned} \langle nlm \mid r^{-k} \mid nlm\rangle &= \int_0^\infty R^2_{nl}(r)r^{-k}dr \\ &= \frac{2^{k-1}a^k n!}{(n+\nu+1)\Gamma(n+2\nu+2)} \int_0^\infty x^{2\nu+2-k}e^{-x} \left[L^{2\nu+1}_n(x) \right]^2 dx \\ &= f(k) \qquad k = 3,4,5,6. \end{aligned} \tag{36}$$

We use the relation between the confluent hypergeometric function $F(-n;\nu+1;x)$ and the associated Laguerre polynomials $L^\nu_n(x)$, namely:

$$L^\nu_n(x) = \frac{\Gamma(n+\nu+1)}{\Gamma(n+1)\Gamma(\nu+1)} F(-n;\nu+1;x), \tag{37}$$

$$\int_0^\infty x^{\nu-1}e^{-x}\left[F(-n;\gamma;x)\right]^2dx = \frac{n!\Gamma(\nu)}{\gamma(\gamma+1)\cdots(\gamma+n-1)}\left\{1+\frac{n(\gamma-\nu-1)(\gamma-\nu)}{1^2\gamma}+\right.$$
$$+\frac{n(n-1)(\gamma-\nu-2)(\gamma-\nu-1)(\gamma-\nu)(\gamma-\nu+1)}{1^2 2^2\gamma(\gamma+1)}+\cdots$$
$$\left.\cdots+\frac{n(n-1)\cdots 1(\gamma-\nu-n)\cdots(\gamma-\nu+n-1)}{1^2 2^2\cdots n^2\gamma(\gamma+1)\cdots(\gamma+n-1)}\right\}. \quad (38)$$

Equation (36) becomes:

$$\begin{aligned}\langle nlm \mid r^{-3} \mid nlm'\rangle &= \int_0^\infty R_{nl}^2(r)r^{-3}dr\delta_{mm'} \\ &= \frac{4a^3 n!}{(n+\nu+1)\Gamma(n+2\nu+2)}\int_0^\infty x^{2\nu-1}e^{-x}\left[L_n^{2\nu+1}(x)\right]^2dx\delta_{mm'} \\ &= \frac{4a^3 n!}{(n+\nu+1)\Gamma(n+2\nu+2)}\left[\frac{\Gamma(n+2\nu+2)}{\Gamma(n+1)\Gamma(2\nu+2)}\right]^2\times \\ &\quad\times\int_0^\infty x^{2\nu-1}e^{-x}\left[F(-n;2\nu+2;x)\right]^2dx\delta_{mm'} \\ &= \frac{2a^3}{\nu(2\nu+1)(n+\nu+1)}\left\{1+\frac{n}{(\nu+1)}\right\}\delta_{mm'} = f(3), \qquad (39) \\ \langle nlm \mid r^{-4} \mid nlm'\rangle &= \frac{4a^4}{(2\nu-1)\nu(2\nu+1)(n+\nu+1)}\left[1+\frac{3n}{(\nu+1)}+\right. \\ &\quad\left.+\frac{3n(n-1)}{(\nu+1)(2\nu+3)}\right]\delta_{mm'} \\ &= f(4), \qquad (40) \\ \langle nlm \mid r^{-5} \mid nlm'\rangle &= \frac{4a^5}{(2\nu-1)(\nu-1)\nu(2\nu+1)(n+\nu+1)}\left[1+\frac{6n}{(\nu+1)}+\right. \\ &\quad\left.+\frac{15n(n-1)}{(\nu+1)(2\nu+3)}+\frac{5n(n-1)(n-2)}{(\nu+1)(2\nu+3)(\nu+2)}\right]\delta_{mm'} \\ &= f(5), \qquad (41) \\ \langle nlm \mid r^{-6} \mid nlm'\rangle &= \frac{8a^6}{(2\nu-3)(2\nu-1)(\nu-1)\nu(2\nu+1)(n+\nu+1)}\left[1+\frac{10n}{(\nu+1)}+\right. \\ &\quad+\frac{45n(n-1)}{(\nu+1)(2\nu+3)}+\frac{35n(n-1)(n-2)}{(\nu+1)(2\nu+3)(\nu+2)}+ \\ &\quad\left.+\frac{35n(n-1)(n-2)(n-3)}{2(\nu+1)(2\nu+3)(\nu+2)(2\nu+5)}\right]\delta_{mm'} \\ &= f(6). \qquad (42)\end{aligned}$$

The first-order correction term is:

$$\langle\psi_{nlm}^0\left|H_{\text{pert}}^{\theta(1)}\right|\psi^0\rangle = 0. \qquad (43)$$

Therefore, the non-commutativity of space-time to the first order has no effect , so we study the non-commutativity effects at second order.

To second order in θ, equation (27) can be written as:

$$2\langle\psi^1\left|H_{\text{pert}}^{\theta(1)}\right|\psi^0\rangle + \langle\psi_{nlm}^0\left|H_{\text{pert}}^{\theta(2)}\right|\psi^0\rangle. \qquad (44)$$

So we have

$$2\langle\psi^1\left|H_{\text{pert}}^{\theta(1)}\right|\psi^0\rangle = \theta^2 8e^8\left[\frac{(B_l^m)^2}{E_{nl}-E_{nl-1}}+\frac{\left(B_{l+1}^m\right)^2}{E_{nl}-E_{nl+1}}\right]\left[E_{nl}f(3)+e^2f(4)\right]^2, \tag{45}$$

and

$$\langle\psi_{nlm}^0\left|H_{\text{pert}}^{\theta(2)}\right|\psi^0\rangle = \theta^2 e^6\left\{\left[(B_l^m)^2+\left(B_{l+1}^m\right)^2\right]\left[5E_{nl}f(5)-4e^2f(6)\right]-E_{nl}f(5)+e^2f(6)\right\}. \tag{46}$$

Putting these results together one gets:

$$\begin{aligned}\Delta E^{\text{nc}}(n,l,m) &= \theta^2\alpha^3\left[8\alpha[\frac{(B_l^m)^2}{E_{nl}-E_{nl-1}}+\frac{\left(B_{l+1}^m\right)^2}{E_{nl}-E_{nl+1}}]\left[E_{nl}f(3)+e^2f(4)\right]^2\right.\\ &\left.+([(B_l^m)^2+\left(B_{l+1}^m\right)^2]\left[5E_{nl}f(5)-4e^2f(6)\right]-E_{nl}f(5)+e^2f(6))\right]. \end{aligned}\tag{47}$$

The energy shift is depends on magnetic quantum number, which clearly reflects the existence of spin. Furthermore it is worth noting that the correction terms containing θ^2 are very similar to the spin–spin coupling, thus the non-commutative parameter θ plays the role of spin and thus the degeneracy of levels is completely removed. The energy levels of the hydrogen atom in the framework of the non-commutative Klein- Gordon equation are:

$$\hat{E} = E_{n,l} + \Delta E^{\text{nc}}(n,l,m). \tag{48}$$

We showed that the energy-levels shift for $1S$ and $2S$ states are:

$$\Delta E_{1\text{S}}^{\text{nc}} = \theta^2\alpha^3\left(-E_{10}f_{1\text{S}}(5)+e^2f_{1\text{S}}(6)\right), \tag{49}$$

$$\Delta E_{2\text{S}}^{\text{nc}} = \theta^2\alpha^3\left(-E_{20}f_{2\text{S}}(5)+e^2f_{2\text{S}}(6)\right). \tag{50}$$

One can obtain a limit for θ by comparing the corrections to transition energies obtained using (49) and (50) with the experimental results from hydrogen spectroscopy. We take as test the levels $1S$ and $2S$ because we have the best experimental precision for the transition between them [20]:

$$f_{1\text{S}-2\text{S}} = (2446061102474851 \pm 34)\,\text{Hz}. \tag{51}$$

The non-commutative correction to this transition reads:

$$\delta E\,(1\text{S}-2\text{S}) = 636,737\theta^2\,(\text{MeV})^3. \tag{52}$$

Comparing with the result of the experimental value, the bound is found to be given by:

$$\theta \leq (8\text{GeV})^{-2}. \tag{53}$$

This is in agreement with other results presented for instance in ref [21]. Thus the experimental signature for space-time non-commutativity differs from that for space-space non-commutativity [11].

5. Conclusions

In this work we started from quantum relativistic charged scalar particle in a canonical non-commutative space-time to find the action which is invariant under the infinitesimal gauge transformation. By using the Seiberg-Witten maps and the Moyal product up to second order in the non-commutativity parameter θ, we derived the deformed Klein-Gorden equation for non-commutative Coulomb potential. By solving the deformed Klein-Gordon equation we found that the energy shift up to the second order of θ, is proportional to θ^2, thus we explicitly accounted for spin effects, resulting from the rotation of the time dimension in space. Hence we can say that the Klein-Gordon equation in non-commutative space-time at the second order in θ describes particles with spin. Thus we came to the conclusion that the non-commutative relativistic energy degeneracy is completely removed. This proofs that the noncommutative space-time is responsible for spin effects.

References

[1] Minwalla S , Van Raamsdonk M and Seiberg N 2000 *JHEP* **0002** 020
[2] Gomis J and Mehen T 2000 *Nucl. Phys. B* **591** 178
[3] Seiberg N, Susskind L and Toumbas N 2000 *JHEP* **0006** 044
[4] Doplicher S, Fredenhagen K and Roberts J E 1994 *Phys. Lett. B* **331** 39
[5] Doplicher S, Fredenhagen K and Roberts J E 1995 *Commun. Math. Phys.* **172** 187
[6] Yoneya T 2000 *Progr. Theor. Phys.* **103** 1081
[7] Doplicher S, Fredenhagen K, Roberts J 1994 *Phys. Lett. B* **331** 39
[8] Doplicher S, Fredenhagen K, Roberts J 1995 *Comm. Math. Phys.* **172** 187
[9] Dayi O F, Yapiskann. B 2002 *JHEP* **10** 022
[10] Bertolami O and Guisado L 2003 *JHEP* **0312** 013.
[11] Zaim S, Khodja L and Delenda Y 2011 *Int. J. Mod. Phys. A* **26** 4133
[12] Chaichian M, Sheikh-Jabbari M M and Tureanu A 2001 *Phys. Rev. Lett.* **86** 2716
[13] Adorno T C, Baldiotti M C, Chaichian M, Gitman D M, Tureanu A 2009 *Phys. Lett. B* **682** 235
[14] Motavalli H, Akbarieh A R 2010 *Mod. Phys. Lett. A* **25** 2523
[15] Seiberg N and Witten E 1999 *JHEP* **032** 9909
[16] Mebarki N, Zaim S, Khodja L and Aissaoui H 2008 *Phys. Scripta* **78** 045101
[17] Stern A 2008 *Phy. Rev. Lett.* **100** 061601
[18] Naudts J 2005 [arXiv:physics/0507193]
[19] Nikiforov A F and Uvarov V B 1988 *Special Functions of Mathematical Physics* (Basel: Birkhauser)
[20] T.W. Hänsch et al 2005 *Phil.Trans.R.Soc. A* **363** 2155
[21] Stern A 2008 *Phys.Rev.D* **78** 065006

Rabi oscillation in a damped rotating magnetic field: A path integral approach

Mekki Aouachria

Laboratoire de Physique Energétique Appliquée (LPEA), Département des Sciences de la Matiére, Faculté des Sciences, Université Hadj Lakhdar, Batna, Algeria

E-mail: mekkiaouachria69@gmail.com

Abstract. In this paper we consider a spin 1/2 particle interacting with a damped rotating magnetic field using path integral formalism. The propagator is first of all written in the standard form by replacing the spin by two fermionic oscillators via the Schwinger's model; then it is determined exactly thanks to the introduction of a particular rotations in coherent state space which has eliminated the rotation angle of the magnetic field and has simplified the Hamiltonian of the considered system. Thus, the Rabi formula are deduced.

1. Introduction

Up to now, a whole class of potentials have been treated successfully within the path-integral formalism, thanks to the use of certain transformations [1]. However, it is known that the most relativistic interactions are those where the spin is taken into account which is a very useful and very important notion in physics. From a practical point of view, the explicit calculus of propagators for such interactions by the path-integral formalism, has been discussed by many authors previously ([2], [3], [4], [5], [6], [7], [8], [9], [10], [11], [12], [13], [14]).

In this paper we are devoted to this type of interaction; by considering a problem treats according to usual quantum mechanics[16]. It acts of a spin 1/2 which interacts with a damped rotating magnetic field.

$$\mathbf{B}(t) = \left(B_1 e^{-\lambda t}\sin\omega t, \left(\frac{\omega}{g} - B_0\right)e^{-\lambda t} - \frac{\omega}{g}, B_1 e^{-\lambda t}\cos\omega t\right) \tag{1}$$

Its dynamics is described by the Hamiltonian

$$H = -\frac{g}{2}\sigma\mathbf{B} \tag{2}$$

where g is the gyromagnetic ratio. Then the Hamiltonian become

$$H = -\omega_1 e^{-\lambda t}\sin\omega t\sigma_x + \left(\left(\frac{\omega}{g} - B_0\right)e^{-\lambda t} - \frac{\omega}{g}\right)\sigma_y - \omega_1 e^{-\lambda t}\cos\omega t\sigma_z \tag{3}$$

where we have put $B_0 = \frac{\omega_0}{g}$ and $B_1 = \frac{2\omega_1}{g}$. The Pauli matrices are the following:

$$\sigma_z = \begin{pmatrix} 1 & 0 \\ 0 & -1 \end{pmatrix} \qquad \sigma_x = \begin{pmatrix} 0 & 1 \\ 1 & 0 \end{pmatrix} \qquad \sigma_y = \begin{pmatrix} 0 & -i \\ i & 0 \end{pmatrix}. \tag{4}$$

Considering this problem by the path integral approach, our motivation is the following. We show that for interaction with the coupling of spin-field type, the propagator is first, by construction, written in the standard form $\sum_{\text{path}} \exp\left(iS\left(path\right)/\hbar\right)$, where S is the action that describes the system, where the discrete variable relative to spin being inserted as the (continuous) path using fermionic coherent states. The knowledge of the propagator is essential to the determination of physical quantities such as the Rabi formula which is the aim of this paper.

The paper is organized as follows. In section 2, we give some notations and the necessary spin coherent state path integral for spin $\frac{1}{2}$ system for our further computations. In section 3, after setting up a path integral formalism for the propagator, we perform the direct calculations. The integration over the spin variables is easy to carry out thanks to simple transformations. The explicit result of the propagator is directly computed and the Rabi formula is then deduced. Finally, in section 4, we present our conclusions.

2. Coherent states formalism

Now, let us focus on some definitions, properties and notations needed for the further developments. As we are interested by the spin field interaction, we shall replace the Pauli matrices σ_i by a pair of fermionic operators (u, d) known as Schwinger fermionic model of spin following the recipe:

$$\sigma \longrightarrow \left(u^{\dagger}, d^{\dagger}\right) \sigma \begin{pmatrix} u \\ d \end{pmatrix} \tag{5}$$

where the pair (u, d) describes a two-dimensional fermionic oscillators.

Incidentally, the spin eigenstates $|\uparrow\rangle$ and $|\downarrow\rangle$ are generated from the fermionic vacuum state $|0,0\rangle$ by the action of the fermionic oscillators u^{+} and d^{+} following the relations

$$u^{+}|0,0\rangle = |\uparrow\rangle \qquad \textit{and} \qquad d^{+}|0,0\rangle = |\downarrow\rangle \tag{6}$$

where the action of u and d on this vacuum state is given by the vanishing results

$$u|0,0\rangle = 0 \qquad \textit{and} \qquad d|0,0\rangle = 0 \tag{7}$$

The pair of the fermionic oscillators (u, d) and its adjoint (u^{+}, d^{+}) satisfy the usual fermionic algebra defined by the following anticommutator relations:

$$\left[u, u^{+}\right]_{+} = 1, \qquad \left[d, d^{+}\right]_{+} = 1 \tag{8}$$

where all other anticommutators vanish.

The notation $[A, B]_{+}$ stands for

$$[A, B]_{+} = AB + BA.$$

Let us now introduce coherent states relative to this fermionic oscillators algebra. These states are generally defined as eigenvectors of the fermionic oscillators u and d:

$$u|\alpha, \beta\rangle = \alpha|\alpha, \beta\rangle, \qquad d|\alpha, \beta\rangle = \beta|\alpha, \beta\rangle \tag{9}$$

where (α, β) is a pair of Grasmmann variables which are anticommuting with fermionic oscillators and with themselves, namely

$$\begin{cases} [\alpha, u]_{+} = [\alpha, u^{+}]_{+} = [\alpha, d]_{+} = [\alpha, d^{+}]_{+} = 0 \\ [\beta, u]_{+} = [\beta, u^{+}]_{+} = [\beta, d]_{+} = [\beta, d^{+}]_{+} = 0 \end{cases} \tag{10}$$

and are commuting with vacuum states $|0,0\rangle\,, \langle 0,0|$:

$$\begin{cases} \alpha\,|0,0\rangle = |0,0\rangle\,\alpha, & \langle 0,0|\,\alpha = \alpha\,\langle 0,0| \\ \beta\,|0,0\rangle = |0,0\rangle\,\beta, & \langle 0,0|\,\beta = \beta\,\langle 0,0| \end{cases} \tag{11}$$

The above definitions are equivalent to the fact that these states are generated from the vacuum state according to the following relation

$$|\alpha,\beta\rangle = \exp\left(-\alpha u^{+} - \beta d^{+}\right)|0,0\rangle \tag{12}$$

The main properties of these states are:

-the completeness relation

$$\int d\bar{\alpha}d\alpha d\bar{\beta}d\beta e^{-\bar{\alpha}\alpha-\bar{\beta}\beta}\,|\alpha,\beta\rangle\ \langle\alpha,\beta| = 1 \tag{13}$$

-non-orthogonality

$$\langle\alpha,\beta \mid \alpha',\beta'\rangle = e^{\bar{\alpha}\alpha'+\bar{\beta}\beta'}. \tag{14}$$

3. Path integral formulation

At this stage we shall provide a path integral expression for the propagator for the Hamiltonian given by the expression (3). This can be readily done by exploiting the above model of the spin by which this Hamiltonian converts to the following fermionic form:

$$H = -\omega_1 e^{-\lambda t}\sin\omega t(u^{\dagger}d + d^{\dagger}u) + \left(\left(\frac{\omega}{g} - B_0\right)e^{-\lambda t} - \frac{\omega}{g}\right)(-iu^{\dagger}d + id^{\dagger}u)$$
$$-\omega_1 e^{-\lambda t}\cos\omega t(u^{\dagger}u - d^{\dagger}d) \tag{15}$$

Moreover, it is convenient to choose the quantum state as $|\alpha,\beta\rangle$ where (α,β) describes the spin variables. According to the habitual construction procedure of the path integral, we define the propagator as the matrix element of the evolution operator between the initial state $|\alpha_i,\beta_i\rangle$ and final state $|\alpha_f,\beta_f\rangle$

$$\mathbf{K}(\alpha_{f,}\beta_f;\alpha_{i,}\beta_i;T) = \langle\alpha_f,\beta_f \mid U(T) \mid \alpha_i,\beta_i\rangle \tag{16}$$

where

$$U(T) = \mathbf{T}_D \exp\left(-\frac{i}{\hbar}\int_0^T H(t)dt\right) \tag{17}$$

and $\mathbf{T}_D$ is the Dyson chronological operator.

To move to path integral representation, we first subdivide the time interval $[0,T]$ into $N+1$ intervals of length ε, intermediate moments, by using the Trotter's formula and we then introduce the projectors according to these intermediate instants N, which are regularly distributes between 0 and T in (17) ,we obtain the discretized path integral form of the propagator

$$\mathbf{K}(\alpha_{f,}\beta_f;\alpha_{i,}\beta_i;T) = \lim_{N\to\infty}\prod_{n=1}^{N}\int d\bar{\alpha}_n d\alpha_n d\bar{\beta}_n d\beta_n e^{-\bar{\alpha}_n\alpha_n-\bar{\beta}_n\beta_n}$$

$$\times\prod_{n=1}^{n=N+1}\exp\left\{\bar{\alpha}_n\alpha_{n-1} + \bar{\beta}_n\beta_{n-1} - i\frac{\epsilon}{2}\omega\left(-i\bar{\alpha}_n\beta_{n-1} + i\bar{\beta}_n\alpha_{n-1}\right)\right.$$

$$+\,i\frac{\epsilon}{2}g\left(\frac{\omega}{g} - B_0\right)e^{-\lambda t}\left(-i\bar{\alpha}_n\beta_{n-1} + i\bar{\beta}_n\alpha_{n-1}\right)$$

$$+i\epsilon\omega_1 e^{-\lambda t_n}\left[\cos\omega t_n\left(\bar{\alpha}_n\alpha_{n-1}-\bar{\beta}_n\beta_{n-1}\right)+\sin\omega t_n\left(\bar{\alpha}_n\beta_{n-1}+\bar{\beta}_n\alpha_{n-1}\right)\right]\Big\} \tag{18}$$

with $(\alpha_0,\beta_0)=(\alpha_i,\beta_i)$ and $\left(\bar{\alpha}_{N+1},\bar{\beta}_{N+1}\right)=\left(\bar{\alpha}_f,\bar{\beta}_f\right)$. This last expression represents the path integral of the propagator which has been the purpose subject of previous papers ([9],[10]), and has the advantage that it permits us to perform explicitly some concrete calculations.

4. Calculation of the propagator

To begin, we first introduce new Grassmann variables via an unitary transformation in spin coherent space which eliminates the angle ωt present in the expression of the magnetic field:

$$\left\{\begin{array}{c}(\alpha_n,\beta_n)\mapsto(\eta_n,\xi_n)\\ \begin{pmatrix}\alpha_n\\ \beta_n\end{pmatrix}=e^{-i\frac{\omega t_n}{2}\sigma_y}\begin{pmatrix}\eta_n\\ \xi_n\end{pmatrix}\end{array}\right. \quad and \quad \left\{\begin{array}{c}(\bar{\alpha}_n,\bar{\beta}_n)\mapsto(\bar{\eta}_n,\bar{\xi}_n)\\ (\bar{\alpha}_n,\bar{\beta}_n)=(\bar{\eta}_n,\bar{\xi}_n)\,e^{i\frac{\omega t_n}{2}\sigma_y}\end{array}\right. \tag{19}$$

Then, it is easy to show that the measure and the infinitesimal action become respectively

$$\prod_{n=1}^{N}\left(d\bar{\alpha}_n d\alpha_n d\bar{\beta}_n d\beta_n e^{-\bar{\alpha}_n\alpha_n-\bar{\beta}_n\beta_n}\right)=\prod_{n=1}^{N}\left(d\bar{\eta}_n d\eta_n d\bar{\xi}_n d\xi_n e^{-\bar{\eta}_n\eta_n d-\bar{\xi}_n\xi_n}\right) \tag{20}$$

$$\bar{\alpha}_n\alpha_{n-1}+\bar{\beta}_n\beta_{n-1}=\left(\bar{\eta}_n,\bar{\xi}_n\right)\left[1+i\frac{\omega}{2}\epsilon\sigma_y+O(\epsilon^2)\right]\begin{pmatrix}\eta_{n-1}\\ \xi_{n-1}\end{pmatrix}$$

$$=\left(\bar{\eta}_n,\bar{\xi}_n\right)\begin{pmatrix}\eta_{n-1}\\ \xi_{n-1}\end{pmatrix}+i\frac{\omega}{2}\epsilon\left(\bar{\eta}_n,\bar{\xi}_n\right)\sigma_y\begin{pmatrix}\eta_{n-1}\\ \xi_{n-1}\end{pmatrix} \tag{21}$$

and

$$i\epsilon\omega_1 e^{-\lambda t_n}\left[\sin\omega t_n\left(\bar{\alpha}_n\beta_{n-1}+\bar{\beta}_n\alpha_{n-1}\right)+\cos\omega t_n\left(\bar{\alpha}_n\alpha_{n-1}-\bar{\beta}_n\beta_{n-1}\right)\right]$$

$$=i\epsilon\omega_1 e^{-\lambda t_n}\left(\bar{\eta}_n\eta_{n-1}-\bar{\xi}_n\xi_{n-1}\right). \tag{22}$$

The propagator in function of the new Grassmann variables η and ξ, becomes

$$\mathbf{K}(\alpha_f,\beta_f;\alpha_i,\beta_i;T)=\lim_{N\to\infty}\prod_{n=1}^{N}\left(d\bar{\eta}_n d\eta_n d\bar{\xi}_n d\xi_n e^{-\bar{\eta}_n\eta_n-\bar{\xi}_n\xi_n}\right)$$

$$\prod_{n=1}^{N+1}\exp\left[\bar{\eta}_n\eta_n+\bar{\xi}_n\xi_n+i\epsilon\omega_1 e^{-\lambda t_n}\left(\bar{\eta}_n\eta_{n-1}-\bar{\xi}_n\xi_{n-1}\right)\right.$$

$$\left.+i\frac{\epsilon}{2}e^{-\lambda t_n}\left(\omega-\omega_0\right)\left(i\bar{\eta}_n\xi_{n-1}+i\bar{\xi}_n\eta_{n-1}\right)\right] \tag{23}$$

Now using the following transformation:

$$\epsilon=e^{\lambda t_n}\tau \quad with \quad \tau=s_n-s_{n-1} \tag{24}$$

The propagator becomes

$$\mathbf{K}(\alpha_f,\beta_f;\alpha_i,\beta_i;T)=\lim_{N\to\infty}\prod_{n=1}^{N}\left(d\bar{\eta}_n d\eta_n d\bar{\xi}_n d\xi_n e^{-\bar{\eta}_n\eta_n-\bar{\xi}_n\xi_n}\right)$$

$$\times\prod_{n=1}^{N+1}\exp\left\{\bar{\eta}_n\eta_n+\bar{\xi}_n\xi_n+i\tau\left(\bar{\eta}_n,\bar{\xi}_n\right)\left[\omega_1\sigma_z+\frac{1}{2}(\omega-\omega_0)\sigma_y\right]\begin{pmatrix}\eta_{n-1}\\ \xi_{n-1}\end{pmatrix}\right\} \tag{25}$$

which can be rewritten in the following form

$$\mathbf{K}(\alpha_f,\beta_f;\alpha_i,\beta_i;T) = \lim_{N\to\infty}\prod_{n=1}^{N}\left(d\bar{\eta}_n d\eta_n d\bar{\xi}_n d\xi_n e^{-\bar{\eta}_n\eta_n-\bar{\xi}_n\xi_n}\right)\prod_{n=1}^{N+1}\exp\left[\bar{\eta}_n\eta_{n-1}\right.$$

$$\left.+\bar{\xi}_n\xi_{n-1}+i\tau\sqrt{\omega_1^2+\frac{1}{4}(\omega-\omega_0)^2}\left(\bar{\eta}_n,\bar{\xi}_n\right)\left[\cos\theta\sigma_z+\sin\theta\sigma_y\right]\begin{pmatrix}\eta_{n-1}\\ \xi_{n-1}\end{pmatrix}\right] \tag{26}$$

with

$$\cos\theta = \frac{2\omega_1}{\sqrt{4\omega_1^2+(\omega-\omega_0)^2}} \tag{27}$$

then we introduce new Grassmann variables γ,δ via an unitary transformation in spin coherent state space defined by

$$\begin{cases}(\eta_n,\xi_n)\longmapsto(\gamma_n,\delta_n)\\ \begin{pmatrix}\eta_n\\ \xi_n\end{pmatrix}=e^{i\frac{\theta}{2}\sigma_x}\begin{pmatrix}\gamma_n\\ \delta_n\end{pmatrix}\end{cases}\quad \textit{and}\quad \begin{cases}(\bar{\eta}_n,\bar{\xi}_n)\longmapsto(\bar{\gamma}_n,\bar{\delta}_n)\\ (\bar{\eta}_n,\bar{\xi}_n)=(\bar{\gamma}_n,\bar{\delta}_n)e^{-i\frac{\theta}{2}\sigma_x}\end{cases} \tag{28}$$

The role of this unitary transformation is twofold. First, it eliminates the angle θ present in the expression of the action , and secondly it diagonalizes the Hamiltonian. In effect the measure and the infinitesimal action become respectively

$$\prod_{n=1}^{N}\left(d\bar{\eta}_n d\eta_n d\bar{\xi}_n d\xi_n e^{-\bar{\eta}_n\eta_n-\bar{\xi}_n\xi_n}\right)=\prod_{n=1}^{N}\left(d\bar{\gamma}_n d\gamma_n d\bar{\delta}_n d\delta_n e^{-\bar{\gamma}_n\gamma_n-\bar{\delta}_n\delta_n}\right) \tag{29}$$

$$\bar{\eta}_n\eta_{n-1}+\bar{\xi}_n\xi_{n-1}=\bar{\gamma}_n\gamma_{n-1}+\bar{\delta}_n\delta_{n-1} \tag{30}$$

and

$$i\epsilon\sqrt{\omega_1^2+\frac{1}{4}(\omega-\omega_0)^2}\left(\bar{\eta}_n,\bar{\xi}_n\right)\left[\cos\theta\sigma_z+\sin\theta\sigma_y\right]\begin{pmatrix}\eta_{n-1}\\ \xi_{n-1}\end{pmatrix}=$$

$$=i\epsilon\sqrt{\omega_1^2+\frac{1}{4}(\omega-\omega_0)^2}(\bar{\gamma}_n,\bar{\delta}_n)\sigma_z\begin{pmatrix}\gamma_{n-1}\\ \delta_{n-1}\end{pmatrix} \tag{31}$$

which modify the expression (27) to the following form

$$\mathbf{K}(\gamma_f,\delta_f;\gamma_i,\delta_i;T) = \lim_{N\to\infty}\prod_{n=1}^{N}d\bar{\gamma}_n d\gamma_n d\bar{\delta}_n d\delta_n e^{-\bar{\gamma}_n\gamma_n-\bar{\delta}_n\delta_n}$$

$$\prod_{n=1}^{N+1}\exp\left\{\bar{\gamma}_n\gamma_{n-1}+\bar{\delta}_n\delta_{n-1}+i\epsilon\sqrt{\omega_1^2+\frac{1}{4}(\omega-\omega_0)^2}(\bar{\gamma}_n,\bar{\delta}_n)\sigma_z\begin{pmatrix}\gamma_{n-1}\\ \delta_{n-1}\end{pmatrix}\right\} \tag{32}$$

Having finished with the diagonalization, we are left with the task of integrating over the Grassmann variables γ,δ. For this, it is useful to write again the propagator in an appropriate form.

We note by $\mathbf{q}_n=\begin{pmatrix}\gamma_n\\ \delta_n\end{pmatrix}$ and $\bar{\mathbf{q}}_n=\left(\bar{\gamma}_n,\bar{\delta}_n\right)$, so the propagator (33) becomes

$$\mathbf{K}(\gamma_f,\delta_f;\gamma_i,\delta_i;T) = \lim_{N\to\infty}\int\prod_{n=1}^{n=N}d\bar{\mathbf{q}}_n d\mathbf{q}_n\exp\left\{\sum_{n=1}^{n=N}-\bar{\mathbf{q}}_n\mathbf{q}_n+\right.$$

$$+\sum_{n=1}^{n=N+1}\bar{\mathbf{q}}_n\mathbf{R}(n)\mathbf{q}_{n-1}\Bigg\}, \tag{33}$$

where $\mathbf{R}(n)$ is a diagonal matrix

$$\mathbf{R}(n)=\begin{pmatrix} 1+i\epsilon\sqrt{\omega_1^2+\frac{1}{4}\left(\omega-\omega_0\right)^2} & 0 \\ 0 & 1-i\epsilon\sqrt{\omega_1^2+\frac{1}{4}\left(\omega-\omega_0\right)^2} \end{pmatrix} \tag{34}$$

we note that the action can be re-arranged as follows

$$\sum_{n=1}^{n=N}-\bar{\mathbf{q}}_n\mathbf{q}_n+\sum_{n=1}^{n=N+1}\bar{\mathbf{q}}_n\mathbf{R}(n)\mathbf{q}_{n-1}=$$

$$=\sum_{j,n=1}^{n=N}-\left[\bar{\mathbf{q}}_j(\delta_{j,n}-\mathbf{R}(n)\delta_{j,n+1})\mathbf{q}_n\right]+\bar{\mathbf{q}}_{N+1}\mathbf{R}(N+1)\mathbf{q}_N+\bar{\mathbf{q}}_1\mathbf{R}(1)\mathbf{q}_0. \tag{35}$$

In order to rewrite the propagator in an adequate form, we introduce the vectors

$$\mathbf{P}=\begin{pmatrix}\mathbf{q}_1\\ \cdot\\ \cdot\\ \cdot\\ \mathbf{q}_N\end{pmatrix},\quad \mathbf{V}=\begin{pmatrix}\mathbf{R}(1)\mathbf{q}_0\\ 0\\ \cdot\\ \cdot\\ \cdot\end{pmatrix},\quad \mathbf{W}^\dagger=\begin{pmatrix}0, & . & . & . & ,\bar{\mathbf{q}}_{N+1}\mathbf{R}(N+1)\end{pmatrix}, \tag{36}$$

the propagator, thus, takes the following form:

$$\mathbf{K}(\gamma_f,\delta_f;\gamma_i,\delta_i;T)=\int d\mathbf{P}^\dagger d\mathbf{P}\exp\left(-\mathbf{P}^\dagger\mathcal{M}\mathbf{P}+\mathbf{P}^\dagger\mathbf{V}+\mathbf{W}^\dagger\mathbf{P}\right) \tag{37}$$

where $\mathcal{M}$ is a matrix defined by

$$\mathcal{M}_{j,n}=\delta_{j,n}-\mathbf{R}(n)\delta_{j,n+1}\qquad n,j\in[1,N] \tag{38}$$

With the help of the following change

$$\mathbf{P}\longrightarrow\mathbf{P}+\mathcal{M}^{-1}\mathbf{V}\ \ and\ \ \mathbf{P}^\dagger\longrightarrow\mathbf{P}^\dagger+\mathbf{W}^\dagger\mathcal{M}^{-1} \tag{39}$$

the integral over the Grassmann variables is reduced to

$$\mathbf{K}(\gamma_f,\delta_f;\gamma_i,\delta_i;T)=\det\mathcal{M}\cdot\exp(\mathbf{W}^\dagger\mathcal{M}^{-1}\mathbf{V})$$

As $\det\mathcal{M}$ equals 1 and

$$\mathbf{W}^\dagger\mathcal{M}^{-1}\mathbf{V}=\lim_{N\longrightarrow\infty}\left[\bar{\mathbf{q}}_f\mathbf{R}(N+1)\mathbf{R}(N)\cdots\mathbf{R}(1)\mathbf{q}_0\right]=\bar{\mathbf{q}}_f\mathbf{R}(T)\mathbf{q}_0, \tag{40}$$

thus the propagator reduce to following form

$$\mathbf{K}(\gamma_f,\delta_f;\gamma_i,\delta_i;T)=\left(\bar{\gamma}_f,\bar{\delta}_f\right)\mathbf{R}(T)\begin{pmatrix}\gamma_i\\ \delta_i\end{pmatrix} \tag{41}$$

with

$$\mathbf{R}(S)=e^{\frac{i}{2}S\sqrt{4\omega_1^2+(\omega-\omega_0)^2}\sigma_z}. \tag{42}$$

Now we come back to the old Grassmann variables (α,β). So, the exact expression of the propagator concerning to our problem is the following

$$\mathbf{K}(\alpha_f,\beta_f;\alpha_i,\beta_i;T)=\exp\left(\bar{\alpha}_f,\bar{\beta}_f\right)M\left(S\right)\begin{pmatrix}\alpha_i\\ \beta_i\end{pmatrix} \tag{43}$$

where M is a matrix defined by

$$M\left(S\right)=e^{-i\frac{\omega T}{2}\sigma_y}e^{i\frac{\theta}{2}\sigma_x}\mathbf{R}(S)e^{-i\frac{\theta}{2}\sigma_x} \tag{44}$$

5. The transition probability

If, for convenience, we specify the initial and final states as the eigenstates of σ_y, the transition amplitude from the initial state $|m_i\rangle = |\uparrow\rangle_y = \frac{1}{\sqrt{2}}(|\uparrow\rangle + i\,|\downarrow\rangle)$ to the final state $|m_i\rangle = |\downarrow\rangle_y = \frac{1}{\sqrt{2}}(|\uparrow\rangle - i\,|\downarrow\rangle)$ is related to $K(f, i; T)$ by :

$$K(\downarrow,\uparrow;T) ={}_y \langle\downarrow|U(T)\,|\uparrow\rangle_y \tag{45}$$

With the help of the completeness relations, this amplitude becomes

$$K(\downarrow,\uparrow;T) = \int d\bar{\alpha}_f d\alpha_f d\overline{\beta}_f d\beta_f d\bar{\alpha}_i d\alpha_i d\overline{\beta}_i d\beta_i e^{-\bar{\alpha}_f\alpha_f-\overline{\beta}_f\beta_f} e^{-\bar{\alpha}_i\alpha_i-\overline{\beta}_i\beta_i}$$

$$\times_y\langle\downarrow|\,\alpha_f,\beta_f\rangle\langle\alpha_i,\beta_i\,|\uparrow\rangle_y\mathbf{K}(\alpha_f,\beta_f,\alpha_i,\beta_i;T). \tag{46}$$

We substitute $|\downarrow\rangle_y$ and $|\uparrow\rangle_y$ in (46) then

$$K(\downarrow,\uparrow;T) = \frac{1}{2}\int d\bar{\alpha}_f d\alpha_f d\bar{\beta}_f d\beta_f d\bar{\alpha}_i d\alpha_i d\bar{\beta}_i d\beta_i e^{-\bar{\alpha}_i\alpha_i-\bar{\beta}_i\beta_i} e^{-\bar{\alpha}_f\alpha_f-\bar{\beta}_f\beta_f}$$

$$\{\langle\uparrow|\,\alpha_f,\beta_f\rangle\,\langle\alpha_i,\beta_i|\uparrow\rangle - \langle\downarrow|\,\alpha_f,\beta_f\rangle\,\langle\alpha_i,\beta_i|\downarrow\rangle +$$
$$i\left(\langle\downarrow|\,\alpha_f,\beta_f\rangle\,\langle\alpha_i,\beta_i|\uparrow\rangle + \langle\downarrow|\,\alpha_f,\beta_f\rangle\,\langle\alpha_i,\beta_i|\downarrow\rangle\right)\}$$
$$\times\exp\left(\bar{\alpha}_f,\bar{\beta}_f\right)\begin{pmatrix} M_{11}(T) & M_{12}(T) \\ M_{21}(T) & M_{22}tT \end{pmatrix}\begin{pmatrix}\alpha_i \\ \beta_i\end{pmatrix} \tag{47}$$

where $M_{ij}(T)$ are the elements of the matrix $M(T)$ which can be written as sum of four terms

$$K(\downarrow,\uparrow;T) = K_1(\downarrow,\uparrow;T) + K_2(\downarrow,\uparrow;T) + K_3(\downarrow,\uparrow;T) + K_3(\downarrow,\uparrow;T) \tag{48}$$

We calculate for instance the first term as [17]:

$$\langle\uparrow|\,\alpha_f,\beta_f\rangle = \alpha_f \quad , \quad \langle\alpha_i,\beta_i\,|\uparrow\rangle = \bar{\alpha}_i \;\; and \qquad \alpha_f\bar{\alpha}_i = e^{-\bar{\alpha}_i\alpha_f} - 1 \tag{49}$$

Thus

$$K_1(\downarrow,\uparrow;T) = \frac{1}{2}\int d\overline{\alpha}_f d\alpha_f d\overline{\beta}_f d\beta_f d\overline{\alpha}_i d\alpha_i d\overline{\beta}_i d\beta_i e^{\overline{\alpha}_f\alpha_f+\overline{\beta}_f\beta_f} e^{\overline{\alpha}_i\alpha_i+\overline{\beta}_i\beta_i}$$

$$\left\{\exp\left[\overline{\alpha}_f\mathbf{M}_{11}\alpha_i + \overline{\alpha}_f\mathbf{M}_{12}\beta_i + \overline{\beta}_f\mathbf{M}_{21}\alpha_i + \overline{\beta}_f\mathbf{M}_{22}\beta_i - \bar{\alpha}_i\alpha_f\right]\right.$$
$$\left. - \exp\left[\overline{\alpha}_f\mathbf{M}_{11}\alpha_i + \overline{\alpha}_f\mathbf{M}_{12}\beta_i + \overline{\beta}_f\mathbf{M}_{21}\alpha_i + \overline{\beta}_f\mathbf{M}_{22}\beta_i\right]\right\}, \tag{50}$$

and with the help of the four-component vector ν is defined by

$$\nu = \begin{pmatrix}\alpha_i \\ \alpha_f \\ \beta_i \\ \beta_f\end{pmatrix}, \qquad \nu^\dagger = \begin{pmatrix}\overline{\alpha}_i & \overline{\alpha}_f & \overline{\beta}_i & \overline{\beta}_f\end{pmatrix} \tag{51}$$

Accordingly, the expression (50) is reduced to

$$K_1(\downarrow,\uparrow;T) = \int d\nu^\dagger d\nu.\left[\exp\nu^\dagger G'\nu - \exp\nu^\dagger G\nu\right]$$

$$= \left[\det G' - \det G\right], \tag{52}$$

where G and G' are the two following 4×4 matrices

$$G=\begin{pmatrix}-1&0&0&0\\ \mathbf{M}_{11}&-1&\mathbf{M}_{11}&0\\ 0&0&-1&0\\ \mathbf{M}_{21}&0&\mathbf{M}_{22}&-1\end{pmatrix}\qquad and\qquad G^{'}=\begin{pmatrix}-1&-1&0&0\\ \mathbf{M}_{11}&-1&\mathbf{M}_{11}&0\\ 0&0&-1&0\\ \mathbf{M}_{21}&0&\mathbf{M}_{22}&-1\end{pmatrix}. \tag{53}$$

A simple computation shows that $\det G'=1+\mathbf{M}_{11}$ and $\det G=1$. Consequently,

$$K_1(\downarrow,\uparrow;T)=\frac{1}{2}\mathbf{M}_{11}. \tag{54}$$

In the same manner, we proceed for the remaining elements. The result takes the following expression

$$K(\downarrow,\uparrow;T)=\frac{1}{2}\left[\mathbf{M}_{11}(T)-\mathbf{M}_{22}(T)+i\left(\mathbf{M}_{12}(T)+\mathbf{M}_{21}(T)\right)\right] \tag{55}$$

By simple calculus the transition probability is then given by

$$P_{\downarrow\uparrow}=\left|K(\downarrow,\uparrow;T)\right|^2$$

$$P_{\downarrow\uparrow}=\frac{4\omega_1^2}{4\omega_1^2+(\omega-\omega_0)^2}\sin^2\frac{S(T)}{2}\sqrt{4\omega_1^2+(\omega-\omega_0)^2},\quad with\ S(T)=\frac{1-e^{-\lambda T}}{\lambda} \tag{56}$$

Thus

$$P_{\downarrow\uparrow}=\frac{4\omega_1^2}{4\omega_1^2+(\omega-\omega_0)^2}\sin^2\frac{1-e^{-\lambda T}}{2\lambda}\sqrt{4\omega_1^2+(\omega-\omega_0)^2} \tag{57}$$

This formula is exactly the well-known Rabi formula given in the literature[16].

6. Conclusion

By using the formalism of the path integral and the fermionic coherent states approach, we are able to calculate the explicit expression of the propagator relative to spin 1/2 interacting with damped rotating magnetic field. To treat the spin dynamics, we have used the Schwinger's recipe which replaces the Pauli matrices by a pair of fermionic oscillators. The introduction of a particular rotations in coherent state space has eliminated the rotation angle of the magnetic field and has then been simplified somewhat the Hamiltonian of the considered system. As a consequence, we are able to integrate over the spin variables described by fermionic oscillators. The exactness of the result is displayed in the evaluation of the Rabi formula.

Acknowledgments

This work was supported by PNR research project 8/u05/4094 N^0 31/13.

References

[1] Grosche C and Steiner F 1998 *Handbook of Feynman Path Integrals* (Verlag: Springer)
[2] Boudjedaa T, Bounames A, Nouicer Kh, Chetouani L and Hammann T F 1995 *J. Math. Phys.* **36** 1602
[3] Boudjedaa T, Bounames A, Nouicer Kh, Chetouani L and Hammann T F 1996 *Phys. Scripta.* **54** 225
[4] Boudjedaa T, Bounames A, Nouicer Kh, Chetouani L and Hammann T F 1998 *Phys. Scripta.* **56** 545
[5] Merdaci A, Boudjedaa T and Chetouani L 2001 *Phys. Scripta.* **64** 15
[6] Merdaci A, Boudjedaa T and Chetouani L 2001 *Czech. J. Phys.* **51** 865
[7] Merdaci A, Boudjedaa T and Chetouani L 2001 *Eur. Phys. J. C* **22** 585
[8] Nouicer Kh and Chetouani L 2001 *Phys. Lett. A* **281** 218
[9] Aouachria M and Chetouani L 2002 *Eur. Phys. J.* **25** 333
[10] Aouachria M and Chetouani L 2002 *Chinese. J. Phys.* **40** 496

[11] Aouachria M and Chetouani L 2009 *Can. J. Phys.* **87** 389
[12] Aouachria M 2011 *Can. J. Phys.* **89** 1141
[13] Aouachria M 2011 *J. Korean. Phys. Soc.* **58** 689
[14] Aouachria M 2011 *Chinese. J. Phys.* **49** 689
[15] Aouachria M and Rekik R 2012 AIP Conf. Proc. **1444** 265
[16] Tahmasebi M J and Sobouti Y 1991 *Mod. Phys. Lett.* **B 5** 1919
[17] Nakamura M and Kazuo Kitahara K 1991 *J. Phys. Soc. of Japan.* **60** 1388

Journal of Physics: Conference Series **435** (2013) 012022 doi:10.1088/1742-6596/435/1/012022

Planar Microstrip Slot Antenna for S & C band Wireless Applications

P A Ambresh[1, a], P M Hadalgi [2, b] and P V Hunagund[3, c]

[1, 2, 3] Microwave Research Laboratory, Department Of P.G. Studies & Research in Applied Electronics,Gulbarga University, Gulbarga-585106 (INDIA).
E-mail: [a]ambreshpa@rediffmail.com, [b]hadalgipm@rediffmail.com, [c]prabhakar_hunagund@yahoo.co.in

Abstract. In this paper, compact planar microstrip antennas comprising of broad slots for enhancing various antenna parameters are discussed and presented. The proposed microstrip slot antennas achieved a compactness of maximum 58 % with a peak gain of 7.58 dB. These antennas can be operated for S band applications such as WiMax, operating in the frequency range of 3.3 – 3.6 GHz, RADAR, WLAN, fixed satellite services and maritime mobile services etc. covering 2 - 6 GHz frequency range. The antennas can be used as a compact antenna system where limited size is a foremost requirement. Results also show the satisfactory performance with the dual band frequency characteristics. Details of the antenna design structure, results of return loss (RL), impedance bandwidth and radiation pattern with co-cross polarization are also given

1.Introduction

In present day scenario, the significance in the design and developing the compact, inexpensive planar antennas with wide bandwidths are fuelled by rapid developments of various communication services and other wireless services. Hence, compact microstrip slot antennas have found increased usage in wireless applications specifically in S and C band wireless frequency range.

Several researchers have focused on improving various antenna parameters such as impedance bandwidth enhancement [1] and size reduction [2] of microstrip antennas. Bandwidth enhancement techniques that have been studied for planar antennas include the use of a thick substrates, parasitic patches, E-slot patch, H-shaped patch, U-shaped slot patch, T-shaped parasitic strip, meandered slot and slit loaded and shorting pins [3]. Also, several methods for reducing the size (compact) of the microstrip antenna have already been studied [4, 5, 6, 7, 8 and 9]. It is known that, the resonant frequency of a patch antenna is inversely proportional to dielectric constant. It is also possible to reduce the resonant frequency by using a substrate with high dielectric constant [10]. The rectangular U-slot patch antenna with a folded patch feed [11] and E-H shaped patch antenna with L-probe feed had been studied practically [12]. By cutting a square slot at the centre of a rectangular microstrip patch, both compactness and dual frequency operation can be achieved [13].

Corresponding author: [1]ambreshpa@rediffmail.com

By loading a pair of narrow slots parallel and close to the radiating edges of a bow-tie patch, dual-frequency operation with tunable frequency-ratio can be achieved [14].

A circularly polarized, dual-frequency, slotted square patch with probe feeding mechanism is reported to operate as the telemetry, telecommand and control (TT&C) antenna for satellite spacecrafts at 2.25 GHz and 3.0 GHz [15]. A compact patch design for circular polarization application based on third iteration minkowski-like pre fractal geometry to be used in dual band application was reported by Ali Jalal [16]. Two short-circuited microstrip patch antennas can radiate at dual frequencies was reported by Roy and Thomos [17]. A trade-off is taken among these techniques and optimized geometries are developed in this study. This paper demonstrates a technique to achieve both bandwidth enhancement and reduction of antenna structural profile. It is achieved by a novel design of wide slots along the radiating edges of the microstrip patch fabricated on low cost FR4 dielectric material.

2. Design structure of planar microstrip slot antennas

Figure 1 (a), (b) and (c) illustrate the fabricated planar microstrip slot antennas (MSSA) with length L = 17.76 mm, width W = 23.28 mm etched on FR4 dielectric material with dielectric constant $\epsilon_r = 4.4$ and thickness h = 1.6 mm were designed using computer software AutoCAD-2010. Because, the dimensions of the patch are finite along the length and width, the fields at the edges of the patch undergo fringing. Most of the electric field lines reside in the substrate and the part of some lines exist in air. Since some of the waves travel in the substrate and some in air, an effective dielectric constant ϵ_{eff} is introduced to account for fringing. The amount of fringing is a function of the dimensions of the patch and the thickness of the substrate. The designed frequency of the patch antenna is 3.85 GHz. The RF power is fed to the radiating element (patch) by using the coax feed technique. It is a very common technique used for exciting microstrip antenna. The inner conductor of the coaxial feed connector extends through the dielectric and is soldered to the radiating element, whereas the outer conductor is connected to the ground plane. The main advantage of this type of feeding technique is easy to fabricate, has low spurious radiation and the feed can be placed at any desired location on the patch to match with its input impedance [18]. The coax feed is placed on the axis of symmetry to provide a good impedance matching. The feed position has been chosen at a distance $X_f = L/2[\epsilon_{eff}(\mathrm{sqrt}(L))]^{1/2}$, as given by M. Kara [19]. The copper plate is used as ground plane with dimension 35 x 35 mm having thickness 0.16 cm. The pair of slots are etched along the radiating and non - radiating edges of the radiating element, which is finally structured as microstrip slot antennas 1, 2 and 3 (MSSA 1, MSSA 2 and MSSA 3) with its slot dimensions in length (L_1, L_2), width (W_1, W_2) and spacing between slots (S_1, S_2, S_3, S_4, S_5). For MSSA 1, $L_1 = L_2 =$ 17.16 mm, $W_1 = W_2 =$ 2.87 mm, $S_1 = S_3 =$ 8.04 mm $S_2 =$ 3.77 mm, $S_4 = S_5 =$ 2.78 mm as shown in figure 1.

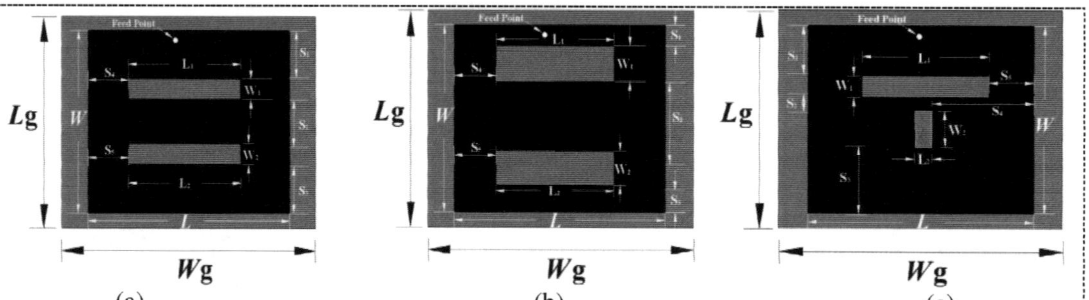

(a) (b) (c)

Figure 1. Microstrip slot antenna (MSSA). (a). MSSA 1, (b). MSSA 2 and (c) MSSA 3.

For MSSA 2 as shown in figure1, the slot dimensions are $L_1 = L_2 =$ 18.14, $W_1 = W_2 =$ 3.86 mm, $S_1 = S_3$ = 5.17 mm, $S_2 =$ 7.54 mm, $S_4 = S_5 =$ 1.68mm and for MSSA 3 as shown in figure 1, the slot dimensions are $L_1 =$ 18.14 mm, $L_2 =$ 3.2 mm, $W_1 =$ 3.86 mm, $W_2 =$ 7.51 mm, $S_1 =$ 5.16 mm, $S_2 =$ 2.01 mm, $S_3 =$ 9.04 mm, $S_4 =$ 8.17 mm, $S_5 =$ 2.68mm. The antennas resonate at two independent frequencies signifying dual band frequency characteristics. Slot lengths, widths and spacing dimensions are functions of operating wavelength λ_o.

3. Results and discussions

Good antenna performance can be achieved by using a thick dielectric substrate and having a low dielectric constant. It is desirable for obtaining improved bandwidth, high efficiency and good radiation characteristics leading to larger antenna size. Designing a compact antenna requires higher dielectric constant, leading to narrower bandwidth, lesser efficiency and higher loss tangents (dissipation factors) [18]. Hence, the final design in this paper required a trade-off between antenna dimensions and antenna performance, depending on the system requirement. In this design, the use of pair of broad slots on the radiating element (patch) are etched and it appear as discontinuity to the microstrip transmission line and provide a transverse component of current, which generates a longitudinal component of the magnetic field and is modelled as series inductance [20], so the electrical length of the patch increases, thus lowering the resonant frequency to achieve compactness of the antenna for fixed frequency operation. The antenna incorporated with the slots also has an advantage that it introduces a capacitance that suppresses some of the inductance introduced by the coax feed due to the thick substrate. The upper conducting layer, i.e., the patch element of the antenna is the source of radiation and it radiates primarily because of the fringing fields between the patch edge and the ground plane. The lower conducting layer acts as a perfectly reflecting ground plane, bouncing energy back through the substrate and into the free space. The various parameters of the fabricated microstrip slot antennas (MSSA) are measured on Vector Network Analyzer (Rohde and Schwarz, Germany make ZVK model 1127.8651) which has a operating frequency range starting from 30 MHz to 40 GHz.

The pair of narrow slots of MSSA 1 as shown in figure 1 resonates at two frequencies; at 3.51 GHz with return loss (RL = – 14.10 dB) and at 4.52 GHz (RL = – 17.14 dB) signifying dual band nature is depicted in figure 2. An impedance bandwidth of 230 MHz (5.8 %) and 140 MHz (3.9 %) are obtained at these two independent resonant frequencies with a moderate gain of 5.6 dB and 4.3 dB. MSSA 2 also resonates at two frequencies; at 3.81 GHz with return loss (RL = – 16.58 dB) and 5.16 GHz (RL = – 15.21 dB) as shown in figure 2 with achievable impedance bandwidth of 280 MHz (7.5 %) and 170 MHz (4.6 %) and a moderate gain of 3.7 dB and 4.9 dB. Due to the pair of wide slots printed on MSSA 2, the impedance bandwidth is increased to 280 MHz compared to impedance bandwidth of MSSA 1. As shown in figure 1 (c), when one horizontal and vertical slots are printed, the MSSA 3 resonates at two lower frequencies compared to MSSA 1 and MSSA 2; i.e., at is 2.22 GHz with return loss (RL = – 24.98 dB) and 3.68 GHz (RL = – 19.89 dB) with achievable impedance bandwidth of 195 MHz (4.9 %) and 260 MHz (5.6 %) with a peak gain of 5.75 dB and 7.58 dB.

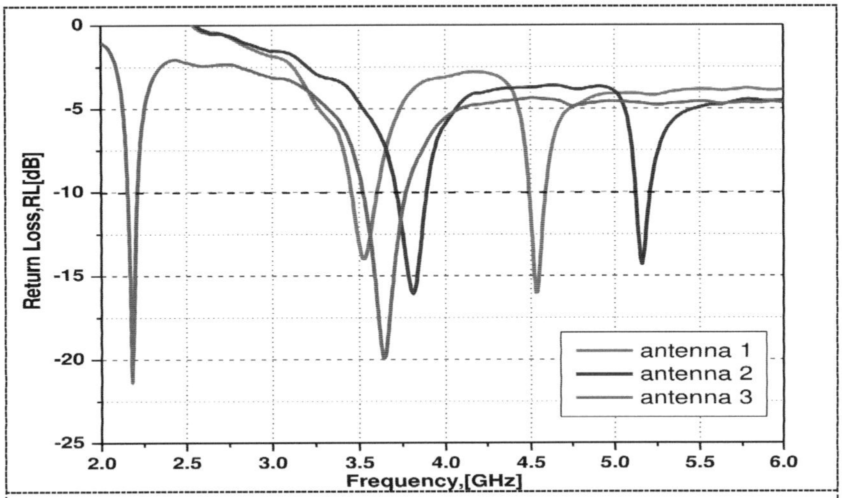

Figure 2. Return loss (RL) characteristics versus frequency

These two independent resonances for MSSA 1, 2 and 3 are produced due to the presence of slots on the radiating element. A compactness of 21 %, 15 % and 58 % are also achieved for MSSA 1, MSSA 2 and MSSA 3. From the return loss characteristics, the proposed antennas showed a lower resonant

Journal of Physics: Conference Series **435** (2013) 012022 doi:10.1088/1742-6596/435/1/012022

frequency with minimum return loss by varying slot length and width and spacing of the slots compared to the standard microstrip antenna. It is observed that there is a reduction in resonant frequency of MSSA 1, 2 but MSSA 3 showed a maximum lower resonant frequency of 2.22 GHz compared to the designed frequency 3.85 GHz making the antenna compact by 58 % with better return loss, improved impedance bandwidth and gain. Figure 3 shows the measured radiation patterns for the proposed microstrip slot antennas 1, 2 and 3 (MSSA 1, MSSA 2 and MSSA 3) with co-polar and cross-polar characteristics. The radiation patterns are measured and plotted at respective resonant frequencies 3.51 GHz and 4.52 GHz for MSSA1, 3.81 GHz and 5.16 GHz for MSSA 2 and 2.22 GHz and 3.68 GHz for MSSA 3. As shown in figure 3, the proposed antennas exhibited nearly omni-directional radiation patterns suitable for various wireless applications. MSSA 3 is the best antenna among the MSSA 1 and MSSA 2 in terms of compactness.

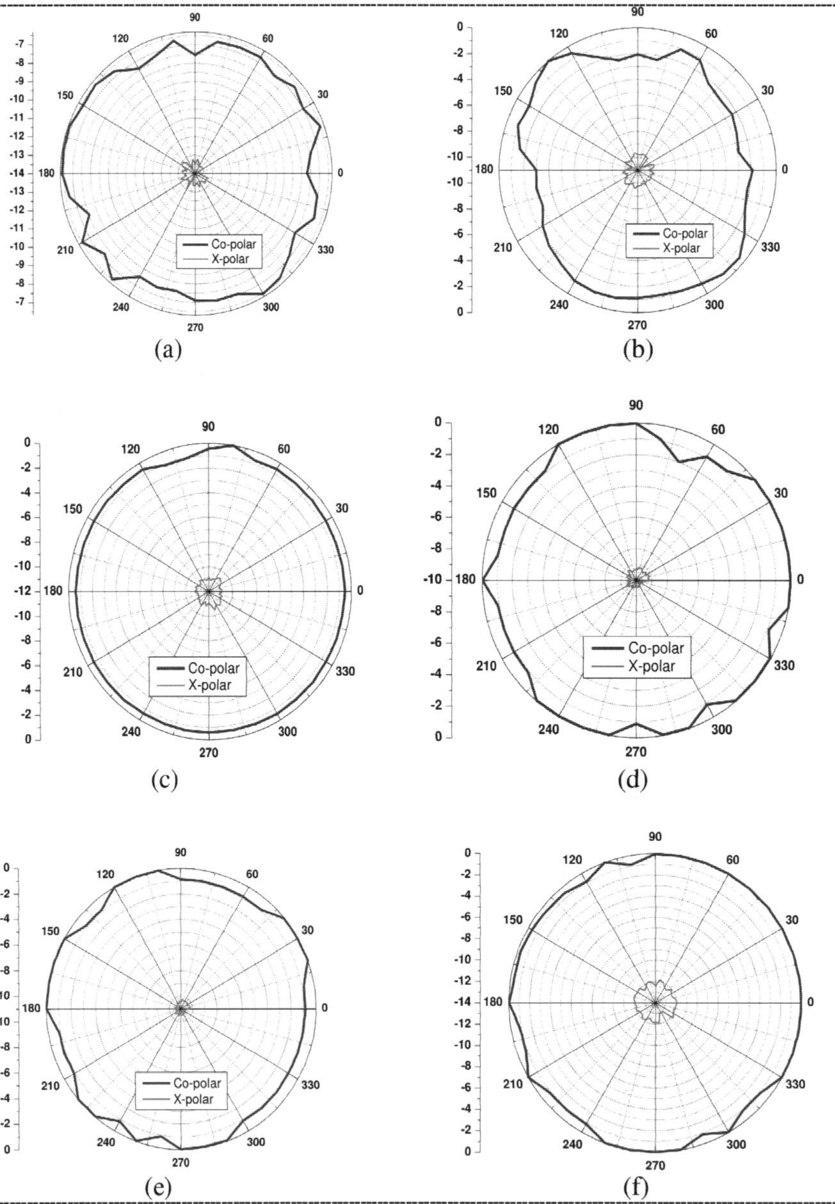

Figure 3. Measured radiation patterns of the microstrip slot antenna (MSSA). (a) MSSA 1 at 3.51 GHz, (b) MSSA 1 at 4.52 GHz, (c) MSSA 2 at 3.81 GHz, (d) MSSA 2 at 5.16 GHz, (e) MSSA 3 at 2.22GHz, (f) MSSA 3 at 3.68 GHz.

Journal of Physics: Conference Series **435** (2013) 012022 doi:10.1088/1742-6596/435/1/012022

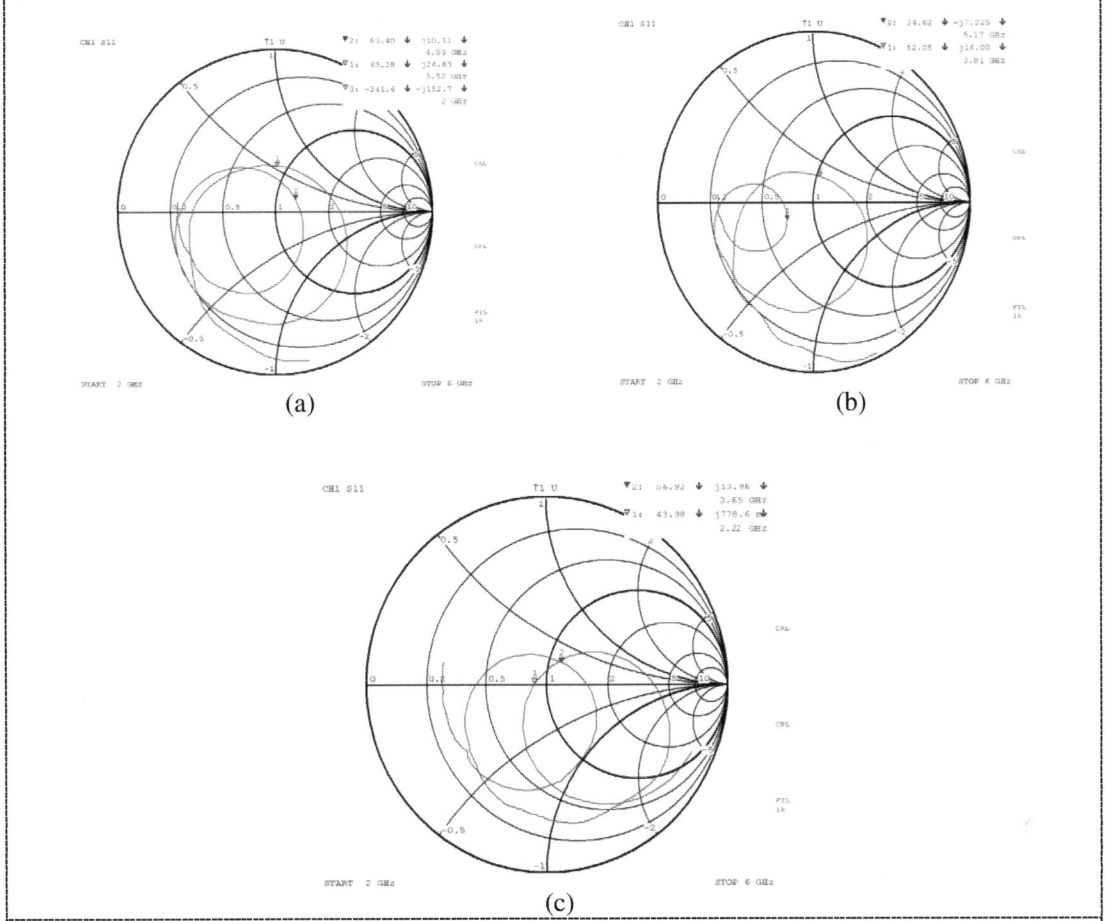

Figure 4. Smith chart plot of (MSSA). (a). MSSA 1, (b). MSSA 2 and (c) MSSA 3.

Figure 4 (a), (b) and (c) shows the Smith chart plots of impedance locus versus frequency for the proposed microstrip slot antennas 1, 2 and 3 (MSSA 1, MSSA 2 and MSSA 3). The presence of two loops on the Smith chart shows dual band characteristics. Hence, the design parameters of the antenna lead to good impedance matching between the input and the load in S and C band frequency range.

4.Conclusion

The proposed antennas are rugged and have low cost and moderate gain suitable for RADAR, WLAN, fixed satellite services, maritime mobile services and space applications covering 2 - 6 GHz frequency range. The microstrip slot antennas are compact, having a size reduction of 21 %, 15% and 58 % with the peak gain of 7.58 dB. The proposed antennas also resonate at two frequencies in S band. This dual frequency behavior increases its applications in radar and satellite communications. The low impedance bandwidth of the antenna is attributed to the low loss tangent of the FR4 substrate material with higher gain.

Acknowledgments

Authors thank the Department of Science and Technology (DST), Government of India, New Delhi, for sanctioning Vector Network Analyzer to this Department under FIST Project and also providing financial assistance to Ambresh P.A under Rajiv Gandhi National fellowship- Junior Research Fellowship & Senior Research Fellowship (RGNF-JRF& RGNF-SRF) [No.F.14 (SC)/2009(SA-III) dated 18 November 2010] scheme by University Grants Commission, New Delhi.

5. References

[1] Kin-Lu Wong, and Wen-Hsiu Hsu 2000 A broadband patch antenna with wide slits *IEEE Trans. on Antennas & Propagation,* **3** 1414–1417.

[2] H.Iwasaki 1996 A circularly polarized small-size microstrip antenna with a cross slot *IEEE Trans. Antennas Propagat.,* **44** 10 1399–1401

[3] K.L.Wong 2002 *Compact and Broadband Microstrip Antennas* (John Wiley and Sons).

[4] M. Sanad 1994 Effect of shorting posts on short circuit microstrip antennas *IEEE Trans Antennas Propag.* **42** 794–797.

[5] N.Kuga and H.Arai 1996 Circular patch antennas miniaturized by shorting posts *Electron. Commun. Jpn* **79**,51–58.

[6] K.L.Wong and S.S Pan 1997 Compact Triangular microstrip antenna," *Electron Letters* **33** 433–434.

[7] L.Desclos 2000 Size reduction of patch by means of slot insertion *Microwave and Optical Technology Letters* **25** 111–113.

[8] J.S.Hong and M.J.Lancaster 1994 Capacitively loaded microstrip loop resonator," *Electron. Letters* **30** 1494–1495.

[9] A.K.Shackelford, K.F.Lee and K.M. Luk 2003 Design of small-size wide-bandwidth microstrip-patch antennas *IEEE Trans. Microwave Theory Tech.* **45** 75–83.

[10] T.K.Lo, C.o.Ho, Y.Hwang, E.K.W. Lam and B., Lee 1997 Miniature aperture-coupled microstrip antenna of very high permittivity *Electron Letters* **33** 9-10.

[11] C. Y.Chiu, H. Wong and C. H.Chan 2007 Study of small wideband folded patch- feed antennas *IET Microwave Antennas Propagation* **1** 2 501–505.

[12] M T Islam, N Misran, and M N Shakib 2009 Broadband E-H shaped microstrip patch antenna for wireless systems *Progress In Electromagnetic Research PIER* **98** 163-167.

[13] Wen-Shyang.Chen 1998 Single-Feed Dual- Frequency Rectangular Microstrip Antenna with Square Slot *Electronics Letters* **34** 3, 231-232.

[14] K. L.Wong and W. S. Che 1998 Slot-Loaded Bow-Tie Microstrip Antenna for Dual-Frequency Operation *Electron. Letters* **34** 1713–1714.

[15] A. M.El-Tager, M. A.Eleiwa and M. I.Salama 2009 A Circularly Polarized Dual frequency Square Patch Antenna for TT&C Satellite Applications *Proceedings of Progress in Electromagnetics Research Symposium* (Beijing: China).

[16] Ali Sadeq Abdulhadi Jalal 2008 A New Compact Patch Design For Circular Polarization Application Based On Third Iteration Minkowski-Like Pre Fractal Geometry *Journal of Engineering and Applied Sciences* **3** 9 729-734.

[17] J. S.Roy, N.Chattoraj and N.Swain 2007 New dual-frequency microstrip antennas for wireless communication *Romanian Journal of Information Science and Technology* **10** 1, 113–119.

[18] C.A.Balanis 1997 *Antenna Theory Analysis and Design* (John Wiley and Sons: New York).

[19] M.Kara 1996 Formulas for the computation of physical properties of rectangular microstrip antenna elements with various substrate thicknesses *Microwave and Optical Technology Letters* **12** 268–272.

[20] W.J.R.Hoefer 1997 Equivalent series inductivity of a narrow transverse slit in microstrip *IEEE Trans. Microwave Theory Tech.* **25** 822– 824.

High Frequency Wave Packet Concentrated near the moving Line on the Liquid Surface

Anton I. Popov[1]
St. Petersburg State University, St. Petersburg, Russia

E-mail: popov239@gmail.com

Abstract. Wave wall type solutions for waves on the surface of an ocean with slowly varying depth are constructed. It is a generalization of the solution of quasi-photon type. Wave wall is a packet of short waves concentrated near the moving line which forms the wave front. The approach is based on the representation of the solution in the form of asymptotic space-time rays (STR) series each term of which is a formal power series. The appropriate form of the ansatz is suggested. Chains of equations for coefficients of formal power series for phase and amplitude terms are derived. These equations are investigated and the solvability of the problem at each step is proved. As a result, asymptotic space-time series for the wave wall moving with group velocity on the surface of a heavy liquid is obtained.

Introduction and statement of the problem

The theory of spatially-localized waves is an interesting problem of wave theory. This problem is important not only for basic researches but also for various engineering applications (from electronic devices in electrodynamics to different constructions in an ocean subjected to wave action). The localized solution construction methods were developed in electrodynamics. However, it is an open problem for surface ocean waves. The purpose of present article is to develop asymptotic methods for localized waves on the surface of a heavy liquid.

We consider waves on the surface of an ocean with slowly varying depth and construct wave wall type solutions. Let us formulate the problem. The equation and the boundary conditions have the following form:

$$\begin{cases} \Delta\Phi = 0 \\ \left(\dfrac{\partial\Phi}{\partial z} + \dfrac{\partial H}{\partial x^1}\dfrac{\partial\Phi}{\partial x^1} + \dfrac{\partial H}{\partial x^2}\dfrac{\partial\Phi}{\partial x^2}\right)\Bigg|_{z=-H} = 0 \\ \left(\dfrac{\partial^2\Phi}{\partial t^2} + g\dfrac{\partial\Phi}{\partial z}\right)\Bigg|_{z=0} = 0 \end{cases} \qquad (1)$$

Here $\vec{v} = -grad\Phi$, Φ is the velocity field potential. $z = -H(\varepsilon x^1, \varepsilon x^2)$ is the bottom.

$$\begin{cases} -H \le z \le 0 \\ -\infty < x^1, x^2 < \infty \end{cases} \qquad (2)$$

We assume the following form of the ansatz (ε is small parameter characterizing the depth variation):

[1] To whom any correspondence should be addressed.

$$\Phi \sim e^{\frac{i\theta(\tau,\xi^1,\xi^2)}{\varepsilon}} \cdot \sum_{j=0}^{\infty} \Phi_j(\tau,\xi^1,\xi^2,z)(i\varepsilon)^j, \tag{3}$$

where $\tau = \varepsilon t$, $\xi^1 = \varepsilon x^1$, $\xi^2 = \varepsilon x^2$.

Here the frequency and the wave vector are, correspondingly,

$$\omega = -\frac{\partial\theta}{\partial\tau} = -\frac{\partial}{\partial t}\left(\frac{\theta}{\varepsilon}\right),\quad \vec{k} = \left(\frac{\partial\theta}{\partial\xi^1}, \frac{\partial\theta}{\partial\xi^2}\right) = \nabla\left(\frac{\theta}{\varepsilon}\right),\quad \nabla = \left(\frac{\partial}{\partial x^1}, \frac{\partial}{\partial x^2}\right). \tag{4}$$

Substitution of the ansatz into the equation $\Delta\Phi = 0$ gives one the following relations (by making equals the coefficients of the terms of the same orders):

$$\mathrm{L}_0\Phi_0 = 0,\ \mathrm{L}_0\Phi_1 + \mathrm{L}_1\Phi_0 = 0, \ldots ,\ \mathrm{L}_0\Phi_j + \mathrm{L}_1\Phi_{j-1} + \mathrm{L}_2\Phi_{j-2} = 0. \tag{5}$$

Here

$$\mathrm{L}_0\Phi \equiv \frac{\partial^2\Phi}{\partial z^2} - k^2\Phi, \tag{6}$$

$$\mathrm{L}_1\Phi = 2\nabla'\theta\nabla'\Phi + \Delta'\theta\cdot\Phi,\ \mathrm{L}_2\Phi = -\Delta'\Phi, \tag{7}$$

$$\nabla' = \left(\frac{\partial}{\partial\xi_1}, \frac{\partial}{\partial\xi_2}\right),\quad \Delta' = \frac{\partial^2}{\partial\xi_1^2} + \frac{\partial^2}{\partial\xi_2^2},\quad k^2 = \left(\frac{\partial\theta}{\partial\xi_1}\right)^2 + \left(\frac{\partial\theta}{\partial\xi_2}\right)^2. \tag{8}$$

Correspondingly, the boundary conditions from (1) lead to the relations:

$$\left.\left(g\frac{\partial\Phi_j}{\partial z} - \omega^2\Phi_j + \frac{\partial^2\theta}{\partial\tau^2}\Phi_{j-1} + 2\frac{\partial\theta}{\partial\tau}\frac{\partial\Phi_{j-1}}{\partial\tau} - \frac{\partial^2\Phi_{j-2}}{\partial^2\tau}\right)\right|_{z=0} = 0, \tag{9}$$

$$\left.\left(\frac{\partial\Phi_j}{\partial z} + \nabla' H\nabla'\theta\cdot\Phi_{j-1} - \nabla' H\nabla'\Phi_{j-2}\right)\right|_{z=-H} = 0, \tag{10}$$

$$j = 0,1,2,\ldots,\qquad \Phi_{-1} = \Phi_{-2} \equiv 0.$$

For $j = 0$ one gets an unconventional spectral problem with ω^2 (eigenvalue) inserted into the boundary condition:

$$\frac{\partial^2\Phi_0}{\partial z^2} - k^2\Phi_0 = 0,\ \left.\left(g\frac{\partial\Phi_0}{\partial z} - \omega^2\Phi_0\right)\right|_{z=0} = 0,\ \left.\frac{\partial\Phi_0}{\partial z}\right|_{z=-H} = 0. \tag{11}$$

Due to (11) one gets

$$\Phi_0 = A_0\left(\tau,\xi^1,\xi^2\right)\cdot ch\left[k\left(z+H\right)\right],\ \omega^2 = gk\cdot th\left(kH\right),\ k = \sqrt{\theta_{\xi^1}^2 + \theta_{\xi^2}^2}. \tag{12}$$

Remark. At this stage A_0 is unknown and will be determined later.

Thus, the dispersion relation has the following form:

$$\omega = \omega\left(k\right) = \sqrt{g}\sqrt{th\left(kH\right)},\ \text{where } \vec{k} = \left(k_1,k_2\right) = \left(\theta_{\xi^1},\theta_{\xi^2}\right). \tag{13}$$

Using the relation $\omega = -\dfrac{\partial\theta}{\partial\tau}$ one transforms (13) to the form of conventional Hamilton-Jacobi equation:

$$\frac{\partial\theta}{\partial\tau} + \mathrm{H}\left(\xi^1, \xi^2, \theta_{\xi^1}, \theta_{\xi^2}\right) = 0,\ \mathrm{H} = \sqrt{g}\sqrt{k \cdot th(kH)}. \tag{14}$$

The canonical system for this Hamilton-Jacobi equation has the form:

$$\frac{d\tau}{ds} = 1,\ \frac{d\xi^i}{ds} = \frac{\partial\mathrm{H}}{\partial\theta_{\xi^i}},\ \frac{d\theta_\tau}{ds} = 0,\ \frac{d\theta_{\xi^i}}{ds} = -\frac{\partial\mathrm{H}}{\partial\xi^i},\ \frac{d\theta}{ds} = -\mathrm{H} + \theta_{\xi^i}\frac{\partial\mathrm{H}}{\partial\theta_{\xi^i}}. \tag{15}$$

The solution $\xi^i = \xi^i(s)$ of (15) gives us a space-time ray.

$$\Phi_0 = A_0\left(\tau, \xi^1, \xi^2\right) \cdot ch\left[k\left(z + H\right)\right],\ \omega^2 = gk \cdot th\left(kH\right),\ k = \sqrt{\theta_{\xi^1}^2 + \theta_{\xi^2}^2}. \tag{12}$$

θ construction

For the following construction it is convenient to introduce new coordinates. We will do this in two steps. At first, we introduce the coordinates τ, s^1, s^2. We start from the basic curve – a line at the ocean surface plane playing a role of wave wall front at initial moment of time. Let us describe new coordinates. Determine the coordinates of a point M. Let l_1 be the basic line on the initial plane. Construct a line parallel to the time axis through M. Its intersection with the initial plane is M_1 and with the surface formed by STR starting from the basic line - M_2. Let l_2 be the space time ray (STR) starting from the point $P_1 \in \ell_1$ and coming to M_2. P is the initial point on the basic curve. s^1 is the length of the basic curve between the points P and P_1. $s^2 = |M_1M_2|$, $\tau = |M_1M|$.

Note that the equations of the STR in new coordinates are:

$$\begin{cases} s^1 = const \\ s^2 = \tau \end{cases} \tag{16}$$

It is convenient to use modified coordinates τ, η^1, η^2:

$$\eta^1 = s^1,\ \eta^2 = s^2 - \tau. \tag{17}$$

We try to find the wave wall type solution of (1), i.e. the solution concentrated near the surface $\eta^2 = 0$, having the exponential decay far from the surface. The Hamilton-Jacobi equation in the coordinates τ, η^1, η^2 has the form:

$$\frac{\partial\theta}{\partial\tau} - \frac{\partial\theta}{\partial\eta^2} + \mathrm{H}\left(\eta^1, \eta^2, \theta_{\eta^1}, \theta_{\eta^2}\right) = 0. \tag{18}$$

We use the same letter for the Hamiltonian as a function of new coordinates. θ is sought in the form of formal series in powers of η^2:

$$\theta = \theta^{(0)} + \theta^{(1)} + \theta^{(2)} + \ldots.., \tag{19}$$

where $\theta^{(0)}=\theta^{0}(\eta^{1},\tau),\ \theta^{(1)}=\theta^{1}(\eta^{1},\tau)\eta^{2},\ \theta^{(2)}=\theta^{2}(\eta^{1},\tau)(\eta^{2})^{2},\ \ldots\ldots\ldots,\theta^{(l)}=\theta^{l}(\eta^{1},\tau)(\eta^{2})^{l}$. We insert series for θ and $\mathrm{H}\left(\eta^{1},\eta^{2},\theta_{\eta^{1}},\theta_{\eta^{2}}\right)$ into (18), equals terms of the same power of η^{2} and comes to the recurrent system of equation for θ^{l}. The solution gives us θ.

Φ_{j} construction

All Φ_{j} are sought in the form of formal series in powers of η^{2}. $\Phi_{j},\ j=0,1,2...$, are determined from (5) consequently. Series for Φ_{0} has the same form as earlier (12), where $A_{0}=A_{0}(\tau,\eta^{1},\eta^{2})$ is got from the solvability condition of the problem (see (5)) for Φ_{1}. The solvability condition is given by the following theorem:

Theorem 1.
Consider homogeneous Sturm-Liouville problem:

$$\begin{cases} Ly=(p(x)y')'-q(x)y=0 \\ l_{0}y=(\alpha_{0}y+\alpha_{1}y')\big|_{x=x_{0}}=0\,, \\ l_{1}y=(\beta_{0}y+\beta_{1}y')\big|_{x=x_{1}}=0 \end{cases} \tag{20}$$

where $p(x)>0,\ \mathrm{Im}\,q=0.$ $\alpha_{j},\ \beta_{j}\ (j=0,1)$ *are real.* $p(x),\ q(x)$ *are formal power series.*
Let there exists a solution $y_{0},\ y_{0}\neq 0,$ *in the form of a formal power series. Then the necessary and sufficient condition for solvability of the inhomogeneous Sturm-Liouville problem for formal power series,*

$$\begin{cases} Ly=-F \\ l_{0}y=A \\ l_{1}y=B \end{cases}$$

is as follows:

$$p(x_{1})\frac{B}{\beta_{1}}y_{0}\bigg|_{x=x_{1}}-p(x_{0})\frac{A}{\alpha_{1}}y_{0}\bigg|_{x=x_{0}}=-\int_{x_{0}}^{x_{1}}F(x)y_{0}(x)dx. \tag{21}$$

Here $F(x),\ A,\ B$ *are formal power series.*

The solvability condition (21) for Φ_{1} can be written in the divergence form:

$$div\vec{A}=0. \tag{22}$$

Here $\vec{A}=\left(\langle\varepsilon\rangle,v_{gr_{1}}\langle\varepsilon\rangle,v_{gr_{2}}\langle\varepsilon\rangle\right)$, where $\langle\varepsilon\rangle=\frac{1}{g}\left(-\frac{\partial\theta}{\partial\tau}\right)\Phi_{0}^{2},\ v_{gr_{i}}\langle\varepsilon\rangle=\int_{-H}^{0}\Phi_{0}^{2}dz\cdot\frac{\partial\theta}{\partial\xi^{i}},\ i=1,2.$

Here $\langle\varepsilon\rangle$ is the mean surface energy density, v_{gr} is the group velocity.
For the next step it is important to introduce preliminary information and Theorem 2.

Preliminary information.
Consider $r+s$ formal power series:

$$\sum_{|\alpha|=0}^{\infty} A_\alpha^j(x)y^\alpha = a^{(j)}, \quad j=1,2,.....,r. \tag{23}$$

$$\sum_{|\alpha|=1}^{\infty} A_\alpha^j(x)y^\alpha = b^{(j-r)}, \quad j=r+1,r+2,.....,r+s. \tag{24}$$

$$\alpha=(\alpha_1,\alpha_2,.....,\alpha_s), \tag{25}$$

$$|\alpha| = \alpha_1+\alpha_2+.......+\alpha_s, \quad \alpha_i \in \mathbb{Z}_+. \tag{26}$$

The coefficients $A_\alpha^j \in C_{loc}^\infty(\Omega)$, $\Omega \subset \mathbb{R}^r$.

We will call variables $x=(x^1,x^2,.......,x^r)$ as finite variables and $y=(y^1,y^2,.......,y^s)$ as small variables.

Let us suppose that x and y are formal power series:

$$x^{(j)} = \sum_{|\alpha|=0}^{\infty} B_\alpha^j(a)b^\alpha, \quad j=1,2,.......,r. \tag{27}$$

$$y^{(j-r)} = \sum_{|\alpha|=1}^{\infty} B_\alpha^j(a)b^\alpha, \quad j=r+1,r+2,.......,r+s. \tag{28}$$

$B_\alpha^j(a) \in C^\infty(\Omega_1)$ $(\Omega_1 \subset \mathbb{R}^r)$ are functions, where $B_{(0,0,.......,0)}^j(a) \in \Omega$.

$a=(a^{(1)},a^{(2)},.......,a^{(r)})$ are finite variables, $b=(b^{(1)},b^{(2)},.......,b^{(s)})$ are small variables.

Then we can substitute $x^{(j)}$ and $y^{(j)}$ from (27), (28) to (23), (24). When one substitutes formal power series for $x^{(j)}$ and $y^{(j)}$, one replaces the functions $A_\alpha^j(x)$ by its Taylor series. The converges are not assumed.

Definition.
The inversion of series (23), (24) is the determination of series (27), (28) such that its substitution into (23), (24) gives us:

$$\sum_{|\alpha|=0}^{\infty} A_\alpha^j(x)y^\alpha = a^{(j)}, \quad j=1,2,.....,r.$$

$$\sum_{|\alpha|=1}^{\infty} A_\alpha^j(x)y^\alpha = b^{(j-r)}, \quad j=r+1,r+2,.....,r+s.$$

Theorem 2.
Consider series:

$$\sum_{|\alpha|=0}^{\infty} A_\alpha^j(x)y^\alpha = a^{(j)}, \quad j=1,2,.....,r. \tag{29}$$

$$\sum_{|\alpha|=1}^{\infty} A_\alpha^j(x)y^\alpha = b^{(j-r)}, \quad j=r+1,r+2,.....,r+s. \tag{30}$$

$$A_\alpha^j \in C_{loc}^\infty(\Omega), \quad \Omega \subset \mathbb{R}^r.$$

Let us suppose that

$$J = \left.\frac{D(A^1_{(0,...0)}(x),...,A^r_{(0,...0)},A^{r+1}_{(1,0,...0)}\cdot y^{(1)},...,A^{r+s}_{(0,...0,1)}\cdot y^{(s)})}{D(x^{(1)},.....,x^{(r)},y^{(1)},.....,y^{(s)})}\right|_{x^{(j)}=x_0^{(j)},y^{(j)}=0} \neq 0, \qquad (31)$$

where

$$x_0^{(j)} = x^{(j)}\Big|_{b^{(1)}=0,b^{(2)}=0,.....,b^{(s)}=0} \qquad (32)$$

Then one can convert the series (29), (30).

Return to our problem. Using Theorem 2, we can rewrite equation (22) in the coordinates (τ,η^1,η^2). It takes essentially more simple form:

$$\frac{1}{J}\frac{\partial}{\partial\tau}\left(\langle\varepsilon\rangle J\right)=0,\ J=\frac{D(\tau,\xi^1,\xi^2)}{D(\tau,\eta^1,\eta^2)}. \qquad (33)$$

It allows one to find A_0:

$$A_0(\tau,\eta^1,\eta^2)=\sqrt{\frac{g\left(\langle\varepsilon\rangle J\right)\Big|_{\tau=0}}{-\frac{\partial\theta}{\partial\tau}\cdot J\Big|_{z=0}}}\cdot\frac{1}{\cosh(kH)}. \qquad (34)$$

Thus, we have determined A_0, hence, Φ_0 too, from the solvability condition of the problem for Φ_1. Analogously, Φ_j is determined from the solvability condition of the problem for Φ_{j+1}.

Thus, full asymptotic expansion for wave wall type solution is constructed. From one side, it is rigorous mathematical result with the justification for all steps. From the other side, it can be applied to the description of natural ocean waves.

Acknowledgment
The work was supported by RFBR grant № 12-01-09337.

References

[1] V.M. Babich and A.I. Popov 2012 An asymptotic solution of the Hamilton-Jacobi equation concentrated near a surface *Journal of Mathematical Sciences, Springer Science* **185** 4, 523-525
[2] V. M. Babich and A. I. Popov 2011 Quasiphotons of waves on the surface of a heavy liquid *Journal of Mathematical Sciences, Springer Science* **173** 3, 243-253
[3] V. M. Babich, V. S. Buldyrev and I. A. Molotkov 1985 *The Space-Time Ray Method: Linear and Nonlinear Waves* (Izd. LGU, Leningrad)
[4] S. Bochner and W.T. Martin 1948 *Several complex variables* (Princeton, Princeton Univ. Press)

Journal of Physics: Conference Series **435** (2013) 012024 doi:10.1088/1742-6596/435/1/012024

Variational analysis of soliton scattering by external potentials

B A Umarov[1], A Messikh[2], N Regaa[1] and B B Baizakov[3]

[1] Department of Computational and Theoretical Sciences, Faculty of Science, International Islamic University Malaysia, P.O Box 141, 25710 Kuantan, Malaysia

[2] Faculty of Information and Communication Technology, International Islamic University Malaysia, P.O. Box 10, 50728 Kuala Lumpur, Malaysia

[3] Physical-Technical Institute of the Uzbek Academy of Sciences, 2-b, Bodomzor str., 100084, Tashkent, Uzbekistan

E-mail: bakhram@iiu.edu.my

Abstract. Dynamics of the width and center-of-mass position of a matter wave soliton subject to interaction with arbitrary external potential is analyzed using the collective coordinates approach. It is shown that approximation of the trial function and external potential only in the interaction region of the spatial domain is sufficient for adequate description of the soliton scattering process. The validity of the developed approach is illustrated for the Gaussian and Pöschl-Teller potentials.

1. Introduction

Soliton is a localized wave packet that can propagate in nonlinear media preserving its shape and velocity and exhibit particle like elastic collision with other solitons [1]. Theoretically studied and experimentally observed in different areas of physics, such as plasma physics, hydrodynamics, nonlinear optics, electric circuits, and more recently in Bose-Einstein condensates, solitons remain in the forefront of modern physics research [2]. Perhaps, the most important practical application of solitons to date pertains to information transfer through optical fibers [3]. Over the years enhancing the capacity of optical fiber communication systems employing solitons as information bit carriers has been the strongest motivation for scientific and engineering efforts in this field [4, 5].

Solitons in Bose-Einstein condensates present a promising candidate system for applications in high precision devices, such as atom interferometers [6], and studies of atom-surface interfaces [7]. A central mechanism here is the coherent interaction of the bright matter wave soliton with a potential barrier or well. The possibility to split the soliton through scattering on a Gaussian barrier with subsequent recombining the fragments was demonstrated in the recent experiment [8]. Theoretical analysis of the experiment and related process of soliton molecule formation was reported in [9].

Variational approximation, also known as the collective coordinates approach, is widely employed theoretical tool for the analysis of non-integrable soliton bearing equations [10, 11]. The success of the variational approximation critically depends of the proper choice of the trial function. The decisive point here is that, analytical evaluation of the averaged/effictive Lagrangian should have been possible to obtain the system of evolution equations for the parameters of the soliton. The choice of the Gaussian profile as a trial function is frequently found in the literature, because it suits the above mentioned purpose in many cases. However, in the matter wave soliton scattering processes under investigation, the Gaussian initial profile gives rise to a breathing wave packet. Moreover, the motion of the packet is accompanied by shedding linear waves, since the Gaussian is not the exact solution to the governing equation without external potential. Both of these processes, breathing motion of the soliton before the scattering and radiation of linear waves, are irrelevant and must be avoided. A possible way around this difficulty is adopting a trial function belonging to the soliton family, and replacing it by an equivalent Gaussian function *only* in the interaction region of the spatial domain. Then the corresponding term of the averaged Lagrangian can be evaluated exactly. Efficiency of this method was demonstrated in [12] for propagation of solitons in liquid crystals.

In the variational approximation developed below we assume the trial function to be of the hyperbolic secant function with time dependent parameters. While the soliton is far away from the scattering potential the interaction between them is effectively zero, and one has a freely propagating matter wave packet with constant parameters, such as the amplitude, width, center-of-mass position and velocity. Time dependence of the parameters become evident when the soliton starts to interact with the potential, namely it experiences deformation and changes velocity. Perturbation of the soliton in the form of oscillating amplitude and width remains long after the interaction. This process is interpreted as excitation of soliton's internal modes. Frequency of oscillations can be estimated from variational equations, as illustration of the predictive power of the variational approach.

The paper is organized as follows. In Sec. II the mathematical model is formulated and basic equations are derived. Sec. III is devoted to numerical simulations of the variational equations and comparison of the results with direct solution of the original Gross-Pitaevskii equation. Finally, in Sec. IV we summarize our findings, and discuss some interesting directions for future studies.

2. The model and main equations

The model is based on the reduced nonlinear Schrödinger/Gross-Pitaevskii equation (GPE)

$$i\psi_t + \frac{1}{2}\psi_{xx} - V(x)\psi + g|\psi|^2\psi = 0, \tag{1}$$

where $\psi(x,t)$ is the macroscopic wave function of the quasi-one dimensional condensate, $V(x)$ is the external potential. We consider positive coefficient of nonlinearity ($g > 0$), which corresponds to attractive interactions between atoms in the condensate, or focusing nonlinearity in optics applications. In absence of external potential ($V(x) = 0$) and $g = 1$ the Eq. (1) possesses fundamental bright soliton solution

$$\psi(x,t) = A\text{sech}[A\,(x-\xi(t))]\exp(i\Theta), \quad \Theta(x,t) = \frac{1}{2}(A^2 - \xi^2)t + vx + \varphi, \quad v = \xi_t, \tag{2}$$

where A, ξ, v, φ are the amplitude, center of mass position, velocity and phase of the soliton respectively. Trial function of the variational approximation will be assumed to be of this form. This means that the shape of the soliton remains close to the shape of hyperbolic secant function, although its amplitude and width may change significantly during the whole period of evolution.

It is easy to verify that the governing Eq. (1) can be obtained from the following Lagrangian density

$$\mathcal{L} = \frac{i}{2}(\psi\psi_t^* - \psi^*\psi_t) + \frac{1}{2}|\psi_x|^2 + V(x)|\psi|^2 - \frac{g}{2}|\psi|^4, \tag{3}$$

by means of the Euler-Lagrange equation. Spatial integration of the Lagrangian density $L = \int_{-\infty}^{\infty} \mathcal{L} dx$ using the trial function

$$\psi(x,t) = A\text{sech}\left(\frac{x-\xi}{a}\right) e^{ib(x-\xi)^2+iv(x-\xi)+i\varphi}, \tag{4}$$

gives rise to following averaged/effective Lagrangian

$$L = N\left[\frac{\pi^2}{12}a^2 b_t + \frac{\pi^2}{6}a^2 b^2 - \frac{1}{2}\xi_t^2 + \varphi_t + \frac{1}{6a^2} - \frac{gN}{6a} + \frac{1}{2a}\int_{-\infty}^{\infty} V(x)\text{sech}^2\left(\frac{x-\xi}{a}\right)dx\right]. \tag{5}$$

The norm of the wave function $N = \int_{-\infty}^{\infty} |\psi|^2 dx = 2A^2 a$ is proportional to the number of atoms in the condensate and represents a conserved quantity.

Evolution equations for variational parameters can be derived from the Euler-Lagrange equations $d/dt(\partial L/\partial \dot{q}_i) - \partial L/\partial q_i = 0$, where q_i are time dependent collective coordinates a, ξ, b, φ. The equation for the phase φ reduces to $dN/dt = 0$ and illustrates the conservation of the norm of the wave function or number of atoms in the condensate. It is decoupled from other equations and can be dropped in further analysis. What remains is a set of coupled equations for the width and center-of-mass position of the soliton

$$a_{tt} = \frac{4}{\pi^2 a^3} - \frac{2gN}{\pi^2 a^2} + \frac{6}{\pi^2 a^2} F_1(V,a,\xi), \tag{6}$$

$$\xi_{tt} = -\frac{1}{a^2} F_2(V,a,\xi), \tag{7}$$

where

$$F_1(V,a,\xi) = \int_{-\infty}^{\infty} V(x)\,\text{sech}^2\left(\frac{x-\xi}{a}\right)\left[1 - \frac{2\,(x-\xi)}{a}\text{th}\left(\frac{x-\xi}{a}\right)\right] dx,$$

$$F_2(V,a,\xi) = \int_{-\infty}^{\infty} V(x)\,\text{sech}^2\left(\frac{x-\xi}{a}\right)\,\text{th}\left(\frac{x-\xi}{a}\right) dx.$$

The coupled system of dynamical equations (6)-(7) for description of soliton scattering by arbitrary external potential $V(x)$ is the main result of this work. Here it is necessary to note that the last term in effective Lagrangian (5) cannot be calculated analytically for arbitrary potentials $V(x)$, therefore Eqs. (6)-(7) are the integro-differential equations. Available integration routines [13, 14] provide fast and high precision numerical simulations for both the GPE (1) and variational (6)-(7) equations.

However, before proceeding to numerical simulations we present analytical approximations for the integral in Eq.(5) to enhance the efficiency of the developed method. Similar technique was reported in [15] where a deep and narrow Pöschl-Teller potential was replaced by the Dirac delta function $V(x) = V_0\,\delta(x)$. Below we consider another case when both the potential and wave function can be approximated by Gaussian functions.

$$V(x) \to V_0 \exp(-x^2/w^2), \qquad \text{sech}^2\left[(x-\xi)/a\right] \to \exp\left[-(x-\xi)^2/a^2\beta^2\right], \tag{8}$$

where w is the width of the potential, and $\beta = 2/\sqrt{\pi}$ is found from the requirement of equal areas beneath the curves

$$\int_{-\infty}^{\infty} \text{sech}^2\left(\frac{x}{a}\right) dx = \int_{-\infty}^{\infty} \exp\left(-\frac{x^2}{\beta^2 a^2}\right) dx. \tag{9}$$

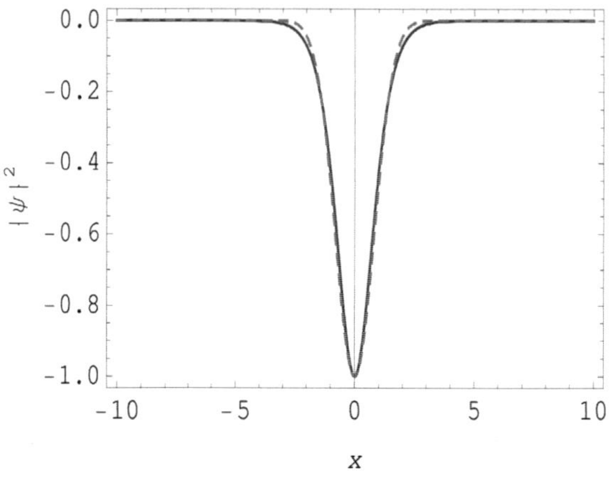

Figure 1. Comparison of the Gaussian (blue solid line) and hyperbolic secant (red dashed line) functions of equal norms as illustration for the approximation (8). Parameters: $V_0 = -1$, $a = 1$, $\beta = 2/\sqrt{\pi}$, $\xi = 0$.

The accuracy of the approximation (8) is illustrated in Fig. 1. Now the integrals in Eq. (5) can be readily calculated with Gaussian functions and the effective Lagrangian (5) takes the form

$$L = N\left[\frac{\pi^2}{12}a^2 b_t + \frac{\pi^2}{6}a^2 b^2 - \frac{1}{2}\xi_t^2 + \varphi_t + \frac{1}{6a^2} - \frac{gN}{6a} + V_0\frac{w}{\beta(\beta^2a^2+w^2)^{1/2}}\exp\left(-\frac{\xi^2}{\beta^2a^2+w^2}\right)\right].$$

Corresponding dynamical equations for the width and center-of-mass position read as follows

$$a_{tt} = \frac{4}{\pi^2 a^3} - \frac{2gN}{\pi^2 a^2} - V_0\frac{12\beta a w(2\xi^2 - \beta^2 a^2 - w^2)}{\pi^2(\beta^2a^2+w^2)^{5/2}}\exp\left(-\frac{\xi^2}{\beta^2a^2+w^2}\right), \quad (10)$$

$$\xi_{tt} = V_0\frac{2w\xi}{(\beta^2a^2+w^2)^{3/2}}\exp\left(-\frac{\xi^2}{\beta^2a^2+w^2}\right), \quad (11)$$

If the scattering potential also has the form $\sim \mathrm{sech}^2(x/w)$, one should substitute $w \to \beta w$ in these equations. In the next section we compare predictions of exact (6)-(7) and approximate (10)-(11) variational equations with the numerical solution of the original GPE (1).

3. Numerical simulations

Numerical solution of the GPE (1) has been performed by the split-step fast Fourier transform method [16] using 2048 Fourier modes within the integration domain of length $L \in [-50 \div 50]$, and the time step was $\delta t = 0.005$. Accuracy of calculations was monitored through conserved quantities of the governing equation. In particular the norm of the wave function was constant with tolerance $\sim 10^{-3}$ through the whole period of evolution. To avoid the adverse effect of linear waves emitted by the soliton during scattering on the potential, we installed absorbing boundary layers at both ends of the integration domain as described in [17]. This technique emulates the condition of infinite spatial domain for the GPE (1).

The dynamical equations of the variational approximation (6)-(7) and (10)-(11) are solved using the Runge-Kutta procedure of 5-th order with adaptive step-size control [13]. Fast evaluation of the integral terms $F_1(V, a, \xi)$, $F_2(V, a, \xi)$ in Eq. (6)-(7) are performed by adaptive Newton-Cotes integration routine, based on eighth-degree polynomial [14].

Journal of Physics: Conference Series **435** (2013) 012024 doi:10.1088/1742-6596/435/1/012024

Arrangement for the soliton scattering by external potential well/barrier is depicted in Fig. 2. Soliton $\psi(x)$ is set in motion with some velocity v towards the potential well/barrier $V(x)$ initially located at some distance from it. Depending on the parameters of the problem soliton is either reflected or transmitted through the potential. We monitor the evolution of the width

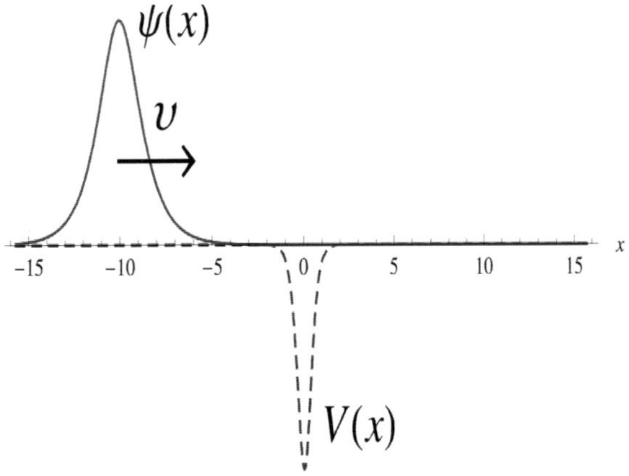

Figure 2. Schematic view for the soliton scattering in numerical experiments.

and center-of-mass position of the soliton via numerical solution of the GPE and variational equations

$$\xi(t) = \frac{\int_{-\infty}^{\infty} x|\psi(x,t)|^2\,dx}{\int_{-\infty}^{\infty} |\psi(x,t)|^2\,dx}, \quad a(t) = \left(\frac{12}{\pi^2} \cdot \frac{\int_{-\infty}^{\infty} (x-\xi(t))^2|\psi(x,t)|^2\,dx}{\int_{-\infty}^{\infty} |\psi(x,t)|^2\,dx}\right)^{1/2}.$$

3.1. Scattering on the Gaussian *potential*

First we consider the scattering of a matter wave soliton by a Gaussian potential barrier and well. In experiments with Bose-Einstein condensates the potential barrier/well for neutral atoms is created using a laser beam, whose frequency ω is slightly shifted from the frequency ω_0 of atomic transitions [18]. The mechanism is based on the dipole force exerted on an atom by an inhomogeneous light field. If the difference $\Delta = \omega - \omega_0$ is negative ($\Delta < 0$, red detuning) the dipole potential is attractive and the atoms are pulled towards the maximum of the laser light intensity. In this case the laser beam acts as a potential well for the atomic wave packet. Similarly, blue detuned ($\Delta > 0$) laser beam acts as a potential barrier for matter wave solitons.

In Fig. 3 the scattering of a soliton by a Gaussian potential barrier is shown through the time dependence of its relative width and center-of-mass trajectory. The soliton at initial position ξ_0 is set in motion, with velocity v, towards the potential barrier at the origin. Until the soliton reaches the interaction range of the potential barrier $\sim w$, it moves freely width constant width and velocity. Reflecting barrier forces the soliton to shrink and expand again, reversing the direction of its motion. As can be seen from the figure, dynamical equations of the variational approximation (6)-(7) and (10)-(11) provide fairly good description of the scattering process, when compared with direct numerical solution of the original GPE (1).

Scattering of the soliton on the potential well is quite different from the previous case of a barrier. Excitation of bound states of the well by approaching matter wave packet is the influential process in this case, because the outcome of the soliton scattering strongly depends on the nature of this interaction, i.e. whether it is repulsive or attractive. In fact, when the bound states of the potential well are excited, the single soliton trial function (4) of the variational approximation becomes inaccurate (see also the next subsection).

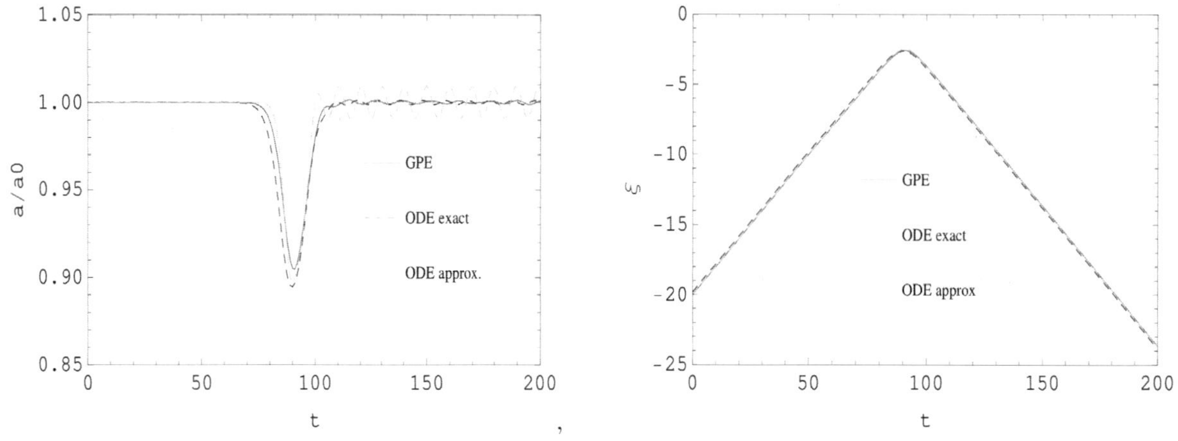

Figure 3. Evolution of the width and center-of-mass position of a soliton $\psi(x,0) = A\text{sech}\left[(x-\xi_0)/a\right]\exp(-vx)$ reflected by a Gaussian potential barrier $V(x) = V_0\exp(-x^2/w^2)$, according to GPE and ODE systems: (6)-(7) - exact, (10)-(11) - approximate. Left panel: At the impact upon the potential barrier soliton shrinks and expands, but small amplitude oscillations remain after the collision. Right panel: Center-of-mass dynamics is perfectly reproduced by the variational approximation. Parameters: $V_0 = 0.5$, $w = 1$, $A = 1$, $a = 1$, $v = 0.2$, $\xi_0 = -20$.

In numerical simulations presented below we consider a wave packet which is narrow and tall compared with the extent and strength of the well. In this case the leakage of matter from the wave packet and excitation of the bound states in the well, are inhibited. Fig. 4 illustrates typical example of soliton scattering by Gaussian potential well.

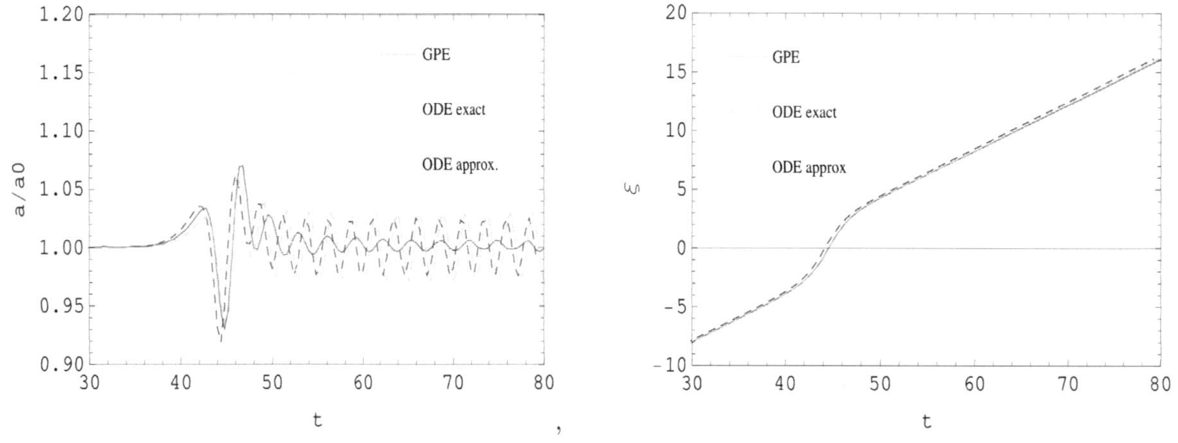

Figure 4. Time dependence of the width and center-of-mass position of the soliton with $A = 2$, $a = 0.5$, $v = 0.4$, $\xi_0 = -20$, transmitted through a Gaussian potential well with $V_0 = -1$, $w = 2$ according to GPE and ODE systems: (6)-(7) - exact, (10)-(11) - approximate. Left panel: In the spatial domain of the potential soliton exhibits strong variations of the width, which transforms into steady oscillations after the interaction. Therefore, traversing the potential well leads to excitation of internal modes of the soliton. Right panel: As in the case of the barrier, center-of-mass dynamics is described very well by both versions of the variational approximation.

A noteworthy feature of the scattering process shown in the right panel of Fig. 4 is the accelerated motion of the soliton within the spatial domain of the potential well (at $t \sim 40 \div 50$). This phenomenon, called "time advance" effect, was first shown for a linear wave packet passing over the reflectionless potential [19], and later reported for solitons traversing the Pöschl-Teller potential [15]. Here we observe the same effect in the numerical simulations of the soliton scattering by a Gaussian potential well. As in the above mentioned cases, on leaving the interaction region the soliton restores its original velocity possessed before the scattering event.

3.2. Scattering on the Pöschl-Teller potential

The Pöschl-Teller potential

$$V(x) = -\frac{\hbar^2}{2mw^2}\frac{\nu(\nu+1)}{\cosh^2(x/w)} \tag{12}$$

has a remarkable property of being fully transparent for linear waves with arbitrary energy, if ν is a positive integer [20]. The absence of reflected waves at any energy, when the dielectric function profile possesses the form (12), has found practical applications in the design of optical coatings permitting the uninterrupted passage of light [21].

Recently it was revealed that the potential (12) exhibits peculiar features also in the interactions with nonlinear waves. Namely, if the soliton approaches the potential well at velocity less than some critical value, it is almost fully reflected. By contrast, solitons with velocity higher than critical, are almost fully transmitted [22]. Particularly, for the potential $V(x) = V_0/\cosh^2(x/w)$ with $V_0 = -4, w = 0.5$, the critical velocity was found to be $v_{cr} = 0.325$ [22]. Similar behavior was observed in the scattering of soliton molecules as well [15]. Matter wave soliton approaching the Pöschl-Teller potential well excites quantum bound states of the well. Interaction of the soliton with emerging localized states in the potential well crucially influences the process of scattering. That is why here the variational approximation with the single component trial function fails, as can be inferred from Fig. 5. The possibility of taking into account the bound states of the well in the collective coordinate approach was considered in Refs. [23, 22], but for the delta function case and without comparison with GPE. Below we provide such comparison for the case of the weak potential well.

In Fig. 6 the results for soliton scattering on the shallow potential (12) is presented. Predictions of the variational approach are compared with the results of direct numerical solution of the GPE (1). As can be admitted, the coupled set of variational equations provide qualitatively correct description for the evolution of the soliton's width and center-of-mass dynamics. One peculiarity to be noted here is that the reflected soliton (middle panel) performs breathing oscillations with less amplitude ($\sim 3\%$ of stationary value) than the transmitted one ($\sim 40\%$). In both cases the variational approximation overestimates the amplitude of oscillations. However, the frequencies of oscillations are in good agreement.

4. Conclusions

We have developed a variational approximation to describe the scattering of matter wave solitons by external potentials of arbitrary shape. Dynamical equations for the parameters of the soliton have the form of integro-differential equations. The possibility to solve these equations quickly and accurately, using standard numerical procedures, has been demonstrated. Quite good agreement between the results of variational equations and direct numerical solution of the original Gross-Pitaevskii equation is found for soliton scattering on potential barriers, while scattering on potential wells is described qualitatively. The proposed approach can be extended to other soliton bearing equations, such as generalized nonlinear Schrödinger equation and nonlocal Gross-Pitaevskii equation for dipolar condensates. The results can be useful in development of new methods aimed at probing the external potentials/defects by scattering solitons on them.

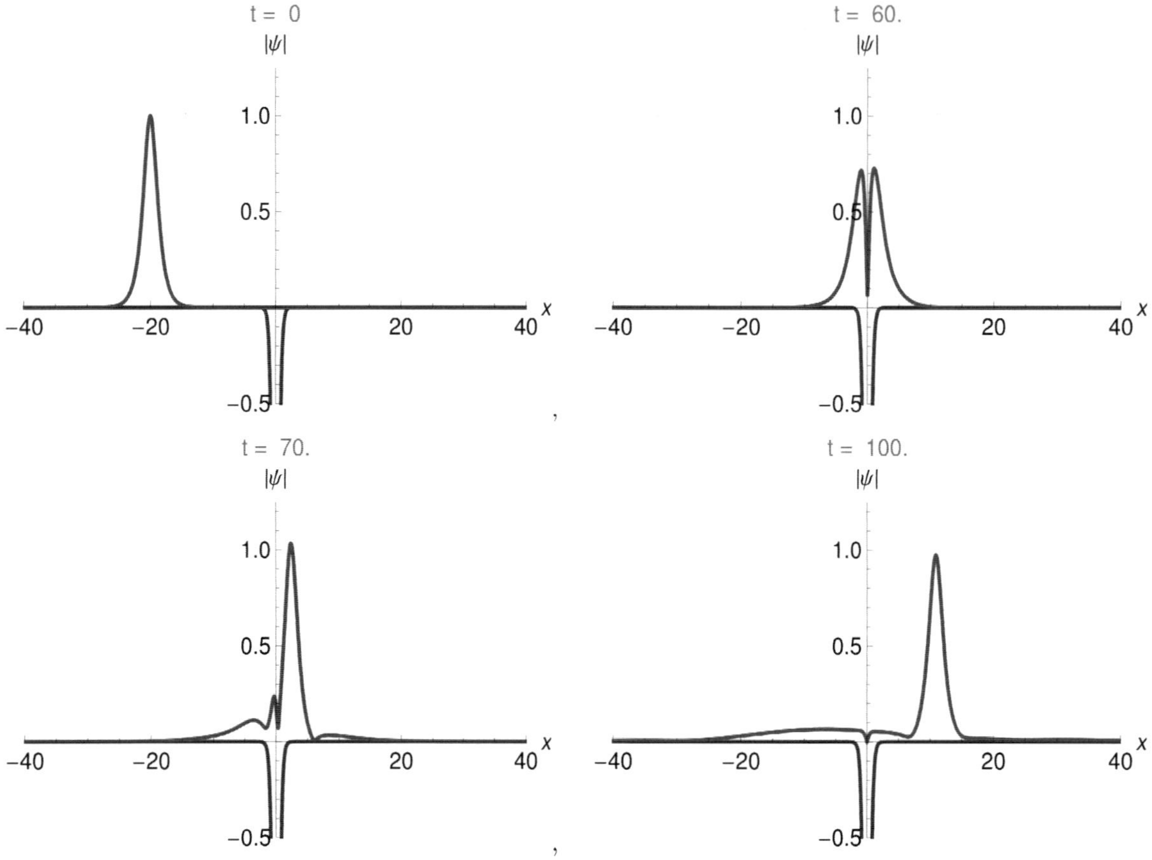

Figure 5. Transmission of the soliton with $A = 1, a = 1, \xi_0 = -20$ through the potential well $V(x) = V_0/\cosh^2(x/w)$, with $V_0 = -4, w = 0.5$ at slightly overcritical velocity $v = 0.35$. Strong deformation of the soliton in the interaction region implies that the variational approximation with the trial function (4) will fail in this case. For visual convenience only the upper part of the potential is shown (blue line).

Acknowledgments
This work has been supported by the research grant No. FRGS 0409-108 of the Ministry of Higher Education, Malaysia and research grant of IIUM EDW B10-089-0428. BBB acknowledges partial financial support from the Fund for Fundamental Research of the Uzbek Academy of Sciences under grant No. $\Phi.5 - 12$.

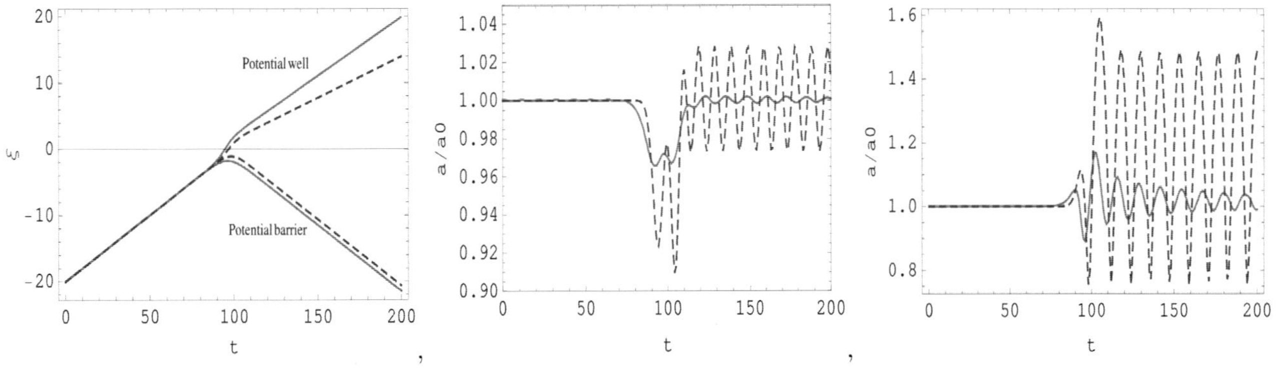

Figure 6. Reflection/transmission of the matter wave soliton with $A = 1, a = 1, v = 0.2$, through a shallow potential barrier/well $V(x) = V_0/\cosh^2(x/w)$ with $V_0 = \pm 0.1$, $w = 1$. Red solid and blue dashed lines correspond to numerical solution of the GPE (1) and predictions of the variational equations (10)-(11). Left panel: Center-of-mass dynamics. Middle and right panels: Dynamics of the width for potential barrier and well respectively

.

References

[1] *Encyclopedia of Nonlinear Science* 2005 ed A Scott (New York: Routledge)
[2] Remoissenet M 2003 *Waves Called Solitons. Concepts and Experiments* (Berlin: Springer)
[3] Hasegawa A and Kodama Y 1995 *Solitons in Optical Communications* (Oxford: Clarendon)
[4] Mollenauer L F and Gordon J P 2006 *Solitons in Optical Fibers. Fundamentals and Applications* (Amsterdam: Elsevier Academic Press)
[5] Mitschke F 2009 *Fiber Optics. Physics and Technology* (Berlin Heidelberg: Springer)
[6] Cronin A D, Schmiedmayer J and Pritchard D E 2009 *Rev. Mod. Phys.* **81** 1051
Billam T P, Cornish S L and Gardiner S A 2011 *Phys. Rev.* A **83** 041602(R)
[7] Cornish S L, Parker N G, Martin A M, Judd T E, Scott R G, Fromhold T M and Adams C S 2009 *Physica* D **238** 1299
[8] Hulet R, Pollack S, Dries D and Olson E 2010 *NLQUGAS: Nonlinear phenomena in degenerate quantum gases* ed Salguero J R (Ourense Spain) p 31
Dries D, Pollack S E, Hitchcock J M and Hulet R G 2010 *Phys. Rev.* A **82** 033603
[9] Al Khawaja U and Stoof H T C 2011 *New J. Phys.* **13** 085003
[10] Anderson D 1983 *Phys. Rev.* A **27** 3135
[11] Malomed B A 2002 *Progress in Optics* Vol **43** ed Wolf E (Amsterdam: Elsevier) pp 69191
[12] Minzoni A A, Smyth N F and Worthy A L 2007 *J. Opt. Soc. Am.* B **24** 1549
[13] Press W H, Teukolsky S A, Vetterling W T and Flannery B P 1996 *Numerical Recipes. The Art of Scientific Computing* (Cambridge: Cambridge University Press)
[14] Forsythe G E, Malcolm M A and Moler C B 1977 *Computer methods for mathematical computations* (New Jersey: Prentice-Hall)
[15] Al-Marzoug S M, Al-Amoudi S M, Al Khawaja U, Bahlouli H and Baizakov B B 2011 *Phys. Rev.* E **83** 026603
[16] Agrawal G P 1995 *Nonlinear Fiber Optics* (New York: Academic Press)
[17] Berg P, If F, Christiansen P L and Skovgaard O 1987 *Phys. Rev.* A **35** 4167
[18] Grimm R, Weidemüller M and Ovchinnikov Y B 2000 *Adv. At. Mol. Opt. Phys.* **42** 95
[19] Crandall R E and Litt B R 1983 *Ann. Phys., NY* **146**, 458
Kiriushcheva N and Kuzmin S 1998 *Am. J. Phys.* **66** 867
Lekner J 2007 *Am. J. Phys.* **75** 1151
[20] Flügge S 1971 *Practical Quantum Mechanics I* (Berlin: Springer)
[21] Lekner J 1987 *Theory of Reflection of Electromagnetic and Particle Waves* (Berlin: Dordrecht)
[22] Lee C and Brand J 2006 *Europhys. Lett.* **73** 321
[23] Goodman R H, Holmes P J and Weinstein M I 2004 *Physica* D **192** 215

Dynamic effects of electromagnetic wave on a damped two-level atom: Exact solution via path integral

Mekki Aouachria and Farida Halimi

Laboratoire de Physique Energétique Appliquée (LPEA), Département des Sciences de la Matiére, Faculté des Sciences, Université Hadj Lakhdar, Batna, Algeria

E-mail: mekkiaouachria@yahoo.fr

Abstract. The spin coherent state path integral describing the dynamics of a two-level system interacting with electromagnetic wave of circular polarization is considered. The propagator is first written in the standard form by replacing the spin by a unit vector aligned along the polar and azimuthal directions. Then it is determined exactly using perturbation methods. Thus, the exact energy spectra with corresponding wave functions are deduced.

1. Introduction

The applications of path-integral formalism have widely increased since a large class of potentials had been resolved [1]. However it is known that the most relativistic interactions are those where the spin, which is a very useful and very important notion in physics, is taken into account. In the framework of non-relativistic theory the phenomena of spin is automatically introduced by the Pauli equation which contains the Schrödinger Hamiltonian and a spin-field interaction. This motivates the research into the solvable Pauli equations which is inevitably useful in applied physics. For instance a well-known example of its direct application is the time-dependent field acting on an atom with two levels whose time-evolution is controlled by the Pauli-type equation. The solution for this equation has made clear the associated transition amplitudes [2]. This and similar [3, 4] types of interaction aside, there are little analytical and exact computations which treat the time-dependent spin-field interaction. Furthermore, if one replaces the time dependence of the exterior field by a space-time dependence or by only space dependence this becomes even more restrictive [5, 6, 7, 8, 9, 10].

Moreover, the problem becomes nearly unsolvable if we try to build these solutions by the path integral formalism because the spin is a discrete quantity. The difficulty here is associated to the fact that the path integral lacks some classical ideas such as trajectories and up to now one does not know how to deal with this kind of technique in this important case. Thus some effort has been made to find a partial solution using the Schwinger's model of spin and some explicit computations are then carried out [11, 12, 13, 14, 15, 16, 17].

A different model for spin is the use of the spin coherent state path integral which turns out to be helpful in visualizing the quantum dynamics in terms of classical ideas, and to our knowledge few explicit and semiclassical calculations are carried out [18, 19, 20]. If on the other hand the time-dependence of the exterior field is replaced by space-time dependence or space dependence, to our knowledge, the exact solution in these cases using spin coherent state path integral are very rare [21, 22, 23, 24]. The aim of this paper is to give an attempt for the case below where we present an explicit spin coherent state path integral. For this reason we are devoted to this type of interaction by considering a problem which has recently been treated according to usual quantum mechanics [25]. It acts on an atom which has two levels and which interacts with electromagnetic wave of circular polarization. The same problem has been considered with damping in ref. [26].

The purpose of this article is to deal with the same problem using the formalism of spin coherent path integral. The two-level atom with a mass m and an angular frequency ω has a dipole moment $\mathbf{D}$. The electromagnetic wave has wave vector k and angular frequency ω_L, propagating along the z-axis, and is described by the electric field

$$\mathbf{E} = (A\cos(\omega_L t - kz), -A\sin(\omega_L t - kz), 0)\,, \tag{1}$$

where A is the amplitude of $\mathbf{E}$. The dynamics of the atom in interaction with the electromagnetic wave is described by the following Hamiltonian:

$$H = \frac{p^2}{2m} + \frac{1}{2}\hbar\omega\sigma_z - \frac{i\hbar}{2}\begin{pmatrix} \gamma_1 & 0 \\ 0 & \gamma_2 \end{pmatrix} + \mathbf{V}, \tag{2}$$

where

- the first term represents the kinetic energy associated with the center–of–mass momentum along the z-direction,
- the second and the third terms describe the internal movement of the atom with, $\gamma_1 = \frac{1}{\tau_1}$ and $\gamma_2 = \frac{1}{\tau_2}$, being the lifetimes of the two atomic levels,
- $\mathbf{V}$ is the interaction energy between the atom and electromagnetic wave represented in the dipole approximation by

$$\mathbf{V} = -\mathbf{DE} = \begin{pmatrix} 0 & -\frac{1}{2}\hbar\Omega e^{-i(\omega_L t - kz)} \\ -\frac{1}{2}\hbar\Omega e^{i(\omega_L t - kz)} & 0 \end{pmatrix}, \tag{3}$$

where $DA = \frac{1}{2}\hbar\Omega$.

The Hamiltonian related to our problem has the following form

$$\begin{aligned} H &= H_0 + H_{int} \\ &= \frac{p^2}{2m} + \frac{1}{2}\hbar(\omega - \frac{i\hbar\gamma_-}{2})\sigma_z - \frac{i\hbar\gamma_+}{4}\mathbf{I} - \frac{1}{2}\hbar\Omega e^{-i(\omega_L t - kz)}\sigma_+ - \frac{1}{2}\hbar\Omega e^{i(\omega_L t - kz)}\sigma_-. \end{aligned} \tag{4}$$

where

$$\sigma_z = \begin{pmatrix} 1 & 0 \\ 0 & -1 \end{pmatrix}, \sigma_+ = \begin{pmatrix} 0 & 1 \\ 0 & 0 \end{pmatrix}, \sigma_- = \begin{pmatrix} 0 & 0 \\ 1 & 0 \end{pmatrix}, \mathbf{I} = \begin{pmatrix} 1 & 0 \\ 0 & 1 \end{pmatrix}, \tag{5}$$

are the usual Pauli matrices.

Considering this problem by the path integral approach, our motivation is the following. We show that for interaction with the coupling of spin-field type, the propagator is first written in the standard form $\sum_{\text{path}} \exp\left(iS\left(path\right)/\hbar\right)$, where S is the action that describes the system, and the discrete variable relative to spin being inserted as the (continuous) path using spin coherent states. With this approach, the formulation that uses the concept of trajectory is more suitable for a discussion of the semiclassical case which is based on the determination of classical paths. Then we show that for this interaction, the propagator is exactly calculable and the wave functions and the energy spectrum can be extracted.

We note that this problem has been recently studied using fermionic coherent state path integral [27, 28], where the authors use the Schulman-Langhlin procedure to estimate and replace the $(\Delta z_n)^2$ terms present in the action by $i\varepsilon\hbar/m$, which is translated into an effective potential. In contrary, in this article, we avoid this procedure and the effective potential naturally arises in the calculation, which is also another motivation for using the spin coherent state path integral formalism. The difference between the two approaches stems simply from the difference between the two distinct coherent states.

Our paper is organized as follows. In the next section we give some notation and the spin coherent state path integral for spin $\frac{1}{2}$ system for our further computations. In section 3, after setting up a path integral formalism for the propagator, we perform the direct calculations. To integrate over the variable of the exterior motion we introduce a particular rotation in spin-coherent state space which eliminates the phase of the electromagnetic field, then we linearise the kinetic energy term using the phase space. Consequently the effective potential $i\varepsilon\hbar/m$ naturally arises, and then we integrate over z. Accordingly, the integration over the spin variables is easy to carry out and the result is given as a perturbation series. These are summed up exactly and the explicit result of the propagator is directly computed and the wave function is then deduced. Finally, in section 4, we present our conclusions.

2. Path-integral Formulation

There are several ways to represent the spin in the path integral formalism [29, 30, 31]. We use the simplest way [32, 19] which consists of:

- replacing σ by a unit vector $\mathbf{n}$ directed along (θ, φ),
- associating a coherent state $|\Omega\rangle$

$$|\Omega\rangle = |\theta, \varphi\rangle = e^{-i\varphi S_z} e^{-i\theta S_y} |\uparrow\rangle \,, \tag{6}$$

obtained from two rotations of the angles θ and ϕ around z and y axes over the state $|\uparrow\rangle$, and whose scalar product and projector are respectively:

$$\langle \Omega | \, \Omega' \rangle = \cos\frac{\theta}{2}\cos\frac{\theta'}{2} e^{\frac{i}{2}(\varphi-\varphi')} + \sin\frac{\theta}{2}\sin\frac{\theta'}{2} e^{-\frac{i}{2}(\varphi-\varphi')}, \tag{7}$$

$$\frac{1}{2\pi}\int d\cos(\theta) d\varphi \, |\Omega\rangle \langle\Omega| = \mathbf{I}. \tag{8}$$

We label by z the real variable that describes the atom position, with the corresponding projector

$$\int |z\rangle \langle z| \, dz = 1, \tag{9}$$

and (θ, φ) the polar variables generating the dynamics of the spin.

The transition amplitude from the initial state $|z_i, \theta_i, \varphi_i\rangle$ at $t_i = 0$ to the final state $|z_f, \theta_f, \varphi_f\rangle$ at $t_f = T$ is given by the matrix elements of the time evolution operator

$$K\left(z_f, \Omega_f, z_i, \Omega_i; T\right) = \langle z_f, \theta_f, \varphi_f \mid \mathbf{T}_D \exp(-\frac{i}{\hbar}\int_0^T H dt) \mid z_i, \theta_i, \varphi_i\rangle, \tag{10}$$

where $\mathbf{T}_D$ is Dyson's time ordering symbol.

Discretizing the time $\varepsilon = T/(N+l)$, using the Trotter formula and inserting N-times the resolution of unity (8) and (9) between each pair of the evolution operator at time ε, we obtain the discretized form of the transition amplitude

$$K\left(z_f,\Omega_f,z_i,\Omega_i;T\right) = \lim_{N\longrightarrow\infty}(m/2\pi i\hbar\varepsilon)^{\frac{N}{2}}\int\prod_{n=1}^{n=N}dz_n\prod_{n=1}^{N+1}$$

$$\times\exp\frac{im}{2\hbar\varepsilon}\left(z_n-z_{n-1}\right)^2K^z\left(\Omega_f,\Omega_i;T\right), \tag{11}$$

with

$$K^z\left(\Omega_f,\Omega_i;T\right) = \lim_{N\longrightarrow\infty}\int\prod_{n=1}^{N}\frac{d\cos(\theta_n)d\varphi_n}{2\pi}$$

$$\times\prod_{n=1}^{N+1}\left[\langle\Omega_n\mid\Omega_{n-1}\rangle - i\varepsilon\left\langle\Omega_n\right|H_{int}\left|\Omega_{n-1}\right\rangle\right], \tag{12}$$

where

$$z_{N+1}=z_f,\quad \Omega_{N+1}=\Omega_f \quad\text{and}\qquad z_0=z_i,\quad \Omega_0=\Omega_i. \tag{13}$$

It is easy to find that the following matrix elements:

$$\left\langle\Omega\right|\sigma_z\left|\Omega'\right\rangle = \cos\frac{\theta}{2}\cos\frac{\theta'}{2}e^{+\frac{i}{2}(\varphi-\varphi')} - \sin\frac{\theta}{2}\sin\frac{\theta'}{2}e^{-\frac{i}{2}(\varphi-\varphi')}, \tag{14}$$

$$\left\langle\Omega\right|\sigma_+\left|\Omega'\right\rangle = \cos\frac{\theta}{2}\sin\frac{\theta'}{2}e^{+\frac{i}{2}(\varphi+\varphi')}, \tag{15}$$

$$\left\langle\Omega\right|\sigma_-\left|\Omega'\right\rangle = \sin\frac{\theta}{2}\cos\frac{\theta'}{2}e^{-\frac{i}{2}(\varphi+\varphi')}, \tag{16}$$

and the propagator related to our problem (11) takes the form of Feynman path integral

$$K = \int Dpath\exp(iAction), \tag{17}$$

which means in our case:

$$K\left(z_f,\Omega_f,z_i,\Omega_i;T\right) = \lim_{N\longrightarrow\infty}(m/2\pi i\hbar\varepsilon)^{\frac{N}{2}}\int\prod_{n=1}^{n=N}dz_n\prod_{n=1}^{N+1}\exp\frac{im}{2\hbar\varepsilon}\left(z_n-z_{n-1}\right)^2$$

$$\times\int\prod_{n=1}^{N}\frac{d\cos(\theta_n)d\varphi_n}{2\pi}\exp\left\{\sum_{n=1}^{n=N+1}\left[\log<\Omega_n\left|\Omega_{n-1}\right\rangle - i\varepsilon\frac{\left\langle\Omega_n\right|H_{int}\left|\Omega_{n-1}\right\rangle}{<\Omega_n\left|\Omega_{n-1}\right\rangle}\right]\right\}. \tag{18}$$

Having obtained the conventional form it remains to integrate it in order to extract the interesting physical properties. We thus proceed to the calculation of $K\left(z_f,\Omega_f,z_i,\Omega_i;T\right)$.

3. The calculation of the propagator

We note that (18) can be rewritten in the form

$$K\left(z_f,\Omega_f,z_i,\Omega_i;T\right) = \lim_{N\longrightarrow\infty}(m/2\pi i\hbar\varepsilon)^{\frac{N}{2}}\int\prod_{n=1}^{n=N}dz_n\prod_{n=1}^{N+1}$$

$$\times e^{-\frac{T\gamma_+}{4}} \exp \frac{im}{2\hbar\varepsilon} (z_n - z_{n-1})^2 \lim_{N \longrightarrow \infty} \int \prod_{n=1}^{N} \frac{d\cos(\theta_n) d\varphi_n}{2\pi}$$

$$\times \prod_{n=1}^{N+1} \begin{pmatrix} \cos\frac{\theta_n}{2} e^{+\frac{i}{2}\varphi_n} & \sin\frac{\theta_n}{2} e^{-\frac{i}{2}\varphi_n} \end{pmatrix} R(z_n, t_n) \begin{pmatrix} \cos\frac{\theta_{n-1}}{2} e^{-\frac{i}{2}\varphi_{n-1}} \\ \sin\frac{\theta_{n-1}}{2} e^{+\frac{i}{2}\varphi_{n-1}} \end{pmatrix}, \tag{19}$$

with

$$R(z_n, t_n) = \begin{pmatrix} 1 - i\varepsilon\frac{1}{2}\left(\omega - \frac{i\hbar\gamma_-}{2}\right) & i\varepsilon\frac{1}{2}\Omega e^{-i(\omega_L t_n - kz_n))} \\ i\varepsilon\frac{1}{2}\Omega e^{i(\omega_L t_{n-1} - kz_{n-1})} & 1 + i\varepsilon\frac{1}{2}\left(\omega - \frac{i\hbar\gamma_-}{2}\right) \end{pmatrix}. \tag{20}$$

To integrate it is necessary to first eliminate the inconvenient terms $e^{-i(\omega_L t_n - kz_n)}$ and $e^{i(\omega_L t_{n-1} - kz_{n-1})}$ with the help of the following change of variable:

$$\varphi_n = \varphi_n' + \omega_L t_n - kz_n. \tag{21}$$

Then, the measure

$$\prod_{n=1}^{N} \frac{d\cos(\theta_n) d\varphi_n}{2\pi} = \prod_{n=1}^{N} \frac{d\cos(\theta_n) d\varphi_n'}{2\pi}, \tag{22}$$

remains unchanged. The expression (19) becomes

$$K\left(z_f, \Omega_f', z_i, \Omega_i'; T\right) = \lim_{N \longrightarrow \infty} (m/2\pi i\hbar\varepsilon)^{\frac{N}{2}} \int \prod_{n=1}^{n=N} dz_n \prod_{n=1}^{N+1}$$

$$\times e^{-\frac{T\gamma_+}{4}} \exp \frac{im}{2\hbar\varepsilon} (z_n - z_{n-1})^2 \lim_{N \longrightarrow \infty} \int \prod_{n=1}^{N} \frac{d\cos(\theta_n) d\varphi_n'}{2\pi}$$

$$\times \prod_{n=1}^{N+1} \begin{pmatrix} \cos\frac{\theta_n}{2} e^{+\frac{i}{2}\varphi_n'} & \sin\frac{\theta_n}{2} e^{-\frac{i}{2}\varphi_n'} \end{pmatrix} R_1(z_n, t_n) \begin{pmatrix} \cos\frac{\theta_{n-1}}{2} e^{-\frac{i}{2}\varphi_{n-1}'} \\ \sin\frac{\theta_{n-1}}{2} e^{+\frac{i}{2}\varphi_{n-1}'} \end{pmatrix}, \tag{23}$$

where

$$R_1(z_n, t_n) = \begin{pmatrix} 1 - i\varepsilon\frac{1}{2}\left(\Delta\omega - \frac{i\hbar\gamma_-}{2}\right) e^{-i\frac{k}{2}\Delta z_n} & i\varepsilon\frac{1}{2}\Omega \\ i\varepsilon\frac{1}{2}\Omega & 1 + i\varepsilon\frac{1}{2}\left(\Delta\omega - \frac{i\hbar\gamma_-}{2}\right) e^{+i\frac{k}{2}\Delta z_n} \end{pmatrix} \tag{24}$$

and $\Delta\omega = \omega - \omega_L$

Then we use the following identity:

$$\int_{-\infty}^{+\infty} \frac{dp_n}{2\pi\hbar} \exp\left[\frac{-i\varepsilon}{2m\hbar} p_n^2 + \frac{i}{\hbar} p_n \left(\Delta z_n \pm \frac{\varepsilon\hbar}{2m} k\right)\right]$$

$$= \sqrt{\frac{m}{2\pi i\varepsilon\hbar}} \exp\left[\frac{im}{2\varepsilon\hbar}\left(\Delta z_n \pm \frac{\varepsilon\hbar}{2m} k\right)^2\right]. \tag{25}$$

Thus the propagator (23) can be rewritten as:

$$K\left(z_f,\Omega'_f,z_i,\Omega'_i;T\right)=\lim_{N\longrightarrow\infty}\int_{-\infty}^{+\infty}\prod_{n=1}^{n=N}dz_n\prod_{n=1}^{n=N+1}\int_{-\infty}^{+\infty}\frac{dp_n}{2\pi\hbar}$$
$$\times\exp\sum_{n=1}^{n=N+1}\left[\frac{-i\varepsilon}{2m\hbar}p_n^2+\frac{i}{\hbar}p_n\Delta z_n-i\frac{\varepsilon\hbar}{8m}k^2\right]\lim_{N\longrightarrow\infty}\int\prod_{n=1}^{N}\frac{d\cos(\theta_n)d\varphi'_n}{2\pi}\prod_{n=1}^{N+1}$$
$$\times\begin{pmatrix}\cos\frac{\theta'_n}{2}e^{+\frac{i}{2}\varphi'_n} & \sin\frac{\theta'_n}{2}e^{-\frac{i}{2}\varphi'_n}\end{pmatrix}R_2(z_n,t_n)\begin{pmatrix}\cos\frac{\theta'_{n-1}}{2}e^{-\frac{i}{2}\varphi'_{n-1}}\\ \sin\frac{\theta'_{n-1}}{2}e^{+\frac{i}{2}\varphi'_{n-1}}\end{pmatrix},\tag{26}$$

where

$$R_2(p_n,t_n)=\begin{pmatrix}1-i\varepsilon(\frac{1}{2}\left(\Delta\omega-\frac{i\hbar\gamma_-}{2}\right)+\frac{p_nk}{2m}) & i\varepsilon\frac{1}{2}\Omega\\ i\varepsilon\frac{1}{2}\Omega & 1+i\varepsilon\left(\frac{1}{2}\left(\Delta\omega-\frac{i\hbar\gamma_-}{2}\right)+\frac{p_nk}{2m}\right)\end{pmatrix}.\tag{27}$$

By integrating over the N variables z_n, we clearly get Dirac functions $\delta\left(\dot{p}\right)$ which reflects the conservation of the atom impulsion during the movement, i.e:

$$p_1=p_2=\cdots=p_{N+1}=p.\tag{28}$$

Hence the propagator (25) takes the following form

$$K\left(z_f,\Omega'_f,z_i,\Omega'_i;T\right)=\int_{-\infty}^{+\infty}\frac{dp}{2\pi\hbar}\exp\left[\frac{i}{\hbar}p(z_f-z_i)-\frac{iTp^2}{2m\hbar}-iT\frac{\hbar k^2}{8m}\right]$$
$$\times\lim_{N\longrightarrow\infty}\int\prod_{n=1}^{N}\frac{d\cos(\theta_n)d\varphi'_n}{2\pi}$$
$$\times\prod_{n=1}^{N+1}\begin{pmatrix}\cos\frac{\theta'_n}{2}e^{+\frac{i}{2}\varphi'_n} & \sin\frac{\theta'_n}{2}e^{-\frac{i}{2}\varphi'_n}\end{pmatrix}R_2(p,t_n)\begin{pmatrix}\cos\frac{\theta'_{n-1}}{2}e^{-\frac{i}{2}\varphi'_{n-1}}\\ \sin\frac{\theta'_{n-1}}{2}e^{+\frac{i}{2}\varphi'_{n-1}}\end{pmatrix},\tag{29}$$

where

$$R_2(p,t_n)=\begin{pmatrix}1-i\varepsilon\left(\frac{1}{2}\left(\Delta\omega-\frac{i\hbar\gamma_-}{2}\right)+\frac{pk}{2m}\right) & i\varepsilon\frac{1}{2}\Omega\\ i\varepsilon\frac{1}{2}\Omega & 1+i\varepsilon\left(\frac{1}{2}\left(\Delta\omega-\frac{i\hbar\gamma_-}{2}\right)+\frac{pk}{2m}\right)\end{pmatrix}.\tag{30}$$

let us integrate on the anguler variables $\theta_n$et φ'_n
the propagateur takes the following form

$$K(f,i;T)=e^{-\frac{T\gamma_+}{4}}\int_{-\infty}^{+\infty}\frac{dp}{2\pi\hbar}\exp\left[\frac{i}{\hbar}p(z_f-z_i)-\frac{iTp^2}{2m\hbar}-iT\frac{\hbar k^2}{8m}\right]$$
$$\times\begin{pmatrix}\cos\frac{\theta_f}{2}e^{\frac{i\varphi'_f}{2}} & \sin\frac{\theta_f}{2}e^{-\frac{i\varphi'_f}{2}}\end{pmatrix}R\left(p,T\right)\begin{pmatrix}\cos\frac{\theta_i}{2}e^{-\frac{i\varphi'_i}{2}}\\ \sin\frac{\theta_i}{2}e^{\frac{i\varphi'_i}{2}}\end{pmatrix}.\tag{31}$$

with

$$R\left(p,T\right)=\lim_{N\longrightarrow\infty}(-1)^N\prod_{\underleftarrow{n=1}}^{n=N+1}R_2(p,t_n),\tag{32}$$

The arrow under the product symbol indicates the time ordering operation. We concentrate on the evaluation of the matrix elements $R_{ij}(p,T)$. To this aim the matrix $R(p,t_n)$ is written as a sum of a diagonal matrix and an off-diagonal matrix. To first order in ε we have

$$R(p,t_n) = e^{-i\varepsilon\omega(j)\sigma_z} + i\varepsilon K(n),$$

where the off-diagonal matrix is given by

$$K(n) = \begin{pmatrix} 0 & u(n) \\ u(n) & 0 \end{pmatrix}, \tag{33}$$

with

$$\omega(n) = \frac{1}{2}\left(\Delta\omega - \frac{i\hbar\gamma_-}{2}\right) + \frac{kp}{2m}, \text{ and } \quad u(n) = \frac{\Omega}{2}, \tag{34}$$

let us pass to the calculation of produce matrix according to [11]

$$\prod_{n=1}^{N}(e^{-i\varepsilon\omega(n)\sigma_z} + i\varepsilon K(n)) =$$

$$= e^{-i\sum\limits_{j=1}^{N}\varepsilon\omega(j)\sigma_z} + \sum_{l=1}^{N}(i\varepsilon)e^{-i\sum\limits_{l+1}^{N}\varepsilon\omega(k)\sigma_z}K(l)e^{-i\sum\limits_{1}^{l-1}\varepsilon\omega(k)\sigma_z}$$

$$+\sum_{l_1=1}^{N}\sum_{l_2=1}^{l_1-1}(i\varepsilon)^2 e^{-i\sum\limits_{l_1+1}^{N}\varepsilon\omega(k)\sigma_z}K(l_1)e^{-i\sum\limits_{l_2+1}^{l_1-1}\varepsilon\omega(k)\sigma_z}K(l_2)e^{-i\sum\limits_{1}^{l_1-1}\varepsilon\omega(k)\sigma_z} + \ldots$$

$$+\sum_{l_1=1}^{N}\sum_{l_2=1}^{l_1-1}\sum_{l_3=1}^{l_2-1}\ldots\sum_{l_{N-1}=1}^{l_{N-2}-1}\sum_{l_N=1}^{l_{N-1}-1}(i\varepsilon)^N e^{-i\sum\limits_{l_1+1}^{N}\varepsilon\omega(k)\sigma_z}K(l_1)e^{-i\sum\limits_{l_2+1}^{l_1-1}\varepsilon\omega(k)\sigma_z}K(l_2)$$

$$e^{-i\sum\limits_{l_3+1}^{l_2-1}\varepsilon\omega(k)\sigma_z}\ldots e^{-i\sum\limits_{l_{N-1}+1}^{l_{N-2}-1}\varepsilon\omega(k)\sigma_z}K(l_{N-1})e^{-i\sum\limits_{l_N+1}^{l_{N-1}-1}\varepsilon\omega(k)\sigma_z}K(l_N)e^{-i\sum\limits_{1}^{l_N-1}\varepsilon\omega(k)\sigma_z}. \tag{35}$$

Taking the limit $N \longrightarrow +\infty$ of (36) leads to

$$R(p,T) = e^{-i\int_0^T ds\omega(p,s)\sigma_z} + i\int_0^T ds_1 e^{-i\int_{s_1}^T ds\omega(p,s)\sigma_z}K(s_1)e^{-i\int_0^{s_1} ds\omega(p,s)\sigma_z}$$

$$+(i)^2\int_0^T ds_1\int_0^{s_1} ds_2 e^{-i\int_{s_1}^T ds\omega(p,s)\sigma_z}K(s_1)e^{-i\int_{s_2}^{s_1} ds\omega(p,s)\sigma_z}K(s_2)$$

$$+\ldots(i)^N\int_0^T ds_1\int_0^{s_1} ds_2\int_0^{s_2} ds_3\ldots\int_0^{s_{N-1}} ds_N e^{-i\int_{s_1}^T ds\omega(p,s)\sigma_z}K(s_1)$$

$$\times e^{-i\int_{s_2}^{s_1} ds\omega(p,s)\sigma_z}K(s_2)\ldots K(s_{N-1})e^{-i\int_{s_N}^{s_{N-1}} ds\omega(p,s)\sigma_z}K(s_N)e^{-i\int_0^{s_N} ds\omega(p,s)\sigma_z} + \ldots \tag{36}$$

a simple calculation shows us that the terms odd respectively even are the element antidiagonaux respectitively diagonal of R.

$$R_{11}(p,T) = e^{-i\int_0^T ds\omega(p,s)} + \sum_{n=1}^{\infty} i^{2n}\int_0^T ds_1\int_0^{s_1} ds_2\int_0^{s_2} ds_3\ldots\int_0^{s_{2n-1}} ds_{2n}\times$$

$$e^{-i\int_{s_1}^T ds\omega(p,s)}u(p(s_1),s_1)e^{+i\int_{s_2}^{s_1} ds\omega(p,s)}u(p(s_2),s_2)$$

$$\times\ldots e^{+i\int_{s_{2n}}^{s_{2n-1}} ds\omega(p,s)}u(p(s_{2n}),s_{2n})e^{-i\int_0^{s_{2n}} ds\omega(p,s)} \tag{37}$$

and

$$R_{12}(p,T) = i\int_0^T ds_1 e^{-i\int_{s_1}^T ds\omega(p,s)} u(p(s_1),s_1)R_{22}(p(s_1),s_1), \quad (38)$$

$$\begin{aligned} R_{22}(p,T) &= R^*_{11}(p,T), \\ R_{21}(p,T) &= -R^*_{12}(p,T). \end{aligned} \quad (39)$$

For instance

$$R_{11}(p,T) =$$
$$\left[1+\sum_{n=1}^{\infty}(i\frac{\Omega}{2})^{2n}\int_0^T e^{i\Delta s_1}ds_1\int_0^{s_1} e^{-i\Delta s_2}ds_2...\int_0^{s_{2n-1}} e^{-i\Delta s_{2n}}ds_{2n}\right]e^{-i\frac{T\Delta}{2}}, \quad (40)$$

with

$$\begin{aligned} \Delta &= 2\omega(n,s). \quad (41) \\ \omega(n,s) &= \frac{\Delta\omega}{2} - \frac{i\gamma_-}{4} + \frac{kp}{2m} \quad (42) \end{aligned}$$

4. Summation of the perturbation series

Let us put

$$F(0,T) = \sum_{n=1}^{\infty}(i\frac{\Omega}{2})^{2n}\int_0^T e^{i\Delta s_1}ds_1\int_0^{s_1} e^{-i\Delta s_2}ds_2...\int_0^{s_{2n-1}} e^{-i\Delta s_{2n}}ds_{2n}$$

The evaluation of this multiple integral which has the form of a repeated convolution product, proceeds by taking the Laplace transform of all functions with respect to the time differences in their arguments and using the convolution theorem of Laplace transform [11]. The result is

$$\widetilde{F}(0,q) = \int_0^{+\infty} dTe^{-qT}F(0,T)dT = .\frac{1}{q}\sum_{n=1}^{\infty}\left[\frac{-\Omega^2/4}{q(q-i\Delta)}\right]^n$$

the sum over N is carried out with the result

$$\widetilde{F}(0,q) = \frac{q-i\Delta}{q(q-i\Delta)+\frac{\Omega^2}{4}} - \frac{1}{q}$$

this result is valid for $\mid \frac{(i\frac{\Omega}{2})^2}{q(q-i\Delta)} \mid < 1$.We notice that is possible to get a contour of the inverse Laplace integration where the condition above is verified. Taking the inverse Laplace transform we obtain for the element $R_{11}(p,T)$ the expression

$$R_{11}(p,T) = \cos(\Omega' T) - \frac{i\Delta}{2\Omega'}\sin(\Omega' T), \quad \text{with} \quad \Omega' = \sqrt{\frac{\Delta^2}{4}+\frac{\Omega^2}{4}}$$

Following the same method of calculations, the remaining elements are calculated and listed below

$$\begin{aligned} R_{12}(p,T) &= \frac{i\Omega}{2\Omega'}\sin(\Omega' T), \quad (43) \\ R_{21}(p,T) &= \frac{i\Omega}{2\Omega'}\sin(\Omega' T), \quad (44) \\ R_{22}(p,T) &= \cos(\Omega' T) + \frac{i\Delta}{2\Omega'}\sin(\Omega' T). \quad (45) \end{aligned}$$

Substituting this result in the equation (31) the propagator takes finally the from:

$$\begin{aligned} K(f,i;T) &= \int_{-\infty}^{+\infty} \frac{dp}{2\pi\hbar} \exp\left[-\frac{iTp^2}{2m\hbar} - \frac{i\hbar k^2 T}{8m} + \frac{i}{\hbar}p(z_f - z_i)\right] \\ &\times \begin{pmatrix} \cos\frac{\theta_f}{2}e^{\frac{i\varphi'_f}{2}} & \sin\frac{\theta_f}{2}e^{-\frac{i\varphi'_f}{2}} \end{pmatrix} R(p,T) \begin{pmatrix} \cos\frac{\theta_i}{2}e^{-\frac{i\varphi'_i}{2}} \\ \sin\frac{\theta_i}{2}\exp(+\frac{i\varphi'_i}{2}) \end{pmatrix}, \end{aligned} \tag{46}$$

with

$$R(p,T) = \begin{pmatrix} \cos(\Omega' T) - \frac{i\Delta}{2\Omega'}\sin(\Omega' T) & -\frac{i\Omega}{2\Omega'}\sin(\Omega' T) \\ -\frac{i\Omega}{2\Omega'}\sin(\Omega' T) & \cos(\Omega' T) + \frac{i\Delta}{2\Omega'}\sin(\Omega' T) \end{pmatrix}. \tag{47}$$

Now we come back to the old angailars variables (θ, φ). So, the exact expression of the propagator concerning to our problem is the following

$$\begin{aligned} K(f,i;T) &= \int_{-\infty}^{+\infty} \frac{dp}{2\pi\hbar} \exp\left[-\frac{iTp^2}{2m\hbar} - \frac{i\hbar k^2 T}{8m} + \frac{i}{\hbar}p(z_f - z_i)\right] e^{-\frac{T\gamma_+}{4}} \\ &\cdot \begin{pmatrix} \cos\frac{\theta_f}{2}e^{\frac{i\varphi_f}{2}} & \sin\frac{\theta_f}{2}e^{-\frac{i\varphi_f}{2}} \end{pmatrix} \mathbf{S}(p,T) \begin{pmatrix} \cos\frac{\theta_i}{2}e^{-\frac{i\varphi_i}{2}} \\ \sin\frac{\theta_i}{2}e^{+\frac{i\varphi_i}{2}} \end{pmatrix}, \end{aligned} \tag{48}$$

Where the elements of matrix $\mathbf{S}(p,T)$ is given by:

$$\mathbf{S}_{11}(p,T) = \left[\cos\Omega' T - \frac{i\Delta}{2\Omega'}\sin\Omega' T\right] e^{-\frac{T\gamma_+}{4}} \exp\frac{i}{2}(kz_f - \omega_L T - kz_i), \tag{49}$$

$$\mathbf{S}_{22}(p,T) = \left[\cos\Omega' T + \frac{i\Delta}{2\Omega'}\sin\Omega' T\right] e^{-\frac{T\gamma_+}{4}} \exp -\frac{i}{2}(kz_f - \omega_L T - kz_i), \tag{50}$$

$$\mathbf{S}_{12}(p,T) = -\frac{i\Omega}{2\Omega'}\sin\Omega' T e^{-\frac{T\gamma_+}{4}} \exp\frac{i}{2}(kz_f - \omega_L T + kz_i), \tag{51}$$

$$\mathbf{S}_{21}(p,T) = -\frac{i\Omega}{2\Omega'}\sin\Omega' T e^{-\frac{T\gamma_+}{4}} \exp -\frac{i}{2}(kz_f - \omega_L T + kz_i). \tag{52}$$

The angles θ, φ are allowed to vary only in the limited domains $[0, 2\pi]$ and $[0, 4\pi]$. Our propagator is the following:

$$\begin{aligned} K(f,i;T) &= \sum_{n=-\infty}^{+\infty} K(z_f, \theta_f + 2n\pi, z_i, \varphi_f + 4n\pi,\ \theta_i, \varphi_i; T) \\ &= K(z_f, \Omega_f, z_i, \Omega_i; T) \end{aligned} \tag{53}$$

It is an expression for the propagator in the spin coherent state representation.

5. The wave functions

Let us now eliminate the coherent states by computing the transition amplitude between the proper states of the spin. We take an example of the matrix element

$$K_{\uparrow\uparrow}(z_f, z_i; T) = \langle\uparrow|K(z_f, z_i; T)|\uparrow\rangle. \tag{54}$$

With the help of the completeness relations, this amplitude becomes

$$K_{\uparrow\uparrow}(z_f, z_i; T) = \frac{1}{2\pi}\int d\cos(\theta_f)d\varphi_f \frac{1}{2\pi}\int d\cos(\theta_i)d\varphi_i$$

$$\times \langle \uparrow |\Omega_f\rangle \, K\left(z_f, \Omega_f, z_i, \Omega_i; T\right) \langle \Omega_i | \uparrow \rangle \tag{55}$$

If we fix the initial state of the atom as $|m_i\rangle = |\uparrow\rangle$, and the final state as $|m_f\rangle = |\uparrow\rangle$, we obtain

$$\langle \uparrow |\Omega_f\rangle = \cos\frac{\theta_f}{2} e^{-\frac{i}{2}\varphi_f}, \qquad \text{and} \qquad \langle \Omega_i | \uparrow\rangle = \cos\frac{\theta_i}{2} e^{+\frac{i}{2}\varphi_i}. \tag{56}$$

Substituting (56) in (55) and integrating over polar coordinates, (55) takes the following form

$$K_{\uparrow\uparrow}(z_f, z_i; T) = \int_{-\infty}^{+\infty} \frac{dp}{2\pi\hbar} \, \exp\left[\frac{i}{\hbar} p(z_f - z_i) - \frac{iTp^2}{2m\hbar} - i\frac{\Delta e_p}{2\hbar} T\right] \mathbf{S}_{11}. \tag{57}$$

In the same manner, we proceed for the remaining elements. The result takes the following matrix form

$$K\left(z_f, z_i; T\right) = \int_{-\infty}^{+\infty} \frac{dp}{2\pi\hbar} \exp\left[\frac{i}{\hbar} p(z_f - z_i) - \frac{iTp^2}{2m\hbar} - i\frac{\Delta e_p}{2\hbar} T\right] \mathbf{S}\left(T\right), \tag{58}$$

which coincide with those obtained in [28] using fermionic coherent states path integral.

From this propagator, the atom state at time T is deduced from the initial state via the evolution equation

$$\Psi(z, T) = \begin{pmatrix} \Psi_1(z, T) \\ \Psi_2(z, T) \end{pmatrix} = \int_{-\infty}^{+\infty} K(z, y; T)\Psi(y, 0) dy. \tag{59}$$

The same physical features can thus be reobtained as has been done in [25] and in the case with out damping done in [26].

6. Conclusion

We have given an exact treatment using the path-integral formalism to the problem of the two-level atom in interaction with an electromagnetic wave. Thanks to the two variables (θ, φ) replacing the spin, the propagator has been written, first in the conventional form $\int \mathcal{D}(path) \exp \frac{i}{\hbar} S(path)$, then determined with exactitude. Thus the wave function has been deduced via the evolution equation. Our results through the path-integral approach are in accordance with those obtained in [25].

Finally let us note that the expression (58) has the form $\sum_{\text{path}} \exp\left(iS\left(path\right)/\hbar\right)$ where the sum runs over all classical paths (parameterized here by the momentum). This form, remarkable for this problem, shows that a traditional semi-treatment can lead to an exact result.

Acknowledgments

This work was supported by PNR research project 8/u05/4094 N^0 31/13.

References

[1] Grosche C and Steiner F 1998 *Handbook of Feynman Path Integrals* (Verlag: Springer)
[2] Rabi I. I. 1936 *Phys. Rev.* **51** 652
[3] Tahmasebi M J and Sobouti Y 1991 *Mod. Phys. Lett.* B **5** 1919
[4] Tahmasebi M J and Sobouti Y 1991 *Mod. Phys. Lett.* B **6** 1255
[5] Lämmerzahl C and Bordé C J 1995 *Phys. Lett.* A **203** 59
[6] Qiong-Gui L 2005 *Phys. Lett.* A **342** 67
[7] Calvo M and Codriansky S 1983 *J. Math. Phys.* **24** 553
[8] Barut A O and Beker H 1991 *Eur. Phys. Lett.* **14** 197
[9] Mijatović M, Ivanovski C, Veljanoski B and Trenčevski K 1993 Z. *Phys.* A **345** 65
[10] Codriansky S, Cordero P and Salamo S 1995 *Z. Phys.* A **353** 341

[11] Boudjedaa T, Bounames A, Nouicer Kh, Chetouani L and Hammann T F 1995 *J. Math. Phys.* **36** 1602
[12] Boudjedaa T, Bounames A, Nouicer Kh, Chetouani L and Hammann T F 1996 *Phys. Scripta.* **54** 225
[13] Boudjedaa T, Bounames A, Nouicer Kh, Chetouani L and Hammann T F 1998 *Phys. Scripta.* **56** 545
[14] Merdaci A, Boudjedaa T and Chetouani L 2001 *Phys. Scripta.* **64** 15
[15] Merdaci A, Boudjedaa T and Chetouani L 2001 *Czech. J. Phys.* **51** 865
[16] Merdaci A, Boudjedaa T and Chetouani L 2001 *Eur. Phys. J. C* **22** 585
[17] Nouicer Kh and Chetouani L 2001 *Phys. Lett.* A **281** 218
[18] Aouachria M and Chetouani L 2009 *Can. J. Phys.* **87** 389
[19] Alscher A and Grabert H 1999 *J. Phys.* A **32** 4907
[20] Alscher A and Grabert H 2001 *Eur. Phys. J.* D **14** 127
[21] Aouachria M 2011 *Can. J. Phys.* **89** 1141
[22] Aouachria M 2011 *J. Korean. Phys. Soc.* **58** 689
[23] Aouachria M 2011 *Chinese. J. Phys.* **49** 689
[24] Aouachria M and Rekik R 2012 *AIP Conf. Proc.* **1444** 265
[25] Zeng G J, Zhou S L, Ao S M and Zeng Z Y 1997 *Phys. Rev.* A **55** 2945
[26] Zeng Z Y, Zeng G J, Kuang L M and Zhang L D 1999 *Phys. Lett.* A **261** 316
[27] Aouachria M and Chetouani L 2002 *Eur. Phys. J.* C **25** 333
[28] Aouachria M and Chetouani L 2002 *Chinese. J. Phys.* **40** 496
[29] Klauder J R 1960 *Ann. Phys.* (NY)**11** 123
[30] Klauder J R and Skagerstam B S 1985 *Coherent states application in physics and mathematical physics* (Word Scientific, Singapore)
[31] Ohnuki Y and Kashiwa T 1978 *Prog. Theor. Phys.* **60** 548
[32] Perelomov A M 1972 *Commun. Math. Phys.* **26** 22

Study of Localized Solutions of the Nonlinear Discrete Model for Dipolar BEC in an Optical Lattice by the Homoclinic Orbit Method

Azeddine Messikh[1] and Bakhram Umarov[2]

[1] Kulliyah of Information and Communication Technology, IIUM, Jalan Gombak 53100, Kuala Lumpur, Malaysia.

[2] Kulliyyah of Science, IIUM, Jalan Sultan Ahmad Shah, Bandar Indera Mahkota, 25200 Kuantan, Pahang, Malaysia.

E-mail: messikh@iium.edu.my

Abstract. We use homoclinic orbits to find solutions of a dynamical system of the dipolar Bose Einstein Condensate (BEC) in a deep optical lattice. The equation of motion is transformed to a two-dimensional map and its homoclinic orbits are computed numerically. Each homoclinic orbit leads to a different solution. These different solutions lead to different types of solitons. We also analyse the stability of the solutions.

1. Introduction

It is well known that the dynamics of the Bose Einstein condensate (BEC) trapped in an optical lattice can be described in tight-binding approximation by the discrete Nonlinear Schrödinger Equation (DNLSE) [1, 2]. This model opens the way to study different aspects of the BEC dynamics in an optical lattice, such as discrete solitons and nonlinear localized modes and their stability and dynamics, modulational instability, superfluid-insulator transition, etc. [3]. Also recently the nonlinearity caused by long range dipole-dipole interaction attracts much interest [4,5]. In this case the corresponding discrete model for dipolar BEC in an optical lattice was introduced in the work [6], and further investigated in [7].

The equation for one dimensional optical lattice in the dimensionless variables was derived in [8] and the nonlinear localized modes has been investigated in details by application of numerical Newton-Raphson method. In this work, following the paper [9] we rather use homoclinic orbits to find the localized solutions. The equation of motion is transformed to a two-dimensional map and the homoclinic orbits for this map are computed numerically. Each homoclinic orbit leads to a different solution. Also the linear stability of the found solutions is investigated.

2. Two-Dimensional map

Consider the dipolar Bose Einstein Condensate (BEC) in a deep optical lattice. It is well known that the governing equation for this system is given by [6–8]

$$i\frac{d}{dt}\Phi_n + \kappa(\Phi_{n+1} + \Phi_{n-1}) + q|\Phi_n|^2\Phi_n + g\left(|\Phi_{n+1}|^2 + |\Phi_{n-1}|^2\right)\Phi_n = 0, \quad (1)$$

where Φ_n is the complex field amplitude at the n-th site of the lattice, and κ is the linear parameter, while q and g correspond to the nolinearity parameters. This equation, which is 1D equation, has no exact analytical solution. Therefore, we use numerical methods and look for localized stationary solutions of the form

$$\Phi_n = u_n e^{i\omega t}. \tag{2}$$

If we substituted this form into the equation of motion (1), we get

$$q\, u_n^3 + \left[g\left(u_{n+1}^2 + u_{n-1}^2\right) - \omega\right] u_n + (u_{n+1} + u_{n-1})\,\kappa = 0. \tag{3}$$

It easy to show that this last equation has at most three fixed points

$$p_0 = 0 \text{ and } p_\pm = \pm\sqrt{\frac{\omega - 2\kappa}{2g + q}}. \tag{4}$$

It is worth mentioning that $p_\pm$ should be real numbers.

With the transformation $v_n = u_{n-1}$, Eq. (3) becomes a two-dimensional map

$$\left\{\begin{array}{lcl} u_{n+1} & = & 2\dfrac{(\omega - g v_n^2)u_n - q u_n^3 - \kappa v_n}{\kappa \pm \sqrt{\kappa^2 + 4g(\omega - g v_n^2)u_n^2 - 4gq u_n^4 - 4g\kappa u_n v_n}}, \\ v_{n+1} & = & u_n. \end{array}\right. \tag{5}$$

In what follows we restrict to the case when $\kappa > 0$ and to the map that corresponds to the denominator $\kappa + \sqrt{\kappa^2 + 4g(\omega - g v_n^2)u_n^2 - 4gq u_n^4 - 4g\kappa u_n v_n}$. Since u_n and v_n are real numbers, the map exists only when

$$\kappa^2 + 4g(\omega - g v_n^2)u_n^2 - 4g(q u_n^3 + \kappa v_n)u_n \geq 0. \tag{6}$$

Some examples for the existence of the map are shown by the dashed area in Fig. 1.

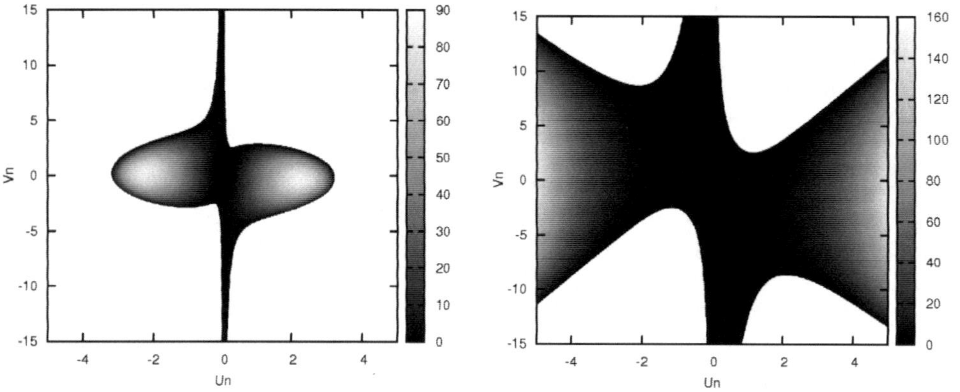

Figure 1. The map exists only in the dashed area. The parameters used are: (left figure) $\kappa = 1$, $q = 1$, $g = 0.8$, $\omega = 10$ and (right figure) $\kappa = 1$, $q = -0.6$, $g = 0.1$ and $\omega = 0.1$. These homoclinic points lead to three different solitons.

To get some solutions of Eq. (1), we fellow the method described in [10] which is based on homoclinic orbits. This method is briefly reviewed in the next section.

3. Homoclinic orbits

There are two interesting manifolds for a two-dimensional map, stable manifold and unstable manifold. The stable (unstable) manifold is the set of points that converge to a saddle fixed point through forward (backward) iterations of the map. It is well know that a fixed point in two-dimensional map is a saddle point if the two eigenvalues of its Jacobian, $\lambda_\pm$, satisfy the following inequalities

$$|\lambda_-| < 1 \quad \text{and} \quad |\lambda_+| > 1. \tag{7}$$

The stable manifold corresponds to the eigenvalue λ_-, while the unstable manifold corresponds to the other eigenvalue λ_+. An intersection point of the stable and unstable manifolds is called a homoclinic point. The orbit that contains this point is called a homoclinic orbit. It is clear that any point in a homoclinic orbit will converge to the saddle fixed point by forward and backward iterations. Therefore, homoclinic orbits give stationary solutions for Eq. (1) with the form given by Eq. (2) and converges to corresponding saddle fixed point.

Returning back to our map, Eq.(5). At the fixed point p_0 , the Jacobian matrix of the map for $\kappa > 0$ becomes

$$J = \begin{bmatrix} \omega' & -1 \\ 1 & 0 \end{bmatrix}, \tag{8}$$

where $\omega' = \omega/\kappa$. The eigenvalues are then given by

$$\lambda_\pm = \frac{\omega' \pm \sqrt{\omega'^2 - 2}}{2}. \tag{9}$$

One can show that the fixed point p_0 is a saddle point if and only if $|\omega'| > 2$. This condition depends only on ratio ω/κ. However, for the two other fixed points, the condition for them to be a saddle point depends on all parameters, κ, q, g and ω. As an example, one can show that the two fixed points are saddle points for $\kappa = 1$, $q = -0.6$, $g = 0.1$, and $\omega = 0.1$,

To compute the growth of the stable and unstable manifolds we use the search circle algorithm [11,12]. This algorithm is implemented in the dynamical system software (DSTOOL) [13].

Let us first look for some solutions of Eq. (1) with $\lim_{n\to\pm\infty} u_n = 0$. To find these solutions, we simply identify homoclinic points correspond to the fixed point p_0. That is, the intersection of the stable and unstable manifolds for the saddle point p_0. In Figure 2 we identify three homoclinic points: dot marker at (2.16,2.24), triangle marker at (-0.66,0.66) and square marker at (0.99,0.11). These homoclinic points leads to three different type of the solitons.

Let us choose for example the soliton obtained from the homoclinic point at (-0.66,0.66). We found that this type of soliton doesn't change much in the range $0 < \kappa < 1$. Figure 3 shows the soliton for $\kappa = 1$ and $\kappa = 0.005$.

Now, we will be interested in solitons with $\lim_{n\to\pm\infty} u_n = p_+$. We choose the parameters $\kappa = 1$, $q = -0.6$, $g = 0.1$ and $\omega = 0.1$. As stated before, the point p_+ is a saddle point. In Fig. 4 we identify three homoclinic points and their corresponding solitons. Next we plot the two manifolds for different values of the parameter κ. One can see from this figure that decreasing κ leads to disappearance of solitons. That is, we start with three homoclinic points and at $\kappa = 0.2$ we get two homoclinic points. By decreasing the value of κ, the two manifolds become tangent to each other and finally they will be separated, see Fig. 5.

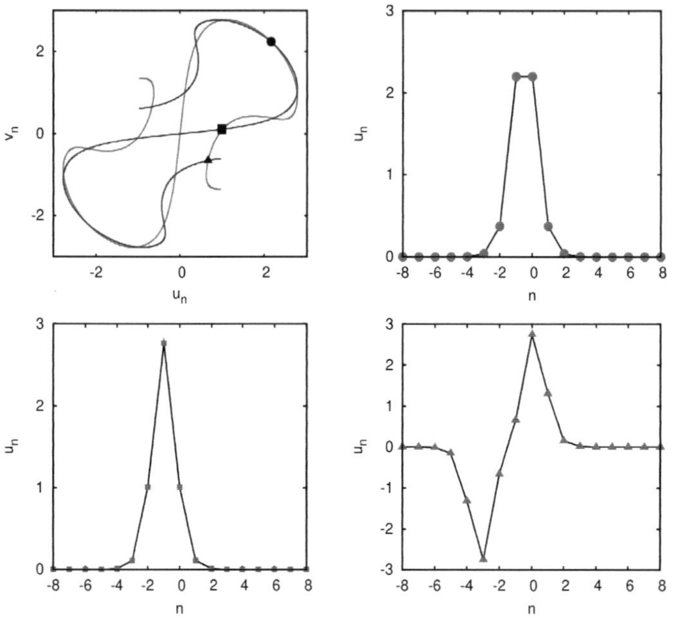

Figure 2. Type of solitons found from the homoclinic points. The stable (blue color) and unstable (red color) manifolds. There are three important homoclic points labeled by •, ▲ and ■. The parameters used are: $\kappa = 1$, $q = 1$, $g = 0.8$, $\omega = 10$.

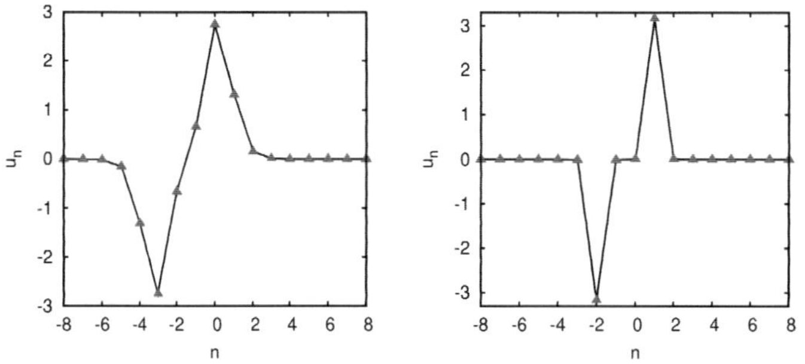

Figure 3. The parameters are $q = 1$, $g = 0.8$, $\omega = 10$ and different values of κ: (left) 1 and (right) 0.005.

In the next section, we study the linear stability of the solitons we found.

4. Stability analysis

Introducing a small perturbation to the solution of Eq. (1) as

$$\bar{\Phi}_n(t) = \left[u_n(t) + \epsilon\left(a_n \mathrm{e}^{-i\phi t} + b_n \mathrm{e}^{i\phi^* t}\right)\right] \mathrm{e}^{i\omega t}. \tag{10}$$

Inserting this last equation into Eq. (1), keeping linear terms with respect to ϵ, we obtain the equation of motion in the rotating frame as a set of linear differential equations

$$\dot{a}_n(t) = -\left(g u_{n-1} u_n + \kappa\right) a_{n-1} - \left(g u_{n-1}^2 + 2q u_n^2 + g u_{n+1}^2 - \omega\right) a_n - \left(g u_n u_{n+1} + \kappa\right) a_{n+1}$$

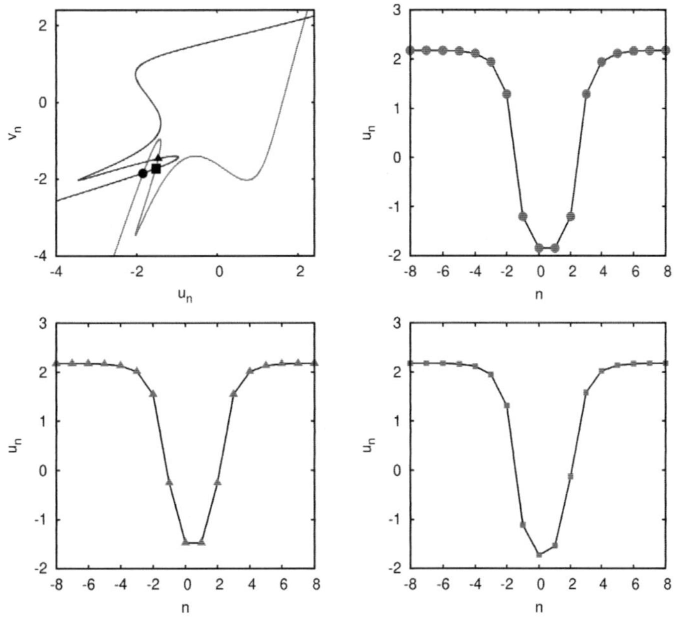

Figure 4. (left figure) Type of solitons found from the homoclinic points. (right figure) The stable (blue color) and unstable (red color) manifolds. Three important homoclic points labeled by •, ▲ and ■ are shown. The parameters used are $\kappa = 1$, $q = -0.6$, $g = 0.1$, $\omega = 0.1$.

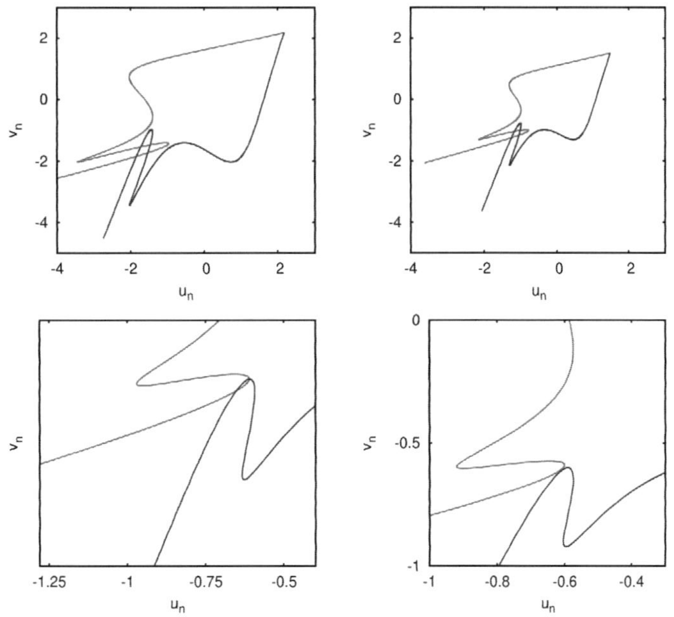

Figure 5. The stable (blue color) and unstable (red color) manifolds. The parameters used are $q = -0.6$, $g = 0.1$, $\omega = 0.1$ and different values of κ: $\{1, 0.5, 0.2, 0.19\}$.

$$- (g u_{n-1} u_n)\, b_{n-1} - \left(q u_n^2\right) b_n - (g u_n u_{n+1})\, b_{n+1}, \tag{11}$$

$$\begin{aligned}\dot{b}_n(t) &= (gu_{n-1}u_n+\kappa)\,b_{n-1}+\left(gu_{n-1}^2+2qu_n^2+gu_{n+1}^2-\omega\right)b_n+(gu_nu_{n+1}+\kappa)\,b_{n+1}\\ &\quad+(gu_{n-1}u_n)\,a_{n-1}+\left(qu_n^2\right)a_n+(gu_nu_{n+1})\,a_{n+1}.\end{aligned} \qquad (12)$$

The eigenvalues of the linear system determine whether the solution is stable or not. Using Eq. (12) we found that in Fig. 2 symmetric odd mode (square marker) is stable however, the symmetric even mode (dot marker) is unstable. This is in accordance with [1]. In addition to that we found that the third soliton (triangle marker) is also stable. However, all the solutions in Fig. 4 are unstable.

5. Conclusion

The objective of this work was to obtain numerically the localized solutions of Eq. (1) using the homoclinic orbits method.The existence of the different type solutonic solutions have been shown and their linear stability have been checked.

Acknowledgments

This work has been supported by the research grant No. FRGS 0409-108 of the Ministry of Higher Education, Malaysia.

References

[1] Abdullaev F Kh, Baizakov B B, Darmanyan S A, Konotop V V, and Salerno M 2001 *Phys. Rev.* A **64** 043606
[2] Trombettoni A and Smerzi A 2001 *Phys. Rev. Lett.* **86** 2353
[3] Morsh O and Oberthaler M 2006 *Rev. Mod. Phys.* **78** 179
[4] Lahaye T, Menotti C, Santos L, Lewenstein M and Pfau 2009 *Rep. Prog. Phys.* **72** 126401
[5] Trefzger C, Menotti C, Capogrosso-Sansone B and Lewenstein M 2011 *J. Phys. B: At. Mol. Phys.* **44** 193001
[6] Fattori M, Roati G, Deissler B, D'Errico C, Zaccanti M, Jona-Lasinio M, Santos L, Inguscio M and G.Modugno 2008 *Phys. Rev. Lett.* **101** 190405
[7] Ai-Xia Zhang and Ju-Kui Xue 2010 *Phys. Rev.* A **82** 013606
[8] Rojas-Rojas S, Vicencio R A, Molina M I, and Abdullaev F Kh 2011 *Phys. Rev.* A **84** 033621
[9] Carretero-Gonzalez R, Talley J D, Chong C and B.A. Malomed 2006 *Physica* D **216** 77
[10] Carretero-Gonzalez R, Talley J D, Chong C and Malomed B A 2006 *Physica D* **216** 77
[11] England J, Krauskopf B and Osinga H M 2004 *SIAM Journal on Applied Dynamical Systems* **3** 161
[12] Krauskopf B and Osinga H M 1998 *J. Comput. Phys.* **146** 404
[13] DSTOOL Exploring dynamical systems, http://www.dynamicalsystems.org/sw/sw/

Bounding the greybody factors for the Reissner-Nordström black holes

Tritos Ngampitipan[1] and Petarpa Boonserm[2]

[1] Department of Physics, Faculty of Science, Chulalongkorn University, Bangkok 10330, Thailand

[2] Department of Mathematics and Computer Science, Faculty of Science, Chulalongkorn University, Bangkok 10330, Thailand

E-mail: `tritos.ngampitipan@gmail.com`

Abstract. A black hole can emit radiation called Hawking radiation. Such radiation seen by an observer outside the black hole differs from the original radiation near the horizon of the black hole by the so-called "greybody factor". In this paper, the bounds of the greybody factors for the Reissner-Nordström black holes are obtained. These bounds can be derived by using the 2×2 transfer matrices. It is found that the charges of black holes act as good barriers.

1. Introduction

A black hole was believed to have been associated with the concept that anything which entered the black hole cannot escape. In 1974, Stephen Hawking, however, showed that a black hole could indeed emit radiation, which became known as the Hawking radiation [1]. He discovered this radiation by studying quantum field theory in a black hole background. Hawking radiation results from one particle from pair production moving out of the black hole. According to general relativity, the spacetime around a black hole behaves as gravitational potential under which particles move. Some of the radiations are reflected back into the black hole and the rest are transmitted out of the black hole. Therefore, Hawking radiation before passing the gravitational potential is different from one after passing the potential. This difference can be measured by the so-called "greybody factor".

There has been a number of studies devoted to calculating the greybody factors [2] - [7]. Moreover, one interesting technique, the 2×2 transfer matrix, which is used to derive a rigorous bound on the greybody factors, was studied in [8] - [10]. By using this method, the bounds of the greybody factors of the four-dimensional Schwarzschild black holes was obtained [11]. In this paper, we derive the bounds of the greybody factors of the four-dimensional Reissner-Nordström black holes by using the 2×2 transfer matrix.

2. Black holes

2.1. The Schwarzschild black holes

The Schwarzschild metric in four dimensions is given by

$$ds^2 = -f(r)dt^2 + f^{-1}(r)dr^2 + r^2 d\Omega^2, \quad (1)$$

where $d\Omega^2 \equiv d\theta^2 + \sin^2\theta d\phi^2$ and

$$f(r) \equiv 1 - \frac{2GM}{r}. \tag{2}$$

The Regge Wheeler equation is

$$\frac{d^2\psi}{dr_*^2} + \left[\omega^2 - V(r)\right]\psi = 0, \tag{3}$$

where

$$V(r) = f(r)\left[\frac{l(l+1)}{r^2} + \frac{2GM}{r^3}\right], \tag{4}$$

and the tortoise coordinate r_* is given by

$$\frac{dr}{dr_*} = 1 - \frac{2GM}{r}. \tag{5}$$

The structure of the Schwarzschild potential with $l = 1$ and $GM = 2$ is shown in Figure 1.

2.2. The Reissner-Nordström black holes

The Reissner-Nordström metric in four dimensions is given by

$$ds^2 = -\Delta dt^2 + \Delta^{-1}dr^2 + r^2 d\Omega^2, \tag{6}$$

where

$$\Delta \equiv 1 - \frac{2GM}{r} + \frac{GQ^2}{r^2}. \tag{7}$$

If there are no electric charges, the Reissner-Nordström metric is reduced to the Schwarzschild metric (1). The only difference between the Reissner-Nordström black holes and the Schwarzschild black holes is the presence of charges. The Schrödinger like equation is

$$\frac{d^2\psi}{dr_*^2} + \left[\omega^2 - V(r)\right]\psi = 0, \tag{8}$$

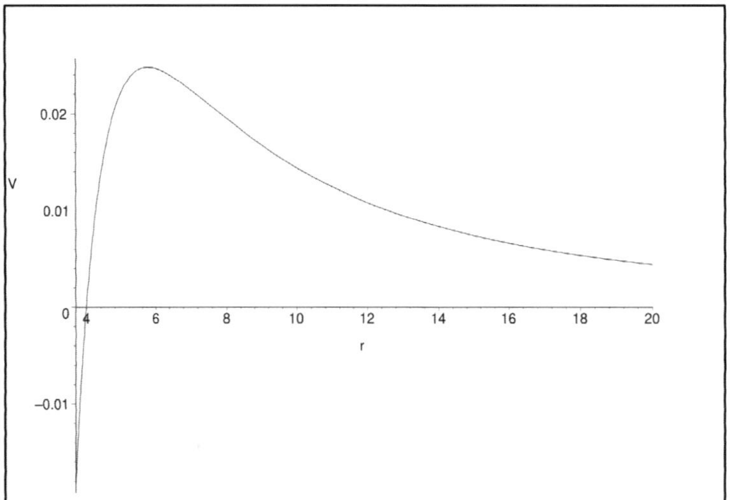

Figure 1. The Schwarzschild potential with $l = 1$ and $GM = 2$.

where

$$V(r) = \frac{l(l+1)\Delta}{r^2} + \frac{\Delta\partial_r\Delta}{r}. \tag{9}$$

The tortoise coordinate r_* is given by

$$r_* = \begin{cases} r + GM\ln\left|u^2 - A^2\right| + \dfrac{G^2M^2 + A^2}{2A}\ln\left|\dfrac{u-A}{u+A}\right|, & GM^2 > Q^2 \\ r + GM\ln\left|u^2 + B^2\right| + \dfrac{G^2M^2 - B^2}{B}\arctan\dfrac{u}{B}, & GM^2 < Q^2 \\ r + GM\ln\left|u^2\right| - \dfrac{G^2M^2}{u}, & GM^2 = Q^2 \end{cases}, \tag{10}$$

where

$$\begin{aligned} u &\equiv r - GM \\ A^2 &\equiv G^2M^2 - GQ^2 \\ B^2 &\equiv -G^2M^2 + GQ^2. \end{aligned} \tag{11}$$

The black holes with $GM^2 < Q^2$ are unreal and unphysical. Thus, we do not consider them. The black holes with $GM^2 = Q^2$ are extreme. The extremal black holes are useful in some theories, especially in supersymmetric theories. We do not consider these black holes in this paper. We focus only on the black holes with $GM^2 > Q^2$.

We can see the structure of the Reissner-Nordström potential with $Q = 1$, $l = 1$, and $GM = 2$ from Figure 2.

3. The bounds of the greybody factors

The bounds of the greybody factors using the 2×2 transfer matrix are given by [8] - [10]

$$T \geq \operatorname{sech}^2\left(\int_{-\infty}^{\infty} \vartheta dr_*\right) \tag{12}$$

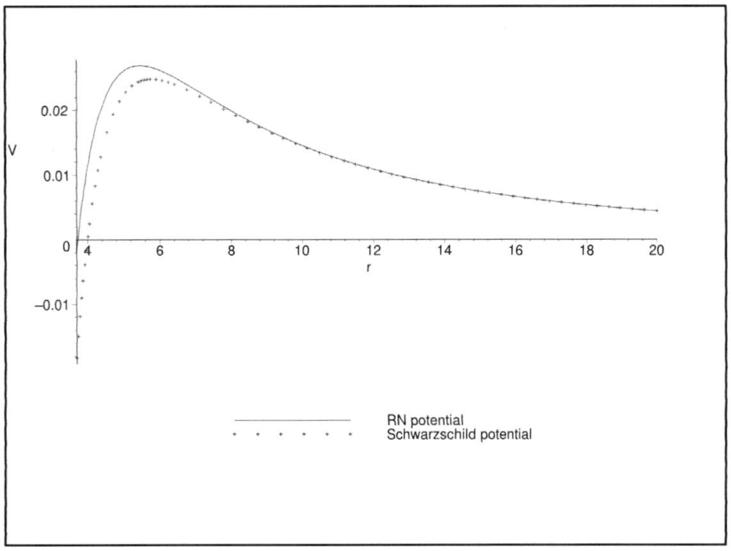

Figure 2. The Reissner-Nordström potential with $Q = 1$, $l = 1$, and $GM = 2$ compared to the Schwarzschild potential with equal mass and angular momentum.

and

$$R \leq \tanh^2\left(\int_{-\infty}^{\infty} \vartheta dr_*\right), \tag{13}$$

where

$$\vartheta \equiv \frac{\sqrt{(h')^2 + (\omega^2 - V - h^2)^2}}{2h}, \tag{14}$$

for some positive function h. We set $h = \omega$, then

$$T \geq \text{sech}^2\left(\frac{1}{2\omega}\int_{-\infty}^{\infty} V(r)dr_*\right) \tag{15}$$

and

$$R \leq \tanh^2\left(\frac{1}{2\omega}\int_{-\infty}^{\infty} V(r)dr_*\right). \tag{16}$$

3.1. The Schwarzschild black holes

For the potential from (4), the bounds of the greybody factors are given by [11]

$$T \geq \text{sech}^2\left[\frac{2l(l+1)+1}{8GM\omega}\right] \tag{17}$$

and

$$R \leq \tanh^2\left[\frac{2l(l+1)+1}{8GM\omega}\right]. \tag{18}$$

3.2. The Reissner-Nordström black holes

For the potential from (9), the bounds of the transmission probabilities are given by

$$T \geq \text{sech}^2\left[\frac{1}{2\omega}\left\{\frac{l(l+1)}{GM+A} + \frac{GM+2A}{3(GM+A)^2}\right\}\right] \tag{19}$$

and

$$R \leq \tanh^2\left[\frac{1}{2\omega}\left\{\frac{l(l+1)}{GM+A} + \frac{GM+2A}{3(GM+A)^2}\right\}\right]. \tag{20}$$

If the black holes have no electric charges, it is found that $A = GM$ and the above bounds are reduced to

$$T \geq \text{sech}^2\left[\frac{2l(l+1)+1}{8GM\omega}\right] \tag{21}$$

and

$$R \leq \tanh^2\left[\frac{2l(l+1)+1}{8GM\omega}\right], \tag{22}$$

which are exactly the bounds for the Schwarzschild black holes (17) and (18). From Figure 3 and 4, the graphs are plotted by setting $GM = 2$, $Q = 1$, and $l = 1$. The graphs show that for the Reissner-Nordström black holes and the Schwarzschild black holes with equal mass and angular momentum, the greybody factor of the Reissner-Nordström black holes is less than one of the Schwarzschild black holes. Since the difference between the Reissner-Nordström black holes and the Schwarzschild black holes with equal mass and angular momentum is the presence of charges only, it follows that the charges of the Reissner-Nordström black holes block the Hawking radiation.

4. Conclusions

In this paper, we obtain the bounds on the greybody factors by using the 2×2 transfer matrix for the Reissner-Nordström black holes compared with ones for the Schwarzschild black holes with equal mass and angular momentum in four dimensions. It is found that the charges of the Reissner-Nordström black holes behave as good barriers.

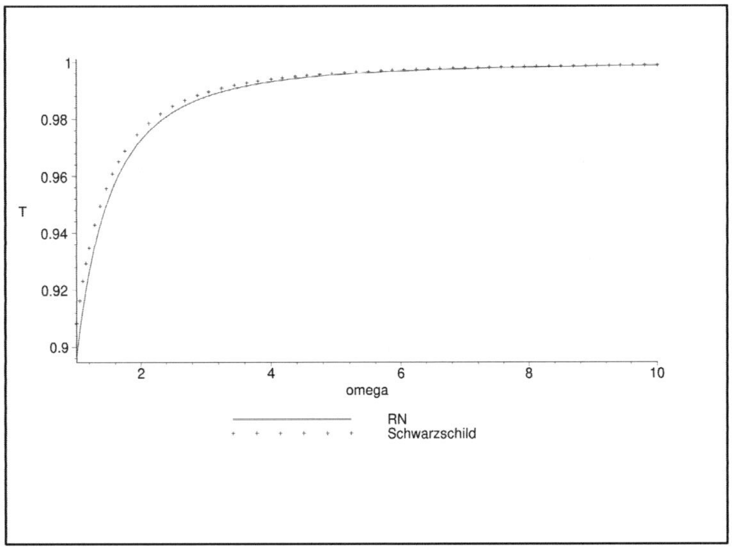

Figure 3. Comparison of the bounds of the transmission probabilities of the Reissner-Nordström black hole and the Schwarzschild black hole with equal mass and angular momentum.

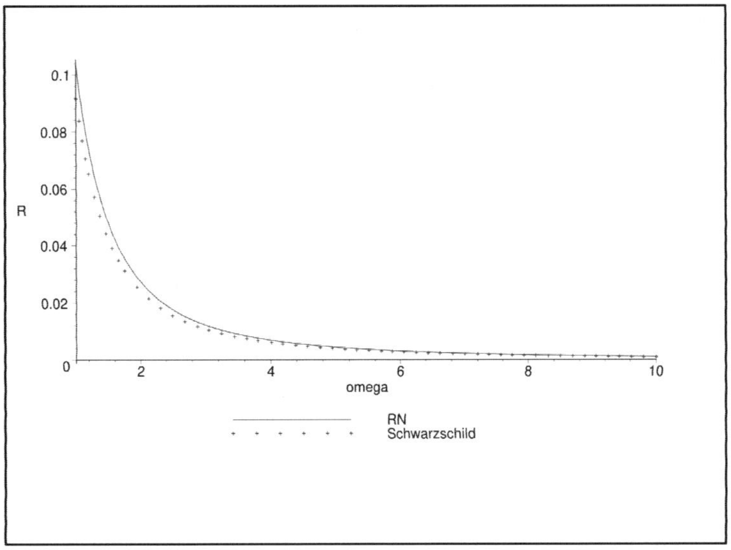

Figure 4. Comparison of the bound of the reflection probabilities of the Reissner-Nordström black hole and the Schwarzschild black hole with equal mass and angular momentum.

Acknowledgments
This research was supported by a grant for the professional development of new academic staff from the Ratchadapisek Somphot Fund at Chulalongkorn University, by the Thailand Toray Science Foundation (TTSF), and by the Research Strategic plan program (A1B1), Faculty of Science, Chulalongkorn University. PB was additionally supported by a scholarship from the Royal Government of Thailand. TN was additionally supported by a scholarship from the Development and Promotion of Science and Technology talent project (DPST). TN gives a special thanks to Dr. Auttakit Chatrabuti for his invaluable advice. We also thank Prof. Matt Visser for his extremely useful suggestions and comments.

References
[1] Hawking S 1975 Particle creation by black holes *Commun. Math. Phys.* **43** 199
[2] Parikh M and Wilczek F 2000 Hawking radiation as tunneling *Phys. Rev. Lett.* **85** 5042-45 (*Preprint* hep-th/9907001)
[3] Fleming C 2005 Hawking radiation as tunneling
[4] Lange P 2007 Calculation of Hawking radiation as quantum mechanical tunneling Thesis, Uppsala Universitet
[5] Fernando S 2005 Greybody factors of charged dilaton black holes in 2 + 1 dimensions *Gen. Relativ. Gravit.* **37** 461 - 81
[6] Kim W and Oh J 2008 Greybody factor and Hawking radiation of charged dilatonic black holes *JKPS* **52** 986 - 91
[7] Escobedo J 2008 Greybody factors Hawking radiation in disguise Master's Thesis, University of Amsterdam
[8] Visser M 1999 Some general bounds for 1-D scattering *Phys. Rev.* A **59** 42738 (*Preprint* quant-ph/9901030)
[9] Boonserm P and Visser M 2008 Bounding the Bogoliubov coefficients *Annals Phys.* **323** 2779 - 98 (*Preprint* quant-ph 0801.0610)
[10] Boonserm P 2009 Rigorous bounds on transmission reflection and Bogoliubov coefficients Ph. D. Thesis, Victoria University of Wellington (*Preprint* math-ph 0907.0045)
[11] Boonserm P and Visser M 2008 Bounding the greybody factors for Schwarzchild black holes *Phys. Rev.* D **78** 101502 (*Preprint* gr-qc 0806.2209)

Computational Simulation of Ultraviolet ZnO Diode Laser

A.Ouerdane[1,2], M.Bouslama[2], A. Abdellaoui[2], M.Ghaffour[2] and Y. Al Douri[3]

[1] Centre Universitaire Khemis Miliana route de Theniet El Had 44225 Ain Defla, Algeria
[2]Laboratoire Matériaux ENSET Mnaouer BP 1523 Oran 31000 , Algeria
[3] Institute of Nanoelectronic Engineering, University Malaysia Perlis, Malaysia

Email : ouerdanea@yahoo.fr

Abstract: This article deals with a mathematical technique for the study of functionality of semiconductor laser, by its rate equations, to be used in the simulation of high speed optical systems. We describe the development of numerical simulation model of a p-ZnO laser emitting in UV optical wavelength. A powerful dynamic algorithmical model is used to study the p-ZnO active region waveguide laser. It is based on time dependant rate equations of a quasi two levels system for the population density and time dependant for the pump and optical signal power. The model is sufficient to account for many of the observed dynamics in a single mode semiconductor laser in response to a dynamic drive current, such as relaxation oscillations and frequency chirping.

1. Introduction

Recently one can observe a great interest to ZnO films possessing nanocrystallites as materials for electronics, optoelectronics and others technological applications. With a wide band gap which is equal at 3.37 eV and a resistivity (10^{-5} -10^{-3} Ω cm) which comes, partly, from stoichiometric shift mainly due to oxygen vacancies and/or zinc interstitials, ZnO is a good materials for ultra violet diode laser (LD) which are widely used for applications in photonics, information storage, biology and medical therapeutics[1,2].

The optical power emitted by the diode laser is proportional to the current supplied to the ZnO semiconductor heterojunction. The laser dynamics can be modeled by coupled rate equations which describe the relation between the carrier number , the photon density and the optical phase[3]. To use these models, values for the rate equations, parameters must be chosen appropriately in order to obtain agreement between simulated and measured results for system performance.

The purpose of this work is to solve the coupled nonlinear rate equations describing the complex electric field and the carrier density in a simple model of the ZnO semiconductor laser. The model is sufficient to account for many of the observed dynamics in a single mode semiconductor laser in response to a dynamic drive current, such as relaxation oscillations and frequency chirping [4].

[1] To whom any correspondence should be addressed

The output of the solver is the time, injection current, and the transformed components of the state vector, computed at discrete time steps. The rate equations are Integrated using the fourth order Runge-Kutta computation[5]

2. Rate equations: steady state

The laser dynamics can be modeled by coupled rate equations which describe the relation between the carrier number $N_p(t)$, the photon density $S_p(t)$ and the optical phase $\varphi(t)$. [1-6].

$$\frac{dN_p(t)}{dt} = \frac{I(t)}{q} - \frac{N_p(t)}{\tau_n} - g(N,T)\,\frac{N_p(t) - N_0}{1+\varepsilon S_p(t)}\,S_p(t) \tag{1}$$

$$\frac{dS_p(t)}{dt} = \Gamma\, g(N,T)\frac{N_p(t) - N_0}{1+\varepsilon S_p(t)}\,S_p(t) - \frac{S_p(t)}{\tau_p} + \Gamma\,\frac{\beta N_p(t)}{\tau_n} \tag{2}$$

$$\frac{d\varphi(t}{dt} = \frac{\alpha_0}{2}\, g(N,T)[N_p(t) - N_0\,]. \tag{3}$$

N_0 is the carrier number at transparency, τ_p is the photon lifetime, τ_n is the carrier lifetime, Γ is the optical confinement factor describing the confinement mode in the active region, β is the spontaneous emission factor, ε is the gain compression factor, $g(N,T)$ is the optical gain coefficient dependent on the carrier density and the temperature function, $I(t)$ is the injected current, α_o is the linewidth enhancement factor and q is the electron charge. The output power and the threshold current are given by: [7].

$$P(t) = \frac{\eta h\upsilon}{\tau_p}\, s_p(t) \tag{4}$$

$$I_{th} = \frac{q}{\tau_n}\left\{n_0 + \frac{1}{g_0\tau_p}\right\} \tag{5}$$

Where η is the quantum efficiency, h the Planck constant and υ the radiation frequency. The steady-state solution to the rate equations is to obtained by setting all the time derivatives $\frac{dN_p(t)}{dt}$ and $\frac{dS_p(t)}{dt}$ to zero. The carrier concentration that satisfies a given steady-state injected current is obtained by iterative self-consistent solutions of the two coupled equations. Hence, the couple rate equations (1) and (2) will be written as follow:

$$\frac{I(t)}{q} - \frac{N_p(t)}{\tau_n} - g(N,T)\frac{N_p(t) - N_0}{1+\varepsilon S_p(t)}\,S_p(t) = 0 \tag{6}$$

$$\Gamma g(N,T)\,\frac{N_p(t) - N_0}{1+\varepsilon S_p(t)}\,S_p(t) - \frac{S_p(t)}{\tau_p} + \Gamma\,\frac{\beta N_p(t)}{\tau_n} = 0 \tag{7}$$

Using equations (6) and (7) the carrier number N will be deduced as follow:

$$N = \frac{\frac{I(t)}{q.V} - \frac{\tau_n S_p}{\tau_p \Gamma}}{\beta - 1} \tag{8}$$

3. Photon density S(f) and Frequency Response of the Semiconductor Laser.

3.1. Photon density

For laser with small signal, the expression of the photon density S(f) and transfer function $R(f)$ above threshold are as follow[8].

$$\mathrm{R(f)} = \frac{\text{photon Density}}{\text{Excitation Current}} = \frac{S_p(f)}{i\,(f)} = \frac{S_p(0)}{i\,(0)}\,\frac{f_0^2}{f_0^2 - f^2 + jff_d} \tag{9}$$

Where:

$$f_0 === \frac{1}{2\pi}\sqrt{\frac{g_0\eta}{\tau_p}(I - I_{th})} = \text{Resonance frequency} \quad (10)$$

$$f_d = \frac{\epsilon\,\eta(I - I_{th})}{2\pi\tau_p} = \text{Damping frequency} \quad (11)$$

$$f_p = \sqrt{f_0^2 - \frac{f_d^2}{4}} = \text{Frequency of peak response.} \quad (12)$$

3.2. Amplitude signal modulation
We consider a power modulation given by:

$$\text{P(t)} = P_0[1 + M.\cos(2\pi F_M\, t)] \quad (13)$$

where P_0 is the unmodulated power and *P(t)* is the modulated power. The parameter M is called the power modulation index. The signal on the photodiode is modulated at a frequency *Fm*. At the output of the Fabry- Perot, we observe the optical spectrum. For weak M, the electric field can be written as follows: [9].

$$E_L(t) = A.\left[1 + \frac{M}{2}.\cos(2\pi f t)\right].\cos(2\pi v_L t + \phi_L) \quad (14)$$

$$\text{E(t)} = \text{A}\cos(2\pi v_L t + \phi_L) + \text{M}\frac{A}{4}\cos\,(2\pi(v_L - f)t + \phi_L)$$
$$+$$
$$\text{M}\frac{A}{4}\cos(2\pi(v_L + f)t + \phi_L) \quad (15)$$

3.3. Frequency signal modulation

Let us now consider a pure sinusoidal frequency modulation at frequency at frequency f:

$$v(t) = v_L + \Delta v.\cos\,(2\pi f t) \quad (16)$$

The phase of the electric field is given by:

$$\int_0^t dt'\, v(t') = v_L t + \frac{\Delta v}{2\pi F_m}\sin\,(2\pi f t) \quad (17)$$

The electric field can therefore be expressed as follows:

$$E_L(t) = A.cos\left[2\pi v_L t + \frac{\Delta v}{f}.sin(2\pi f t + \phi_L)\right] \quad (18)$$

We define $m = \frac{\Delta v}{f}$ as the frequency modulation index:
This field can be expanded in a sum of sinusoidal terms using Fourier techniques as follows:

$$E_L = A.J_O\, cos.(2\pi v_L t\, \phi_L) - A.J_1(m).\cos(2\pi(v_L - f)t + \phi_L) + A.J_1(m).\cos\,(2\pi(v_L - f)t + \phi_L)$$
$$+$$
$$A.J_2(m).\cos(2\pi(v_L - 2f)t + \phi_L) + A.J_1(m).\cos\,(2\pi(v_L - f)t + \phi_L - A.J_3 m.\cos(2\pi(v_L - 3f)t + \phi_L) + A.J_1 m.\cos\,(2\pi(v_L - f)t + \phi_L)$$
$$+A.J_4(m).\cos(2\pi(v_L - 4f)t + \phi_L) + A.J_1(m).\cos(2\pi(v_L - f)t + \phi_L). \quad (19)$$

Frequency modulation is therefore characterized by the presence of sidebands at harmonics of the modulation frequency generated by Bessel function$J_i(m)$ for $i = 1,2,3 \ldots$. Bessel functions can evaluate to zero for specific values of the modulation index. For example, $J_0(2.4) = 0$: the optical carrier at v_L is suppressed for the modulation index equal to m=2.4

4. Numerical Simulation and Results.

4.1 Steady state solution

A desired modulated laser signal can be generated by direct injection of the laser diode by an electric signal I(t) of suitable format. The carrier number Np *(t),* the photon density or optical signal Sp*(t*) will carry the same information as the modulating electric signal I(t) does. Therefore, in the absence of current modulation, the laser diode emits a constant-power monochromatic light. The current modulation induces a variation of the power and frequency of the beam. The deriving current of a modulated laser diode is the sum of two terms. The first term represents the bias current, I_b, which sets a value above the threshold level of the laser while the second term, which determines the modulation level, is noted I_m. Both currents are adjusted to achieve the desired variable power. Hence, the injection current *I(t)*, which is assumed to represent the modulating electrical signal is given as follow: [10].

$$I(t) = I_b + I_m f(t) \tag{20}$$

Where $f(t)$ is a time function witch describe the signal current format.
The ZnO laser diode is formed by phosphorus doping a very thin layer of n-ZnO wafer. The cleaved facets of the semiconductor ZnO active region serve as mirrors for the laser cavity figure.1. A recombination between the electrons of the n-junction with the holes of the p-junction is obtained by an electric current. Photons emitted into a mode of the waveguide will travel along the waveguide and be reflected several times in the mirrors. When the gain is higher than the losses of the cavity the laser starts to oscillate . The energy of the emitted photons is therefore approximately equal to the band gap energy of the semiconductor.

Using the laser parameters listed in Table 1, the threshold current for the modeled laser diode is 50mA. Together with a parasitic time-constant of 0.5 ns, a typical plot of the transient carrier density and photon number from the deterministic single-mode model is recorded on figure.2.

We note that the coupled relaxation oscillation occurs between the carrier density and photon number before reaching steady state.
The damping rate of laser relaxation oscillations is determined by the non linear gain and parasitic time-constant. As illustrated in figure.3a and figure.3b directly modulated lasers are derived by different value of the current source *I(t)*, which modulates the injected carrier density in the active layer.
The laser operation and dynamics are influenced by property of gain suppression when the laser is biased above threshold, which originates from intraband relaxation processes of injected carriers. [11].

Table.1: diode laser parameters.

Parameter	Symbol	Value
Optical Confinement factor	Γ	0.05
Active region volume	V	$3.\ 10^{-11}\ cm^3$
Spontaneous emission factor	β	10^{-5}
Photon Lifetime	τ_p	$32\ .\ 10^{-12}\ s$
Carrier lifetime	τ_n	$2\ .\ 10^{-10}\ s$
Slope gain constant	g_0	$5\ 10^{-7}\ cm^{-3}\ s^{-1}$
Carrier density at transparency	N_0	$5\ 10^{19}\ cm^{-3}$
Quantum efficiency	η	0.5
Gain compression factor	ε	$4\ 10^{-17}$
Linewidth enhancement factor	α_o	5

Journal of Physics: Conference Series **435** (2013) 012028 doi:10.1088/1742-6596/435/1/012028

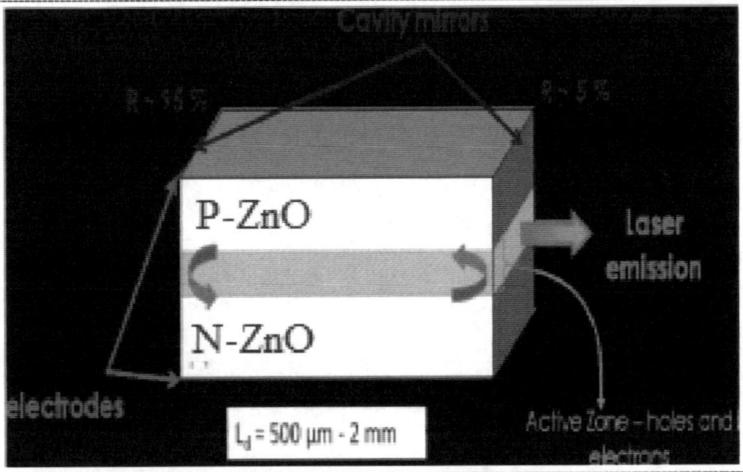

Figure.1 :diode laser structure showing the active region , the P-doped ZnO, the N-doped ZnO and the two cavity mirrors.

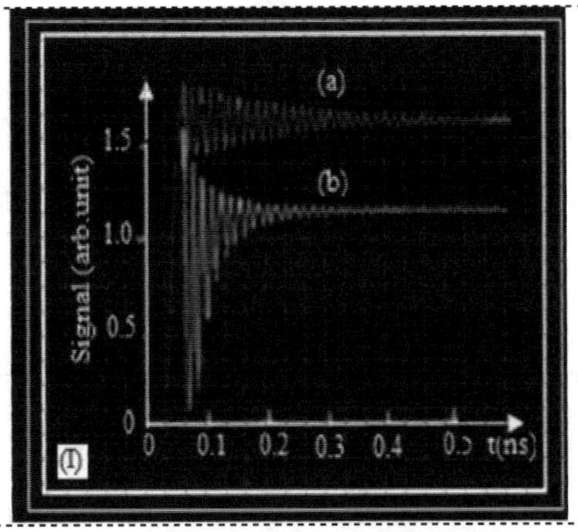

Figure 2.I. modulated laser derived by the current source I(t) = 40 mA

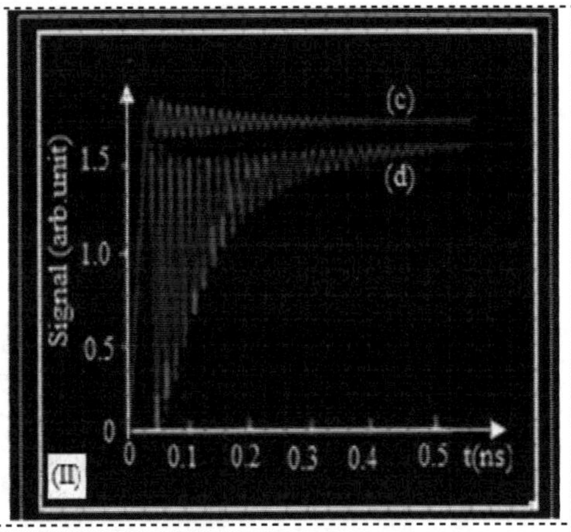

Figure 2.II. modulated laser derived by the current source I(t) =30 mA

4.2. Transient solution to the rate equations

To obtain transient solutions to the rate equations we have to solve numerically the rate equations to obtain the optical power and the phase of the electrical field at the laser output. For this, we use a fourth-order Runge-Kutta algorithm, we proceed first by a number of simplifying assumptions such as (1) the photon and electron distributions are spatially uniform, (2) the refraction index is spatially uniform and the effect of its variations with time is neglected, (3) the optical confinement factor and the spontaneous emission factor are treated as constant. [12].The computational result is given on figure.3

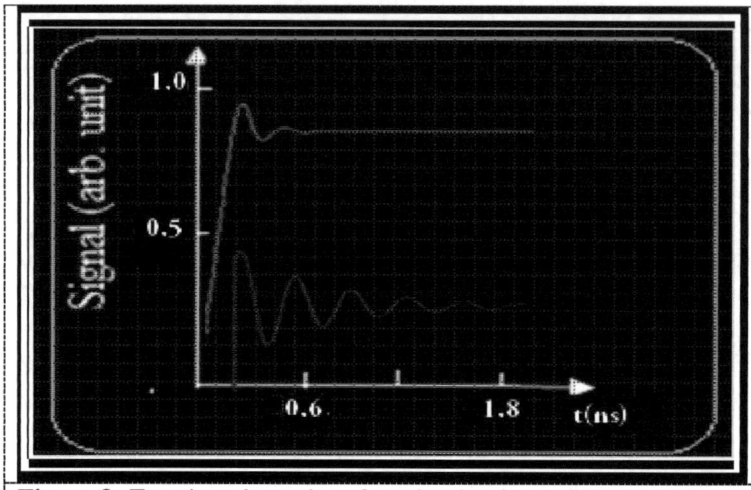

Figure.3: Transient dynamics of carrier density and photon number for a deterministic single mode rate-equation model with bias current of 50 mA.

It will be possible to obtain very short optical pulses, suitable for high-speed light wave telecommunication systems, from a quasi-rectangular modulation current. In these operation regime the small signal laser model, defining the modulation index as the ratio of the modulation current and the difference between the bias and the threshold current, we can obtain different pulses just by varying the bias current

In figure.4 we can see the simulated I(t) and the laser signal. It becomes clear that the optical signal can carry the same information as the modulating electric signal does. There exists a significant difference between the up and down levels of the optical signal such that one can distinguish the 1 and 0 bits. In addition, the bits duration of both the electric signals and the optical ones are identical.: [13].

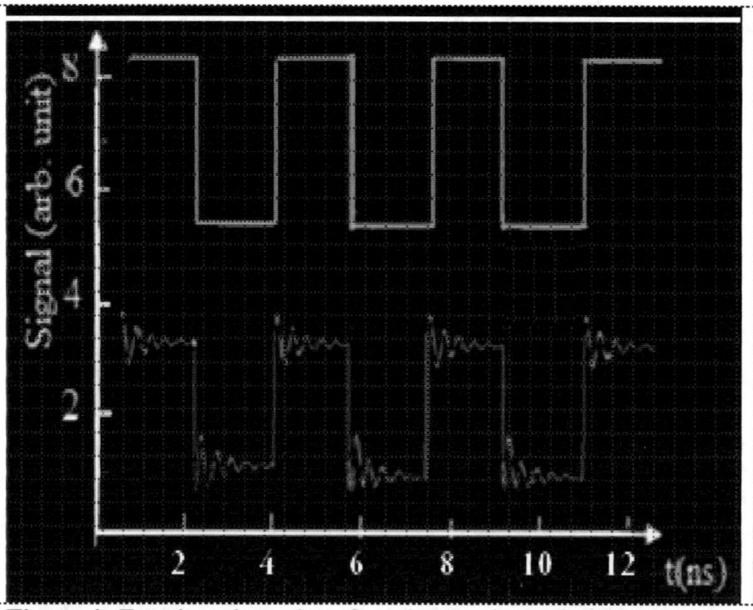

Figure.4: Transient dynamics of carrier density and photon number for a deterministic single mode suitable for high-speed light wave telecommunication systems.

4.3. Frequency response in optical modulation

In practice, optical modulation is a way to minimize the effects of electrical parasitics. Within the model, this is equivalent to adding an optical perturbation to optical signal generated by the carriers. The frequency response of the semiconductor laser, given by the equation13,with the frequency response of the mount fixture.

On figure 5 we plot the simulated optical frequency responses of the laser obtained using the full model with parasite effects. We note that the bandwidth is mainly controlled by the resonance frequency figure.5 The damping due to the injected current variation is simulated and recorded on figure.6, the latter being significantly influenced by parasitics and adiabatic phenomenon. We note also that the bandwidth increases when the bias current increases [7,14].

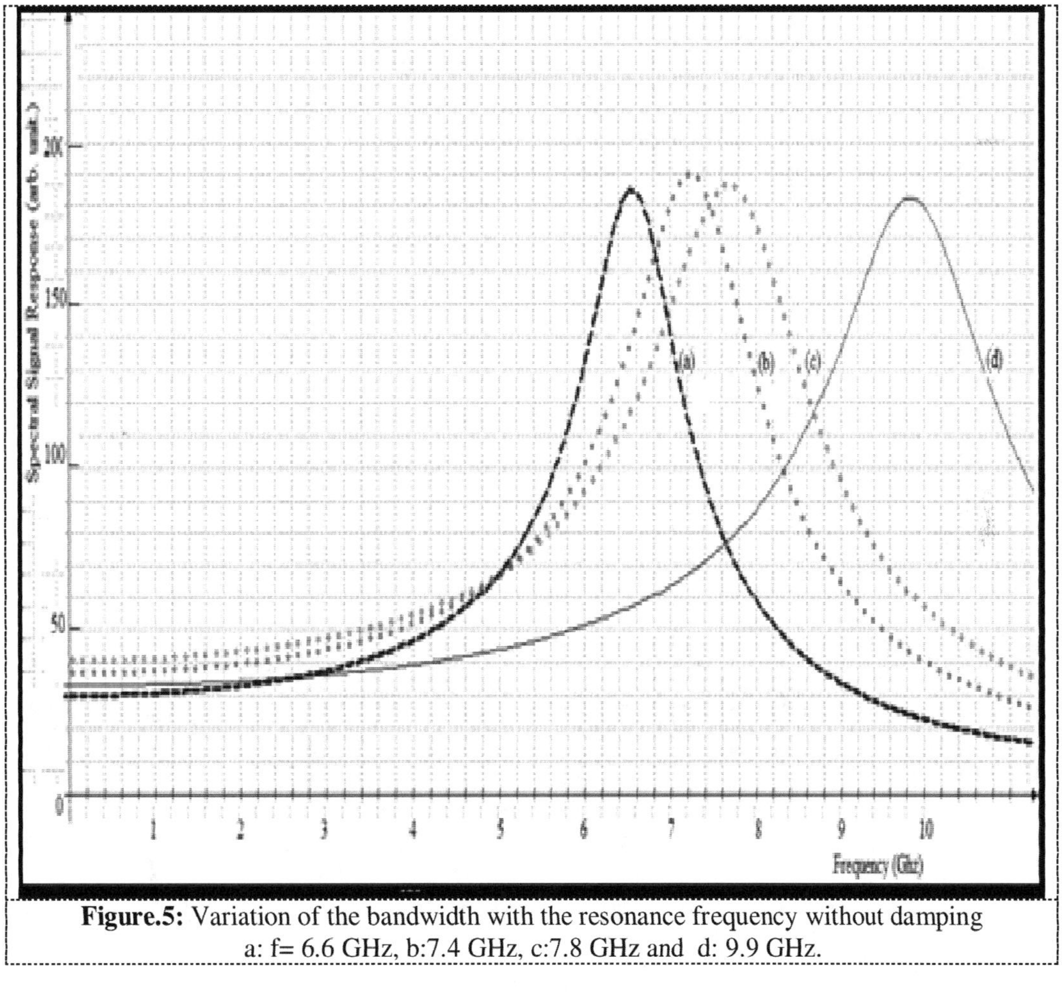

Figure.5: Variation of the bandwidth with the resonance frequency without damping
a: f= 6.6 GHz, b:7.4 GHz, c:7.8 GHz and d: 9.9 GHz.

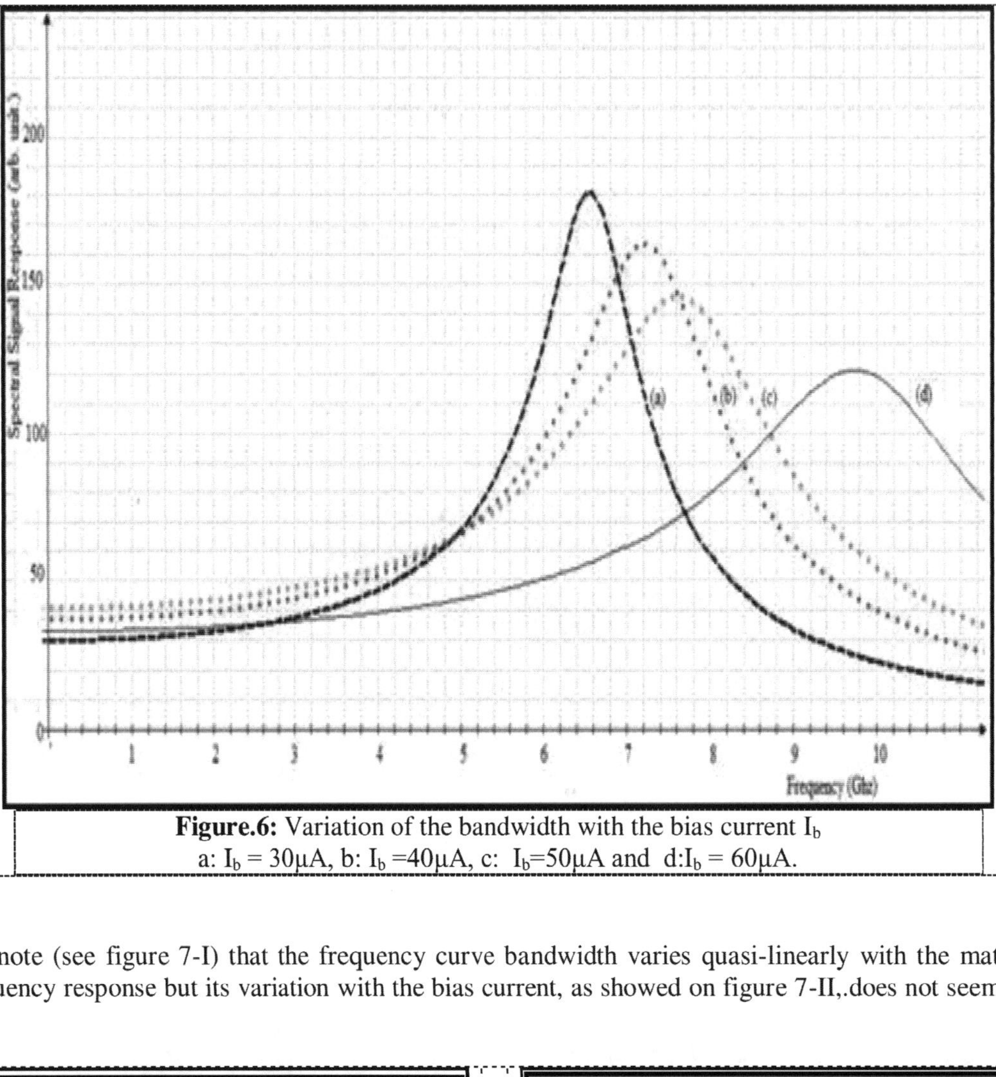

Figure.6: Variation of the bandwidth with the bias current I_b
a: $I_b = 30\mu A$, b: $I_b = 40\mu A$, c: $I_b = 50\mu A$ and d: $I_b = 60\mu A$.

We note (see figure 7-I) that the frequency curve bandwidth varies quasi-linearly with the material frequency response but its variation with the bias current, as showed on figure 7-II,.does not seem one .

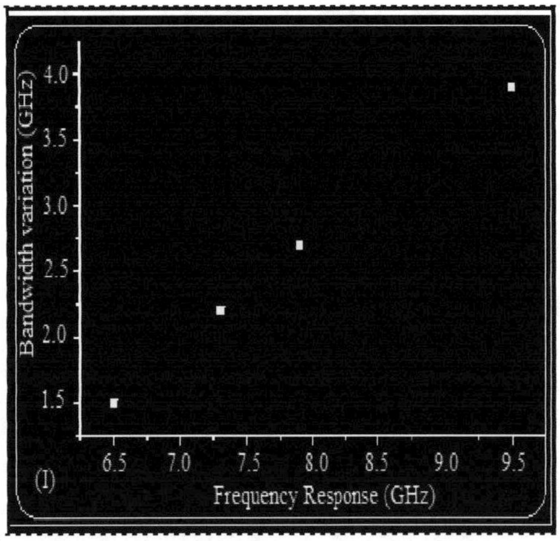

Figure 7.I. Bandwidth variation as function of frequency response.

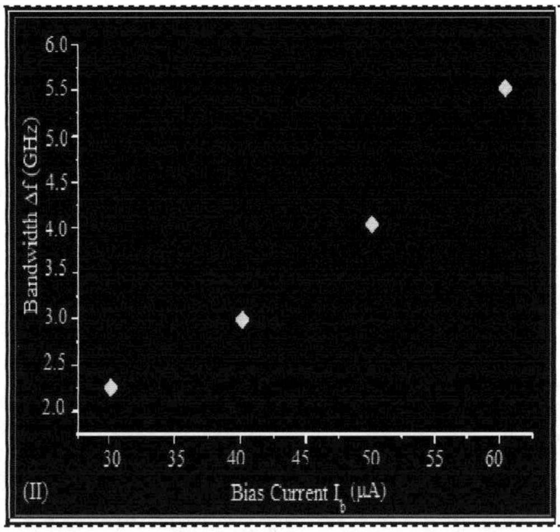

Figure 7.II. Bandwidth variation as function of the bias current.

4.4. Frequency chirping

Diode lasers suffer from the frequency drift caused by changes in the junction temperature, by current noise, and by the perturbation of the external cavity length used for narrowing the laser spectrum. We can observe the effects of chirp on the spectrum of the signal modulated directly by amplitude modulation given by equation (15). This theoretical result is simulated and recorded on figure.8-I while the frequency modulation given by equation (19) is simulated and represented on figure.8-II.

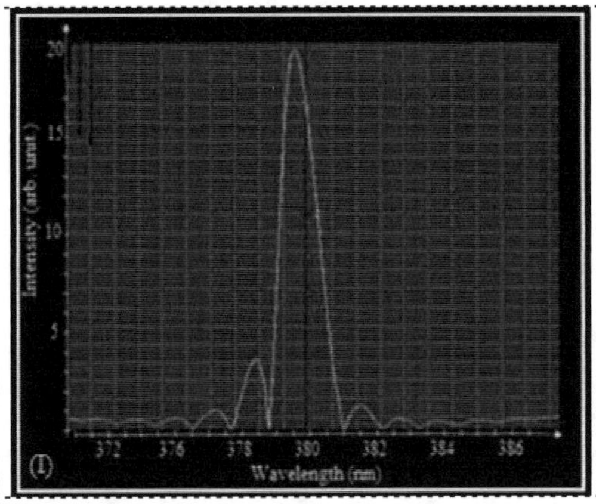

Figure 8-I Effects of chirp on the spectrum of the signal modulated directly by amplitude modulation.

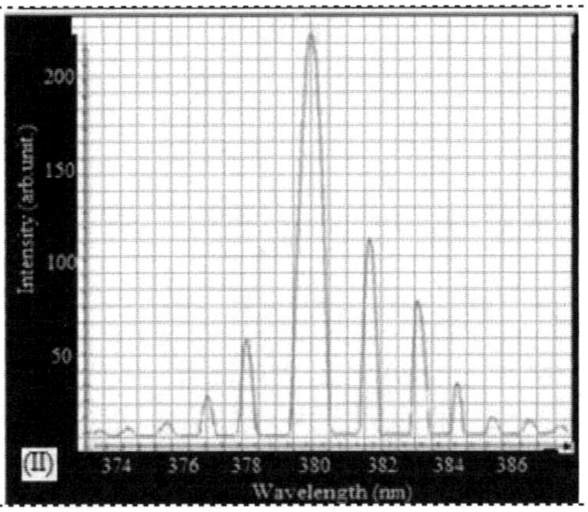

Figure 8-II. Effects of chirp on the spectrum of the signal modulated directly by frequency modulation.

The deformation of the carrier wave is the most visible effect occurring in the signal. This effect is quite larger at lower bit rates, as the adiabatic phenomenon is dominant. This is related to the transient chirp, and reveals that the frequency excursion, for the photon density, is different in minimum signal (low level) to the maximum signal (high level) figure.9[3]

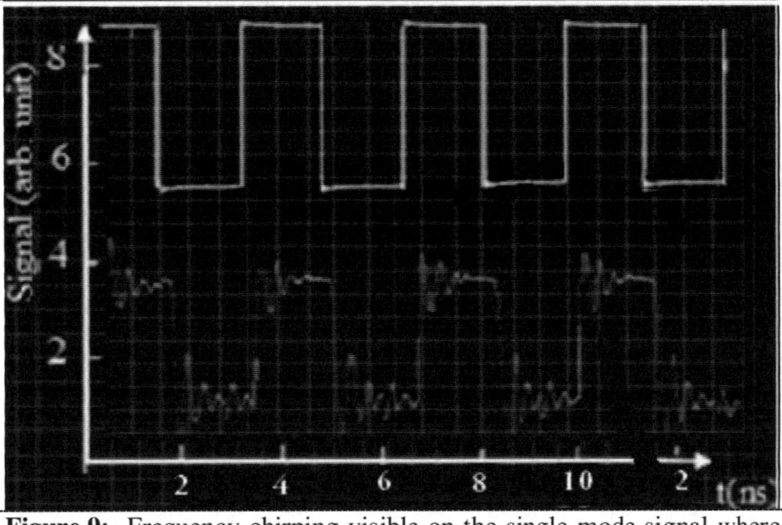

Figure.9: Frequency chirping visible on the single mode signal where the frequency excursion, for the photon density, is different in minimum signal (low level) compared to the maximum signal (high level).

Conclusion

We have simulated a ZnO laser diode using an efficiency computational program and our results are in good agreement with others theoretical and experimental works. This program has been used to study the modulation response of laser diode emitting in the 380 nm ultraviolet wavelength. The simulation program developed can be used to illustrate the performance of a waveguide laser diode as function of device parameters for all semiconductors electrically pumped laser diode.

Acknowledgement: This work is supported by the PNR (National Research Program) Algiers (Algeria).

References

[1] R.A. Abdullah ,K. Ibrahim 2010 Optoelectronics and Advanced Materials–Rapid Communication **4** 568
[2] Jeffrey O. White, George Rakulene, Carl Mungen 2011 ARL-TN0451 **1** 12
[3] Safwat W.Z. Mahmoud 2007 Egypt Journal Solids **30** 277
[4] M. Aleshams 2009 Progress In Electromagnetic Research **8** 121
[5] H.K. Liang, S. F. Yu and H.Y. Yang (2010) Applied Physics letters **97** 241107
[6] Matthew Lam, Ben Lunderwille, Matthew Von Schilling Applied Sciences 459 University of British Colombia
[7] Vitezslav Jerabek, Ivan Huttel 2011 Radioengineering **20** 486
[8] Linh V.T. Nguyen 2002 Systems Sciences Laboratory DSTO RR02
[9] Krishna Myneni. Semiconductor 2011 Laser Rate Equation Solver 44 **1** 37
[10] T homas Erneux, Evgeny A.Viktorix and Paul Mandal 2007 Physical Review **A76** 0238
[11] Ovidio H. Anton, Dinesh Patel Carmen S. Menoni, Jeng-Ya Ych, T.T. Van Roy, L.J. Mawst, J.M.Pikal and Nelson Tansu 2005 Quantum Electronic **11** 1079
[12] Asier Villafranca, Javier Lasobras and Ignacio Garcès Member, IEEE. TOYBA Lab. - I3A University of Zaragoza, PT Walqa, Ed. 1 22197 30 Janv-2Fev 2007 6th Spanish Conference on Electronic Devices
[13] Sheng Chu, Guoping Wang , Weihang Zhou, Yuqing Kong, Lin Li, Jingjian Ren and Jianlin Liu 2011 Nature Technology Doi:10.1038 **6** 506
[14] Sheng Chu ,Mario Olmedo, Zheng Yang , Jieying Kong and Jianlin Liu 2008 Apllied Physics Letters **93** 181106

Effect of Internal Heat Generation on Benard-Marangoni Convection in Micropolar Fluid with Feedback Control

N F M Mokhtar[1], I K Khalid[2], N M Arifin[3]

[1,3]Centre of Foundation Studies, Universiti Putra Malaysia, 43400 UPM, Serdang, Malaysia

[2,3]Department of mathematics, Universiti Putra Malaysia, 43400 UPM, Serdang, Malaysia

norfadzillah.mokhtar@gmail.com

Abstract. The effect of uniform distribution of internal heat generation on the linear stability analysis of the Benard-Marangoni convection in an Eringen's micropolar fluids with feedback control is investigated theoretically. The upper free surface is assumed to be non-deformable and the lower boundary is taken to be rigid and isothermal with fixed temperature and span-vanishing boundaries. The eigenvalue is solved numerically using the Galerkin method. The influence of the internal heat generation; Q and feedback control; K in micropolar fluids with various parameters on the onset of stationary convection has been analysed.

1. Introduction

The theory of micropolar fluid was developed by Eringen [1]-[3], in which the fluid elements possess rotator motion, in addition to unusual translator motion. The translator motion is characterized by the velocity vector, while the rotator motion is characterized by the gravitation vector. Physically speaking, micropolar fluids may be thought of the fluids containing dumb-bell shaped molecule, e.g., polymeric fluids suspension, blood, muddy fluids like crude oils etc. These fluids differ from the classical fluids in the sense that they support couple stresses due to rotation motion, in addition to force stress due to translator motion and they are asymmetric in nature. The former is absent in the classical fluids. Rayleigh-Benard situation in Eringen's [4]-[8] micropolar fluids has been investigated by many authors. The main result from all of these studies is that for heating from below stationary convection is the preferred mode. The instability of convection driven by buoyancy is referred to as the Rayleigh-Benard convection. The classical Rayleigh problem on the onset of convective instabilities in a horizontal layer of fluid heated from below has its origin in the experimental observations of Benard [9] and [10]. Rayleigh's paper is the pioneering work for almost all modern theories of convection. Rayleigh [11] showed that Benard convection, which is caused by buoyancy effects, will occur when the Rayleigh number exceeds a critical value.

The problem of convective instability of a micropolar fluids layer confined between two rigid boundaries and pointed out that the analysis of the instability finds applications in the area of geophysics was studied by Walzer [12]. The onset of convection for a heat conducting micropolar fluids layer between two rigid boundaries is investigated by Rama Rao [13]. The instability of a rotating micropolar fluids also has been investigated by Sastry & Rao [14] and Qin & Kaloni [15]. The universal stability of magneto-micropolar fluids was studied by Ahmadi & Shahinpoor [16]. The effect of throughflow on Marangoni convection in micropolar fluids is investigated by Murty & Rao [17]. The magnetoconvection in micropolar fluids is studied Siddheshwar and Pranesh [18]. Tang and Bau [19] and Howle [20] have shown that the critical Rayleigh number for the onset of Rayleigh-Benard convection can be delayed.

The nonlinear temperature distribution in a horizontal fluid layer arising due to internal heat generation has been studied theoretically by Sparrow et. al [21] and Roberts [22]. The effect of quadratic basic state temperature gradient caused by uniform internal heat generation was first addressed by Char and Chiang [23] for Benard-Marangoni convection. Wilson [24] used a combination of analytical and numerical techniques to analyze the effect of internal heat generation on the onset of Marangoni convection.

Or et al. [25] studied theoretically the use of feedback control strategies to stabilize long wavelength instabilities in the Marangoni-Benard convection. Bau [26] has shown independently how such a feedback control can delay the onset of Marangoni-Benard convection on a linear basis with no-slip boundary conditions at the bottom. Arifin et al. [27] have shown that a feedback control also can delay the onset of Marangoni-Benard convection with free-slip boundary conditions at the bottom. In the presence of strong electric field, the electric conductivity is affected by the magnetic field. Bachok and Arifin [28] studied the feedback control of the Marangoni-Benard convection in the presence of internal heat generation.

To the best of our knowledge, there has been no work on Benard-Marangoni convection in the presence of internal heat generation in micropolar fluids with feedback control. In an attempt to rectify this situation, we seek to investigate the effect of uniform internal heat generation on the onset of Benard-Marangoni convection in micropolar fluids with feedback control and employed the stability analysis. This analysis is based on the linear stability theory and the resulting eigenvalue problem is solved using the Galerkin method.

2. Mathematical formulation

Consider an infinite horizontal layer of Boussinesquian micropolar fluids layer of depth d, heated from below with the internal heat generation exists within the fluid system. The stability of a horizontal layer of micropolar fluids in the presence of thermal feedback control is examined. The no-spin boundary condition is assumed for micro rotation. Let ΔT be the temperature difference between the lower and upper surfaces with the lower boundary at a higher temperature than the upper boundary and these boundaries are maintained at constant temperatures. A Cartesian coordinate system (x, y, z) is used with the origin at the bottom of the surface and the z-axis vertically upward. We assumed that the upper free is non-deformable and the surface tension σ^*at the free surface is assumed to vary linearly with temperature in the form

$$\sigma^* = \sigma_0^* - \sigma_T^* \nabla T , \tag{1}$$

where σ_0^* is the unperturbed value and $-\sigma_T^*$ is the rate of change of surface tension with temperature T. The fluid density ρ is assumed to vary linearly with temperature in the form

$$\rho = \rho_0 \left[1 - \alpha_t \left(T - T_0 \right) \right], \tag{2}$$

where α_t is the thermal expansion coefficient and ρ_0 is the density at $T = T_0$. The governing equations for the Benard-Marangoni situation in a Boussinesq micropolar fluids are

$$\nabla \cdot \boldsymbol{q} = 0, \tag{3}$$

$$\rho_0 \left[\frac{\partial \boldsymbol{q}}{\partial t} + (\boldsymbol{q} \cdot \nabla) \boldsymbol{q} \right] = -\nabla p - \rho g \hat{k} + (2\zeta + \eta) \nabla^2 \boldsymbol{q} + \zeta (\nabla \times \boldsymbol{\omega}) \tag{4}$$

$$\frac{\partial T}{\partial t}+(\boldsymbol{q}\cdot\nabla)T=\frac{\beta}{\rho C_v}(\nabla\times\boldsymbol{\omega})\cdot\nabla T+\kappa\nabla^2 T+h_g, \tag{5}$$

$$\rho_0 I\left(\frac{\partial\boldsymbol{\omega}}{\partial t}+(\boldsymbol{q}\cdot\nabla)\boldsymbol{\omega}\right)=(\lambda'+\eta')\nabla(\nabla\cdot\boldsymbol{\omega})+\eta'\nabla^2\boldsymbol{\omega}+\zeta(\nabla\times\boldsymbol{q}-2\boldsymbol{\omega}), \tag{6}$$

where $\boldsymbol{q}$ is the velocity, $\boldsymbol{\omega}$ is the microrotation, p is the pressure, ρ_0 is the density, T is the temperature, g is the acceleration due to gravity, $\hat{k}$ is the unit vector in the z-direction, ζ is the coupling viscosity coefficient or vortex viscosity, β is the micropolar heat conduction coefficient, C_v is the specific heat, λ and η are the bulk and shear kinematic viscosity coefficients, λ' and η' are the bulk and shear spin-viscosity coefficients, I is the moment of inertia, κ is the thermal conductivity and h_g is the overall uniformly distributed volumetric internal heat generation within the micropolar fluid layer. The basic state of the fluid is quiescent and is described by

$$\boldsymbol{q}_b=(0,0,0),\ \boldsymbol{\omega}_b=(0,0,0),\ p=p_b(z)\ \text{and}\ T=T_b(z) \tag{7}$$

where the subscript b denotes the basic state. Substituting equation (7) into equations (4) and (5), we get the basic state governing equations as

$$\frac{dp_b}{dz}=-\rho_b g\ , \tag{8}$$

$$\frac{d^2T_b}{dz^2}=-\frac{h_g}{\kappa}, \tag{9}$$

with

$$\rho_b=\rho_0\left[1-\alpha_t(T_b-T_0)\right]. \tag{10}$$

Subject to the boundary conditions $T_b = T_0$ at $z = 0$ and $T_b = T_0 - \Delta T$ at $z = d$, Eq. (9) is solved and we obtained

$$T_b(z)=-\frac{h_g}{2\kappa}z^2+\left(\frac{h_g d}{2\kappa}-\frac{\Delta T}{d}\right)z+T_0. \tag{11}$$

Take note that equation (11) is a parabolic distribution with the liquid layer height due to the existence of the internal heat generation. Without the internal heat generation; $Q = 0$, the basic state temperature distribution in the fluid layer is linear. Let the basic state be disturbed by an infinitesimal thermal perturbation and we now have

$$\boldsymbol{q}=\boldsymbol{q}_b+\boldsymbol{q}',\ \boldsymbol{\omega}=\boldsymbol{\omega}_b+\boldsymbol{\omega}',\ p=p_b+p'\ \text{and}\ T=T_b+T', \tag{12}$$

where the primes indicate that the quantities are infinitesimal perturbations. Substituting equation (12) into equations (3) – (6) and linearized in the usual manner, we obtained the linearised equations in the form

$$\nabla\cdot\boldsymbol{q}'=0, \tag{13}$$

$$0=-\nabla p'-\rho' g\hat{k}+(2\zeta+\eta)\nabla^2\boldsymbol{q}'+\zeta(\nabla\times\boldsymbol{\omega}'), \tag{14}$$

$$-W\frac{\Delta T}{d}=\frac{\beta}{\rho_0 C_v}\left((\nabla\times\boldsymbol{\omega}')\cdot\left[-\frac{\nabla T}{d}\right]\hat{k}\right)\cdot\Delta T+\kappa\nabla^2 T'+\left(\frac{zh_g}{\kappa}+\frac{h_g d}{2\kappa}+\frac{\nabla T}{d}\right)W \tag{15}$$

$$0=(\lambda'+\eta')\nabla(\nabla\cdot\boldsymbol{\omega}')+\eta'\nabla^2\boldsymbol{\omega}'+\zeta(\nabla\times\boldsymbol{q}'-2\boldsymbol{\omega}') \tag{16}$$

In the present problem, we assume the principle of exchange of stability is valid and deal only with stationary convection. Hence, the time derivatives have been dropped in equations (13) – (16). The perturbation equations are non-dimensionalized using the following definitions

$$\left(x^*,y^*,z^*\right)=\left(\frac{x}{d},\frac{y}{d},\frac{z}{d}\right), W^*=\frac{W'}{x/d}, \Omega^*=\frac{(\nabla\times\omega')z}{x/d^3} \text{ and } T^*=\frac{T'}{\Delta T}. \tag{17}$$

Substituting equation (17) into equations (14), (15) and (16), eliminating the pressure term by operating curl twice on the resulting equation of (14), operating curl once on equation (16) and non-dimensionalising we get

$$(1+N_1)\nabla^4W+N_1\nabla^2\Omega+R_a\left(\frac{\partial^2T}{\partial x^2}+\frac{\partial^2T}{\partial y^2}\right)=0, \tag{18}$$

$$\nabla^2T+\left[1-Q(1-2z)\right]W-N_5\Omega=0, \tag{19}$$

$$N_3\nabla^2\Omega-2N_1\Omega-N_1\nabla^2W=0, \tag{20}$$

where the asterisks have been dropped for simplicity. Here,

$N_1=\dfrac{\zeta}{\eta+\zeta}$, is the coupling parameter,

$N_3=\dfrac{\eta'}{(\eta+\zeta)d^2}$, is the couple stress parameter,

$N_5=\dfrac{\beta}{\rho_0C_v d^2}$, is the micropolar heat conduction parameter,

$R_a=\dfrac{\alpha g\Delta T\rho_0d^3}{(\eta+\zeta)\chi}$, is the Rayleigh number,

and $Q=\dfrac{h_gd^2}{2\kappa\Delta T}$ is the heat source strength.

The perturbation quantities in a normal mode form are

$$(W,T,\Omega)=\left[W(z),\Theta(z),G(z)\right]\exp\left[i\left(a_xx+a_yy\right)\right], \tag{21}$$

where $W(z)$, $\Theta(z)$, $G(z)$ are amplitudes of the perturbations of vertical velocity, temperature and spin, and $a=\sqrt{a_x^2+a_y^2}$ is the wavenumber of the disturbances at the liquid layer. Substituting equation (21) into equations (18) – (20) we get

$$(1+N_1)\left(D^2-a^2\right)^2W+N_1\left(D^2-a^2\right)G=R_aa^2\Theta, \tag{22}$$

$$\left(D^2-a^2\right)\Theta=\left[Q(1-2z)-1\right]W+N_5G, \tag{23}$$

$$N_1\left[\left(D^2-a^2\right)W+2G\right]-N_3\left(D^2-a^2\right)G=0, \tag{24}$$

where $D = d/dz$.

Following the proportional feedback control in (1999), the continuously distributed actuators and sensors are arranged in a way that for every sensor, there is an actuator positioned directly beneath it. The determination of a control; $q(t)$ can be accomplished using the proportional-integral-differential (PID) controller of the form

$$q(t) = r + K\left[e(t)\right] \text{ where } e(t) = \hat{m}(t) + m(t), \tag{25}$$

where r is the calibration of the control, $e(t)$ an error or deviation from the measurement, $\hat{m}(t)$ from some desired or reference value, $m(t)$, $K = K_p + K_D\, d/dt + K_I \int_0^t dt$ with K_p is the proportional gain, K_D differential gain and K_I integral gain. Based on equation (25), for one sensor plane and proportional feedback control, the actuator modifies the heated surface temperature using a proportional relationship between the upper, $z=1$ and the lower, $z=0$ thermal boundaries for perturbation field

$$T'(x, y, 0, t) = -KT'(x, y, 1, t), \tag{26}$$

where T' denotes the deviation of the fluid's temperature from its conductive state and K is the scalar controller gain in which it will be used to control our system.

Equations (21) - (23) are solved subject to appropriate boundary conditions that are

$$W = DW = G = \Theta(0) + K\Theta(1) = 0 \text{ at } z = 0. \tag{27}$$

Equations (22) – (24) are solved subject to appropriate boundary conditions that are

$$W = D\Theta = G = D^2W = 0 \text{ at } z = 0, \tag{28}$$

$$W = D\Theta = G = D^2W + a^2\Theta M_a = 0 \text{ at } z = 1, \tag{29}$$

where $M_a = \sigma_T^* \Delta T d / \mu\chi$ is the Marangoni number. Equation (25) indicates the use of rigid and isothermal for lower boundary, and for equation (26) the upper surface is assumed to be stress free and insulating. The condition on G is the no-spin boundary condition for both boundaries.

3. Method of Solution

Equations (22)–(24) together with the boundary conditions (25) and (26) constitute a Sturm-Liouville problem with the Marangoni number M_a or the Rayleigh number R_a as an eigenvalue while keeping other physical parameters fixed. The Galerkin method is used to solve the resulting eigenvalue problem. Accordingly, the variables are written in a series of basis function as

$$W = \sum_{i=1}^{n} A_i W_i(z), \quad \Theta = \sum_{i=1}^{n} C_i \Theta_i(z) \text{ and } G = \sum_{i=1}^{n} D_i \Gamma_i(z), \tag{30}$$

where the trial functions $W_i(z)$, $\Theta_i(z)$ and $\Gamma_i(z)$ will be chosen in such a way that they satisfy the respective boundary conditions and A_i, C_i and D_i are constants. Substituting equation (27) into equations (22) – (24), multiplying the resulting equation (22) by $W_j(z)$, equation (23) by $\Theta_j(z)$ and equation (24) by $\Gamma_j(z)$; performing the integration by parts with respect to z between $z = 0$ and 1 and using the boundary conditions (25) and (26), we obtain the following system of linear homogeneous algebraic equations

$$C_{ji}A_i + D_{ji}C_i + E_{ji}D_i = 0, \tag{31}$$

$$F_{ji}A_i + G_{ji}C_i + H_{ji}D_i = 0, \tag{32}$$

$$I_{ji}A_i + J_{ji}D_i = 0. \tag{33}$$

The coefficients $C_{ji} - J_{ji}$ involve the inner products of the basis functions and are given by

$$C_1 = -N_1\left[\langle DW\rangle\langle DG\rangle + a^2\langle WG\rangle\right],$$

$$C_2 = \langle G^2\rangle\left(2N_1 + N_3a^2\right) + N_3\left\langle (DG)^2\right\rangle,$$

$$C_3 = -\left(1+N_1\right)\left[\left\langle \left(D^2W\right)^2\right\rangle + 2a^2\left\langle (DW)^2\right\rangle + a^4\langle W^2\rangle\right],$$

$$C_4 = \left\langle (D\Theta)^2\right\rangle + a^2\langle \Theta^2\rangle,$$

$$C_5 = -N_5\langle G\Theta\rangle,$$

$$C_6 = \left(1+N_1\right)\langle\Theta\rangle\langle DW\rangle,$$

$$C_7 = N_1\left[\langle DG\rangle\langle DW\rangle + a^2\langle GW\rangle\right],$$

$$C_8 = \left[1 - Q(1-2z)\right]\langle W\Theta\rangle,$$

$$C_9 = \langle \Theta W\rangle,$$

$$C_{10} = K\Theta(1),$$

$$C_{11} = C_4 + C_{10}, \tag{34}$$

where the angle bracket $\langle\cdots\rangle$ denotes the integration with respect to z from 0 to 1.

The eigenvalue has to be extracted from the characteristic equation (32). Now, we choose the trial functions as

$$W_i = z^2\left(1-z^2\right)T_{i-1}^*,\ G_i = z^2(1-z)T_{i-1}^* \text{ and } \Theta_i = z(2-z)T_{i-1}^* \tag{35}$$

where T_i^* are the Chebyshev polynomials, such that W_i, Θ_i and G_i satisfy the corresponding boundary conditions except $D^2W + a^2M_a\Theta = 0$ at z = 1 but the residuals from the equations are included as residuals from the differential equations.

4. Result and Discussion

The criterion for the onset of Benard-Marangoni convection in micropolar fluids in the presence of feedback control and internal heat generation is investigated theoretically. The sensitiveness of critical Rayleigh number, R_{ac} and Marangoni number, M_{ac} to the changes in the micropolar fluids parameters; N_1, N_3 and N_5 are also studied.

Table 1 shows the comparison of the critical Marangoni number, M_{ac} for different values of Rayleigh number, R_a and internal heat generation effect, Q when K = 0, N_1 = 0.5, N_3 = 2 and N_5 = 1. Our results are compared with Siddheshwar and Pranesh [18] in the absence of internal heat generation and feedback control, and the results are in a good agreement. Thus, this validates our numerical procedure. From this table, we found that the critical Marangoni number decreases as the value of internal heat generation increase for all R_a values considered. This shows that the internal heat generation parameter has a significant impact on the thermal and surface tensile modes, making the system unstable. Besides that, we also notice that the M_{ac} decreases with the increase in R_a.

The variation of R_a and M_a with wavenumber, a for different values of Q are illustrate in Figure 1 and Figure 2 respectively. We choose $N_1 = 0.5$, $N_3 = 2$, $N_5 = 1$ and $K = 2$ for both cases, and we found that the internal heat generation has a rapid impact on the stability of the system. This can be proved by looking at the difference in the R_a and M_a number, where the R_a and M_a number recorded between $Q = 0$ and $Q = 2$ are much larger compare with the value of $Q = 4$ and $Q = 6$. It is evident that the internal heat generation is a destabilizing factor for the system to be more unstable.

Figure 5 and Figure 6 show the plot of R_{ac} and M_{ac} versus the coupling parameter, N_1 respectively for values of $Q = 0$ and $Q = 10$ when $N_3 = 2$, $N_1 = 1$ and $K = 2$. In each of these plots, the critical number increases with increasing of N_1 for all values of Q considered. N_1 indicates the concentration of microelements, and increasing of N_1 is to elevate the concentration microelements number. When this happened, a greater part of the energy of the system is consumed by these elements in developing gyrational velocities of the fluid and thus delayed the onset of convection.

Figure 7 and Figure 8 show the plot of R_{ac} and M_{ac} versus the coupling parameter, N_1 respectively for values of $K = 0$ and $K = 10$ when $N_3 = 2$, $N_1 = 1$ and $Q = 2$. In each of these plots, the R_{ac} and M_{ac} increases with increasing of N_1 for all values of K considered. It is found that N_1 can delay the onset of convection.

The illustration of the couple stress parameter, N_3 in both thermal buoyancy and surface tension can be seen in Figure 9 and Figure 10 respectively when $N_1 = 0.1$, $N_5 = 10$ and $K = 2$. From these graphs, it can be clearly seen that in the presence of internal heat heating, an increased of N_3 decrease the values of R_{ac} and M_{ac} for $Q = 10$ and $Q = 15$ considered. These situations revealed that the system become more unstable much faster when the couple stress parameter increasing.

Figure 11 and Figure 12 show the plot of R_{ac} and M_{ac} versus the couple stress parameter, N_3 when $N_1 = 0.1$, $N_5 = 10$ and $Q = 2$. From the graph, it can be clearly seen that in the presence of controller gain, K, an increased of N_3 decrease the values of R_{ac} and M_{ac} for all K considered. Therefore, the system become more unstable when the couple stress parameter, N_3 increasing.

Figure 13 and Figure 14 show the plot of R_{ac} and M_{ac} respectively, versus micropolar heat conduction parameter, N_5 when $N_1 = 0.1$, $N_3 = 2$ and $K = 2$. Contrast with the effect of internal heat generation, Q, increasing the micropolar heat conduction, N_5 increased the critical Rayleigh number and critical Marangoni number as well. The reason behind this is, when N_5 increases, the heat induced into the fluid due to the microelements is also increased an thus reducing the heat transfer from the bottom to the top of the system. The decrease in heat transfer is responsible for delaying the onset of instability for both thermal buoyancy and surface tension system. Thus, increasing N_5 promotes stability in the micropolar system in the presence of internal heat generation.

Figure 15 and Figure 16 indicate the variation values of R_{ac} and M_{ac} for $K = 0$ and $K = 5$ respectively, when $N_1 = 0.1$, $N_3 = 2$ and $Q = 2$. It is found that as the number of R_{ac} and M_{ac} increases, the value of gain controller, K also increases for both thermal buoyancy and surface tension considered and thus stabilized the system.

Table 1. Comparison of M_{ac} for different values of R_a and Q when $K = 0$

R_a	M_{ac}			
	Siddheshwar & Pranesh [18]	Present Analysis		
	$Q = 0$	$Q = 0$	$Q = 2$	$Q = 3$
0	98.5736	98.5736	54.146	44.188
100	94.9228	94.9228	50.495	40.537
500	80.3196	80.3196	35.892	25.933
1000	62.0657	62.0657	17.638	7.680

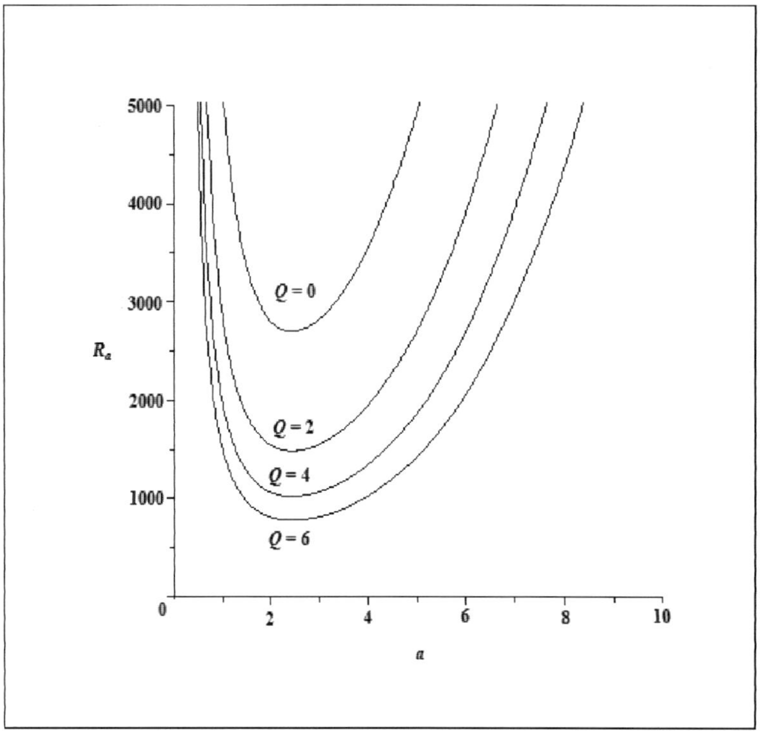

Figure 1. Variation of R_a with a for different values of Q.

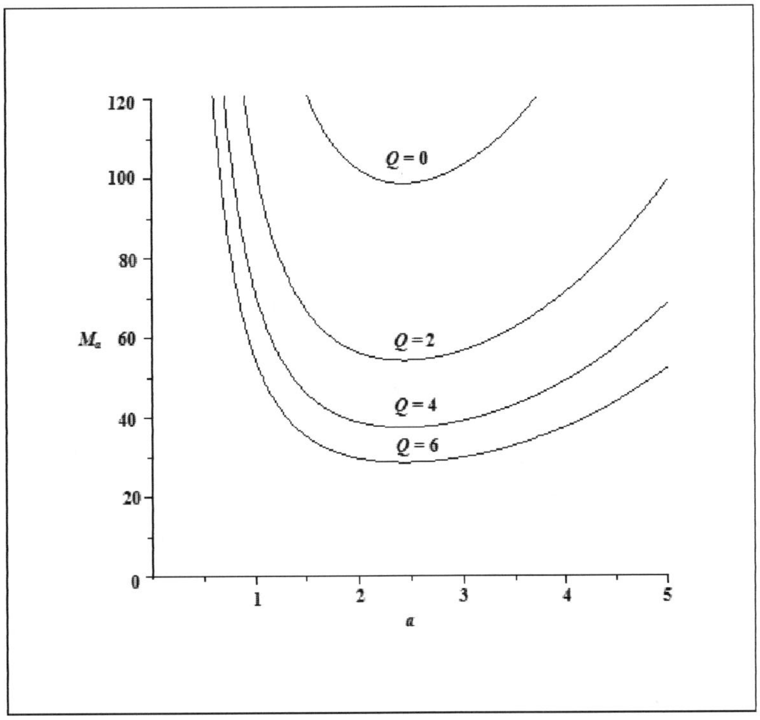

Figure 2. Variation of M_a with a for different values of Q.

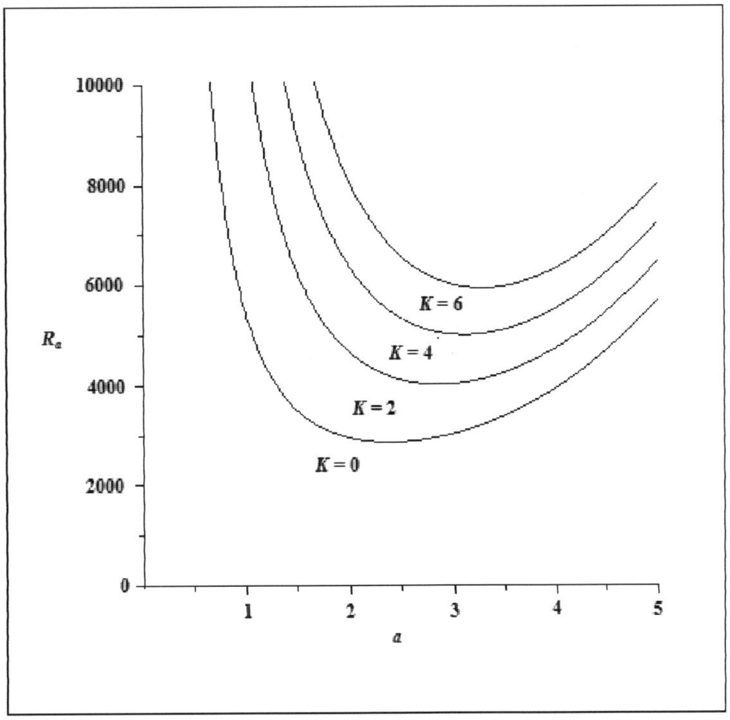

Figure 3. Variation of R_a with a for different values of K.

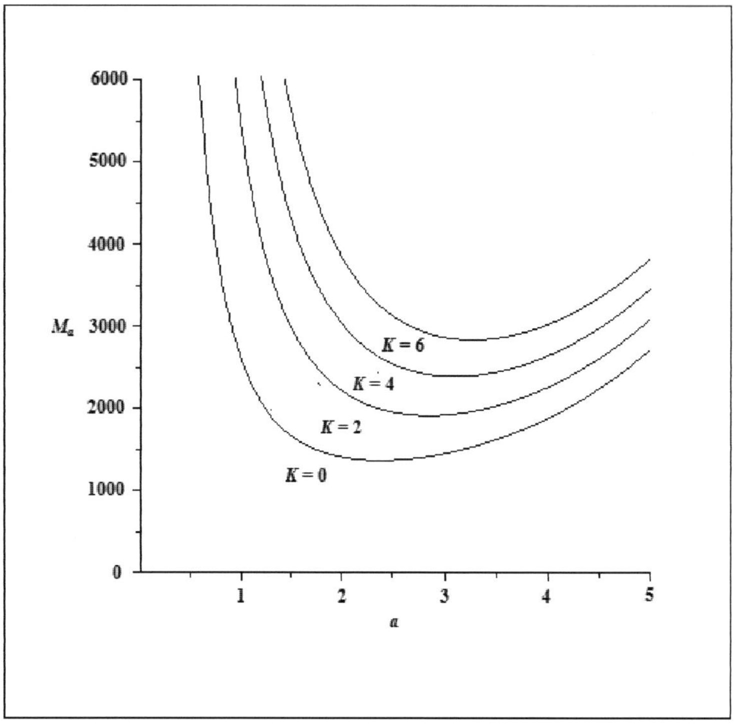

Figure 4. Variation of M_a with a for different values of K.

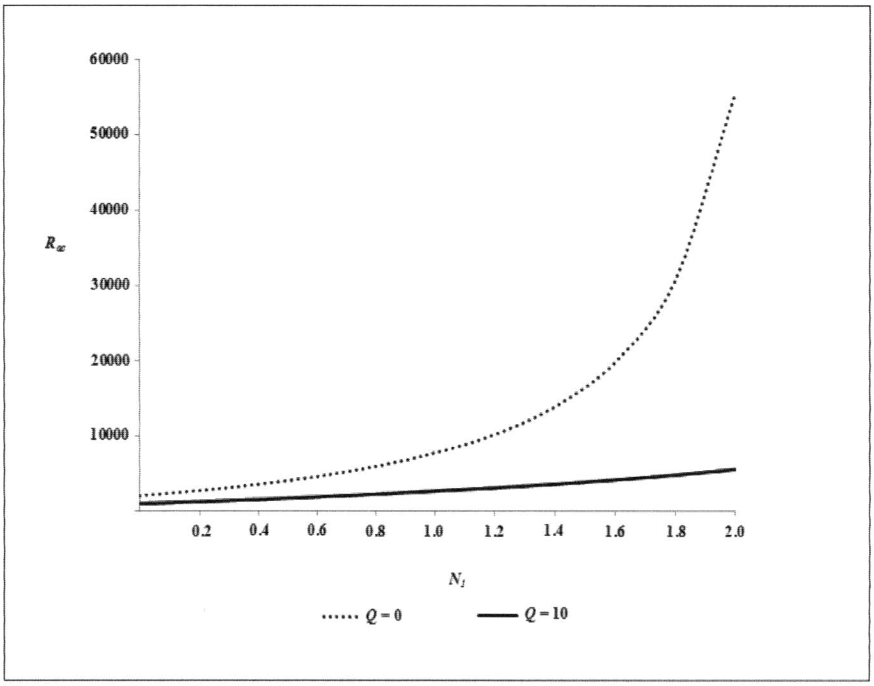

Figure 5. Plot of R_{ac} versus N_1 for different values of Q.

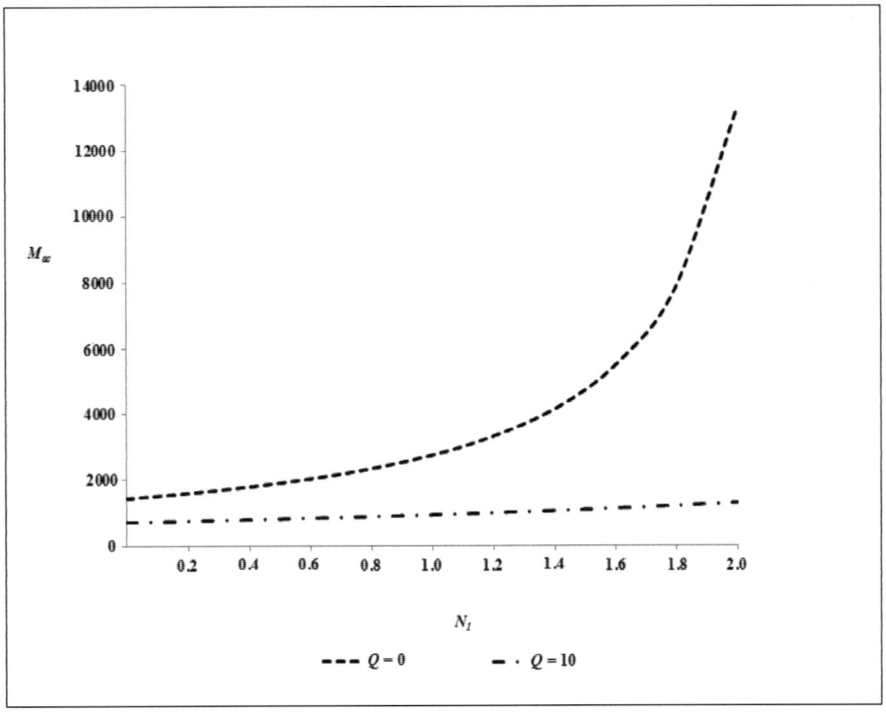

Figure 6. Plot of M_{ac} versus N_1 for different values of Q.

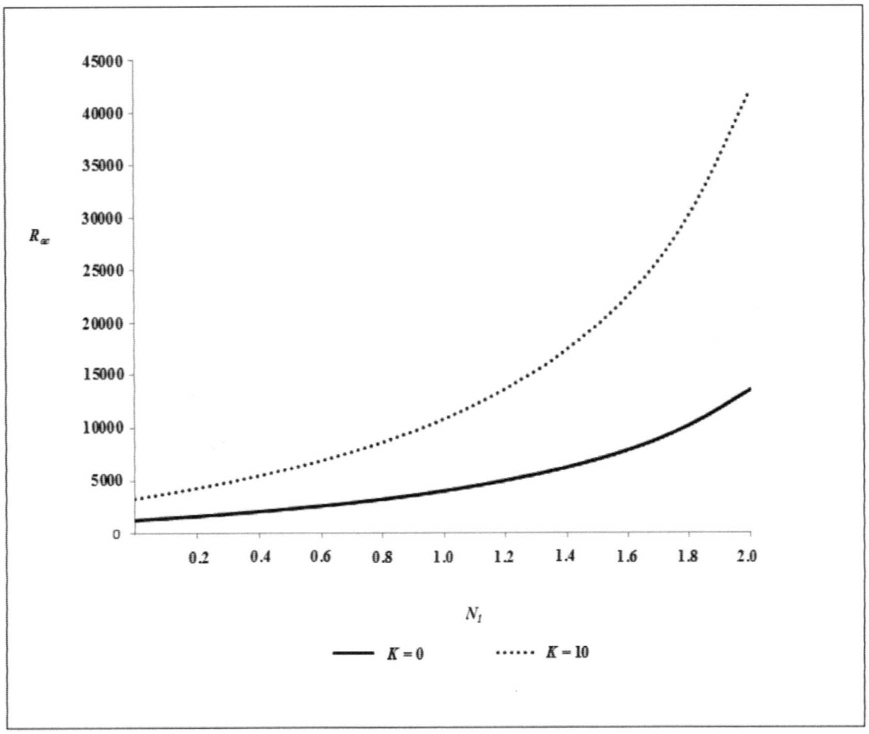

Figure 7. Plot of R_{ac} versus N_1 for different values of K.

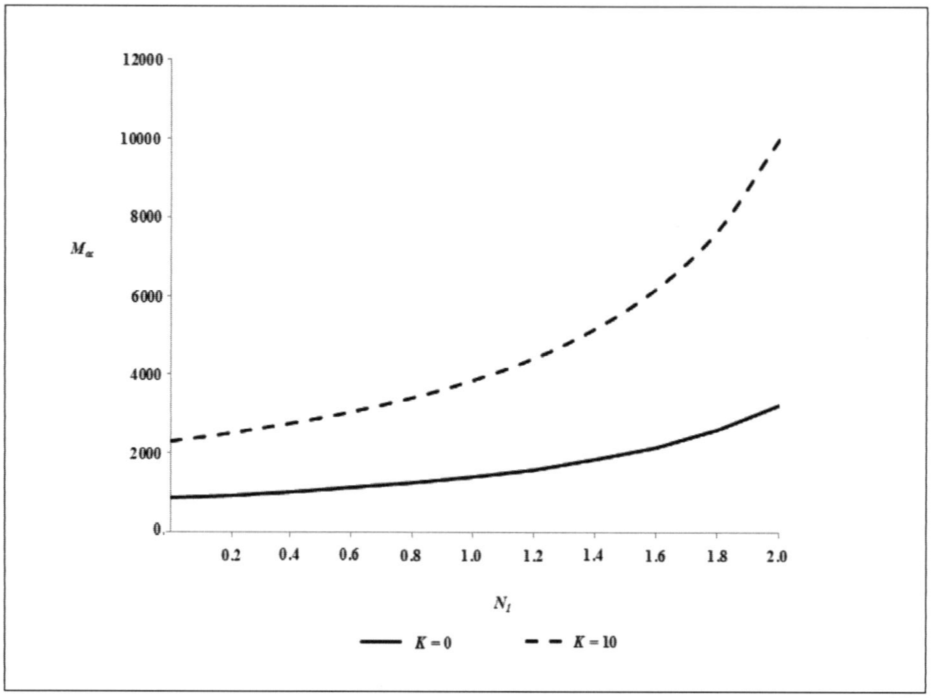

Figure 8. Plot of M_{ac} versus N_1 for different values of K.

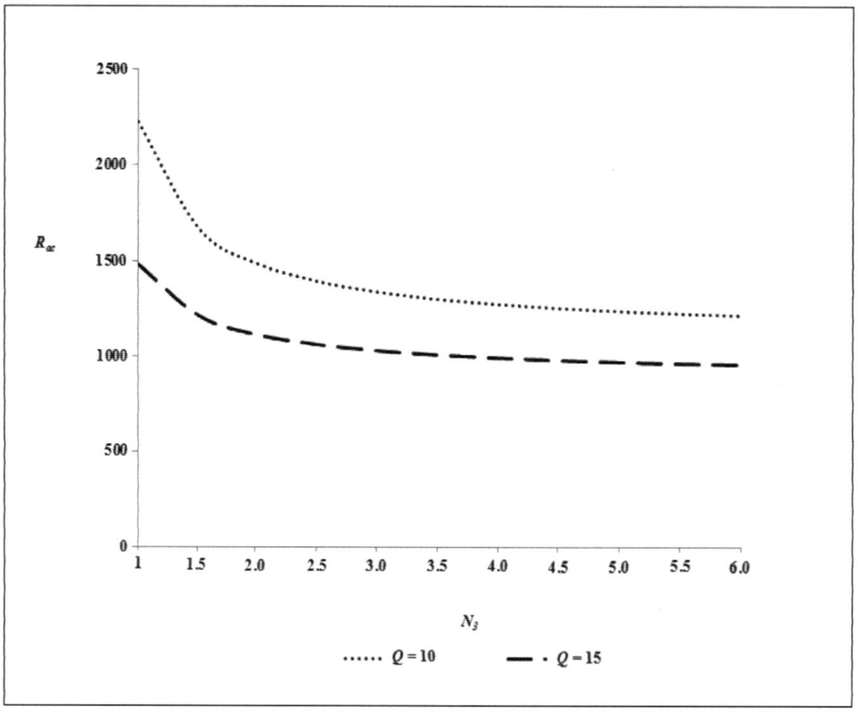

Figure 9. Plot of R_{ac} versus N_3 for different values of Q.

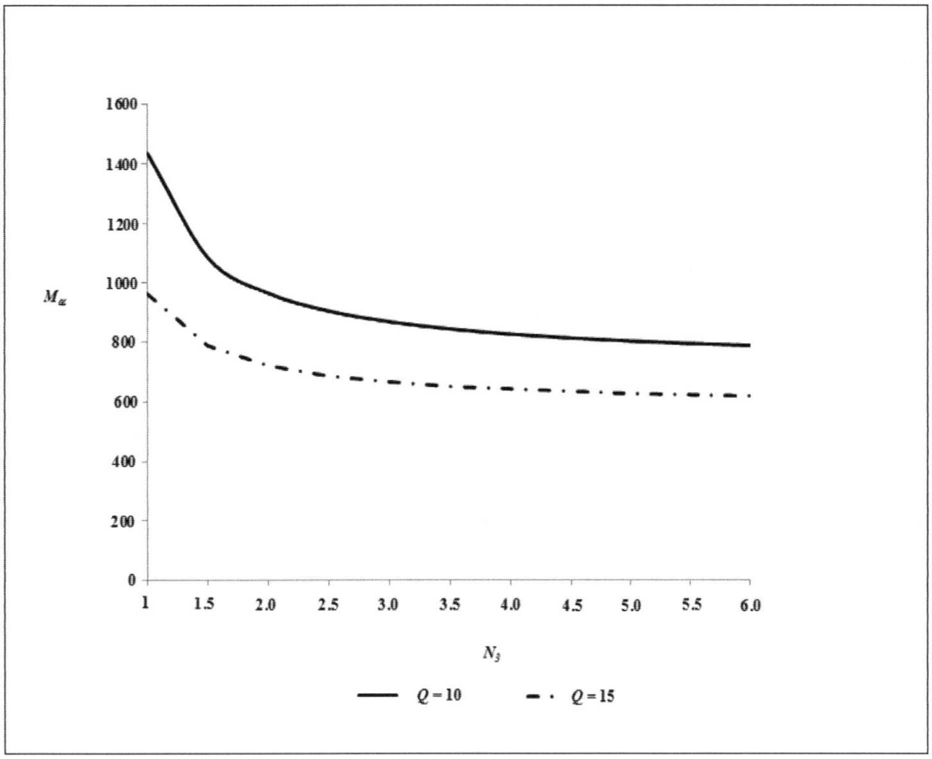

Figure 10. Plot of M_{ac} versus N_3 for different values of Q.

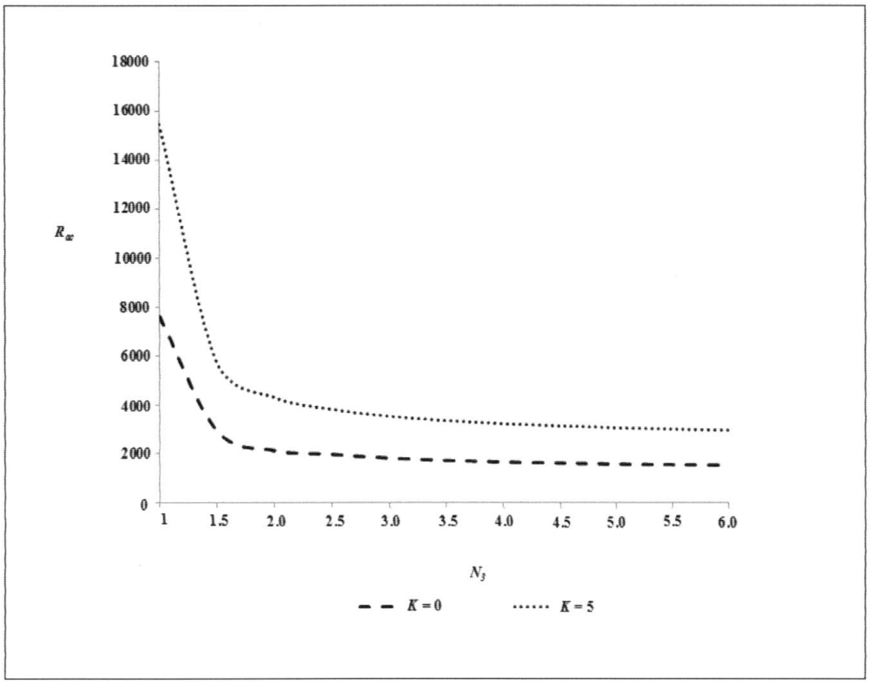

Figure 11. Plot of R_{ac} versus N_3 for different values of K.

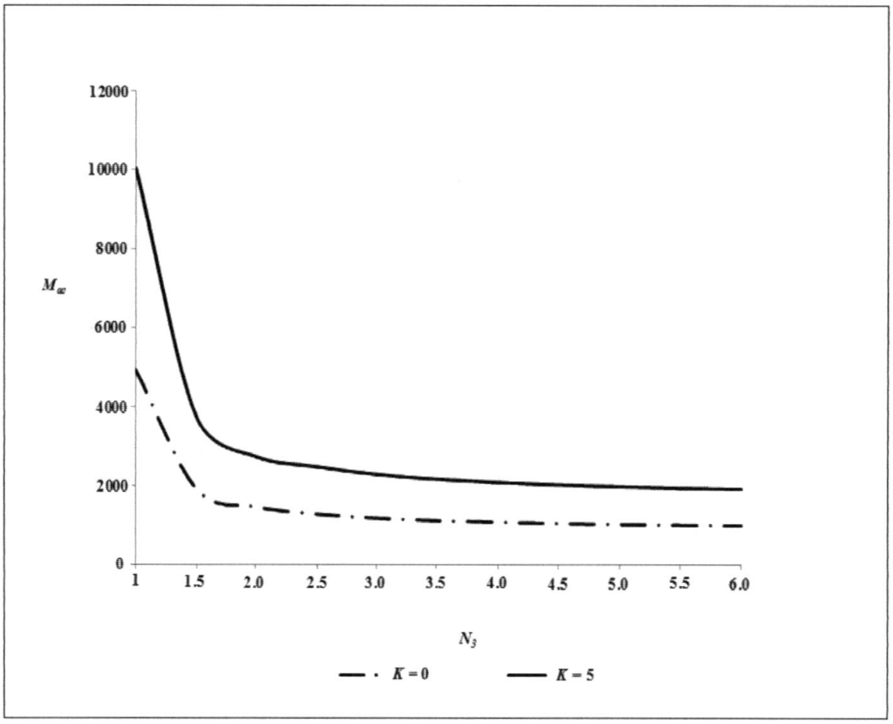

Figure 12. Plot of M_{ac} versus N_3 for different values of K.

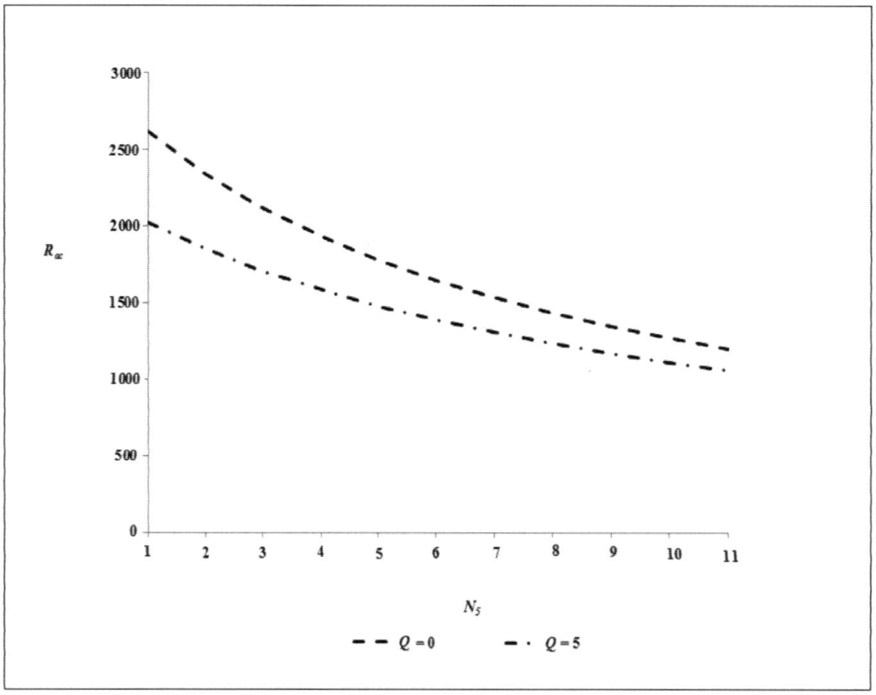

Figure 13. Plot of R_{ac} versus N_5 for different values of Q.

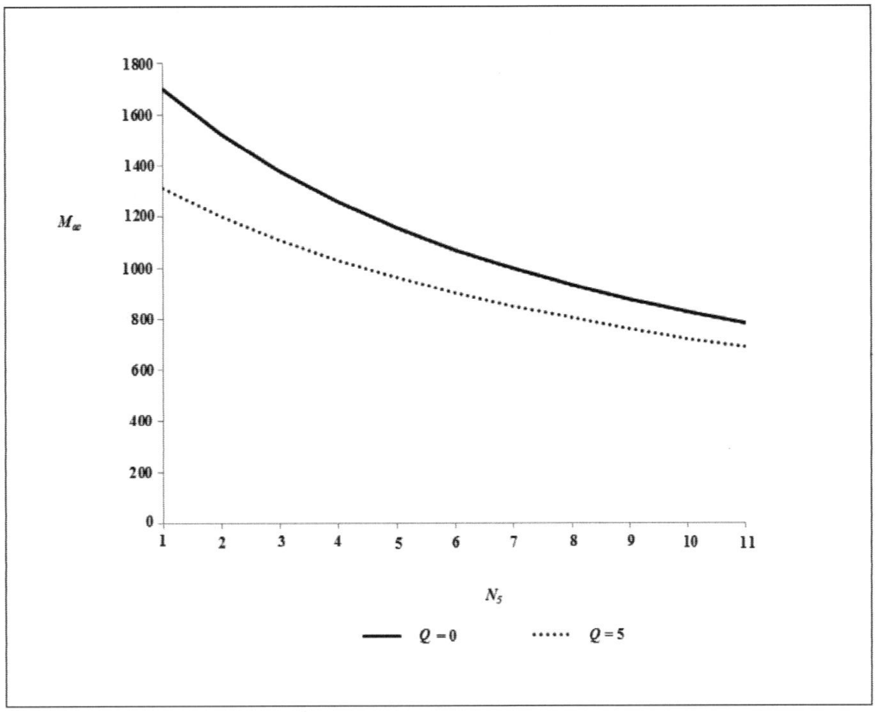

Figure 14. Plot of M_{ac} versus N_5 for different values of Q.

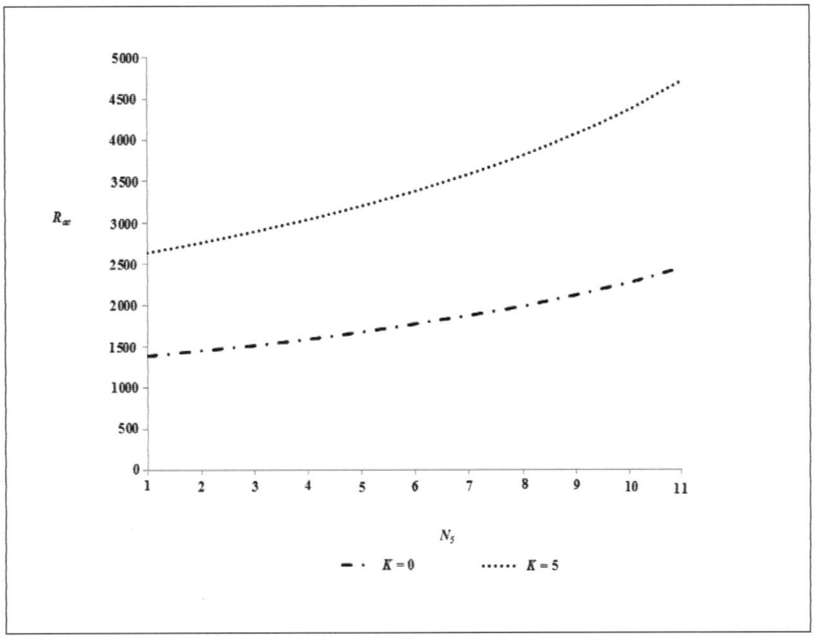

Figure 15. Plot of R_{ac} versus N_5 for different values of K.

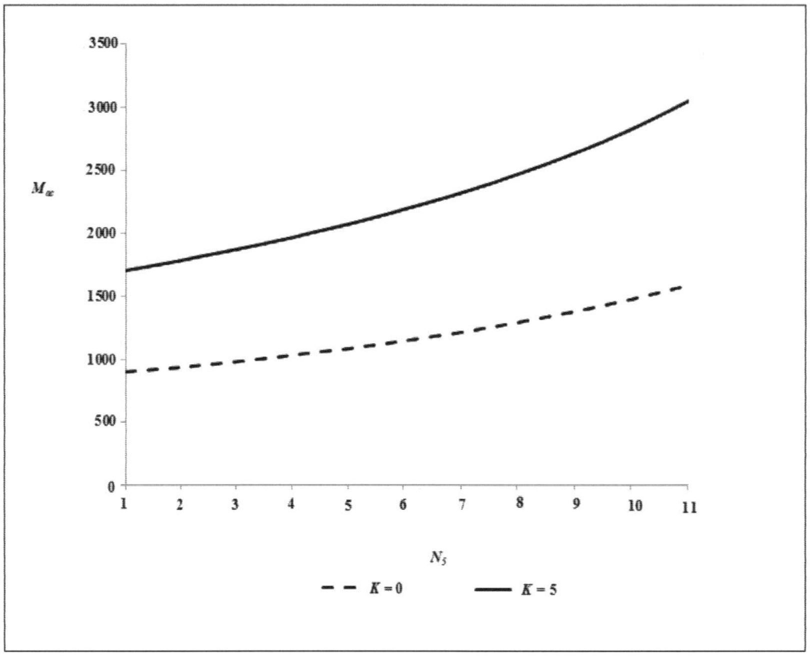

Figure 16. Plot of M_{ac} versus N_5 for different values of K.

5. Conclusion

The stability analysis of feedback control of the Benard-Marangoni convection in micropolar fluids with internal heat generation is investigated theoretically. We found that the effect of internal heat generation, Q in the micropolar fluid has a significant influence on the Benard-Marangoni convection where increasing the internal heating will destabilize the system. As for the effect of controller gain, K in the micropolar fluid, increasing the values of K, elevate the critical Marangoni number and Rayleigh number, and thus promote stability in the system. Although the effect of Q is to destabilize the system, the increase of the microelement concentration, N_1 and N_5 help to slow down the process of destabilizing. The effect of increasing the N_3, decrease the couple stress of the fluid and hence promote the destabilizing effect in the system. Therefore, the coupling parameter, N_1, couple stress parameter, N_3 and micropolar heat conduction parameter, N_5 has a significant effect on the onset of Benard-Marangoni convection.

Acknowledgments

Authors gratefully acknowledge the financial support received from Universiti Putra Malaysia under RUGS scheme.

References

[1] Eringen A C 1966 Theory of micropolar fluids *J. Math. Mech.* **16** 1-18
[2] Eringen A C 1972 Theory of thermo-microfluids *J. Math. Anal Appl.* **38** 480-496
[3] Eringen A C 2001 Microcontinuum Field Theories II: Fluent Media *Springer-Verlag, New York* **71** 1-10
[4] Eringen A C 1964 Simple microfluids *Int. J. Engng Sci.* **2** 205-217
[5] Eringen A C 1969 Micropolar fluids with stretch *Int. J. Engng Sci.* **7** 115-127
[6] Eringen A C 1980 Theory of anistropic micropolar fluids *Int. J. Engng Sci.* **18** 15-17

[7] Eringen A C 1990b Theory of thermo-microstrecth fluids and bubbly liquids *Springer-Verlag* **28** 133-143
[8] Eringen A C 1991 Memory-dependent orientable nonlocal micropolar fluids *Int. J. Engng. Sci.* **29** 1515-1529
[9] Benard H 1900 Les tourbillons cellulaires dans une nappe liquid *Revue Generale des* entries *Sciences Pures et Appliquees* **11** 1261-1271
[10] Benard H 1901 Les tourbillons cellulaires dans une nappe liquide transportant de la chaleur par convection en regime permanent *Ann. Chem. Phys.* **23** 62-144
[11] Rayleigh L 1984 On convection currents in a horizontal layer of fluid when higher temperature is on the under side *Philos. Mag.* **32** 529-546
[12] Walzer W 1976 Convective instability of a micropolar fluid layer *Ger. Beitr. Geophysik. Leipzig* **85** 137 – 143
[13] Rama Rao K V 1979 Onset of instability in a heat conducting micropolar fluid layer *Acta Mech.* **32** 79 – 84
[14] Sastry V U K and Ramamohan R V 1983 Numerical study of thermal instability of a rotating micropolar fluid layer *Int. J. Engng Sci.* **21** 449 – 461
[15] Qin Y and Kaloni P N 1992 A thermal instability problem in a rotating micropolar fluid *Int. J. Engng Sci.* **30** 1117 – 1126
[16] Ahmadi G and Shahinpoor M 1974 Universal stability of magneto micropolar fluid motion *Int. J. Engng Sci.* **12** 657 – 663
[17] Murty Y N and Ramana R V V 1999 Effect of throughflow on Marangoni convection in micropolar fluids *Acta Mech.* **138** 211 – 217
[18] Siddheshwar P G and Pranesh S 2002 Magnetoconvection in fluids with suspended particles under 1g and μg *Aero. Sci. Tech.* **6** 105 – 114
[19] Tang J and Bau H H 1993 Stabilization of the no-motion state in Rayleigh-Benard convection through the use feedback control *Phys. Rev. Lett.* **70** 1795-1798
[20] Howle L E 1997 Linear stability analysis of controlled Rayleigh-Benard convection using shadowgraphic measurement *Phys. Fluids* **9** 3111-3113
[21] Sparrow A, Goldstein R J and Jonsson V K 1964 Thermal stability in a horizontal fluid layer: Effect of boundary conditions and non-linear temperature *J. Fluid Mech.* **18** 513-528
[22] Roberts P H 1967 Convection in horizontal layers with internal heat generation: Theory *J. Fluid Mech.* **30** 33-49
[23] Char M I and Chiang K T 1994 Stability analysis of Benard-Marangoni convection in fluids with internal heat generation *Biometrika* **71** 1-10
[24] Wilson S K 1997 The effect of uniform internal heat generation on the onset of steady Marangoni convection in a horizontal layer of fluid *Acta Mech* **124**. 63-78
[25] Or A C, Kelly R E, Cortelezzi L and Speyer J L 1999 Control of long wavelength Benard-Marangoni convection *J. Fluid Mech.* **387** 321-341
[26] Bau H H 1999 Control of Marangoni-Benard convection *Int. J. Heat Mass Transfer* **42** 1327-1341
[27] Arifin N M, Nazar R and Senu N 1999 Feedback control of the Marangoni-Benard instability in a fluid layer with free-slip bottom . *J. Phys. Soc. Jpn.* **76** 1-4
[28] Bachok N and Arifin N M 2010 Feedback control of the Marangoni-Benard convection in a horizontal fluid layer with internal heat generation *Micro. Sci. Technol.* **22** 97-105

Journal of Physics: Conference Series **435** (2013) 012030 doi:10.1088/1742-6596/435/1/012030

Magnetic properties of rotational bands in ^{160}Dy and $^{170-174}Yb$ isotopes

A A Okhunov[1,3], **A A M Al-Sammarrae**[2], **and H Abu Kassim**[2]

[1] Department of Science in Engineering, International Islamic University Malaysia, 57528, P.O. 10, Kuala Lumpur, Malaysia

[2] Department of Physics, Faculty of Science, University of Malaya, 50603, Kuala Lumpur, Malaysia

[3] Institute of Nuclear Physics, Academy of Science of Uzbekistan, 100132, Tashkent, Uzbekistan

E-mail: abdurahimokhun@iium.edu.my , aaokhunov@gmail.com

Abstract. In this work, non - adiabatic effects manifested in the magnetic properties of low-lying states of even-even deformed nuclei are studied. A simple phenomenological model which takes into account the Coriolis mixing of states of the ground and $K^{\pi} = 2_n^+$ and $K^{\pi} = 1_{\nu}^+$ bands are proposed.

The Calculations for isotopes ^{160}Dy and $^{170,172,174}Yb$, are carried out. The calculated g_R- factor for the states of ground band which are compared with experimental data. Decreases of g_R- factor with increasing angular momentum I have been discussed. The probability of $M1-$ transition from 1_{ν}^+ and coefficients of multipole mixture $\delta(E2/M1)$ from 0_2^+, 0_3^+, 2_1^+, and 2_2^+ bands are calculated and compared with the experimental data. Decreases of g_R- factor with increasing angular momentum I have been discussed.

That these is an obvious inverse relation between g_R- factor and angular momentum I of the ground band states. This has been explained by a mixing ground and $K^{\pi} = 1^+$ bands which have a strong $B(M1)$ to ground state.

1. Introduction

The interest to investigation in the properties of deformed nuclei, especially increased in recent years with the opening of a new collective isovector magnetic dipole mode [1, 2].

The measured values of energies of the excited states such as modes indicate that they are not so high in the excitation spectrum, and consideration of mixing the isovector magnetic modes with low-lying exciting states can lead to interesting physical phenomena [3, 4].

Using ^{85}Kr ions with energy of 350 MeV by Coulomb excitation has been studied ground band state in the flesh to the spin $I = 16\hbar$ isotopes ^{160}Dy and $^{170,174}Yb$ in paper [5]. Non-adiabaticity of magnetic characteristics of low-spin states of ground bands has been observed. It is shown that, with an increase of angular momentum I, g_R- factor decreases in these three nuclei.

The nuclei $^{170,172,174}Yb$ are among the most well studied. These isotopes are studied experimentally in a number of ways: such as radioactive decay of $^{170,172,174}Lu$, and also different nuclear reactions. In the study of nuclear reactions of bands with the small K were supplemented bands with high spin [6]-[8]. In these isotopes detected many 1^+ states and also bands with $K^{\pi} = 0_n^+, 2_n^+$, and some of these bands include the levels with rather high spins.

The aim of the present paper is applied to investigating the properties of positive parity low-lying states of ^{160}Dy and $^{170,172,174}Yb$ isotopes. To studying the properties of positive parity low-lying states of these isotopes used phenomenological model [4] which takes into account the Coriolis mixing all of the experimentally known low-lying rotational bands with $K^{\pi} \leq 2^{+}$. The behavior of g_R- factor of ground band has been discussed by the growth of angular momentum. The energy spectra and the reduced probability of $M1-$ transition and the value of multipole-mixture $\delta\,(E2/M1)$ are calculated and compared with the experimental data.

2. The Model

To analyze the properties of low-lying positive parity states in Dy and Yb isotopes, the phenomenological model of [4] is exploited. This model takes into account the mixing of states of the $K^{\pi} = 0^{+}, 2^{+}$ and 1^{+} bands. The Hamiltonian model is

$$H = H_{rot}(I^2) + H_{K,K'} \tag{1}$$

$$H^{\sigma}_{K'K}(I) = \omega_K \delta_{K,K'} - \omega_{rot}(I)(j_x)_{K,K'}\zeta(I,K)\delta_{K,K'\pm 1} \tag{2}$$

where ω_K- bandhead energy of rotational band, $\omega_{rot}(I)-$ an angular frequency of rotational nucleus, $(j_x)_{K,K'}-$ matrix elements which describe Coriolis mixture between rotational bands and

$$\zeta(I,0) = 1 \qquad \zeta(I,2) = \left[1 - \frac{2}{I(I+1)}\right]^{\frac{1}{2}}$$

The eigenfunction of Hamiltonian model (1) is

$$\begin{aligned} |IMK> &= \sqrt{\frac{2I+1}{16\pi^2}}\left\{\sqrt{2}\Psi^I_{gr,K}D^I_{MK}(\theta)\right. \\ &+ \sum_K \frac{\Psi^I_{K',K}}{\sqrt{1+\delta_{K',0}}}\left[D^I_{M,K'}(\theta)b^+_{K'}\right. \\ &+ \left.\left.(-1)^{I+K'}D^I_{M,-K'}(\theta)b^+_{-K'}\right]\right\}|0> \end{aligned} \tag{3}$$

here $\Psi^I_{K',K}$ is the amplitude of mixture of basis states.

Solving the Shrödinger equation

$$H^{\sigma}_{K,K'}\Psi^I_{K,K'} = \mathcal{E}_K(I)\Psi^I_{K,K'}. \tag{4}$$

we define eigne function and energy of a Hamiltonian. The total energy of state is defined by

$$E_K(I) = E_{rot}(I) + \mathcal{E}_K(I) \tag{5}$$

In figure 1,2,3 and 4 the comparison of calculation values of energy with experimental dates for isotopes ^{160}Db and $^{170-174}Yb$ are presented. From figure we can see that in high spin values $I \geq 12\hbar$ noticeable deviation has been observed in calculated values of energy from experimental data, and this deviation increases with growth of angular momentum I.

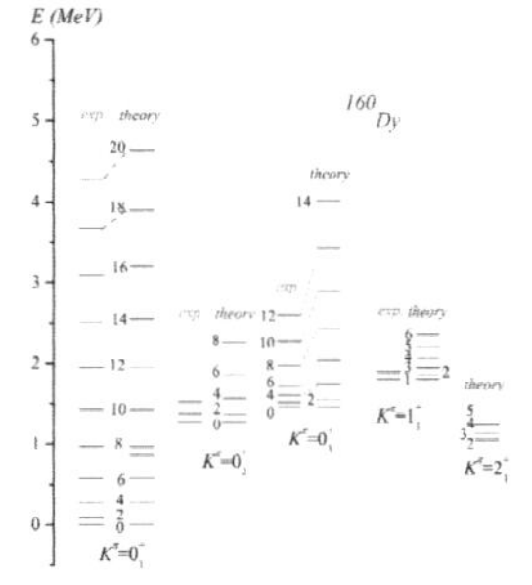

Figure 1. Comparison of the calculated and experimental energy spectra of positive-parity states for ^{160}Dy.

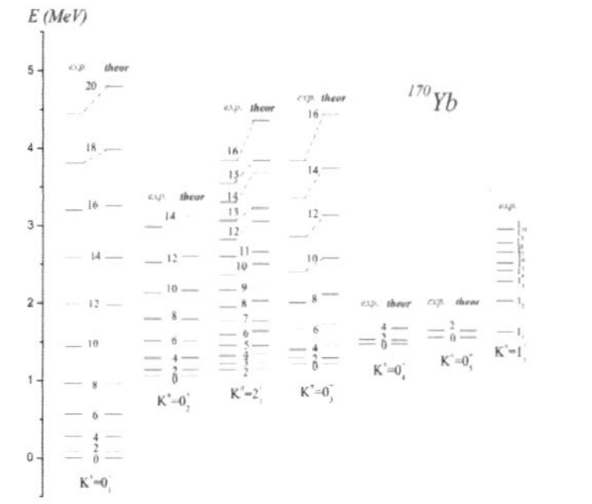

Figure 2. Comparison of the calculated and experimental energy spectra of positive-parity states for ^{170}Yb.

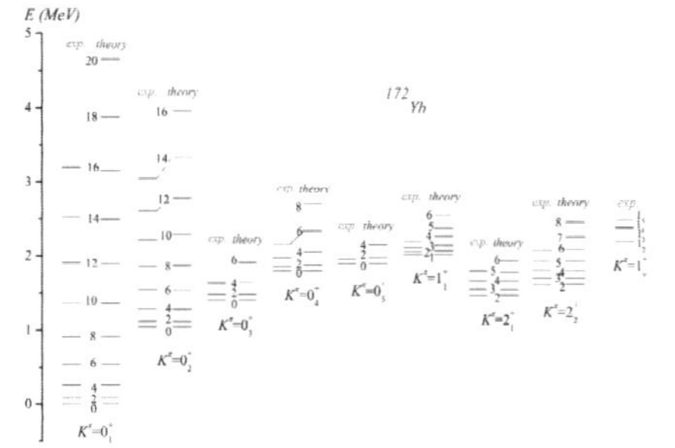

Figure 3. Comparison of the calculated and experimental energy spectra of positive-parity states for ^{172}Yb.

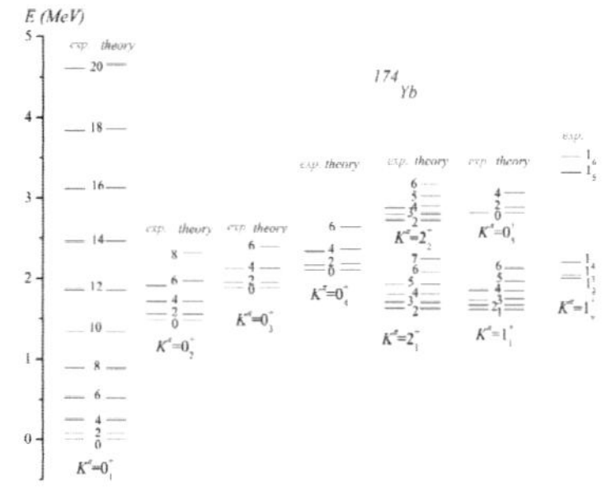

Figure 4. Comparison of the calculated and experimental energy spectra of positive-parity states for ^{174}Yb.

3. Magnetic Characteristics

The equation for the reduced probability of $M1-$ transitions from the states I_iK_i to the level I_f 0_1^+ band by phenomenological model is as follows [4]:

$$
\begin{aligned}
B \quad (\quad & M1; I_iK_i \rightarrow I_f 0_1) = \\
= \quad & \frac{3}{4\pi} | \sum_{n=1}^{\ell+\nu} \Psi^{I_i}_{K_n,K_i} \Psi^{I_f}_{K_n,0_1} K_n C^{I_f K_n}_{I_i K_n;10} \left(g_{K_n} - g_R \right) \\
+ \quad & \frac{\sqrt{6}}{10} C^{I_f 0}_{I_i 1;1-1} \sum_{\nu} m'_{1_\nu} \left(\Psi^{I_f}_{0_1,0_1} \Psi^{I_i}_{1_\nu,K_i} + \Psi^{I_i}_{0_1,0_1} \Psi^{I_f}_{1_\nu,0_1} \right) |^2
\end{aligned} \tag{6}
$$

where $g_R \approx Z/A$, g_K- intrinsic $g-$ factor of $K \neq 0$ bands. Here $m'_{1_\nu} =< 0_1^+ | \hat{m}(M1) | 1_\nu^+ >-$ matrix elements between intrinsic wave functions of 0_1^+ and $K^\pi = 1_\nu^+$ bands, the value which is estimated from the experimental data. In adiabatic approximation of the equation (6) yields

$$
B(M1; 11_\nu^+ \rightarrow 00_1^+) = \frac{3}{4\pi} \times 0.02 \left(m'_{1_\nu} \right)^2 \mu_N^2 \tag{7}
$$

From the known experimental value of probability such is $M1-$ transitions m'_{1_ν} can be calculate. Using experimental values of $B(M1)$ [12], m'_{1_ν} are determined by formula (7) which are presented in Table 1. However, formula (7) does not allow to define sign of the parameters m'_{1_ν}.

Usually, the coefficient of multipole mixture δ is experimentally investigated which is found by the following ratio.

$$\delta\left(I_iK_i \to I_fK_f\right) = 0.834E_\gamma(MeV)\times \times \frac{<I_fK_f\|\hat{m}(E2)\|I_iK_i>}{<I_fK_f\|\hat{m}(M1)\|I_iK_i>}\left(\frac{eb}{\mu_N}\right). \quad (8)$$

Note, that $M1-$ transitions from the $K^\pi = 0^+$ and 2^+ to the level ground bands are equal to the zero (absent) in adiabatic approximation. Thus, $M1-$ transitions appear due to 1^+_ν components $\Psi^{I_i}_{1_\nu,K_i}$ in wave functions of this states.

In case, the calculations use identical signs for parameters m'_{1_ν} the calculated values of coefficients of multipole mixture δ in many cases have smaller values than experimental data. In paper [12], 1^+_ν states have been investigated and illustrated values of ratio $R_{11_\nu} = B(M1;11 \to 20_1)/B(M1;11 \to 00_1)$ for $\gamma-$ transitions from 1^+_ν states to ground state in which adiabatic approximation is ≈ 0.5. As an axiom, for the $\gamma-$ transitions from 1^+_ν states having $R_{11_\nu} > 0.5$ identical sings of m'_{1_ν} are taken, and for $R_{11_\nu} < 0.5$, opposite signs are taken (see Table 1).

Comparison of measured experimental data [10, 11] and calculated value of coefficients of multipole mixture δ for $^{172174}Yb$ are illustrated in Table 2. Table 2, shows that calculated values have good correspondence with experimental data and precisely reproduces a sign of δ. The values of δ are decreasing with the growth of spin I, i.e. with a growth I the probability of $B(M1)$ is increasing.

Experimental data [9] for the probability of $M1-$ transitions from $K^\pi = 0^+_2, 0^+_3, 2^+_1$ and 2^+_2 bands has been compared with calculated values in Table 3. Experimental values of $B(M1)$ for transitions from $I = 2$ and $I = 4$ states of $K^\pi = 2^+_1$ increases with rise of I. Here, it is worth mentioning that this finding confirms our calculated results.

In [5], ground band states up to spin $I = 16\hbar$ of $^{170,174}Yb$ nuclei are investigated by Coulomb excitation through 350 MeV ^{85}Kr ions. In these nuclei, it is observed that g_R- factor of ground band states decreases with an increase of angular momentum I.

A g_R- factor of rotational K band determined within our model by the following formula:

$$g^K_R(I) = g_R + \sum_n^{\ell+\nu} |\Psi^I_{n,K}|^2\left(g_{K_n} - g_R\right)\frac{K^2_n}{I(I+1)} + \frac{\sqrt{3}}{5}\frac{\Psi^I_{gr,K}}{\sqrt{I(I+1)}}\sum_\nu m'_{1_\nu}\Psi^I_{1_\nu,K}. \quad (9)$$

The first non-adiabatic correction is negligible in the calculation of the g_R factors of ground state. But, the correction is considerable for g_R- factor for states with $K \neq 0$.

Theoretical values of magnetic characteristics excited states are obtained for the case $m'_1 = m'_{1_\nu} = 14\ \mu_N$.

4. Conclusion

In the present work, non-adiabatic effects in energies and electromagnetic characteristics of excited states are studied within the phenomenological model which taking into account Coriolis mixing of all experimentally known rotational bands for isotopes ^{160}Dy and $^{170,172,174}Yb$, with $K^\pi \leq 2^+$.

Table 1. Magnetic characteristics of 1^+_ν states in isotopes $^{172,174}Yb$.

ν	$E_{1_\nu}(MeV)$	R_{11_ν}	$B(M1)\uparrow(\mu_N^2)$	m'_{1_ν} (μ_N)
		^{172}Yb [12]		
1	2.010	1.06±0.14	$1.27{\cdot}10^{-2}$	+1.63
2	2.573	0.51±0.09	0.93±0.10	-8.06±0.87
3	2.612	0.70±0.13	0.33±0.09	-4.80±1.31
4	3.002	0.51±0.10	0.34±0.09	-4.87±1.29
5	3.096	0.46±0.12	0.11±0.04	+2.77±1.01
6	3.253	0.46±0.12	0.09±0.03	+2.51±0.84
7	3.604	0.76±0.13	0.49±0.12	-5.85±1.43
8	3.863	1.14±0.24	0.45±0.14	-5.61±1.75
		^{174}Yb [12]		
1	1.625	2.31±0.28	$3.1{\cdot}10^{-2}$	-0.80
2	2.037	0.64±0.40	0.15±0.11	-3.24±2.38
3	2.068	0.67±0.34	0.20±0.12	-3.83±2.30
4	2.338	0.74±0.20	0.28±0.10	-4.34±1.55
5	2.500	0.60±0.16	0.35±0.11	-5.01±1.58
6	2.581	0.46±0.14	0.21±0.08	+3.83±1.46
7	2.815	0.90±0.38	0.16±0.009	-3.24±1.62
8	2.920	0.41±0.07	0.44±0.11	+5.61±1.40
9	3.050	1.06±0.42	0.14±0.07	-3.24±1.62
10	3.122	0.50±0.27	0.10±0.06	-2.51±1.51
11	3.145	0.76±0.29	0.13±0.06	-2.89±1.33
12	3.349	0.58±0.18	0.33±0.14	-4.80±2.04
13	3.485	0.68±0.09	0.24±0.08	+4.09±1.36
14	3.562	0.47±0.08	0.41±0.10	-5.42±1.32
15	3.695	0.48±0.16	0.33±0.13	+4.58±1.80

The energy and structure of wave functions of excited states are calculated. The reduced probabilities of $M1-$ transitions and coefficients of multipole mixture $\delta(E2/M1)$ from $K^\pi = 0^+_m, 2^+_\ell$ and 1^+_ν bands are also calculated and compared with experimental data which gives the satisfactory result are also calculated.

Thus within our theoretical analysis it is possible to explain the $M1-$ transition from the state β_1-, β_2-, γ_1- and γ_2- bands to the ground bands level, which is forbidden by adiabatic approximation.

The experimentally observed K forbidden $M1-$ transitions from state $2^+_\ell-$ bands are explained by the presence of $K^\pi = 1^+_\nu$ components in wave functions of these bands. With the growth of the angular moment I, the coefficient of $\delta(E2/M1)$ for transitions from 0^+_m and 2^+_ℓ bands decrease. Experimental data of $B(M1)$ for transitions $I = 2$ and $I = 4$ state $K^\pi = 2^+_1$ bands shows an increase of $B(M1)$ with the of growth I. This confirms our results.

The result of calculation show that the decrease of g_R- factor with an increase of angular momentum I of rotational states of ground band is associated with the mixing in these states with 1^+_ν bands which have strong $B(M1)$ to the ground state in isotope ^{170}Yb.

Table 2. Multipole mixture coefficients of $\delta(E2/M1)$ in isotopes Yb.

A	$I_iK_i \rightarrow I_fK_f$	E_i (MeV)	E_γ (MeV)	δ [10, 9]	$\delta^{theor.}$
^{172}Yb	$22_1^+ \rightarrow 20_1^+$	1.4659	1.3871	$-5.1^{+1.1}_{-1.6}$	-8.8
	$32_1^+ \rightarrow 40_1^+$	1.5492	1.2888	$+2.8^{+0.7}_{-1.0}$	-9.0
	$\rightarrow 20_1^+$		1.4704	-7.6^{+19}_{-36}	-2.1
	$42_1^+ \rightarrow 40_1^+$	1.6578	1.3975	$-1.1^{+0.2}_{-0.5}$	-1.9
	$22_2^+ \rightarrow 20_1^+$	1.6085	1.5297	+10(3)	+12.6
	$32_2^+ \rightarrow 40_1^+$	1.7006	1.4404	$+6.5^{+2.2}_{-1.4}$	+7.8
	$\rightarrow 20_1^+$		1.6219	+17.0(4)	+11.3
	$42_2^+ \rightarrow 40_1^+$	1.8031	1.5429	$+9.0^{+11}_{-3}$	+8.0
	$52_2^+ \rightarrow 40_1^+$	1.9270	1.6664	$+6.9^{+1.9}_{-1.2}$	+6.8
	$20_2^+ \rightarrow 20_1^+$	1.1179	1.0393	$+2.3^{+0.5}_{-0.3}$	+1.0
	$40_2^+ \rightarrow 40_1^+$	1.2868	1.0265	+0.87(13)	+0.49
	$60_2^+ \rightarrow 60_1^+$	1.5375	0.9977	+0.63(7)	+0.30
^{174}Yb	$22_1^+ \rightarrow 20_1^+$	1.6340	1.1456	-0.48(8) or $-32^{+20}_{-\infty}$[11]	-15.0
	$32_1^+ \rightarrow 20_1^+$	1.7094	1.6329	$-3.8^{+1.4}_{-0.8}$[11]	-17.5
	$20_2^+ \rightarrow 20_1^+$	1.5610	1.4845	+1.7(4) or $+0.12^{+0.12}_{-0.10}$[11]	+0.82
	$40_2^+ \rightarrow 40_1^+$	1.7160	1.4630	$-0.03^{+0.25}_{-0.16}$[11]	+0.40
	$20_3^+ \rightarrow 20_1^+$	1.9582	1.8823	+0.01(8) or +2.2(5)[11]	-0.16
	$21_1^+ \rightarrow 20_1^+$	1.6750	1.5983	-0.23(5) or $+5.2^{+1.8}_{-1.1}$[11]	-0.21
	$31_1^+ \rightarrow 20_1^+$	1.7338	1.6573	+0.50(5) or $+4.6^{+1.4}_{-1.0}$[11]	+0.65

Table 3. Probability of $M1-$ transitions of ^{172}Yb.

$I_iK_i \rightarrow I_f0_f$	$B(M1; I_iK_i \rightarrow I_f0_f)$ μ_N^2	
	exp. [9]	theor.
$20_2^+ \rightarrow 20_1^+$	$0.6 \cdot 10^{-3}(2)$	$1.8 \cdot 10^{-3}$
$20_3^+ \rightarrow 20_1^+$	$1.2 \cdot 10^{-4}(7)$	$1.1 \cdot 10^{-3}$
$22_1^+ \rightarrow 20_1^+$	$1.7 \cdot 10^{-2}(2)$	$0.2 \cdot 10^{-3}$
$42_1^+ \rightarrow 40_1^+$	$9.0 \cdot 10^{-2}(6)$	$3.1 \cdot 10^{-3}$
$22_2^+ \rightarrow 20_1^+$	$5.2 \cdot 10^{-5}(27)$	$4.0 \cdot 10^{-5}$

Acknowledgments

Specially authors would like to thank to Prof. Dr. Usmanov Ph.N. for consultation and discussion of results. And also we thank for financial supporting this work the University of Malaya under the UMRG Grant No.RG 120/10AFR, and the Islamic Development Bank (IDB).

References

[1] Mikhailov I N, Neergor K, Pashkevich V V, Frauendorf C 1977 *Fiz. Elem. Chastits At. Yadra* **8**, No.6 1338
[2] Jungclaus A, Binder B, Dietrich A, et al. 2002 *Phys.Rev.* C **66** 014312.

[3] Usmanov P N, Adam I, Salikhbaev U S, and Solnyshkin A A 2010 *Physics of Atomic Nuclei* **73** No.12 1990
[4] Usmanov Ph N, I.N.Mikhailov IN 1997 *Phys. Part. Nucl. Lett.* **28**, 348
[5] Andrews H R, Hausser O, Ward D and et al. 1980 *Phys.Rev.Lett.* **43**, No.23 1835
[6] Riech C W 2005 *Nucl. Data Sheets* **105** 557
[7] Baglin M 2002 *Nucl. Data Sheets* **96** 611
[8] Bengtsson R, Frauendorf S 1979 *Nucl. Phys.* A **327** No.1 139
[9] Singh B 1995 *Nucl.Data Sheets.* Vol **75** 199
[10] Kracikova T I, Davaa S, Finger M and et al. 1984 *J. Phys.* Vol. G**10** 1115
[11] Youhana H M, Al-Obeidi S R, Al-Amili M A and et al. 1986 *Nucle. Phys.* Vol. A**458** 51
[12] Zilges A, Brentano P von, C.Wesselborg C and et al 1990 *Nuc.Phys.* A**507** 399

Strange Attractors in the Vannimenus Model on an Arbitrary Order Cayley Tree

Nasir Ganikhodjaev[1], **Hasan Akin**[2], **Seyit Temir**[3], **Selman Uğuz**[3], **and Ashraf Mohamed Nawi**[1]

[1]Department of Computational and Theoretical Sciences, Faculty of Science, IIUM, 25200 Kuantan, MALAYSIA
[2]Faculty of Education, Zirve University, 27260, Gaziantep, TURKEY
[3]Department of Mathematics, Harran University, 63120, Sanliurfa, TURKEY

E-mail: [1]gnasir@iium.edu.my, [2]hasanakin69@gmail.com, [3]temirseyit@harran.edu.tr, [4]selmanuguz@gmail.com, [5]algebra-student@yahoo.com

Abstract. We consider the Vannimenus model on a Cayley tree of arbitrary order k with competing nearest-neighbour interactions J_1 and next-nearest-neighbour interactions J_2 and J_3 in the presence of an external magnetic field h. In this paper we study the phase diagram of the model using an iterative scheme for a renormalized effective nearest-neighbour coupling K_r and effective field per site X_r for spins on the rth level; it recovers, as particular cases, previous works by Vannimenus, Inawashiro *et al*, Mariz *et al* and Ganikhodjaev and Uğuz. Each phase is characterized by a particular attractor and the phase diagram is obtained by following the evolution and detecting the qualitative changements of these attractors. These changements can be either continuous or abrupt, respectively characterizing second- or first- order phase transitions. We present a few typical attractors and at finite temperatures, several interesting features (evolution of reentrances, separation of the modulated region into few disconnected pieces, etc) are exhibited for typical values of parameters.

1. Introduction

The anisotropic next-nearest-neighbour Ising (ANNNI) model, which consists of an Ising model with nearest-neighbour interactions augmented by competing next-nearest-neighbour couplings acting parallel to a single axis direction, is one of the simplest model displaying a rich phase diagram with a Lifshitz point and many spatially modulated phases [1]-[8]. There has been a considerable theoretical effort to obtain the structure of the global phase diagram of the ANNNI model in the $T-p$ space, where T is temperature and $p=-J_2/J_1$ is the ratio between the competing exchange interactions. On the basis of numerical mean-field calculations, Bak and von Boehm [7] suggested the existence of an infinite succession of commensurate phases, the so-called devil's staircase, at low temperatures. This mean-field picture has been supported by low-temperature series expansions performed by Fisher and Selke [8]. An Ising model with competing interactions on the Cayley tree has recently been studied extensively because of the appearance of nontrivial magnetic orderings (see [9] - [19] and references therein).

The Cayley tree is not a realistic lattice; however, its amazing topology makes the exact calculation of various quantities possible. For many problems the solution on a tree is much simpler than on a regular lattice and is equivalent to the standard Bethe-Peierls theory [20].

On the Cayley tree one can consider two type of next-nearest-neighbours: prolonged and one-level next-nearest-neighbours (definitions see below). In the case of the Ising model with competing nearest-neighbor interactions J and prolonged next-nearest-neighbour interactions J_p Vannimenus [9] was able to find new modulated phases, in addition to the expected paramagnetic and ferromagnetic ones. From this result follows that Ising model with competing interactions on a Cayley tree is real interest since it has many similarities with models on periodic lattices. In fact, it has many common features with them, in particular the existence of a modulated phase, and shows no sign of pathological behavior - at least no more than mean-field theories of similar systems [9]. Moreover a detailed study of its properties was carried out with essentially exact results, using rather simple numerical methods.

Later Mariz *et al* [10] extended this results assuming existence also interaction J_3 of the one-level nearest-next-neighbours with $k = 2$. C.R. da Silva and S. Coutinho [16] studied the Ising model on a general Cayley tree of arbitrary order with competing interactions between the first-, second-, and third-next-nearest-neighbour spins belonging to the same branch, and in the presence of an external magnetic field. Note that da Silva and Coutinho [16] have generalized the approach used by Thompson [11] for the Ising model on the Cayley tree with only nearest-neighbour interactions and an external field. Yokoi *et al* [14] extended the calculations of Vannimenus [9] for a tree of arbitrary order and gave strong numerical evidence for the existence of chaotic phases associated with strange attractors.

In this paper we extend the model considered by Mariz *et al* [10] for a tree of arbitrary order using the procedures of Inawashiro, Thompson, and Honda [12] to write first-order recursion relations and give strong numerical evidence for the existence of chaotic phases associated with strange attractors. We consider the Vannimenus model on a Cayley tree of arbitrary order k with competing nearest-neighbor interactions J_1, prolonged next-nearest-neighbor interactions J_2 and one-level k-tuple neighbours interaction J_3 in the presence of magnetic field h. Note that the inclusion of the k- tuple neighbours competing interaction J_3 is essential for the presence of different stable modulated phases at $T = 0$. Apparently the one-level k tuple neighbour interactions represent interest not only for physical and biological models but also for models in sociology [22],[23].

This paper has been organized in the following way. In Section 2 the model Hamiltonian is discussed. In Section 3 the recursion equations are defined. Section 4 is devoted to the discussion of the attractors of this dynamical system and finally, the conclusions are given in Section 5.

2. The Vannimenus model

Cayley Tree. A Cayley tree Γ^k of order $k \geq 1$ is an infinite tree, i.e., a graph without cycles with exactly $(k+1)$ edges issuing from each vertex. Let denote the Cayley tree as $\Gamma^k = (V, \Lambda)$, where V is the set of vertices of Γ^k, Λ is the set of edges of Γ^k. Two vertices x and y, $x, y \in V$ are called *nearest-neighbors* if there exists an edge $l \in \Lambda$ connecting them, which is denoted by $l =< x, y >$. The distance $d(x, y), x, y \in V$, on the Cayley tree Γ^k, is the number of edges in the shortest path from x to y. For a fixed $x^0 \in V$ we set

$$W_n = \{x \in V | d(x, x^0) = n\}, \quad V_n = \{x \in V | d(x, x^0) \leq n\}$$

and L_n denotes the set of edges in V_n. The fixed vertex x^0 is called the 0-th level and the vertices in W_n are called the n-th level. For the sake of simplicity we put $|x| = d(x, x^0)$, $x \in V$. Two vertices $x, y \in V$ are called *the next-nearest-neighbours* if $d(x, y) = 2$. The next-nearest-neighbour vertices x and y are called *prolonged next-nearest-neighbours* if $|x| \neq |y|$ and is denoted by $> \widetilde{x, y} <$. The next-nearest-neighbour vertices $x, y \in V$ that are not prolonged are called *one-level next-nearest-neighbours* since $|x| = |y|$ and are denoted by $> \bar{x, y} <$. Two vertices $x, y \in V$ are called the third-next-nearest-neighbour [16] if $d(x, y) = 3$ with $x \in W_n, y \in W_{n+3}$ for some n.

We write $x \prec y$ if the path from x^0 to y goes through x. We call the vertex y a direct successor of x, if $y \succ x$ and x, y are nearest neighbours. The set of the direct successors of x is denoted by $S(x)$, i.e., if $x \in W_n$, then

$$S(x) = \{y_i \in W_{n+1} | d(x, y_i) = 1, i = 1, \cdots, k\}.$$

We observe that for any vertex $x \neq x^0, x$ has k direct successors and x^0 has $k+1$.

The collection $S(x) = \{y_1, \cdots, y_k\}$ we will call one-level k tuple of neighbours. Note that if $k = 2$ then $S(x)$ is a one-level next-nearest-neighbours.

Below we will consider a semi-infinite Cayley tree Γ_+^k of $k-$th order, i.e. an infinite graph without cycles with $(k+1)$ edges issuing from each vertex except for x^0 which has only k edges. In this case $|S(x)| = k$ for any $x \in V$.

The Model For the Vannimenus model with spin values in $\Phi = \{-1, 1\}$, the relevant Hamiltonian with competing nearest-neighbour and prolonged next-nearest-neighbour binary interactions has the form

$$H(\sigma) = -J_1 \sum_{<x,y>} \sigma(x)\sigma(y) - J_2 \sum_{\widetilde{>x,y<}} \sigma(x)\sigma(y) - J_3 \sum_x \prod_{y \in S(x)} \sigma(y) - h \sum_x \sigma(x), \qquad (1)$$

where the sum in the first term ranges over all prolonged next-nearest-neighbours , the second sum ranges over all one-level k- tuple neighbours, the third sum ranges over all nearest neighbours and the spin variables $\sigma(x)$ assume the values ± 1. Here $J_1, J_2, J_3, h \in R$ are coupling constants. This model recovers that of Vannimenus for $k = 2, J_3 = 0$, that of Inawashiro *et al* for $k = 2, J_2 = J_3$, that Mariz *et al* for $k = 2$ and that Ganikhodjaev *et al* for $h = 0$. As noted above the one-level k tuple neighbour interaction J_3 represent interest not only for physical and biological models but also for models in sociology [22], [23].

3. Recurrence Relations

In order to set up the iterative scheme we sum successively over spins as shown schematically in Fig.1.

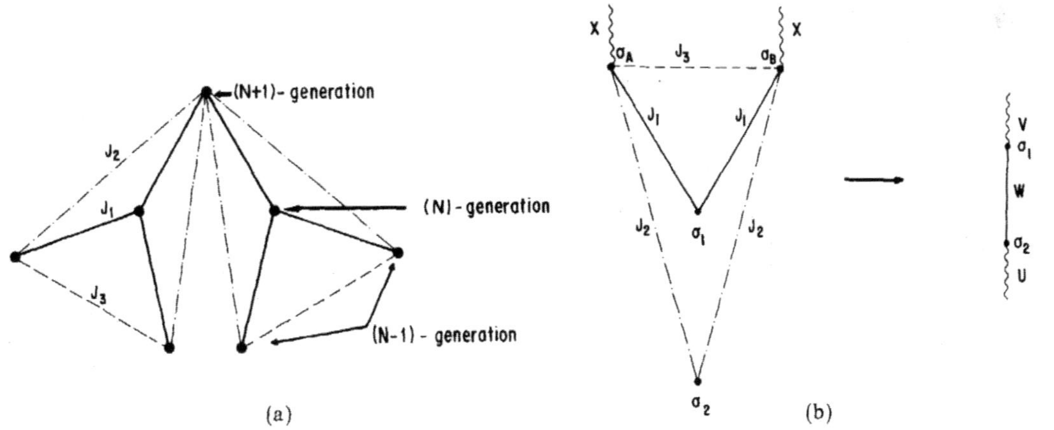

Fig. 1. (a) Three successive generations of a Cayley tree; (b) Schematic illustration of selective summation of spins

Simple but tedious algebra gives

$$\sum_{\sigma_{A_i}=\pm 1} \exp\left[(K_1\sigma_1 + K_2\sigma_2) \cdot \sum_{i=1}^{k} \sigma_{A_i} + K_3 \sum_{i=1<j}^{k} \sigma_{A_i}\sigma_{A_j} + X \sum_{i=1}^{k} \sigma_{A_i}\right] = C \exp[W\sigma_1\sigma_2 + U\sigma_2 + V\sigma_1] \qquad (2)$$

where $K_i \equiv J_i/k_BT, i = 1,2,3$ and

$$U(X,K_1,K_2,K_3) \equiv U = \frac{1}{4}\ln\frac{\omega(1,1)\omega(1,-1)}{\omega(-1,1)\omega(-1,-1)} \tag{3}$$

$$V(X,K_1,K_2,K_3) \equiv V = \frac{1}{4}\ln\frac{\omega(1,1)\omega(-1,1)}{\omega(1,-1)\omega(-1,-1)} \tag{4}$$

$$W(X,K_1,K_2,K_3) \equiv W = \frac{1}{4}\ln\frac{\omega(1,1)\omega(-1,-1)}{\omega(-1,1)\omega(1,-1)} \tag{5}$$

$$C(X,K_1,K_2,K_3) \equiv C = [\omega(1,1)\omega(-1,-1)\omega(1,-1)\omega(-1,1)]^{\frac{1}{4}} \tag{6}$$

where the value ω depends from evenness of k.

If k is odd, then

$$\omega(\sigma_1,\sigma_2) = 2\sum_{i=0}^{\frac{k-1}{2}}\binom{k}{i}\exp\left[\frac{k^2-4ki-k+i^2+3i}{2}\cdot K_3\right]\cosh[(k-2i)(K_1\sigma_1+K_2\sigma_2+X)] \tag{7}$$

and if k is even, then

$$\begin{aligned}\omega(\sigma_1,\sigma_2) &= 2\textstyle\sum_{i=0}^{\frac{k}{2}-1}\binom{k}{i}\exp\left[\frac{k^2-4ki-k+i^2+5i}{2}\cdot K_3\right]\cosh[(k-2i)(K_1\sigma_1+K_2\sigma_2+X)]\\ &+ \binom{k}{k/2}\exp\left[\frac{6k-3k^2}{8}\cdot K_3\right]\end{aligned} \tag{8}$$

Beginning with $X^{(1)} = B = h/k_BT$ and $K_1^{(1)} = K_1 = J_1/k_BT$ where h is the initial applied magnetic field per site and J_1 is the initial nearest-neighbour coupling, we obtain from the above equalities the iteration scheme, for $r = 2,3,\cdots,N$,

$$X^{(r)} = B + k\cdot U(X^{(r-2)},K_1^{(r-2)},K_2,K_3) + V(X^{(r-1)},K_1^{(r-1)},K_2,K_3) \tag{9}$$

and

$$K_1^{(r)} = K_1 + W(X^{(r-1)},K_1^{(r-1)},K_2,K_3) \tag{10}$$

Equations (9) and (10) with functions $U, V,$ and W defined by equations (3–5) and either (7) or (8) constitute basic recursion relations with initial conditions $X^{(0)} = K_1^{(0)} = 0$ and $X^{(1)} = B, K_1^{(0)} = K_1$.

4. The Phase Diagram and Attractors

It is convenient to know the broad features of the phase diagram before discussing the different transitions in more detail. This can be achieved numerically in a straightforward fashion. The recursion relations (9-10) provide us the numerically exact phase diagram in $(T/J_1, -J_2/J_1, J_3/J_1, h/J_1)$ space. Let $k_BT/J_1 = \alpha$, $-J_2/J_1 = \beta$, $J_3/J_1 = \gamma, h/J_1 = \delta$ and respectively $K_1 = 1/\alpha; K_2 = -\beta/\alpha; K_3 = \gamma/\alpha$ and $B = \delta/\alpha$. By iterating equations (9) and (10) numerically for selected values of $\alpha,\beta,\gamma,\delta$ we consider limiting behaviour of the sequence of points $\{(X^{(n)};K_1^{(n)}) : n = 2,3,\cdots\}$ in the $X - K_1$ plane and describe phases. Each phase: Paramagnetic (P), Ferromagnetic (F), Modulated (M), Antiphase ($< 2 >$), Antiferromagnetic (AF) is characterized by a particular attractor in the $X - K_1$ plane, where $X^{(n)}$ plays the role of an effective field. The phase diagram is obtained by following the evolution and detecting the qualitative changements of these attractors. These changements can be either continuous or abrupt, respectively, characterizing second- or first- order transitions. We find firstly that for any k when $J_1 > 0, J_2 > 0$ and J_3 is arbitrary, $X^{(n)}$ and $K_1^{(n)}$ approach a stable fixed point $(\tilde{X}, \tilde{K_1})$ corresponding to a ferromagnetic phase when $\tilde{X} > 0$ and paramagnetic phase when $\tilde{X} = 0$. The same phases result when J_2 is negative and small in magnitude relative to $J_1 > 0$ where this magnitude is smaller and smaller with increasing k. On the other hand when J_2 is negative and sufficiently large in magnitude relative to $J_1 > 0$, $X^{(n)}$ and $K_1^{(n)}$ iterate to a four cycle of points in the $X - K_1$ plane yielding a modulated antiphase $< 2 >$ of the same form discussed previously by Katsura and Takizawa [20] and later on by Vannimenus [9].

The interesting region occurs for intermediate negative values of J_2 and both positive and negative values of J_1 where frustration takes hold and the system cannot decide whether it wants to be ferromagnetic or antiferromagnetic. Below we plot the attractors associated with the mappings (9)-(10) for some selected values of $\alpha, \beta, \gamma, \delta$ with $\delta = +0.1, \delta = 0$, and $\delta = -0.1$ to clarify the role of external magnetic field. The first 10^4 iterations are discarded; the subsequent 10^4 iterations are plotted.

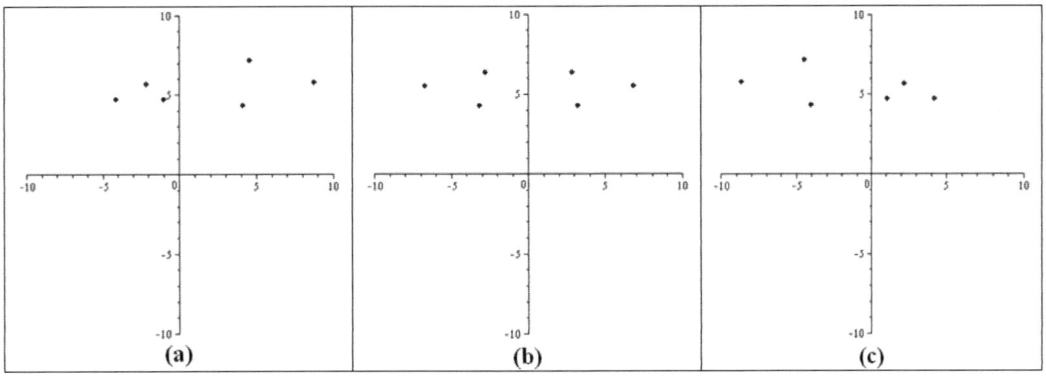

Figure 2. The attractors in $X - K_1$ plane for $k = 2$, $\gamma = -0.35$, $\beta = 0.35$, and $\alpha = 0.11$ with external magnetic for (a) $\delta = +0.1$, (b) $\delta = 0$ and (c) $\delta = -0.1$

The phase corresponding to the attractors in Fig. 2 is a periodic with period 6 and one can see that under external magnetic field the shape of attractor consisting from 6 points a little bit deformed preserving the number of points.

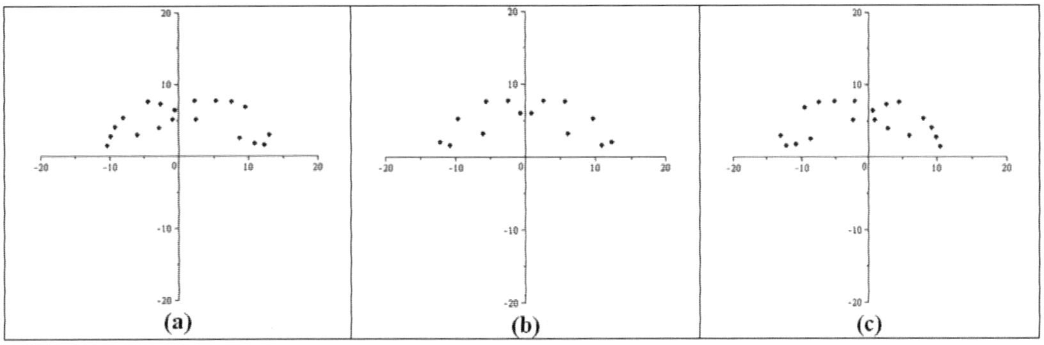

Figure 3. The attractors in $X - K_1$ plane for $k = 2$, $\gamma = -0.475$, $\beta = 0.475$, and $\alpha = 0.13$ with external magnetic for (a) $\delta = +0.1$, (b) $\delta = 0$ and (c) $\delta = -0.1$

The phase corresponding to the attractors in Fig.3 also is a periodic, however one can see that under external magnetic field the shape and cardinality, i.e., the number of points, of attractor consisting from 14 points for $h = 0$ a little bit changed. For $d = \pm 0.1$ the cardinality of corresponding attractor is 19.

The attractors in Fig.4 are finite set, and one can see that under external magnetic field the cardinality of corresponding attractor essentially increase.

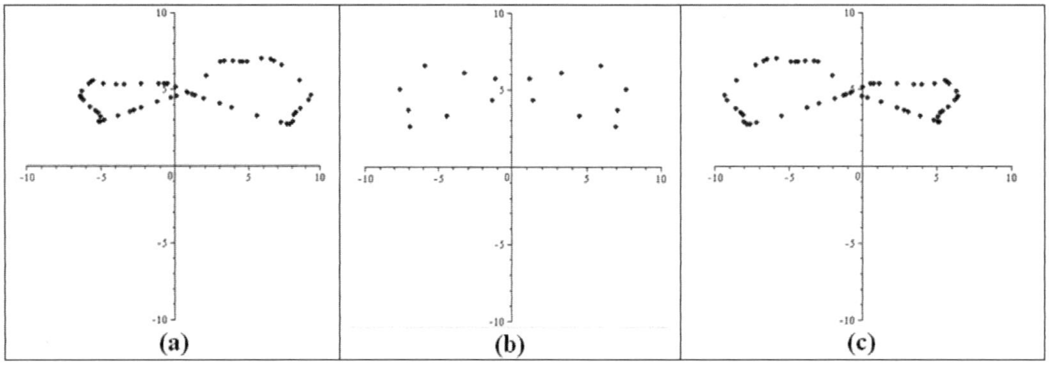

Figure 4. The attractors in $X - K_1$ plane for $k = 2$, $\gamma = -0.4$, $\beta = 0.4$, and $\alpha = 0.13$ with external magnetic for (a) $\delta = +0.1$, (b) $\delta = 0$ and (c) $\delta = -0.1$

In Fig. 5. the attractors are "strange" and one can see that the shape of attractor preserved under external magnetic field.

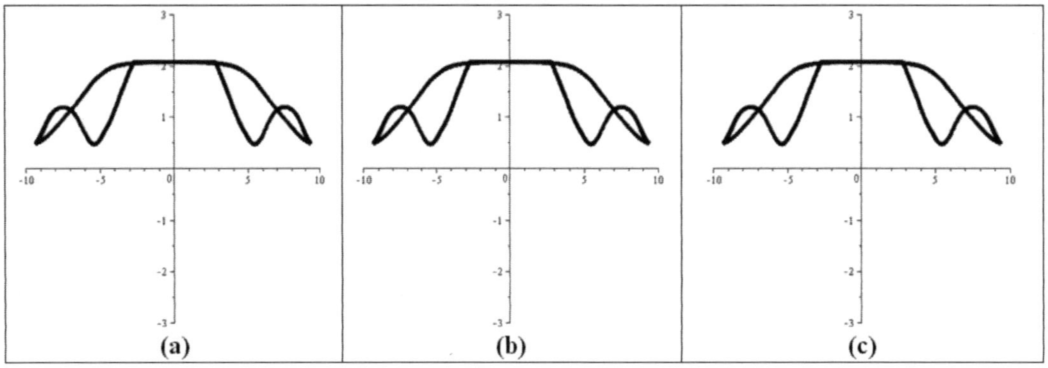

Figure 5. The attractors in $X - K_1$ plane for $k = 2$, $\gamma = -1.35$, $\beta = 0.9$, and $\alpha = 0.48$ with external magnetic for (a) $\delta = +0.1$, (b) $\delta = 0$ and (c) $\delta = -0.1$

Let us consider the attractors in Fig. 6. Here for $h = 0$ the corresponding phase is incommensurate, however in the presence of external magnetic field the corresponding phase transforms into commensurate phase.

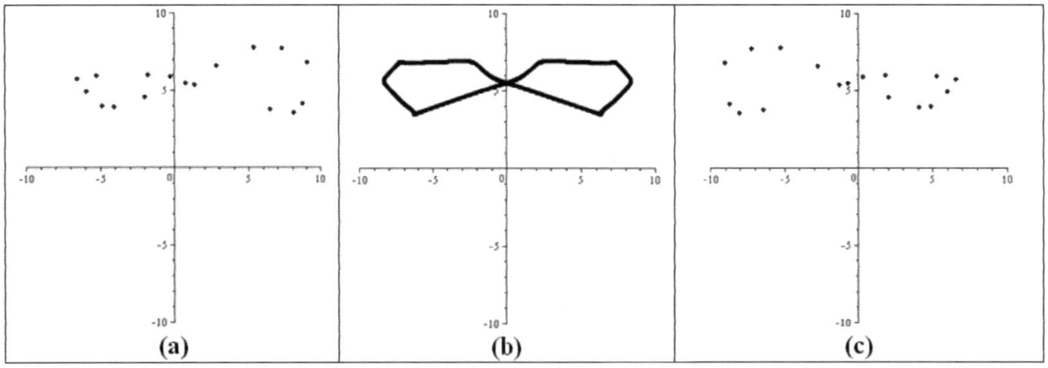

Figure 6. The attractors in $X - K_1$ plane for $k = 2$, $\gamma = -0.38$, $\beta = 0.38$, and $\alpha = 0.11$ with external magnetic for (a) $\delta = +0.1$, (b) $\delta = 0$ and (c) $\delta = -0.1$

The study of additional aspects of this model is planned to be the subject of forthcoming publications.

5. Conclusions

In this paper we have studied the Vannimenus model on an arbitrary order Cayley tree with competing nearest-neighbour and next-nearest-neighbour interactions in the presence of external magnetic field. We obtain recursion relations for effective fields X_r and nearest-neighbour interactions K_r which are rigorous, but which must in general be analyzed numerically. We clarify the role of external magnetic field.

Acknowledgments

This work is supported by the FRGS Grant Phase 2/2010, FRGS 10-022-0141 and by the Scientific and Technological Research Council of Turkey- TUBITAK, Project Number: 109T678.

References

[1] Bak P 1982 *Rep.Prog.Phys.* **45** 587
[2] Elliott R J 1961 *Phys.Rev.* **124** 346
[3] Ostlund S 1981 *Phys.Rev. B* **24** 398
[4] Huse D A 1981 *Phys.Rev. B* **24** 5180
[5] Bak P 1981 *Phys.Rev. Lett.* **46** 791
[6] Jensen M H and Bak P 1983 *Phys.Rev.B* **27** 6853
[7] Bak P and von Boehm J 1980 *Phys.rev.B* **21** 5297
[8] Fisher M E and Selke W 1980 *Phys.Rev.Lett.* **44** 1502
[9] Vannimenus J 1981 *Z.Phys. B* **43** 141
[10] Mariz A M , Tsalis C and Albuquerque A L 1985 *Jour. Stat. Phys.* **40** 577
[11] Thompson C J 1982 *J. Stat. Phys.* **27** 441
[12] Inawashiro S, Thompson C J and Honda G 1983 *Jour. Stat. Phys.* **33** 419
[13] Inawashiro S and Thompson C J 1983 *Physics Letters* **97A** 245
[14] Yokoi C S O, Oliveira M J and Salinas S R 1985 *Phys. Rev. Lett.* **54** 163
[15] Tragtenberg M H R and Yokoi C S O 1995 *Phys. Rev. E* **52** 2187
[16] da Silva C R and Coutinho S 1986 *Physical review B* **34** 7975
[17] Monroe J L 1992 *J. Stat. Phys.* **67** 1185
[18] Monroe J L 1994 *Phys.Lett. A* **188** 80
[19] Ganikhodjaev N and Uğuz S 2011 *Physica A* **390** 4160
[20] Katsura S and Takizawa M 1974 *Prog. Theor. Phys.* **51** 82
[21] Baxter R J 1982 *Exactly Solved Models in Statistical Mechanics* (Academic Press, London/New York)
[22] Kindermann R and Snell J L 1980 *Markov Random Fields and Their Applications*, (Contemporary Mathematics, Providence, R.I.)
[23] Weidlich W 1971 *Br.J.Math.Statist.Psychol.* **24** 251

Journal of Physics: Conference Series **435** (2013) 012032 doi:10.1088/1742-6596/435/1/012032

Ising Model with Competing Interactions on Cayley Tree of Order 4: An Analytic Solution.

Rukiah bin Ali, Farrukh Mukhamedov, Chin Hee Pah

Department of Computational and Theoretical Sciences, Kulliyyah of Science, International Islamic University Malaysia, 25200 Kuantan, Pahang, Malaysia.

pahchinhee@iium.edu.my

Abstract. We investigate an Ising model with two restricted competing interactions (nearest neighbors, and one-level neighbors) on the Cayley tree of order four. We derive a recurrent equation for the Cayley tree of order k. We found an analytic solution for the given interactions in the case of order 4. Our result of the critical curve shows the existence of the phase transition occurs in this model. We also give the calculation of the free energy from the description of Gibbs measure of the given Hamiltonian on Cayley tree of order four.

1. Introduction

There are many mathematical models that can describe the phenomenon of phase transition such as in the Potts model [1], Heisenberg model [2], and Ising model [3, 4]. The Cayley tree Ising model was originally works of Hans Bethe [6]. Unlike the original Ising model, the Cayley tree Ising model takes the basic graph to be a tree rather than set the graph into a lattice point. It is an approximation to Ising model and an alternative model for researchers to mathematically understand about this model especially it's qualitative properties.

The study of the phase transition of the Cayley tree Ising model with competing interactions of ternary and binary on a Cayley tree of order 2, was investigated in [3]. This study is continued by considering a zero external magnetic field and restricted interactions on nearest neighbors and second neighbors to find the exact solution which is also using a recursive method as in [3,4]. A paper on the Ising model with three competing interactions on a Cayley tree [5,10] is investigated with considerations on the external magnetic field and restricted interactions and proved in different cases.

In this paper, a general part is devoted to the Ising model on Cayley tree of order k (see also [8]) and a specific part is provided on finding phase transition curve for the model on Cayley tree of order 4. Note that analytic solution is rarely seen in the study of phase transition, there are only numerical results in [9,11] for studying the Ising model with competing interactions on Cayley tree. Some numerical results for this model are found earlier in [12] that phase transition occurs in Ising model with competing interactions which corresponds to nearest neighbors and one-level neighbors on semi-infinite Cayley tree of order 3 and 4. Our analytic result shall confirm the earlier one.

2. Gibbs Measures & Phase Transition

Let Λ be a finite subset of an infinite lattice, I. Given that $\Omega(\Lambda)$ is the set of all configurations, $\sigma(\Lambda)$ on Λ and $\bar{\sigma}(I/\Lambda)$ be a boundary condition on I/Λ. Assume a Hamiltonian, H is defined on state space, Φ. For any finite $\Lambda \subset I$, a Gibbs measure, P is given by

$$P(\sigma(\Lambda)) = \int_{\Omega} P(d\bar{\sigma}(I/\Lambda)u_{\Lambda}(\sigma(\Lambda)) \qquad (2.1.1)$$

if it satisfies a DLR equation [10]. The conditional Gibbs measure, u_{Λ} in Λ volume under the boundary condition, $\bar{\sigma}(V/\Lambda)$ is given as

$$u_{\Lambda}(\sigma(\Lambda)) = \frac{\exp(-\beta H(\sigma(\Lambda)|\bar{\sigma}(I/\Lambda))}{Z_{\Lambda}(\bar{\sigma}(V/\Lambda))}$$

and the normalizing constant, Z or partition function, is defined by

$$Z_{\Lambda}(\bar{\sigma}(V/\Lambda)) = \sum_{\sigma(\Lambda)\in\Omega(\Lambda)} e^{-\frac{1}{kT}H(\sigma(\Lambda)|\bar{\sigma}(V/\Lambda))}$$

where k is the universal constant and T is the temperature.

The non-uniqueness of Gibbs measure of a given Hamiltonian selected from an equilibrium state will produce a phase transition in physical system. We can say that a phase transition occurs with the existence of the non-uniqueness of Gibbs measure. Indeed, phase transition usually occurs at low temperature. In other words, if it is possible to find the exact value of temperature, T^* such that phase transition occurs for all $T < T^*$, then T^* is called a critical value of temperature.

3. The Recurrent Equations for Partition Functions in the Ising Model with Competing Interactions

A semi-infinite Cayley tree of order k (figure 3.1) is a graph with no cycles, each vertex emanates $k+1$ edges. The root, x^0 in the graph is called the 0th level. From the root, x^0 each vertex emanates exactly k edges. From the vertex at these edges, we called the vertices as the vertex of the first level. The vertices from the edges at first level called as the vertex of the second level and so on. A configuration on a tree is assigned as "+" or "−" to each point.

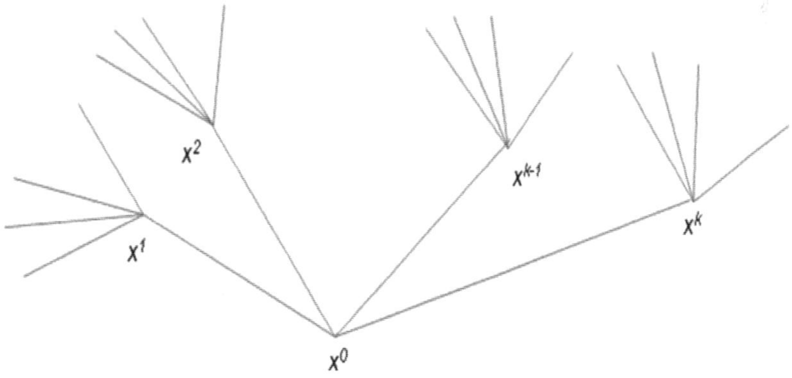

Figure 3.1: Semi-infinite Cayley tree of order k

Let Cayley tree, $\Gamma^k = (V, L), k \geq 1$ where V is given by the set of vertices of the graph, Γ^k and L is the set of edges of Γ^k. Let $x, y \in V$, then the vertices x and y are called *nearest neighbors* ($<x, y>$) if there exists an edge between them. Note that a collection of pairs $<x, x_1>, <x_1, x_2>, \ldots, <x_{d-1}, y>$ are neighboring vertices which called a *path* from x to y. Thus, the distance $d(x, y), x, y \in V$ is the shortest path on the Cayley tree.

For the fixed $x^0 \in V$, we set

$$W_n = \{x \in V | d(x, x^0) = n\}, \; V_n = \bigcup_{k=1}^{n} W_k, \quad L_n = \{l = <x, y> \in L | x, y \in V_n\}.$$

Note that y is the direct successor of x, and x, y are nearest neighboring vertices. In other words, any vertex $x \neq x^0$ has k direct successor and from vertex x^0 has $k+1$ direct successor. Thus, we define the *set of direct successor* of x, $S(x)$ as follows

$S(x) = \{y \in W_{n+1}: d(x, y) = 1\}, x \in W_n$.

The vertices $x, y \in V$ are called second neighbor $(> x, y <)$ when there exist vertex $z \in V$ such that $x, y \in S(z)$.

In this model, we consider the spins values in the set $\Phi = \{-1, +1\}$. A configuration, σ on V is defined as a function $x \in V \to \sigma(x) \in \Phi$ where the set of all configurations coincides with $\Omega = \Phi^V$.

Definition 3.1. *Ising model with competing interactions on Cayley tree is defined by the following Hamiltonian*

$$H(\sigma) = -J \sum_{<x,y>} \sigma(x)\sigma(y) - J_1 \sum_{>x,y<} \sigma(x)\sigma(y) - h \sum_{x \in V} \sigma(x). \tag{3.1}$$

The first term is the total sum for all nearest neighbors, and the second term is the total sum for all second nearest neighbors. The third term is zero since external magnetic field, h in this case is considered zero. The values of its spin variables, $\sigma(x)$ is considered as ± 1. Both signs refer to the state of the spin "up" or "down". Second nearest neighbor $> x, y <$ is called one-level neighbors $(> \overline{x, y} <)$ if x and y are on the same level, n.

Definition 3.2. *One-level Ising model with competing interactions on Cayley tree is defined by the following Hamiltonian*

$$H(\sigma) = -J \sum_{<x,y>} \sigma(x)\sigma(y) - J_1 \sum_{>\overline{x,y}<} \sigma(x)\sigma(y) - h \sum_{x \in V} \sigma(x) \tag{3.2}$$

where the sum in the second term is only taken over one-level second neighbors and assume $h = 0$.

In this study, we consider $J > 0$, and $J_1 \in \mathbb{R}$. Let us define the set of all configurations on $V_1 = \{x^0, x^1, \ldots, x^k\}$ (see figure 3.1). We consider $\bar{\sigma}(V \backslash V_n)$ as $\bar{\sigma}_n$. The conditional Gibbs measure, $\mu_n(\sigma^n | \bar{\sigma}_n)$ is defined on volume V_n as

$$\mu_n(\sigma^n | \bar{\sigma}_n) = \frac{\exp\{-\beta H(\sigma^n | \bar{\sigma}_n)\}}{Z^n(\bar{\sigma}_n)} \tag{3.3}$$

where

$$H(\sigma^n | \bar{\sigma}_n) = -J \sum_{<x,y>: x,y \in V_n} \sigma^n(x)\sigma^n(y) - J_1 \sum_{>x,y<: x,y \in V_n} \sigma^n(x)\sigma^n(y)$$
$$-J \sum_{<x,y>: x \in V_n, y \bar{\in} V_n} \sigma^n(x)\bar{\sigma}^n(y) - J \sum_{>x,y<: y \in V_n, x \bar{\in} V_n} \bar{\sigma}^n(x)\sigma^n(y) \tag{3.4}$$

and partition function is defined by

$$Z^n(\bar{\sigma}_n) = \sum_{\sigma^n \in \Omega_n} \exp\{-\beta H(\sigma^n | \bar{\sigma}_n)\} \tag{3.5}$$

where $\beta = \frac{1}{kT}$, and T is the inverse temperature.

For Cayley tree, Γ^k of order k, the partition function is divided into two terms

$$Z^n(\bar{\sigma}_n) = Z^n{}_+(\bar{\sigma}_n) + Z^n{}_-(\bar{\sigma}_n) \tag{3.6}$$

where

$Z^n{}_+(\bar{\sigma}_n) = \sum_{\sigma^0 = +1} \exp\{-\beta H(\sigma^n | \bar{\sigma}_n)\}$ and $Z^n{}_-(\bar{\sigma}_n) = \sum_{\sigma^0 = -1} \exp\{-\beta H(\sigma^n | \bar{\sigma}_n)\}$.

To have general equations relating the partition function of kth order on Cayley tree, all possible configurations of the spins is considered. Let r be the number of spin down $(-)$, $\sigma(y) = -1$ and $k - r$ be the number of spins up (+), $(y) = +1$. The following diagram illustrates spins up and spins down on V_1.

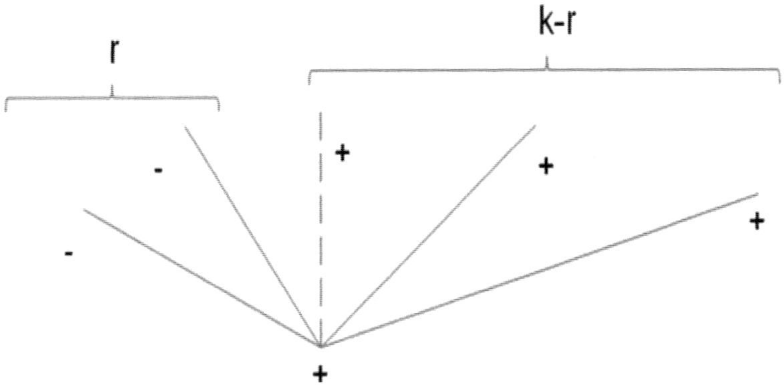

Figure 3.2: Number of spin up (+) and spin down (-) on V_1 with k vertices.

Since we are considering ferromagnetic cases where $J, J_1 > 0$, we let $\theta = \exp(\beta J)$ which corresponds to nearest neighbours, and $\theta_1 = \exp(\beta J_1)$ which corresponds to one-level neighbors.

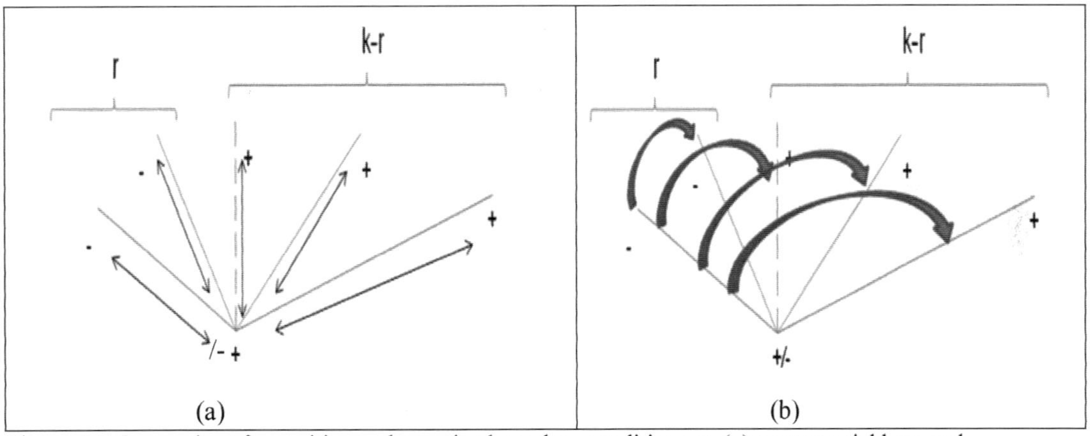

Figure 3.3: Interactions for positive and negative boundary conditions on (a) nearest neighbors and (b)one-level neighbors

Lemma 3.1. *For a given configuration with r number of spin "$-$" and $(k-r)$ of spin "+" on W_1, one-level interactions give an interaction factor as*

$$\theta_1^v = \theta_1^{\frac{1}{2}k^2 - 2kr - \frac{1}{2}k + 2r^2} . \tag{3.7}$$

Proof. Let $p = C_2^r$ which corresponds to a combination of choosing any two sites with spin down (-), $q = C_2^{k-r}$ corresponds to combination of choosing any two sites with spin up (+), and $s = C_2^k$ is the total combination of spins at one-level neighbors. The total number of interactions, denoted by v is equal to

$$v = p + q - (s - (p + q))v = 2(C_2^r + C_2^{k-r}) - C_2^k = \frac{1}{2}k^2 - 2kr - \frac{1}{2}k + 2r^2.$$

Theorem 3.1. *For any positive integer n and a fixed k^{th} order of Cayley tree,Γ^k the recurrent equation* (3.6) *can be rewritten as the following*

$$Z^n = \sum_{r=0}^{k} C_r^k (Z_+^{n-1})^{k-r} (Z_-^{n-1})^r \, \theta^{k-2r} \theta_1^{\frac{1}{2}k^2-2kr-\frac{1}{2}k+2r^2} + \sum_{r=0}^{k} C_r^k (Z_+^{n-1})^{k-r} (Z_-^{n-1})^r \, \theta^{2r-k} \theta_1^{\frac{1}{2}k^2-2kr-\frac{1}{2}k+2r^2}. \quad (3.8)$$

Proof. Interactions with nearest neighbors and one-level neighbors are illustrated by the figure 3.3.

From Hamiltonian (3.1), exponent of total interactions of nearest neighbors on V_1 can be straightaway calculated as:
$\theta^{-r}\theta^{k-r} = \theta^{-r+k-r} = \theta^{k-2r}$ (with +ve boundary) and
$\theta^{r}\theta^{-(k-r)} = \theta^{-r-k+r} = \theta^{2r-k}$ (with −ve boundary).

Meanwhile, exponent of total interactions of one-level neighbors on V_1 are taking over two competing interactions on the same level (figure 3.2 b). Therefore, from lemma above, we have

$$Z_+^n = \sum_{r=0}^{k} C_r^k (Z_+^{n-1})^{k-r} (Z_-^{n-1})^r \, \theta^{k-2r} \theta_1^{\frac{1}{2}k^2-2kr-\frac{1}{2}k+2r^2}$$

$$Z_-^n = \sum_{r=0}^{k} C_r^k (Z_+^{n-1})^{k-r} (Z_-^{n-1})^r \, \theta^{2r-k} \theta_1^{\frac{1}{2}k^2-2kr-\frac{1}{2}k+2r^2}$$

Substituting into equation (3.4), immediately we have the equation (3.5).

Now, let the ratio be

$$u_n(x^0) = \frac{Z_+^n}{Z_-^n}. \quad (3.9)$$

Theorem 3.2. *For any positive integer n, the ratio* (3.9) *will be as follows*:

$$u_n(x^0) = \frac{\sum_{r=0}^{k} C_r^k u_{n-1}^{k-r} \theta^{k-2r} \theta_1^{\frac{1}{2}k^2-2kr-\frac{1}{2}k+2r^2}}{\sum_{r=0}^{k} C_r^k u_{n-1}^{k-r} \theta^{2r-k} \theta_1^{\frac{1}{2}k^2-2kr-\frac{1}{2}k+2r^2}}. \quad (3.10)$$

Proof. Straightforward from Theorem 3.1 and definition of (3.9).

4. Gibbs Measure and Analysis of Recurrent Equation of Order 4

4.1. Gibbs Measure for Ising model on Cayley Tree of order 4

Let $t: x \to \mathbb{R}$ be a real valued function of $x \in V$. The probability measure, μ_n is defined as follows:

$$\mu_n(\sigma^n) = \frac{\exp\{-\beta H(\sigma^n) + \sum_{x\in W_n} t_x \sigma(x)\}}{Z^n} \qquad (4.1.1)$$

where

$$H(\sigma^n) = -J \sum_{<x,y>:x,y\in V_n} \sigma^n(x)\sigma^n(y) - J_1 \sum_{>x,y<:x,y\in V_n} \sigma^n(x)\sigma^n(y) \qquad (4.1.2)$$

for any $n = 1,2,3,\ldots$ and partition function for V_n is defined as

$$Z^n = Z^n(\beta, f) = \sum_{\tilde{\sigma}^n \in \Box V_n} \exp\{-\beta H(\tilde{\sigma}^n) + \sum_{x\in W_n} t_x \tilde{\sigma}(x)\}. \qquad (4.1.3)$$

where $\beta = \frac{1}{kT}$, and T is inverse temperature.

Let consider consistency condition be as follows: for $\mu_n(\sigma^n), n \geq 1$, the consistency condition is given by

$$\sum_{\sigma_n} \mu_n(\sigma^{n-1}, \sigma_n) = \mu_{n-1}(\sigma^{n-1}), \qquad (4.1.4)$$

where $\sigma_n = \{\sigma(x), x \in W_n\}$.

Let $V_1 \subset V_2 \subset \cdots \bigcup_{n-1}^{\infty} W_n$ and $\mu_1, \mu_2, \ldots$ be a sequence of the probability measures on $\Phi^{V_1}, \Phi^{V_2}, \ldots$ where $\Phi = \{-1, +1\}$ for which it satisfies (4.1.4). Based on Kolmogorov theorem (Kolmogorov, 1956), a unique limiting Gibbs measure, μ_n on Ω existed for positive integer n and $\sigma^n \in \Phi^{V_n}$, the equality is given as

$\mu(\{\sigma | V_n = \sigma^n\}) = \mu_n(\sigma^n)$.

We describe the statement below for the conditions on t_x guaranteeing consistency condition of measures, $\mu_n(\sigma^n)$. In this case we will consider for $k = 4$.

Theorem 4.1. *Probability measure, $\mu_n(\sigma^n)$, for $n = 1,2,\ldots$ satisfying the consistency condition (4.1.4) if and only if for any $x \in V$, it holds the following equation:*

$$t_x = \frac{1}{2}\log \frac{\begin{array}{c}\theta^8\theta_1^8 e^{2(t_p+t_q+t_r+t_s)} + \theta^6\theta_1^2\left(e^{2(t_p+t_q+t_r)} + e^{2(t_p+t_q+t_s)} + e^{2(t_p+t_r+t_s)}\right.\\ \left.+e^{2(t_q+t_r+t_s)}\right) + \theta^2\theta_1^2\left(e^{2t_p} + e^{2t_q} + e^{2t_r} + e^{2t_s}\right) + \theta^4\left(e^{2(t_p+t_q)}\right.\\ \left.+e^{2(t_p+t_r)} + e^{2(t_q+t_r)} + e^{2(t_p+t_s)} + e^{2(t_q+t_s)} + e^{2(t_r+t_s)}\right) + \theta_1^8\end{array}}{\begin{array}{c}\theta_1^8 e^{2(t_p+t_q+t_r+t_s)} + \theta^2\theta_1^2\left(e^{2(t_p+t_q+t_r)} + e^{2(t_p+t_q+t_s)} + e^{2(t_p+t_r+t_s)}\right.\\ \left.+e^{2(t_q+t_r+t_s)}\right) + \theta^6\theta_1^2\left(e^{2t_p} + e^{2t_q} + e^{2t_r} + e^{2t_s}\right) + \theta^4\left(e^{2(t_p+t_q)}\right.\\ \left.+e^{2(t_p+t_r)} + e^{2(t_q+t_r)} + e^{2(t_p+t_s)} + e^{2(t_q+t_s)} + e^{2(t_r+t_s)}\right) + \theta^8\theta_1^8\end{array}}. \qquad (4.1.5)$$

Proof. Necessity. Based on the consistency condition (4.1.4), it gives us the following equation:

$$(Z^n)^{-1}\sum_{\sigma_n} \exp\{-\beta H(\sigma^{n-1}) + \beta J \sum_{x\in W_{n-1}} \sigma(x)\left(\sigma(p) + \sigma(q) + \sigma(r) + \sigma(s)\right) +$$

$$\beta J_1 \sum_{x\in W_{n-1}} \{\sigma(p)\left(\sigma(q) + \sigma(r) + \sigma(s)\right) + \sigma(q)\left(\sigma(r) + \sigma(s)\right) + \sigma(r)\sigma(s)\} +$$

$$\sum_{x\in W_{n-1}}\sum_{p\in S(x)} t_p\sigma(p) + \sum_{x\in W_{n-1}}\sum_{q\in S(x)} t_q\sigma(q) + \sum_{x\in W_{n-1}}\sum_{r\in S(x)} t_r\sigma(r) + \sum_{x\in W_{n-1}}\sum_{s\in S(x)} t_s\sigma(s)\} = (Z^{n-1})^{-1}\exp\{-\beta H(\sigma^{n-1}) + \sum_{x\in W_{n-1}} \sigma(x)t_x\}$$

$$\frac{Z^{n-1}}{Z^n}\exp\{-\beta H(\sigma^{n-1})\}\sum_{\sigma_n}\exp\{\beta J\sum_{x\in W_{n-1}}\sigma(x)\big(\sigma(p)+\sigma(q)+\sigma(r)+\sigma(s)\big) + \beta J_1\sum_{x\in W_{n-1}}\{\sigma(p)\big(\sigma(q)+\sigma(r)+\sigma(s)\big)+\sigma(q)\big(\sigma(r)+\sigma(s)\big)+\sigma(r)\sigma(s)\} + \sum_{x\in W_{n-1}}\sum_{p\in S(x)} t_p\sigma(p) + \sum_{x\in W_{n-1}}\sum_{q\in S(x)} t_q\sigma(q) + \sum_{x\in W_{n-1}}\sum_{r\in S(x)} t_r\sigma(r) + \sum_{x\in W_{n-1}}\sum_{s\in S(x)} t_s\sigma(s)\} = (Z^{n-1})^{-1}\exp\{-\beta H(\sigma^{n-1})\}\exp\sum_{x\in W_{n-1}}\sigma(x)t_x\}. \quad (4.1.6)$$

From (4.1.6), we have

$$\frac{Z^{n-1}}{Z^n}\sum_{\sigma_n}\prod_{x\in W_{n-1}}\exp\{\beta J\ \sigma(x)\big(\sigma(p)+\sigma(q)+\sigma(r)+\sigma(s)\big)+\beta J_1\{\sigma(p)\big(\sigma(q)+\sigma(r)\ +\sigma(s)\big) + \sigma(q)\big(\sigma(r)+\sigma(s)\big)+\sigma(r)\sigma(s)\}+t_p\sigma(p)+t_q\sigma(q)+t_r\sigma(r)+t_s\sigma(s)\} = \prod_{x\in W_{n-1}}\exp\{\sigma(x)t_x\}. \quad (4.1.7)$$

When $\sigma_n = \bigcup_{x\in W_{n-1}}\sigma_n^{\,x}$, equation (4.1.7) could be derived as follow. Let $x \in W_{n-1}$ and $S(x) = \{\sigma(p), \sigma(q), \sigma(r), \sigma(s)\}$, thus

$$\frac{Z^{n-1}}{Z^n}\prod_{x\in W_{n-1}}\sum_{\sigma_n^x}\exp\{\beta J\,\sigma(x)\big(\sigma(p)+\sigma(q)+\sigma(r)+\sigma(s)\big)+\beta J_1\{\sigma(p)\big(\sigma(q)+\sigma(r)\ +\sigma(s)\big) + \sigma(q)\big(\sigma(r)+\sigma(s)\big)+\sigma(r)\sigma(s)\}+t_p\sigma(p)+t_q\sigma(q)+t_r\sigma(r)+t_s\sigma(s)\} = \prod_{x\in W_{n-1}}\exp\{\sigma(x)t_x\}. \quad (4.1.8)$$

For $\sigma(x) = 1$ and $\sigma(x) = -1$, equation (4.1.8) can be rewritten as

$$\frac{\sum_{\sigma_n^x=\{\sigma(p),\sigma(q),\sigma(r),\sigma(s)\}}\exp\{\beta J\,\sigma(x)\big(\sigma(p)+\sigma(q)+\sigma(r)+\sigma(s)\big)+\beta J_1\{\sigma(p)\big(\sigma(q)+\sigma(r)\ +\sigma(s)\big)+\sigma(q)\big(\sigma(r)+\sigma(s)\big)+\sigma(r)\sigma(s)\} + t_p\sigma(p)+t_q\sigma(q)+t_r\sigma(r)+t_s\sigma(s)\}}{\sum_{\sigma_n^x=\{\sigma(p),\sigma(q),\sigma(r),\sigma(s)\}}\exp\{\beta J\big(\sigma(p)+\sigma(q)+\sigma(r)+\sigma(s)\big)+\beta J_1\{\sigma(p)\big(\sigma(q)+\sigma(r)\ +\sigma(s)\big)+\sigma(q)\big(\sigma(r)+\sigma(s)\big)+\sigma(r)\sigma(s)\} + t_p\sigma(p)+t_q\sigma(q)+t_r\sigma(r)+t_s\sigma(s)\}}$$
$$= \exp\{2t_x\} \quad (4.1.9)$$

where $x \in W_{n-1}$ is fixed.

Below is the basic diagram for V_1 of Cayley tree of order four

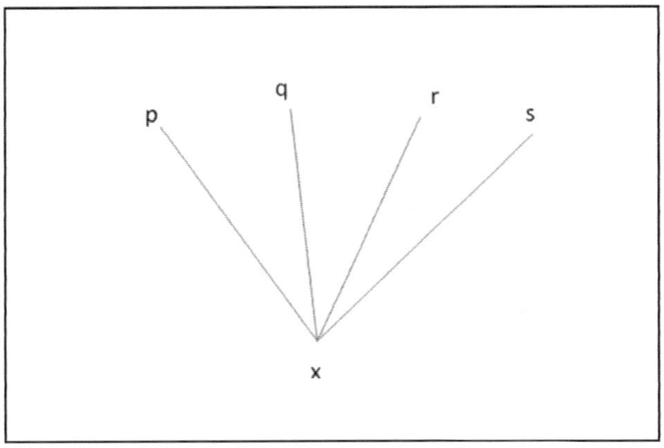

Figure 4.1.1: Basic diagram for Cayley tree of order 4

Based on this diagram, we denote both negative and positive condition as follows:

$$\begin{aligned}
Y_+ = {} & \exp(4\beta J + 6\beta J_1 + t_p + t_q + t_r + t_s) + \exp(2\beta J + t_p + t_q + t_r - t_s) + \\
& \exp(2\beta J + t_p + t_q - t_r + t_s) + \exp(2\beta J + t_p - t_q + t_r + t_s) + \\
& \exp(2\beta J - t_p + t_q + t_r + t_s) + \exp(-2\beta J + t_p - t_q - t_r - t_s) + \\
& \exp(-2\beta J - t_p + t_q - t_r - t_s) + \exp(-2\beta J - t_p - t_q + t_r - t_s) + \\
& \exp(-2\beta J - t_p - t_q - t_r + t_s) + \exp(-2\beta J + t_p + t_q - t_r - t_s) + \\
& \exp(-2\beta J + t_p - t_q + t_r - t_s) + \exp(-2\beta J - t_p + t_q + t_r - t_s) + \\
& \exp(-2\beta J + t_p - t_q - t_r + t_s) + \exp(-2\beta J - t_p + t_q - t_r + t_s) + \\
& \exp(-2\beta J - t_p - t_q + t_r + t_s) + \exp(-4\beta J + 6\beta J_1 - t_p - t_q - t_r - t_s) \\
Y_- = {} & \exp(-4\beta J + 6\beta J_1 + t_p + t_q + t_r + t_s) + \exp(-2\beta J + t_p + t_q + t_r - t_s) + \\
& \exp(-2\beta J + t_p + t_q - t_r + t_s) + \exp(-2\beta J + t_p - t_q + t_r + t_s) + \\
& \exp(-2\beta J - t_p + t_q + t_r + t_s) + \exp(2\beta J + t_p - t_q - t_r - t_s) + \\
& \exp(2\beta J - t_p + t_q - t_r - t_s) + \exp(2\beta J - t_p - t_q + t_r - t_s) + \\
& \exp(2\beta J - t_p - t_q - t_r + t_s) + \exp(-2\beta J + t_p + t_q - t_r - t_s) + \\
& \exp(-2\beta J + t_p - t_q + t_r - t_s) + \exp(-2\beta J - t_p + t_q + t_r - t_s) + \\
& \exp(-2\beta J + t_p - t_q - t_r + t_s) + \exp(-2\beta J - t_p + t_q - t_r + t_s) + \\
& \exp(-2\beta J - t_p - t_q + t_r + t_s) + \exp(4\beta J + 6\beta J_1 - t_p - t_q - t_r - t_s).
\end{aligned}$$

From equation (4.1.9), the following equation could be obtained

$$\exp(2t_x) = \frac{Y_+}{Y_-}$$

$$\Rightarrow \qquad t_x = \frac{1}{2}\log\frac{Y_+}{Y_-}. \tag{4.1.10}$$

Let $\theta = \exp(\beta J)$ and $\theta_1 = \exp(\beta J_1)$, thus

$$\begin{aligned}
Y_+ = {} & \theta^4\theta_1{}^6 e^{t_p+t_q+t_r+t_s} + \theta^2 e^{t_p+t_q+t_r-t_s} + \theta^2 e^{t_p+t_q-t_r+t_s} + \theta^2 e^{t_p-t_q+t_r+t_s} \\
& \theta^2 e^{-t_p+t_q+t_r+t_s} + \theta^{-2} e^{t_p-t_q-t_r-t_s} + \theta^{-2} e^{-t_p+t_q-t_r-t_s} + \theta^{-2} e^{-t_p-t_q+t_r-t_s} \\
& \theta^{-2} e^{-t_p-t_q-t_r+t_s} + \theta^{-2} e^{t_p+t_q-t_r-t_s} + \theta^{-2} e^{t_p-t_q+t_r-t_s} + \theta^{-2} e^{-t_p+t_q+t_r-t_s} \\
& \theta^{-2} e^{t_p-t_q-t_r+t_s} + \theta^{-2} e^{-t_p+t_q-t_r+t_s} + \theta^{-2} e^{-t_p-t_q+t_r+t_s} + \theta^{-4}\theta_1{}^6 e^{-t_p-t_q-t_r-t_s} \\
Y_- = {} & \theta^{-4}\theta_1{}^6 e^{t_p+t_q+t_r+t_s} + \theta^{-2} e^{t_p+t_q+t_r-t_s} + \theta^{-2} e^{t_p+t_q-t_r+t_s} + \theta^{-2} e^{t_p-t_q+t_r+t_s} \\
& \theta^{-2} e^{-t_p+t_q+t_r+t_s} + \theta^{2} e^{t_p-t_q-t_r-t_s} + \theta^{2} e^{-t_p+t_q-t_r-t_s} + \theta^{2} e^{-t_p-t_q+t_r-t_s} \\
& \theta^{2} e^{-t_p-t_q-t_r+t_s} + \theta^{-2} e^{t_p+t_q-t_r-t_s} + \theta^{-2} e^{t_p-t_q+t_r-t_s} + \theta^{-2} e^{-t_p+t_q+t_r-t_s} \\
& \theta^{-2} e^{t_p-t_q-t_r+t_s} + \theta^{-2} e^{-t_p+t_q-t_r+t_s} + \theta^{-2} e^{-t_p-t_q+t_r+t_s} + \theta^{4}\theta_1{}^6 e^{-t_p-t_q-t_r-t_s}.
\end{aligned}$$

$t_x = \frac{1}{2}\log\frac{Y_+}{Y_-}$ multiply with $e^{t_p+t_q+t_r+t_s}$ for both numerator and denominator. Hence, it implies (4.1.5).

Sufficiency. To show that the consistency condition is sufficient in this case, let assume equation (4.1.5) is accepted, then it gives us as in (4.1.10) and hence (4.1.9). From (4.1.9), equation (4.1.11) could be obtained as follows:

$$\begin{aligned}
& \sum_{\sigma_n^x=\{\sigma(p),\sigma(q),\sigma(r),\sigma(s)\}} \exp\{\beta J\,\sigma(x)\big(\sigma(p)+\sigma(q)+\sigma(r)+\sigma(s)\big) + \beta J_1\{\sigma(p) \\
& \qquad \big(\sigma(q)+\sigma(r)+\sigma(s)\big) + \sigma(q)\big(\sigma(r)+\sigma(s)\big) + \sigma(r)\sigma(s)\} \\
& \qquad + t_p\sigma(p) + t_q\sigma(q) + t_r\sigma(r) + t_s\sigma(s)\} \\
& = b(x)\exp\{\sigma(x)t_x\}
\end{aligned} \tag{4.1.11}$$

where $\sigma(x) = \pm 1$. Equation (4.1.11) implies the following equation

$$\begin{aligned}
& \prod_{x\in W_{n-1}} \sum_{\sigma_n^x=\{\sigma(p),\sigma(q),\sigma(r),\sigma(s)\}} \exp\{\beta J\,\sigma(x)\big(\sigma(p)+\sigma(q)+\sigma(r)+\sigma(s)\big) + \beta J_1\{\sigma(p) \\
& \qquad \big(\sigma(q)+\sigma(r)+\sigma(s)\big) + \sigma(q)\big(\sigma(r)+\sigma(s)\big) + \sigma(r)\sigma(s)\} \\
& \qquad + t_p\sigma(p) + t_q\sigma(q) + t_r\sigma(r) + t_s\sigma(s)\} \\
& = \prod_{x\in W_{n-1}} b(x)\exp\{\sigma(x)t_x\}.
\end{aligned} \tag{4.1.12}$$

Let $B^n = \sum_{x\in W_{n-1}} b(x)$. From equation (4.1.12) we have

$$Z^{n-1}B^{n-1}\mu_{n-1}(\sigma^{n-1}) = Z^n \sum_{\sigma^n} \mu^n(\sigma^{n-1},\sigma_n).$$

It is known that probability measure occurred when each $\mu_n, n \geq 1$. Thus, we have

$$\sum_{\sigma^{n-1}}\sum_{\sigma_n} \mu_n(\sigma^{n-1},\sigma_n) = 1, \quad \sum_{\sigma^{n-1}} \mu_{n-1}(\sigma^{n-1}) = 1.$$

Finally, from these equalities, it is obtained that

$$Z^{n-1}B^{n-1} = Z^n \tag{4.1.13}$$

where it can be concluded that consistency condition (4.1.4) holds.

Based on Theorem 3.1, problem on Gibbs measure is done by describing the solutions of functional equation (4.1.5). According to proposition 3.1, we note that any transformation $\tilde{S}$ of the group G_k induces a shift transformation $\tilde{S}: \Omega \to \Omega$ by

$(\tilde{S}\sigma)(t) = \sigma(St), t \in G_k, \sigma \in \Omega$

By G_k, we denote the set of all shifts on Ω. A Gibbs measure,μ on Ω is said to be translation-invariant if for any $T \in G_k$ equality $\mu(T(D)) = \mu(D)$ is valid for all $D \in F$.

The solution of the functional equation (4.1.5) is tricky to be analyzed. Hence, it is natural to begin with $t_x = t$ is constant for all $x \in V$. In this case, Gibbs measure corresponding to this solution is clearly to be translation-invariant one [8]. Thus, from (4.1.5), we have

$$u = \frac{\theta^8\theta_1^8u^4 + 4\theta^6\theta_1^2u^3 + 6\theta^4u^2 + 4\theta^2\theta_1^2u + \theta_1^8}{\theta_1^8u^4 + 4\theta^2\theta_1^2u^3 + 6\theta^4u^2 + 4\theta^6\theta_1^2u + \theta^8\theta_1^8} \tag{4.1.14}$$

where $u = e^{2t}$.

Corollary 4.1. *Equation 4.1.14 is coincide with the classical result for Ising model when* $\theta_1 = 1$, *and* $J_1 = 0$, *i.e.,*

$$u_n(x^0) = \left(\frac{\theta^2u_{n-1}(x^0) + 1}{u_{n-1}(x^0) + \theta^2}\right)^4. \tag{4.1.15}$$

4.2. Analysis of the Recurrent Equation

The general recurrent equation (3.5) and its ratio (3.8) are identified in previous section. The recurrent equation obtained for Cayley tree of order 4 is given as below:

$$u_n(x^0) = \frac{u_{n-1}{}^4\theta^4\theta_1^6 + 4u_{n-1}{}^3\theta^2 + 6u_{n-1}{}^2\theta_1^{-2} + 4u_{n-1}\theta^{-2} + \theta^{-4}\theta_1^6}{u_{n-1}{}^4\theta^{-4}\theta_1^6 + 4u_{n-1}{}^3\theta^{-2} + 6u_{n-1}{}^2\theta_1^{-2} + 4u_{n-1}\theta^2 + \theta^4\theta_1^6}. \tag{4.2.1}$$

Equation (4.1.14) describes the fixed points of the equation (4.1.5). If there is more than one solution in equation (4.2.1), then we can say that there is more than one translation-invariant Gibbs measure existed. Thus, based on Theorem 3.1 a phase transition occurs in our case study. Here, we give the analytic solution for the restricted interactions given in our case.

Let the functional equation, $f(u)$ be

$$f(u) = \frac{u^4\theta^4\theta_1^6 + 4u^3\theta^2 + 6u^2\theta_1^{-2} + 4u\theta^{-2} + \theta^{-4}\theta_1^6}{u^4\theta^{-4}\theta_1^6 + 4u^3\theta^{-2} + 6u^2\theta_1^{-2} + 4u\theta^2 + \theta^4\theta_1^6}. \tag{4.2.2}$$

From (4.2.2), we have

$$u = \frac{u\theta^4\theta_1^6 + 4u^3\theta^2 + 6u^2\theta_1^{-2} + 4u\theta^{-2} + \theta^{-4}\theta_1^6}{u^4\theta^{-4}\theta_1^6 + 4u^3\theta^{-2} + 6u^2\theta_1^{-2} + 4u\theta^2 + \theta^4\theta_1^6}. \tag{4.2.3}$$

We simplify equation (4.2.3), thus we have

$$u^4\theta^4\theta_1^6 + 4u^3\theta^2 + 6u^2\theta_1^{-2} + 4u\theta^{-2} + \theta^{-4}\theta_1^6 - u^5\theta^{-4}\theta_1^6 - 4u^4\theta^{-2} - 6u^3\theta_1^{-2} - 4u^2\theta^2 - u\theta^4\theta_1^6 = 0$$

$$(u-1)(u^4 + \beta u^3 + (\alpha+\beta)u^2 + \beta u + 1) = 0 \tag{4.2.4}$$

where $\alpha = \frac{6\theta^4 - 4\theta^6\theta_1^2}{\theta_1^8}$, $\beta = \frac{4\theta^2 - \theta^8\theta_1^6 + \theta_1^6}{\theta_1^6}$.

From the equation (4.2.4), we can straightaway see that $u_2^* = 1$ is a fixed point. Thus, it is worthwhile to solve the second part of the equation. If $\alpha > 0$, this implies $\gamma \geq 0$, then $u^4 + \beta u^3 + (\alpha + \beta)u^2 + \beta u + 1 = 0$ only has one solution, i.e. $u_2^* = 1$.

Let $t = u + \frac{1}{u}$, $t^2 = u^2 + \frac{1}{u^2}$. Hence, $u^4 + \beta u^3 + (\alpha + \beta)u^2 + \beta u + 1 = 0$ becomes
$t^2 + \beta t + (\alpha + \beta) - 2 = 0$

From this quadratic equation, it is found that

$$t = \frac{-\beta \pm \sqrt{\beta^2 - 4(\alpha + \beta - 2)}}{2} \geq 2$$

Given that value of $t = \gamma_1, \gamma_2$. To solve the equation, we let $t = \gamma_1$ or $t = \gamma_2$. It is showed that γ_2 is rejected since

$$\gamma_2 = \frac{-\beta - \sqrt{\beta^2 - 4(\alpha + \beta - 2)}}{2} \leq 2.$$

Now, let γ_1 equals to

$$\gamma_1 = \frac{-\beta + \sqrt{\beta^2 - 4(\alpha + \beta - 2)}}{2} \geq 2$$

If $\alpha \leq 0$, and we have

i) $\beta^2 - 4\alpha - 4\beta + 8 = (\beta - 2)^2 + 4 - 4\alpha > 0$ and

ii) $4 + \beta > -\sqrt{\beta^2 - 4\alpha - 4\beta + 8}$

Therefore,

$$\sqrt{\beta^2 - 4\alpha - 4\beta + 8} \geq 4 + \beta > -\sqrt{\beta^2 - 4\alpha - 4\beta + 8}$$
$$(4 + \beta)^2 < \beta^2 - 4\alpha - 4\beta + 8$$
$$0 \geq 8\beta + 16 + 4(\alpha + \beta - 2)$$
$$0 \geq 3\beta + \alpha + 2.$$

Substituting α, and β, we have

$$0 \geq 12\theta^2\theta_1^2 - 3\theta^8\theta_1^8 + 5\theta_1^8 + 6\theta^4 - 4\theta^6\theta_1^2. \tag{4.2.5}$$

From (4.2.5), we find another two fixed points, u_1^* and u_3^*. Hence, we can say that phase transition occurs when there exist more than one fixed point. In this case, according to equation (4.2.3), we have a non-negative values given by u_1^*, $u_2^* = 1$, and u_3^*. The critical curve we intend to find is

$$0 = 12\theta^2\theta_1^2 - 3\theta^8\theta_1^8 + 5\theta_1^8 + 6\theta^4 - 4\theta^6\theta_1^2. \tag{4.2.6}$$

Let $C(\theta, \theta_1) = 12\theta^2\theta_1^2 - 3\theta^8\theta_1^8 + 5\theta_1^8 + 6\theta^4 - 4\theta^6\theta_1^2 = 0$ be the implicit function of (θ, θ_1). It can be shown that $C(\theta, \theta_1) = 0$ is a one to one function of (θ, θ_1) for the domain of (θ, θ_1) such that,

$$\frac{\partial}{\partial \theta} C(\theta, \theta_1) = -24\theta(\theta\theta_1 - 1)(\theta\theta_1 + 1)(\theta^4\theta_1^6 + \theta^2\theta_1^4 + \theta^2 + \theta_1^2) < 0$$

where $\alpha \leq 0$, i.e., $\theta_1^2 > \frac{3}{2\theta^2}$ which implies $(\theta\theta_1 - 1)$ always positive. Since $C(\theta, \theta_1)$ is always monotone decreasing respect to θ, $C(\theta, \theta_1) = 0$ is a function of θ to θ_1. Assume there is multiple solution of θ_1 for a given θ in (4.2.6), we rewrite (4.2.6) into
$(-3\theta^8 + 5)\theta_1^8 + (12\theta^2 - 4\theta^6)\theta_1^2 + 6\theta^4 = 0$

and the necessity condition for this equation to have more than one solution for θ_1 is $-3\theta^8+5>0$ and $12\theta^2-4\theta^6<0$ is impossible.

Hence, we could summarize our result as follows:

Theorem 4.2. *For the Ising model* (3.2) *with h=0, the curve* $C(\theta,\theta_1)=0$ *in the plane* (θ,θ_1) *is a critical curve for phase transitions, namely, for an arbitrary pair of parameters* (θ,θ_1) *above the critical curve the phase transition takes place and for any pair of parameters* (θ,θ_1) *below the critical curve there occurs a single Gibbs state.*

5. The Free Energy

In this section, we derive the free energy for Ising model on Cayley tree of order 4 with restricted competing interactions. Let the partition function $Z_n(\beta,t)$ of the state μ_β^t (which corresponds to the solution of $t=\{t_x, x\epsilon V\}$) of the equation (4.1.5) be as follow:

$$Z_n(\beta,t)=\sum_{\tilde{\sigma}_n\epsilon\Omega_{V_n}}\exp\{-\beta H(\tilde{\sigma}_n)+\sum_{x\epsilon W_n}t_x\tilde{\sigma}(x)\}.$$

Then, the free energy [12, 13] is defined as

$$F(\beta,t)=-\lim_{n\to\infty}\frac{3}{\beta(4^{n+1}-1)}\ln Z_n(\beta,t). \qquad (5.2.1)$$

The aim of this section is to prove the following Theorem.

Theorem 5.2.1.(i) *The free energy exists for all t, and is given by the formula*

$$F(\beta,t)=-\lim_{\beta\to\infty}\lim_{n\to\infty}\frac{3}{(4^{n+1}-1)}\sum_{k=0}^{n}\sum_{x\epsilon W_{n-k}}A_\beta(J,J_1,t_p,t_q,t_r,t_s), \qquad (5.2.2)$$

where $p=p(x), q=q(x).r=r(x), s=s(x)$ *are direct successors of* x;

$$A_\beta(J,J_1,t_p,t_q,t_r,t_s)=a_{\beta J}(2\beta J_1+p-s)+a_{\beta J}(-2\beta J_1+p-s)+a_{\beta J}(2\beta J_1+q-s)$$
$$+a_{\beta J}(-2\beta J_1+q-s)+a_{\beta J}(2\beta J_1+r-s)+a_{\beta J}(-2\beta J_1+r-s)+\tilde{a}(\beta J,\beta J_1);$$
$$a_\beta(x)=\frac{1}{4}\ln[4\cosh(x-\beta)\cosh(x+\beta); \qquad (5.2.3)$$
$$\tilde{a}_{\beta J,\beta J_1}(D)=\frac{1}{2}\ln D.$$

(ii) *For any solution* $t=\{t_x, x\epsilon V\}$ *of (4.1.5)*

$$F(\beta,t)=F(\beta,-t),$$

where $-t=\{-t_x, x\epsilon V\}$.

6. Conclusion

We have calculated a recurrent equation of partition function for a ferromagnetic Ising model with competing interactions on Cayley tree of arbitrary order *k*. For the case of order *k*=4, we show that phase transition occurs, i.e., Theorem 4.2, in this model and the critical curve is given by equation (4.2.6). Free energy for the same model is also calculated.

References

[1] Salas J and Sokal A D (1997) *Jou. Stat. Phys.* **86** 551.
[2] Stanley H E and Kaplan T A (1966) *Phys. Rev. Lett.* **17** 913
[3] Ganikhodjaev N N (2002) *Theor. Math. Phys.* **130**(3), 419
[4] Ganikhodjaev N N, Rozikov U A (2009) *Math. Phys. Anal. Geom.* **12** **(**2) 141
[5] Ganikhodjaev N N, Pah C H and Wahiddin M R B (2003) *Jour. Phys.A: Math.Gen.*, **36**, 4283.

[6] Bethe H A (1935) *Proc. Roy. Soc. London Ser A*, **150** 552.
[7] Ganikhodjaev N N, Pah C H, Wahiddin M R B (2004) *Jour. Mah. Phys.* **45** (9), 3645.
[8] Ganikhodjaev N N, Uguz S (2011) *Physica A* **390** 4160
[9] Mariz M, Tsalis C, Albuquerque A L (1985) *Jour. Stat. Phys* **40** 577.
[10] Mukhamedov F, Rozikov U A (2004) *Jour. Stat. Phys.* **114** 825
[11] Tragtenberg M H R, Yokoi C S O (1995) Phys. Rev. E **52** 2187.
[12] Rukiah A. (2010). *Phase Transition for Ising Model with Two Competing Interactions on Cayley Tree of Order Four*. Unpublished Degree Thesis, IIUM.
[13] Sinai G Ya *Theory of Phase Transitions: Rigorous Results* (Pergamon, Oxford, 1982).
[14] Georgii H O (1988), *Gibbs Measures and Phase Transitions* New York, De Gruyter.
[15] Bleher P M, Ruiz J, Zagrebnov V A (1998) *Jour. Stat. Phys.* **93** 1.
[16] Mukhamedov F, Rozikov U (2005). arxiv: 0510022

Journal of Physics: Conference Series **435** (2013) 012033 doi:10.1088/1742-6596/435/1/012033

Two Potts Models on Cayley Tree of Arbitrary Order

Nasir Ganikhodjaev[1], **Ashraf Mohamed Nawi, and**
Mohd Hirzie Mohd Rodzhan

[1]Department of Computational and Theoretical Sciences, Faculty of Science, IIUM, 25200 Kuantan, Malaysia.

E-mail: gnasir@iium.edu.my

Abstract. On a Cayley tree of arbitrary order k we consider two different Potts models with competing nearest-neighbour interactions J_1 and next-nearest-neighbour interactions J_p and J_o, where coupling J_p corresponds to interaction of spins belonging to the same branch of the tree (prolonged) and coupling J_o corresponds to interaction of spins belonging to the same shell of the tree (one-level) and find for each model in addition to the expected paramagnetic, ferromagnetic and antiferromagnetic phases, an intermediate range of coupling values where the local magnetization has chaotic oscillatory glass-like behaviour. We also show that the ranges corresponding two models are different.

1. Introduction

Consideration of spin models with multispin interactions has proved to be fruitful in many fields of physics, ranging from the determination of phase diagrams in metallic alloys and exhibition of new types of phase transition, to site percolation. Systems exhibiting spatially modulated structures, commensurate or incommensurate with the underlying lattice, are of current interest in condensed matter physics [1]. Among the idealized systems for modulated ordering, the axial next-nearest-neighbour Ising (ANNNI) model, originally introduced by Elliot [2] to describe the sinusoidal magnetic structure of Erbium, and the chiral Potts model, introduced by Ostlund [3] and Huse [4] in connection with monolayers adsorbed on rectangular substrates, have been studied extensively by a variety of techniques. A particularly interesting and powerful method is the study of modulated phases through the measure-preserving map generated by the mean-field equations, as applied by Bak [5] and Jensen and Bak [6] to the ANNNI model. The main drawback of the method lies in the fact that thermodynamic solutions correspond to stationary but unstable orbits. However, when these models are defined on a Cayley tree (or on a the Bethe lattice; see [7] for terminology), as in the case of the Ising model with competing interactions examined by Vannimenus [8], it turns out that physically interesting solutions correspond to the attractors of the mapping. This simplifies the numerical work considerably, and detailed study of the whole phase diagram becomes feasible. Apart from the intrinsic interest attached to the study of models on trees, it is possible to argue that the results obtained on trees provide a useful guide to the more involved study of their counterparts on crystal lattices.

The ANNNI model, which consists of an Ising model with nearest-neighbour interactions augmented by competing next-nearest-neighbour couplings acting parallel to a single axis direction, is perhaps the simplest nontrivial model displaying a rich phase diagram with a Lifshitz point and many spatially modulated phases. There has been a considerable theoretical effort to

Journal of Physics: Conference Series **435** (2013) 012033 doi:10.1088/1742-6596/435/1/012033

obtain the structure of the global phase diagram of the ANNNI model in the $T-p$ space, where T is temperature and $p = -J_2/J_1$ is the ratio between the competing exchange interactions. On the basis of numerical mean-field calculations, Bak and von Boehm [9] suggested the existence of an infinite succession of commensurate phases, the so-called devil's staircase, at low temperatures. This mean-field picture has been supported by low-temperature series expansions performed by Fisher and Selke [11]. At the paramagnetic-modulated boundary analytic mean-field calculations show that the critical wave number varies continuously and vanishes at the Lifshitz point.

A phase diagram of a model describes a morphology of phases, stability of phases, transitions from one phase to another and corresponding transition lines. The Cayley tree is not a realistic lattice; however, its amazing topology makes the exact calculation of various quantities possible. For many problems the solution on a tree is much simpler than on a regular lattice and is equivalent to the standard Bethe-Peierls theory [20]. On the Cayley tree one can consider two type of next-nearest-neighbours: prolonged and one-level next-nearest-neighbours (definitions see below). In the case of the Ising model with competing nearest-neighbor interactions J and prolonged next-nearest-neighbour interactions J_p Vannimenus [8] was able to find new modulated phases, in addition to the expected paramagnetic and ferromagnetic ones. From this result follows that Ising model with competing interactions on a Cayley tree is real interest since it has many similarities with models on periodic lattices. In fact, it has many common features with them, in particular the existence of a modulated phase, and shows no sign of pathological behavior - at least no more than mean-field theories of similar systems [8]. Moreover a detailed study of its properties was carried out with essentially exact results, using rather simple numerical methods.

This suggest that more complicated models should be studied on trees, with the hope to discover new phases or unusual types of behavior. The important point is that statistical mechanics on trees involve nonlinear recursion equations and are naturally connected to the rich world of dynamical systems, a world presently under intense investigation [8].

The Potts model (with $q \geq 2$ spin values) was introduced as a generalization of the Ising model [26]. At present the Potts model encompasses a number of problems in statistical physics and lattice theory (see for example [27]). A Potts model just as an Ising model with competing interactions has recently been studied extensively because of the appearance of nontrivial magnetic orderings (see [8], [12],[13],[15],[17],[18],[19],[16],[21],[22],[23], [28],[29] and references therein). The Potts model with competing interactions J_1 and prolonged next-nearest neighbours interactions J_p on Cayley tree of second order have been studied in detail in [19], and Potts model that include also interaction of one-level next-nearest-neighbour interaction J_o have been studied in [28].

In this paper we define a single-trunk Cayley tree, produce the recursion equations for model with competing interactions on the Cayley tree and for the same model on the single-trunk Cayley tree, and show how to reduce the recursion equations on Cayley tree to the simpler recursion equations on the single-trunk Cayley tree. Note that this approach is mathematical justification only of the method developed in [8] and later generalized in [12]. We will consider also the symmetry group of the corresponding model [10].

The aim of this paper is to extend the results of [28] to the Potts model with competing interactions on a Cayley tree of arbitrary order k and to clarify the role of the order k. The paper has been organized in the following way. In Section 2, the model Hamiltonian is discussed. In Section 3, the recursion equations are defined. Section 4 is devoted to the discussion of the phase diagram features. Finally, the conclusions are given in Section 5.

2. The Model Hamiltonian

2.1. Cayley Tree

A Cayley tree Γ^k of order $k \geq 1$ (or the Bethe lattice; see [7] for terminology) is an infinite tree, i.e., a graph without cycles with exactly $k+1$ edges issuing from each vertex. Let denote the Cayley tree as $\Gamma^k = (V, \Lambda)$, where V is the set of vertices of Γ^k, Λ is the set of edges of Γ^k. Two vertices x and y, $x, y \in V$ are called *nearest-neighbors* if there exists an edge $l \in \Lambda$ connecting them, which is denoted by $l =< x, y >$. The distance $d(x,y), x, y \in V$, on the Cayley tree Γ^k, is the number of edges in the shortest path from x to y. For a fixed $x^0 \in V$ we set

$$W_n = \{x \in V | d(x, x^0) = n\}, \quad V_n = \{x \in V | d(x, x^0) \leq n\}$$

and L_n denotes the set of edges in V_n. The fixed vertex x^0 is called the 0-th level and the vertices in W_n are called the n-th level. For the sake of simplicity we put $|x| = d(x, x^0)$, $x \in V$. Two vertices $x, y \in V$ are called *the next-nearest-neighbours* if $d(x,y) = 2$. The next-nearest-neighbour vertices x and y are called *prolonged next-nearest-neighbours* if $|x| \neq |y|$ and is denoted by $> \widetilde{x,y} <$. The next-nearest-neighbour vertices $x, y \in V$ that are not prolonged are called *one-level next-nearest-neighbours* since $|x| = |y|$ and are denoted by $> \bar{x,y} <$.

We write $x \prec y$ if the path from x^0 to y goes through x. We call the vertex y a direct successor of x, if $y \succ x$ and x, y are nearest neighbours. The set of the direct successors of x is denoted by $S(x)$, i.e., if $x \in W_n$, then

$$S(x) = \{y_i \in W_{n+1} | d(x, y_i) = 1, i = 1, \cdots, k\}.$$

We observe that for any vertex $x \neq x^0, x$ has k direct successors and x^0 has $k+1$.

The collection $S(x) = \{y_1, \cdots, y_k\}$ we will call one-level k tuple of neighbours. Note that if $k = 2$ then $S(x)$ is a one-level next-nearest-neighbours.

Below we will consider a semi-infinite Cayley tree Γ^k_+ of k-th order, i.e. an infinite graph without cycles with $k+1$ edges issuing from each vertex except for x^0 which has only k edges. In this case $|S(x)| = k$ for any $x \in V$. Let $< x^0, x >= l \in L$ be an edge of semi-infinite Cayley tree Γ^k_+.

Definition 2.1 The infinite subtree $\Gamma^k_+(l) = (V^l, L^l)$ is called a *single-trunk Cayley tree*, if from vertex x^0 a single edge l emanates and from any other vertex $x \in V^l, x \neq x^0$ exactly $k+1$ edges emanate.

Let $W_1 = \{x_1, x_2, \cdots x_k\}$ and $< x^0, x^1 >= l_1, < x^0, x^2 >= l_2, \cdots < x^0, x^k >= l_k$ be k edges emanating from x^0. It is evident that semi-infinite Cayley tree Γ^k_+ splits into k components - k single-trunk Cayley trees $\Gamma^k_+(l_i), i = 1, 2 \cdots, k$. Let V^{l_i} is the set of vertices of single-trunk Cayley tree $\Gamma^k_+(l_i)$ and $V^{l_i}_n = V_n \cap V^{l_i}$ is the set of vertices $x \in V^{l_i}$ with $d(x^o, x) \leq n$.

2.2. The Models

Let Γ^k_+ be a semi-finite Cayley tree of order k. In addition to nearest-neighbours interaction J_1 and prolonged next-nearest -neighbours interaction J_p we consider also interactions of sites in the set $S(x) = \{y_1, \cdots, y_k\}$ of direct successors for any $x \in V$. Here one can introduce two type interactions:

1) jointly interaction of all sites in $S(x)$, called one-level k-tuple interaction ;

2) binary interaction of any two sites in $S(x) = \{y_1, \cdots, y_k\}$, called one-level next-nearest neighbours interaction.

For the Potts model with spin values in $\Phi = \{1, 2, 3\}$, i.e., the spin variables $\sigma(x), x \in V$ assume the values $\{1, 2, 3\}$, the relevant Hamiltonians with competing nearest-neighbor interactions J_1, prolonged next-nearest-neighbor interactions J_p and one-level k-tuple neighbours or one-level next-nearest neighbors interaction J_o have the forms:

1) Model with one-level k-tuple neighbours interactions

$$H(\sigma) = -J_o \sum_{y_i \in S(x)} \delta_{\sigma(y_1)\sigma(y_2)\cdots\sigma(y_k)} - J_p \sum_{\widetilde{>x,y<}} \delta_{\sigma(x)\sigma(y)} - J_1 \sum_{<x,y>} \delta_{\sigma(x)\sigma(y)}, \tag{1}$$

2) Model with one-level next-nearest neighbor interactions

$$H(\sigma) = -J_o \sum_{y_i, y_j \in S(x): i \neq j} \delta_{\sigma(y_i)\sigma(y_j)} - J_p \sum_{\widetilde{>x,y<}} \delta_{\sigma(x)\sigma(y)} - J_1 \sum_{<x,y>} \delta_{\sigma(x)\sigma(y)}, \tag{2}$$

where $J_0, J_p, J_1 \in R$ are coupling constant, δ in second and third sum in (1) and all sums in (2) is the usual Kronecker symbol and δ in first sum in (1) is the generalized Kronecker symbol, that is defined as follow:

$$\delta_{\sigma(y_1)\sigma(y_2)\cdots\sigma(y_q)} = \begin{cases} 1 & \text{if } \sigma(y_1) = \sigma(y_2) = \cdots = \sigma(y_k) \\ 0 & \text{otherwise.} \end{cases}$$

Note that for $k = 2$ generalized Kroneker's symbol coincides with usual one and one-level 2-tuple of neighbours coincide with one-level next-nearest neighbours. These models recover that in [19] for $k = 2$ with $J_o = 0$ and in [28] for $k = 2$ with $J_o \neq 0$ and they coincide if $J_o = 0$. The generalized Kroneker's symbol was introduced in [24]. The Ising model with one-level k-tuple interactions was considered in [23] and Ising model with one-level next-nearest-neighbours interactions was considered in [30].

2.3. Conditional Gibbs measures with symmetries

Recall that in the case of the Ising model with competing interactions examined by Vannimenus [8], it turns out that physically interesting solutions correspond to the attractors of the recurrence relations, where initial conditions chosen randomly. The main drawback of the method lies in the fact that thermodynamic solutions correspond to stationary but unstable orbits. In this paper we produce recurrence relations fixing boundary configuration $\bar{\sigma}(V \setminus \Lambda)$.
Let Λ be a finite subset of V. We will denote by $\sigma(\Lambda)$ the restriction of a configuration $\sigma : V \to \Phi = \{1, 2, 3\}$ to Λ. Let $\bar{\sigma}(V \setminus \Lambda)$ be a fixed boundary configuration. As usual, one can introduce the notions of total energy $H(\sigma(\Lambda)|\bar{\sigma}(V \setminus \Lambda))$ of configuration $\sigma(\Lambda)$ under boundary condition $\bar{\sigma}(V \setminus \Lambda)$ and partition function $Z_\Lambda(\bar{\sigma}(V \setminus \Lambda))$ in volume Λ under boundary condition $\bar{\sigma}(V \setminus \Lambda)$ that is defined as

$$Z_\Lambda(\bar{\sigma}(V \setminus \Lambda)) = \sum_{\sigma(\Lambda) \in \Omega(\Lambda)} \exp(-\frac{1}{k_B T} H(\sigma(\Lambda)| \ \bar{\sigma}(V \setminus \Lambda))), \tag{3}$$

where $\Omega(\Lambda)$ is the set of all configurations in volume Λ, k_B is Boltzmann constant and T is the absolute temperature. Then conditional Gibbs measure μ_Λ of a configuration $\sigma(\Lambda)$ is defined as

$$\mu_\Lambda(\sigma(\Lambda)| \ \bar{\sigma}(V \setminus \Lambda)) = \frac{\exp(-\beta H(\sigma(\Lambda)| \ \bar{\sigma}(V \setminus \Lambda))}{Z_\Lambda(\bar{\sigma}(V \setminus \Lambda))}$$

We consider the configuration $\sigma(V_n)$, the partition function $Z_{V_n}(\bar{\sigma}(V \setminus V_n))$ and conditional Gibbs measure $\mu_{V_n}(\sigma(V_n)|\bar{\sigma}(V \setminus V_n))$ in volume V_n and for brevity denote them as σ_n, $Z^{(n)}$ and μ_n respectively. Let S_3 be a group of all permutations of a set $\Phi = \{1, 2, 3\}$. **Definition 2.2** For any permutation $\pi \in S_3$ let us define transformation $T_\pi : \Omega \to \Omega$ by the following way: for any $\sigma \in \Omega$ assume $T_\pi \sigma(x) = T_\pi(\sigma(x))$ for any $x \in V$.
It is evident the following Proposition.

Proposition 2.1 The Hamiltonians (1) and (2) are S_3 invariant, i.e., for any $\pi \in S_3$ we have $H(T_\pi\sigma) = H(\sigma)$ for any $\sigma \in \Omega$.
In order to produce the recurrence equations, we consider the relation of the partition function on V_{n+1} to the partition function on subsets of V_n. Given the initial condition $\sigma(x^0) = i_0$ on $V_0 = \{x^0\}$, where x^0 is the root of the Cayley tree, the recurrence equations indicate how their influence propagates down the tree. Below we consider following partition functions.

Let $Z_i^{(n)}$ be a partition function on V_n with the spin i in the root x^0, $i = 1, 2, 3$, i.e.,

$$Z_i^{(n)} = \sum_{\sigma_n\in\Omega_n:\ \sigma_n(x^0)=i} \exp(-\beta H(\sigma_n|\ \bar{\sigma}(V \setminus V_n))), \tag{4}$$

with

$$Z^{(n)} = Z_1^{(n)} + Z_2^{(n)} + Z_3^{(n)},$$

where $n = 0, 1, 2, \cdots$.

Let

$$\sigma_S = \begin{pmatrix} \sigma(y_1), \sigma(y_2), \cdots, \sigma(y_k) \\ \sigma(x^0) \end{pmatrix}$$

be a configuration on the set $S = x^0 \cup S(x^0)$ and $\Omega(S)$ be the set of all such configurations. Assume

$$Z^{(n)}(\sigma_S) = Z^{(n)} \begin{pmatrix} \sigma(y_1), \sigma(y_2), \cdots, \sigma(y_k) \\ \sigma(x^0) \end{pmatrix}$$

be the partition function on V_n with fixed configuration σ_S on V_1. There are a priori 3^{k+1} different partition functions $Z^{(n)}(\sigma_S)$ and the partition function $Z^{(n)}$ in volume V_n can be written as follows:

$$Z^{(n)} = \sum_{\sigma_S\in\Omega(S)} Z^{(n)}(\sigma_S). \tag{5}$$

Lastly let $Z^{(n)}(i, j)$ be a partition function on the single-trunk Cayley tree V_n^l with the configuration (i, j) on an edge $l =< x^0, y >$, where $y \in W_1$ and $i, j = 1, 2, 3$.

For each model there exists corresponding correlations between the partition functions $Z^{(n)}(i, j)$ and $Z^{(n)}(\sigma_S)$.
1) For first model with one-level k-tuple interactions

$$Z^{(n)}(\sigma_S) = \exp\left(\frac{J_0}{k_BT}\delta_{\sigma(y_1)\sigma(y_2)\cdots\sigma(y_k)}\right) \cdot \prod_{i=1}^{k} Z^{(n)}(\sigma(x^0), \sigma(y_i)); \tag{6}$$

2) For second model with one-level next-nearest neighbor interactions

$$Z^{(n)}(\sigma_S) = \exp\left(\frac{J_0}{k_BT}\sum_{i<j}\delta_{\sigma(y_i)\sigma(y_j)}\right) \cdot \prod_{i=1}^{k} Z^{(n)}(\sigma(x^0), \sigma(y_i)). \tag{7}$$

Assume that

$$a = \exp\left(\frac{J_o}{k_BkT}\right); b = \exp\left(\frac{J_p}{k_BT}\right); c = \exp\left(\frac{J_1}{k_BT}\right);$$

where k_B is Boltzmann constant and T is the absolute temperature.
We may assume that the different branches are equivalent, as is usual done for models on tree. Let us fix boundary configuration $\bar{\sigma}_n$ on $V \setminus V_n$ as follows: $\bar{\sigma}_n(x) \equiv i$ for any $x \in V \setminus V_n$, where $i = 1, 2, 3$. It is evident the following Proposition.
Proposition 2.2 For fixed boundary configuration $\sigma(x) \equiv 1$ for any $x \in V\setminus V_n$, the Hamiltonian

$H(\sigma(\Lambda)|\bar{\sigma}(V\setminus\Lambda))$ is S_3^* invariant, where $S_3^*=\{\pi\in S_3:\pi(1)=1\}$ is subgroup of group S_3.
Corollary 2.1 The partition functions $\{Z^{(n)}(i,j):i,j=1,2,3\}$ satisfy following relations:

$$\begin{aligned} Z^{(n)}(1,2)&=Z^{(n)}(1,3);\\ Z^{(n)}(2,1)&=Z^{(n)}(3,1);\\ Z^{(n)}(2,2)&=Z^{(n)}(3,3);\\ Z^{(n)}(2,3)&=Z^{(n)}(3,2); \end{aligned} \tag{8}$$

i.e., the following 5 variables

$$Z^{(n)}(1,1);Z^{(n)}(1,2);Z^{(n)}(2,1);Z^{(n)}(2,2);Z^{(n)}(2,3)$$

are independent. Then from (6) and (7) follow that we can select only 5 independent variables from 3^{k+1} partition functions, the same for both models (1) and (2), namely:

$$Z^{(n)}(1,1,1,\cdots,1),\ Z^{(n)}(1,2,2,\cdots,2),\ Z^{(n)}(2,1,1,\cdots,1),\\ Z^{(n)}(2,2,2,\cdots,2),\ Z^{(n)}(2,3,3,\cdots,3).$$

We note that, in the paramagnetic phase (high symmetry phase, i,e., symmetry with respect to S_3) we have

$$Z^{(n)}(1,1)=Z^{(n)}(2,2);$$

and

$$Z^{(n)}(1,2)=Z^{(n)}(2,1)=Z^{(n)}(2,3)$$

and respectively

$$Z^{(n)}(1,1,1,\cdots,1)=Z^{(n)}(2,2,2,\cdots,2)$$

and

$$Z^{(n)}(1,2,2,\cdots,2)=Z^{(n)}(2,1,1,\cdots,1)=Z^{(n)}(2,3,3,\cdots,3).$$

3. Basic Equations

In order to set up our basic equations in a recurrence scheme relating the partition function of an n- generation tree to the partition functions of its subsystems, we should take into account the partial partition functions for all the possible configurations of the spins in two successive generations. Below we produce the recurrent equations relating the partition functions $Z^{(n)}(i,j)$ for Hamiltonian considered on the single-trunk Cayley tree.

3.1. First Model

Firstly we consider Potts model (1) with competing prolong next-nearest-neighbour interactions, nearest-neighbour interactions and one level k-tuple neighbour interactions. We introduce 5 new variables

$$\begin{aligned} u_1^{(n)}&=\sqrt[q]{Z^{(n)}(1,1,1,\cdots,1)}=a\left[Z^{(n)}\left(1,1\right)\right]\\ u_2^{(n)}&=\sqrt[k]{Z^{(n)}(1,2,2,\cdots,2)}=a\left[Z^{(n)}\left(1,2\right)\right]\\ u_3^{(n)}&=\sqrt[k]{Z^{(n)}(2,1,1,\cdots,1)}=a\left[Z^{(n)}\left(2,1\right)\right]\\ u_4^{(n)}&=\sqrt[k]{Z^{(n)}(2,2,2,\ldots,2)}=a\left[Z^{(n)}\left(2,2\right)\right]\\ u_5^{(n)}&=\sqrt[k]{Z^{(n)}(2,3,3,\ldots,3)}=a\left[Z^{(n)}\left(2,3\right)\right] \end{aligned}$$

where k is the order of the Cayley tree.
Let $Z^{(n)}(i_0, i_1, i_2, \cdots, i_k)$ be a partition function and let m_1 be the number of spins $\{1\}$ and m_2 be the number of spins $\{2\}$ on first level W_1. Then we have

$$Z^{(n)}(1, i_1, i_2, \cdots, i_k) = a^{-k}\left(u_1^{m_1} u_2^{k-m_1}\right)$$
$$Z^{(n)}(2, i_1, i_2, \cdots, i_k) = a^{-k}\left(u_3^{m_1} u_4^{m_2} u_5^{q-(m_1+m_2)}\right)$$

One gets the following system of recurrent equations through direct calculation:

$$\begin{aligned} u_1^{(n+1)} &= a^{1-k}c\left[\left(a^k-1\right)\left(b^k u_1^k + 2u_2^k\right) + (bu_1 + 2u_2)^k\right], \\ u_2^{(n+1)} &= a^{1-k}\left[\left(a^k-1\right)\left(b^k u_3^k + u_4^k + u_5^k\right) + (bu_3 + u_4 + u_5)^k\right], \\ u_3^{(n+1)} &= a^{1-k}\left[\left(a^k-1\right)\left(u_1^k + \left(b^k+1\right)u_2^k\right) + (u_1 + (b+1)\,u_2)^k\right], \\ u_4^{(n+1)} &= a^{1-k}c\left[\left(a^k-1\right)\left(u_3^k + b^k u_4^k + u_5^k\right) + (u_3 + bu_4 + u_5)^k\right], \\ u_5^{(n+1)} &= a^{1-k}\left[\left(a^k-1\right)\left(u_3^k + u_4^k + b^k u_5^k\right) + (u_3 + u_4 + bu_5)^k\right]. \end{aligned} \tag{9}$$

The total partition function is given in terms of (u_i) by

$$Z^{(n)} = \left(a^k-1\right)\left[u_1^k + 2\left(u_2^k + u_3^k + u_4^k + u_5^k\right)\right] + (u_1 + 2u_2)^k + 2(u_3 + u_4 + u_5)^k.$$

For discussing the phase diagram, the following choice of reduced variables is convenient (see (8)):

$$\begin{aligned} x^{(n)} &= \frac{2u_2^{(n)}+u_3^{(n)}+u_5^{(n)}}{u_1^{(n)}+u_4^{(n)}} = \frac{Z^{(n)}(1,2)+Z^{(n)}(1,3)+Z^{(n)}(2,1)+Z^{(n)}(2,3)}{Z^{(n)}(1,1)+Z^{(n)}(2,2)} \\ y_1^{(n)} &= \frac{u_1^{(n)}-u_4^{(n)}}{u_1^{(n)}+u_4^{(n)}} = \frac{Z^{(n)}(1,1)-Z^{(n)}(3,3)}{Z^{(n)}(1,1)+Z^{(n)}(2,2)} \\ y_2^{(n)} &= \frac{u_2^{(n)}-u_3^{(n)}}{u_1^{(n)}+u_4^{(n)}} = \frac{Z^{(n)}(1,3)-Z^{(n)}(3,1)}{Z^{(n)}(1,1)+Z^{(n)}(2,2)} \\ y_3^{(n)} &= \frac{u_2^{(n)}-u_5^{(n)}}{u_1^{(n)}+u_4^{(n)}} = \frac{Z^{(n)}(1,2)-Z^{(n)}(3,2)}{Z^{(n)}(1,1)+Z^{(n)}(2,2)} \end{aligned} \tag{10}$$

As shown above, in the paramagnetic phase (high symmetry phase, i,e., symmetry with respect to S_3) we have

$$u_1^{(n)} = u_4^{(n)};$$

and

$$u_2^{(n)} = u_3^{(n)} = u_5^{(n)}.$$

The variable x is just a measure of the frustration of the nearest-neighbour bonds and is not an order parameter like y_1, y_2 and y_3. The relations (6) now have following form:

$$\begin{aligned}
x' &= 1cD[2\{(a^k-1)[b^k(x-3y_2+y_3)^k+2^k(1-y_1)^k+(x+y_2+y_3)^k] \\
&\quad +[(b+1)x+(1-3b)y_2+(b-3)y_3+2(1-y_1)]^k\} \\
&\quad +\{(a^k-1)[(2(1+y_1))^k+(b^k+1)(x+y_2+y_3)^k] \\
&\quad +[2(1+y_1)+(b+1)(x+y_2+y_3)]^k\} \\
&\quad +\{(a^k-1)[(x-3y_2+y_3)^k+(2(1-y_1))^k+(b(x+y_2-3y_3))^k] \\
&\quad +[(1+b)x+(b-3)y_2+(1-3b)y_3+2(1-y_1)]^k\}]; \\
y_1' &= 1D[\{(a^k-1)[b^k(1+y_1)^k+2^{k+1}(x+y_2+y_3)^k]+[b(1+y_1)+x+y_2+y_3]^k\} \\
&\quad -\{(a^k-1)[(x-3y_2+y_3)^k+2^kb^k(1-y_1)^k+(x+y_2-3y_3)^k] \\
&\quad +[2(x-y_2-y_3)+b(1-y_1)]^k\}]; \\
y_2' &= 1cD[\{(a^k-1)[b^k(x-3y_2+y_3)^k+2^k(1-y_1)^k+(x+y_2+y_3)^k] \\
&\quad +[(b+1)x+(1-3b)y_2+(b-3)y_3+2(1-y_1)]^k\} \\
&\quad -\{(a^k-1)[2^k(1+y_1)^k+(b^k+1)(x+y_2+y_3)^{]} \\
&\quad +[2(1+y_1)+(b+1)(x+y_2+y_3)]^k\}] \\
y_3' &= 1cD[\{(a^k-1)[b^k(x-3y_2+y_3)^k+2^k(1-y_1)^k+(x+y_2+y_3)^k] \\
&\quad +[(b+1)x+(1-3b)y_2+(b-3)y_3+2(1-y_1)]^k\} \\
&\quad -\{(a^k-1)[(x-3y_2+y_3)^k+2^k(1-y_1)^k+b^k(x+y_2+y_3)^k] \\
&\quad +[(b+1)x+(b-3)y_2+(1-3b)y_3+2(1-y_1)]^k\}]
\end{aligned} \tag{11}$$

where

$$\begin{aligned}
D &= \{(a^k-1)[b^k(1+y_1)^k+2^{k+1}(x+y_2+y_3)^k]+[b(1+y_1)+x+y_2+y_3]^k\} \\
&\quad +\{(a^k-1)[(x-3y_2+y_3)^k+(2b(1-y_1))^k+(x+y_2-3y_3)^k] \\
&\quad +[2(x-y_2-y_3)+b(1-y_1)]^k\}]
\end{aligned}$$

3.2. Second Model

Now we consider Potts model (2) with competing prolong next-nearest-neighbour interactions, nearest-neighbour interactions and one level next nearest neighbour interactions. Over again, we select only five variables with introducing new variables, then we have

$$u_1^{(n)} = \sqrt[k]{Z^{(n)}(1,1,1,\ldots,1)} = a^{\frac{k-1}{2}}\left[Z^{(n)}(1,1)\right],$$

$$u_2^{(n)} = \sqrt[k]{Z^{(n)}(1,2,2,\ldots,2)} = a^{\frac{k-1}{2}}\left[Z^{(n)}(1,2)\right],$$

$$u_3^{(n)} = \sqrt[k]{Z^{(n)}(2,1,1,\ldots,1)} = a^{\frac{k-1}{2}}\left[Z^{(n)}(2,1)\right],$$

$$u_4^{(n)} = \sqrt[k]{Z^{(n)}(2,2,2,\ldots,2)} = a^{\frac{k-1}{2}}\left[Z^{(n)}(2,2)\right],$$

$$u_5^{(n)} = \sqrt[k]{Z^{(n)}(2,3,3,\ldots,3)} = a^{\frac{k-1}{2}}\left[Z^{(n)}(2,3)\right].$$

In this case assume

$$a = \exp\left(\frac{J_0}{k_BT}\right), b = \exp\left(\frac{J_p}{k_BT}\right), c = \exp\left(\frac{J_1}{k_BT}\right).$$

As above let m_1 be the number of spins $\{1\}$ and m_2 be the number of spins $\{2\}$ on first level W_1, then we we have

$$Z^{(n)}(1,i_1,i_2,\ldots,i_k) = A\left[u_1^{m_1}u_2^{k-m_1}\right]$$

$$Z^{(n)}(2,i_1,i_2,\ldots,i_k) = A\left[u_3^{m_1}u_4^{m_2}u_5^{k-(m_1+m_2)}\right]$$

$$Z^{(n)}(3,i_1,i_2,\ldots,i_k) = A\left[u_3^{m_1}u_4^{k-(m_1+m_2)}u_5^{m_2}\right]$$

where $A = a^{m_1^2+m_2^2+m_1m_2-k(m_1+m_2)}$.

Next, we produce the following recurrence system through a direct calculation:

$$u_1' = a^{\frac{k-1}{2}}c\left[\sum_{i=0}^{k}\binom{k}{i}\left(a^{-i}bu_1\right)^{k-i}u_2^i\left(\sum_{j=0}^{i}\binom{k}{i}\left(a^{-j}\right)^{i-j}\right)\right]$$

$$u_2' = a^{\frac{k-1}{2}}\left[\sum_{i=0}^{k}\binom{k}{i}\left(a^{-i}bu_3\right)^{k-i}\left(\sum_{j=0}^{i}\binom{i}{j}\left(a^{-j}u_4\right)^{i-j}u_5^j\right)\right]$$

$$u_3' = a^{\frac{k-1}{2}}\left[\sum_{i=0}^{k}\binom{k}{i}\left(a^{-i}u_1\right)^{k-i}u_2^i\left(\sum_{j=0}^{i}\binom{i}{j}\left(a^{-j}\right)^{i-j}b^j\right)\right]$$

$$u_4' = a^{\frac{k-1}{2}}c\left[\sum_{i=0}^{k}\binom{k}{i}\left(a^{-i}u_3\right)^{k-i}\left(\sum_{j=0}^{i}\binom{i}{j}\left(a^{-j}u_5\right)^{i-j}(bu_4)^j\right)\right]$$

$$u_5' = a^{\frac{k-1}{2}}\left[\sum_{i=0}^{k}\binom{k}{i}\left(a^{-i}u_3\right)^{k-i}\left(\sum_{j=0}^{i}\binom{i}{j}\left(a^{-j}u_4\right)^{i-j}(bu_5)^j\right)\right]$$

Then, total partition function is given in terms of (u_i) by

$$Z^{(n)} = \sum_{i=0}^{k}\binom{k}{i}\left(a^{-i}\right)^{k-i}\left[\sum_{j=0}^{i}\binom{i}{j}\left(a^{-j}\right)^{i-j}\left[u_1^{k-i}u_2^i + u_3^{k-i}\left(u_4^ju_5^{i-j} + u_4^{i-j}u_5^j\right)\right]\right]$$

Choosing the same reduced variables as above we produce following recurrent equations (11′):

$$\begin{aligned}
x' &= \frac{\sum_{i=0}^{k}\binom{k}{i}(a^{-i})^{k-i}\left(\sum_{j=0}^{i}\binom{i}{j}(a^{-j})^{i-j}A_1(i,j,x,y_1,y_2,y_3)\right)}{c\sum_{i=0}^{k}\binom{k}{i}(a^{-i})^{k-i}\left(\sum_{j=0}^{i}\binom{i}{j}(a^{-j})^{i-j}D(i,j,x,y_1,y_2,y_3)\right)}\\
y_1' &= \frac{\sum_{i=0}^{k}\binom{k}{i}(a^{-i})^{k-i}\left(\sum_{j=0}^{i}\binom{i}{j}(a^{-j})^{i-j}A_2(i,j,x,y_1,y_2,y_3)\right)}{\sum_{i=0}^{k}\binom{k}{i}(a^{-i})^{k-i}\left(\sum_{j=0}^{i}\binom{i}{j}(a^{-j})^{i-j}D(i,j,x,y_1,y_2,y_3)\right)}\\
y_2' &= \frac{\sum_{i=0}^{k}\binom{k}{i}(a^{-i})^{k-i}\left(\sum_{j=0}^{i}\binom{i}{j}(a^{-j})^{i-j}A_3(i,j,x,y_1,y_2,y_3)\right)}{c\sum_{i=0}^{k}\binom{k}{i}(a^{-i})^{k-i}\left(\sum_{j=0}^{i}\binom{i}{j}(a^{-j})^{i-j}D(i,j,x,y_1,y_2,y_3)\right)}\\
y_3' &= \frac{\sum_{i=0}^{k}\binom{k}{i}(a^{-i})^{k-i}\left(\sum_{j=0}^{i}\binom{i}{j}(a^{-j})^{i-j}A_4(i,j,x,y_1,y_2,y_3)\right)}{c\sum_{i=0}^{k}\binom{k}{i}(a^{-i})^{k-i}\left(\sum_{j=0}^{i}\binom{i}{j}(a^{-j})^{i-j}D(i,j,x,y_1,y_2,y_3)\right)}
\end{aligned} \tag{12}$$

where

$$\begin{aligned}
A_1(i,j,x,y_1,y_2,y_3) &= 2[b(x-3y_2+y_3)]^{k-i}[2(1-y_1)]^{i-j}(x+y_2-3y_3)^j \\
&\quad +b^j[2(1+y_1)]^{k-i}(x+y_2+y_3)^i \\
&\quad +(x-3y_2+y_3)^{k-i}[2(1-y_1)]^{i-j}[b(x+y_2-3y_3)]^j; \\
A_2(i,j,x,y_1,y_2,y_3) &= [2b(1+y_1)]^{k-i}(x+y_2+y_3)^i \\
&\quad -[2b(1-y_1)]^j(x-3y_2+y_3)^{k-i}(x+y_2-3y_3)^{i-j}; \\
A_3(i,j,x,y_1,y_2,y_3) &= [b(x-3y_2+y_3)]^{k-i}[2(1-y_1)]^{i-j}(x+y_2-3y_3)^j \\
&\quad -b^j[2(1+y_1)]^{k-i}(x+y_2+y_3)^i; \\
A_4(i,j,x,y_1,y_2,y_3) &= [b(x-3y_2+y_3)]^{k-i}[2(1-y_1)]^{i-j}(x+y_2-3y_3)^j \\
&\quad -(x-3y_2+y_3)^{k-i}[2(1-y_1)]^{i-j}[b(x+y_2-3y_3)]^j
\end{aligned} \tag{13}$$

and

$$\begin{aligned}
D(i,j,x,y_1,y_2,y_3) &= [2b(1+y_1)]^{k-i}(x+y_2+y_3)^i \\
&\quad +[2b(1-y_1)]^j(x-3y_2+y_3)^{k-i}(x+y_2-3y_3)^{i-j}.
\end{aligned}$$

3.3. The Average Magnetization

In order to investigate the phase diagram we have to look at local properties, namely the local magnetization or the magnetization of the root x^0. The average magnetization m for the n-th generation is given by,

$$m = \left\langle \sigma(x^0) \right\rangle_n = \frac{1 \cdot Z_1^{(n)} + 2 \cdot Z_2^{(n)} + 3 \cdot Z_3^{(n)}}{Z^{(n)}} = 2 - \left(\frac{Z_1^{(n)} - Z_3^{(n)}}{Z^{(n)}} \right)$$

Since the form of spins of the Potts model is unessential one can replace the set of spin values {1,2,3} by the centered set {-1,0,1} [31] and then we have

$$\tilde{m} = \left\langle \sigma(x^0) \right\rangle_n = -\left(\frac{Z_1^{(n)} - Z_3^{(n)}}{Z^{(n)}} \right)$$

Below we will consider the average magnetization $\tilde{m}$ to study the wavevectors.
Average magnetization for first model is given by

$$\begin{aligned}
\tilde{m} &= \frac{1}{E}[(a^k-1)\{2^{-k}((x-3y_2+y_3)^k + (x+y_2-3y_3)^k - 2(x+y_2+y_3)^k)\} \\
&\quad +(1-y_1)^k - (1+y_1)^k + [1+x-(y_1+y_2+y_3)]^k - [1+x+(y_1+y_2+y_3)]^k],
\end{aligned}$$

where

$$\begin{aligned}
E &= [(a^k-1)\{(1+y_1)^k + 2(1-y_1)^k + 2^{1-k}[((x+y_2+y_3)^k + (x-3y_2+y_3)^k \\
&\quad +(x+y_2-3y_3)^k]\} + (1+x+y_1+y_2+y_3)^k + 2(1+x-y_1-y_2-y_3)^k]
\end{aligned}$$

and average magnetization for second model is given by

$$\tilde{m} = -\frac{2^k \sum_{i=0}^k \binom{k}{i} \left(2a^{k-i}\right)^{k-i} \left[\sum_{j=0}^i \binom{i}{j} \left(a^{-j}\right)^{i-j} M(i,j,x,y_1,y_2,y_3)\right]}{\sum_{i=0}^k \binom{k}{i} \left(a^{-i}\right)^{k-i} \left[\sum_{j=0}^i \binom{i}{j} \left(a^{-j}\right)^{i-j} N(i,j,x,y_1,y_2,y_3)\right]}$$

where

$$\begin{aligned} M(i,j,x,y_1,y_2,y_3) &= (1+y_1)^{k-i}(x+y_2+y_3)^i \\ &\quad -2^{2i-k-j}(x-3y_2+y_3)^{k-i}(1-y_1)^{i-j}(x+y_2-3y_3)^j \end{aligned}$$

and

$$\begin{aligned} N(i,j,x,y_1,y_2,y_3) &= 2^{k-i}(1+y_1)^{k-i}(x+y_2+y_3)^i \\ &\quad +2^j(x-3y_2+y_3)^{k-i}(1-y_1)^j(x+y_2-3y_3)^{i-j} \\ &\quad +2^{i-j}(x-3y_2+y_3)^{k-i}(1-y_1)^{i-j}(x+y_2-3y_3)^j \end{aligned}$$

4. Morphology of phase diagrams

It is convenient to know the broad features of the phase diagram before discussing the different transitions in more detail. Below we will consider phase diagram for $k=3$. This can be achieved numerically in a straightforward fashion. The recursion relations (11) for the first model and (12) for the second model provide us the numerically exact phase diagram in $(-J_p/J_1, J_0/J_1, T/J_1)$ space. Let $\beta=-J_p/J_1$, $\alpha=T/J_1$ and $\gamma=J_0/J_1$, by first model and second model, for some fixed values of , starting from initial conditions

$$x^{(1)}=\frac{2b^k+c^k+1}{c\left(c^kb^k+1\right)},\qquad y_1^{(1)}=\frac{c^kb^k-1}{c^kb^k+1},\qquad y_2^{(1)}=\frac{b^k-c^k}{c\left(c^kb^k+1\right)},\qquad y_3^{(1)}=\frac{b^k-1}{c\left(c^kb^k+1\right)}$$

with the parameters $a=\exp\left((k\alpha)^{-1}\gamma\right)$, $b=\exp\left(-\alpha^{-1}\beta\right)$, $c=\exp\left(\alpha^{-1}\right)$ and for second model's as with the parameters $a=\exp\left(\alpha^{-1}\gamma\right)$, $b=\exp\left(-\alpha^{-1}\beta\right)$, $c=\exp\left(\alpha^{-1}\right)$ corresponding to boundary condition $\bar{\sigma}\equiv 1$, one iterates the recurrence relations (11) and (12), we observe their behaviour after a large number of iterations. In the simplest situation a fixed point $(x^*, y_1^*, y_2^*, y_3^*)$ is reached. It corresponds to paramagnetic phase if $y_1^*=0$, $y_2^*=0$, $y_3^*=0$ to a ferromagnetic phase if $y_1^*, y_2^*, y_3^*\neq 0$.

The system may be periodic with period p, where case $p=2$ correspond to antiferromagnetic phase and case $p=4$ correspond to so-called antiphase that denoted $< 2 >$ for compactness. Finally, the system may remain aperiodic. The distinction between a truly aperiodic case and one with a very long period is difficult to make numerically. Below we consider periodic phase with period p where $p\leq 4$. All periodic phase with period $p>4$ and aperiodic phase we will consider as modulated phase.

The resultant phase diagram on the plane (β,α) for both models with some fixed values of γ and k are shown in Figs. 1 and 3. (Here: P – paramagnetic phase, F – ferromagnetic phase, AF – antiferromagnetic phase, P3 – phase with period 3, <2> – antiphase, and M – modulated phase)

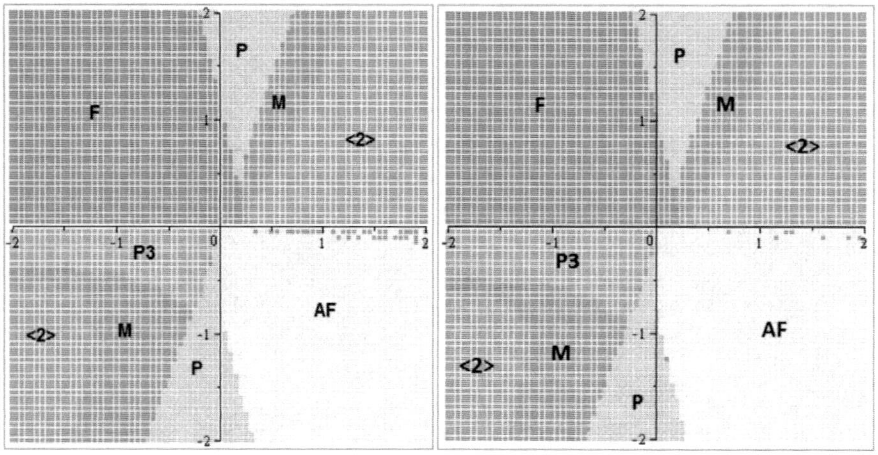

(a) (b)

Fig. 1. Phase diagram of the (a) first model and (b) second model for $\gamma = \pm 0.1$

One can observe the significance of order k with nonzero J_0 of Cayley tree in Figs. 1 – 6 where the models with competing interactions' (1) and (2) phase diagram is illustrated. Here, the systems for both models (11) and (12) was iterated for order $k = 3$ and 4 with $\gamma = \pm 0.1$, $\gamma = \pm 1$ and $\gamma = \pm 10$. In Fig. 1, consider both models with $k = 3$ and $\gamma = \pm 0.1$, contain in these diagrams are six phases which are ferromagnetic, paramagnetic, antiferromagnetic, phase of period 3, antiphase and modulated phases.

Journal of Physics: Conference Series **435** (2013) 012033 doi:10.1088/1742-6596/435/1/012033

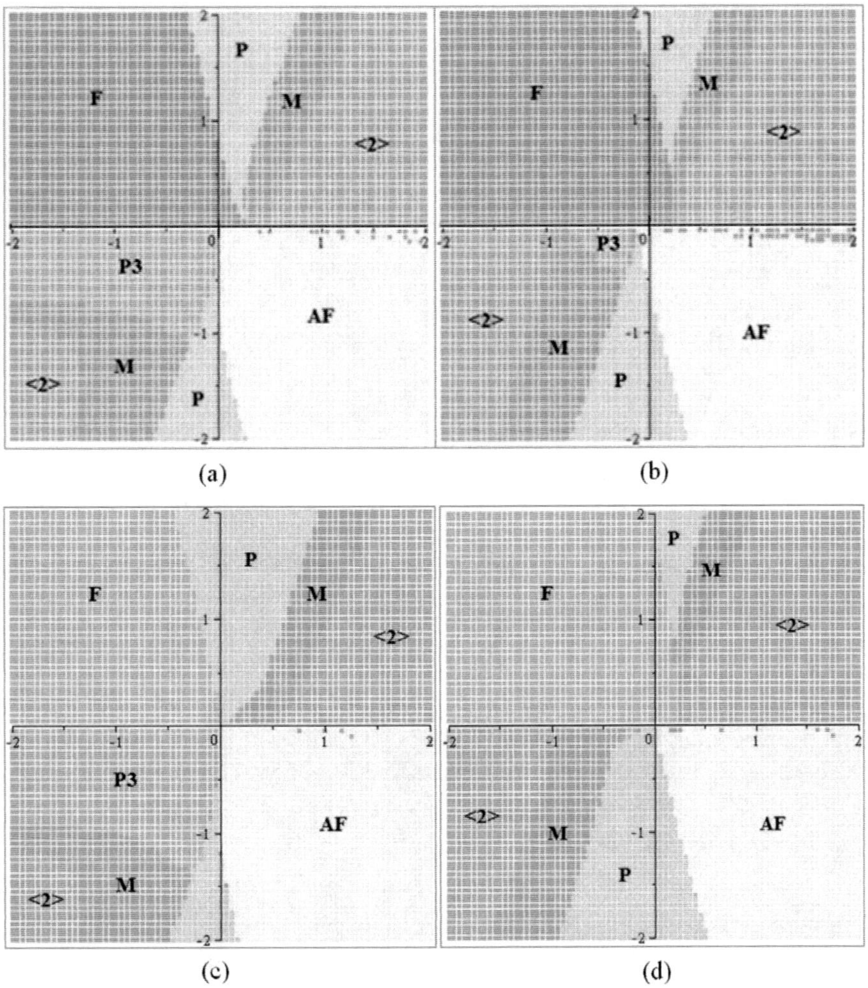

Fig. 2. Phase diagram of the first model for (a) $k = 3$ and $\gamma = -1$ and (b) $k = 3$ and $\gamma = 1$, and for second model with (c) $k = 3$ and $\gamma = -1$ and (d) $k = 3$ and $\gamma = 1$.

The phase diagrams are described in terms of its quadrant. When these phase diagrams are observed, it can be seen that ferromagnetic, paramagnetic, modulated and antiphase phases are situated in the first quadrant $(-J_p/J_1 > 0, T/J_1 > 0)$. On the other hand, the second quadrant $(-J_p/J_1 < 0, T/J_1 > 0)$ holds phase diagrams consisting of ferromagnetic and paramagnetic phases. Moving on, the third quadrant $(-J_p/J_1 < 0, T/J_1 < 0)$ is observed to have four phases namely paramagnetic, phase of period 3, modulated and antiphase phases, except for the case in Fig. 2(d), the phase of period 3 vanish in the this quadrant. Finally in quadrant four $(-J_p/J_1 > 0, T/J_1 < 0)$, the diagrams contain paramagnetic and antiferromagnetic with a small island consists of different phases.

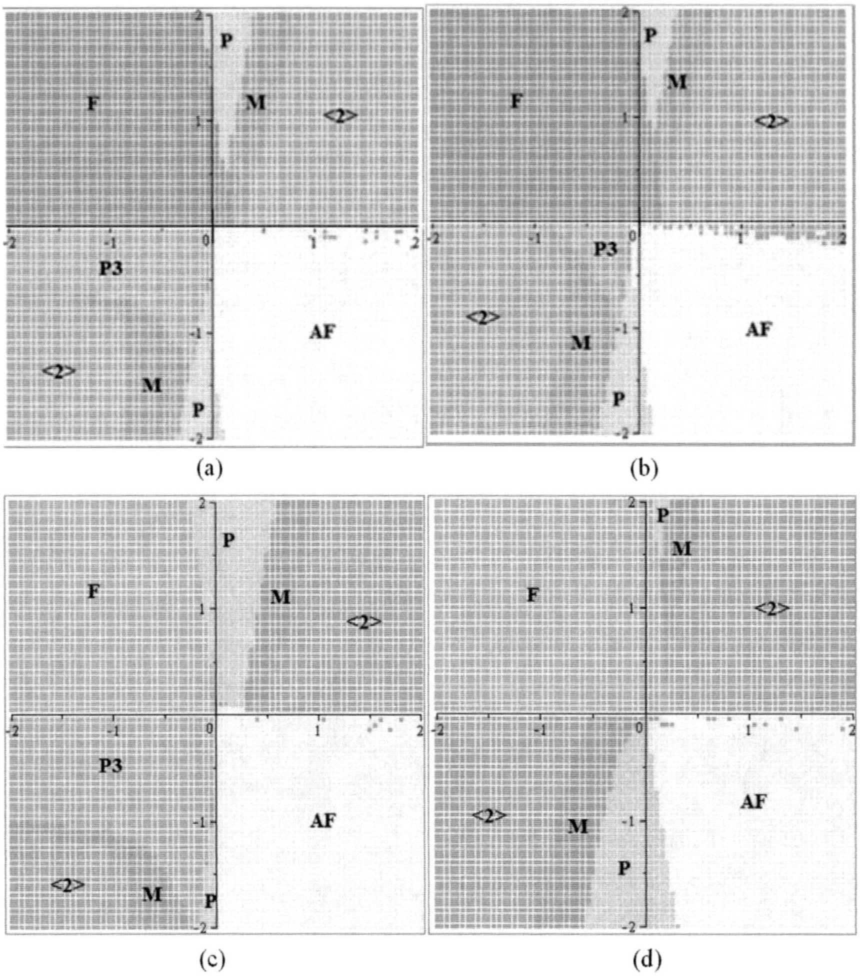

Fig. 3. Phase diagram of the first model for (a) $k = 4$ and $\gamma = -1$ and (b) $k = 4$ and $\gamma = 1$, and for second model with (c) $k = 4$ and $\gamma = -1$ and (d) $k = 4$ and $\gamma = 1$.

Order of Cayley tree with nonzero one-level next nearest neighbour interactions has played a vital role in the phase diagram on the both models (1) and (2) with competing interactions' result. Since the phase diagram was produced in the case $k > 2$, difference on the non-existing of paramodulated phase is observed. This is because this said phase exists in the case of Cayley tree order 2 (Ganikhodjaev et. al [19]). Also, the phase of antiphase exists in the third quadrant when a comparison is done with the case of Cayley tree order 2.

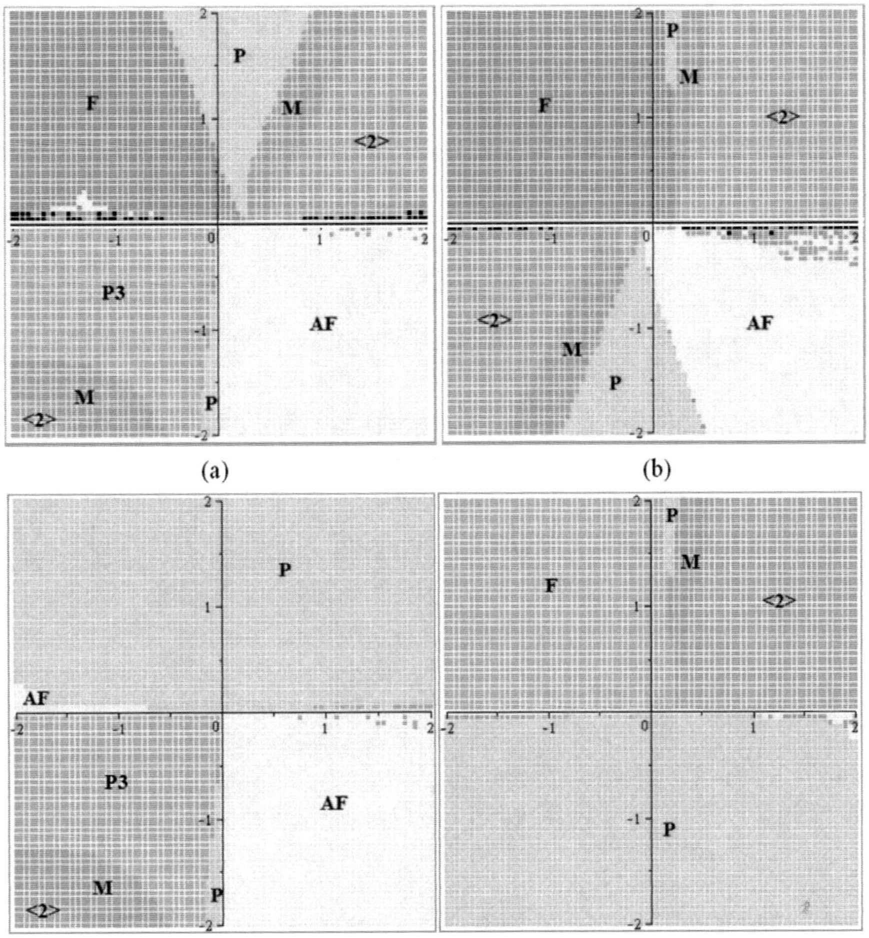

Fig. 4. Phase diagram of the first model for (a) $k = 3$ and $\gamma = -10$ and (b) $k = 3$ and $\gamma = 10$, and for second model with (c) $k = 3$ and $\gamma = -10$ and (d) $k = 3$ and $\gamma = 10$.

The phase diagrams in Figs. (4) – (5) were produced with different order k and $\gamma = \pm 10$ give difference in sizes of some phases. In Fig. 4(c), we found that the paramagnetic phase cover all in first and second quadrant, as we consider $J_1 > 0$ in the case of $k = 3$ and $\gamma = -10$ for second model. Similar to Fig. 4(d), where the paramagnetic phase cover all in third and fourth quadrant ($J_1 < 0$) in the case of $k = 3$ and $\gamma = 10$. The results were different as we see in the case for first model (see Fig. 4(a) and (b)).

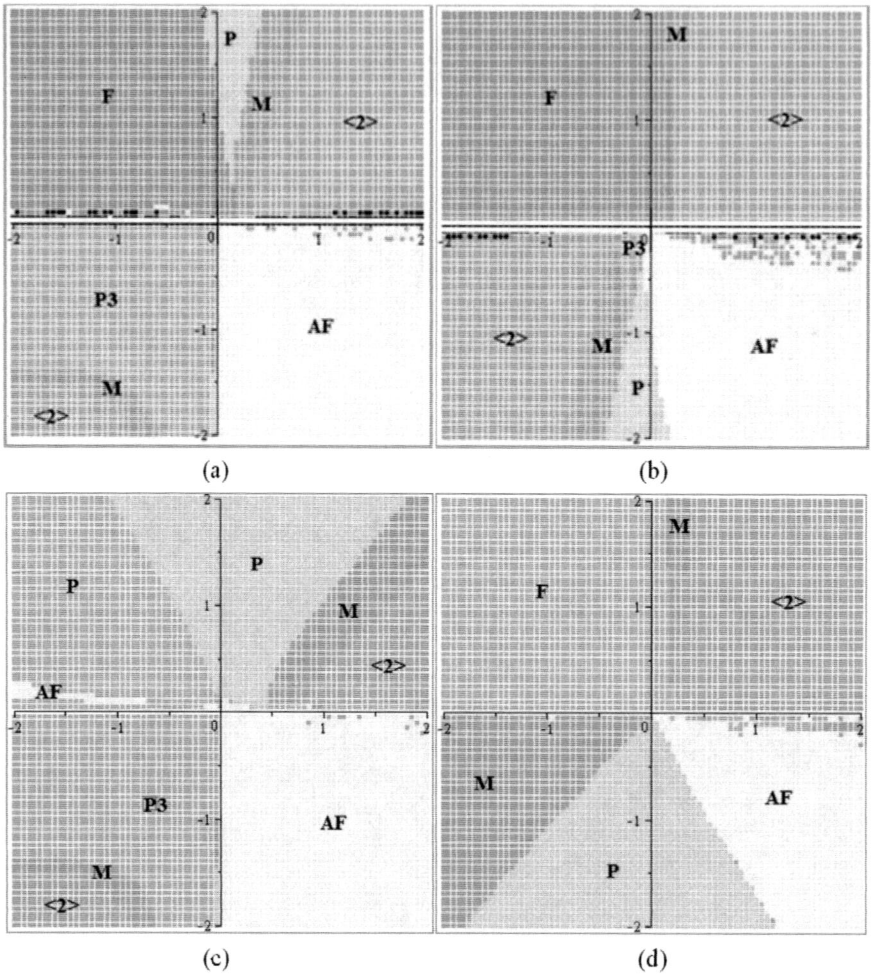

Fig. 5. Phase diagram of the first model for (a) $k = 4$ and $\gamma = -10$ and (b) $k = 4$ and $\gamma = 10$, and for second model with (c) $k = 4$ and $\gamma = -10$ and (d) $k = 4$ and $\gamma = 10$.

Looking on the fig. 5, one can see that the difference between both models occur in the case of $k = 4$ and $\gamma = 10$, as phase of period 3 and antiphase not exist in the third quadrant of our investigation area. (Figs. 5 (c) and (d)). Furthermore, in fig. 6, the phase diagrams were produced with order $k = 10$ and $\gamma = \pm 5$ give an additional in sizes of some phases. In Fig. 6(c), we found that a small phase of antiferromagnetic occur in second quadrant for the second model. Compare to first model, as in first quadrant for second model, we have more region of paramagnetic and modulated phases (see Fig. 6(c)). In addition, one can see in Fig. 6(d), as third quadrant of second model, same region of paramagnetic and modulated phases occur and increase as compare to the first model.

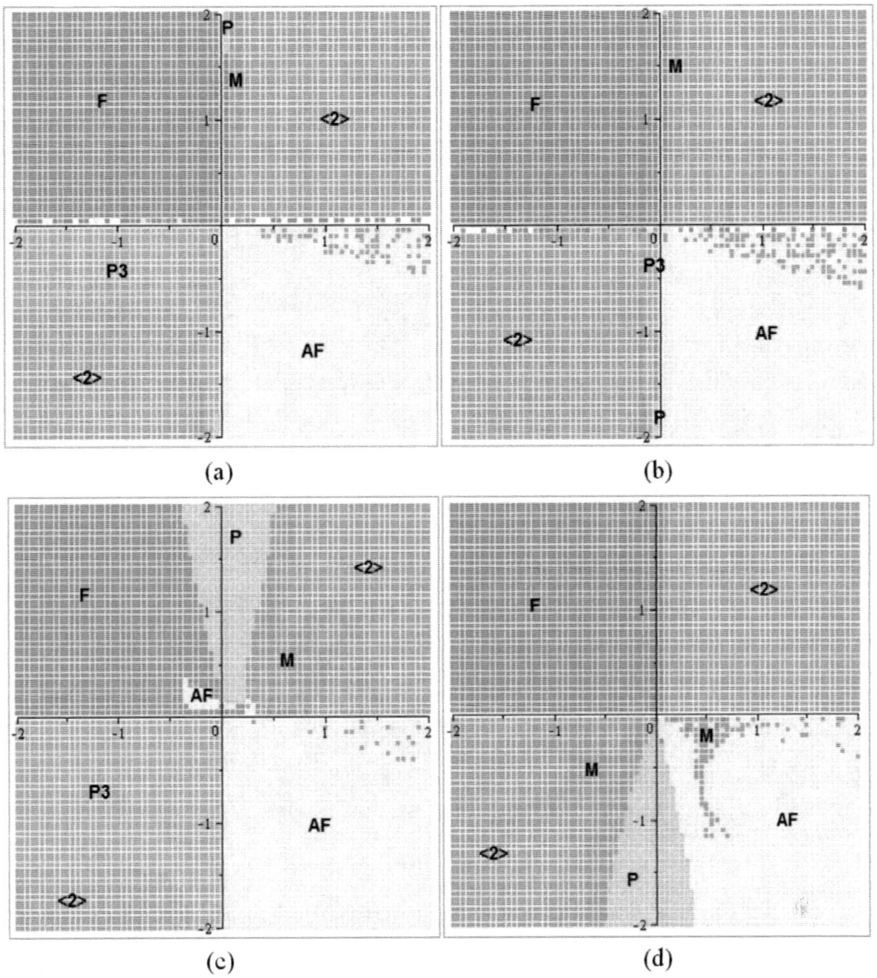

Fig. 6. Phase diagram of the first model for (a) $k = 10$ and $\gamma = -5$ and (b) $k = 10$ and $\gamma = 5$, and for second model with (c) $k = 10$ and $\gamma = -5$ and (d) $k = 10$ and $\gamma = 5$.

From the previously presented resultant phase diagrams, one can see that an investigation on the behaviour of the system (11) and (12) can be performed by applying the numerical approach. The systems were iterated in a large number before the detailed behaviour was studied and visualized on the (α, β) space. As been discussed, differences in the values of k results in differences and changes in the phase. Hence it is very difficult for us to study the behaviour of the systems analytically. Yet, the transition line separating the phases is suggested to be solved analytically. This can be done by linearizing both systems (11) and (12). Later on, the linearized systems and analytical justification of the phase diagrams will be discussed in more detailed.

5. Conclusion

In this paper, we study on relationship between two types of one-level neighbour interactions of Potts model on arbitrary order Cayley tree. One can consider for one-level neighbour interactions, k-tuple interactions and binary interactions. We study on the differences for both types of model on Cayley tree order 3 and order 4 and order 10.

We have found the phase diagram of Potts model (1) and (2) and show that there consist of six phases: ferromagnetic, paramagnetic, antiferromagnetic, period 3, antiphase and modulated phase. Moreover, we study on the difference of phase diagram for both types of model by fixing γ equal to $\pm 0.1, \pm 1, \pm 5$, and ± 10 . From considerations above, we can see that role of k rather significant because there show the phase diagram quite different for each k in first and second model.

Indeed the first and second model show us the difference, even though we only considering two different way in competing the one level interactions. We can say that this two model are differ to each other. They may give an unique significant to any physical phenomena occur. Further investigation can be consider to understand how this difference attribute appear.

6. Acknowledgments

This work was supported by Fundamental Research Grant Scheme FRGS10-022-0141.

7. References

[1] Bak P 1982 *Rep.Prog.Phys.* **45** 587
[2] Elliott R J 1961 *Phys.Rev* **124** 346
[3] Ostlund S 1981 *Phys.Rev. B* **24** 398
[4] Huse D A 1981 *Phys.Rev. B* **24** 5180
[5] Bak P 1981 *Phys.Rev. Lett.* **46** 791
[6] Jensen M H and Bak P 1983*Phys.Rev.B* **27** 6853
[7] Baxter R J 1982 *Exactly Solved Models in Statistical Mechanics*, Academic Press, London/New York
[8] Vannimenus J 1981*Z.Phys. B* **43**, 141
[9] Bak P and von Boehm J 1980 *Phys.rev.B* **21** 5297
[10] Georgii H-O, 1988 *Gibbs Measures and Phase Transitions* de Gruyter, Berlin
[11] Fisher M E and Selke W 1980 *Phys.Rev.Lett.* **44** 1502
[12] Mariz M, Tsalis C, Albuquerque A L 1985 *Jour. Stat. Phys.* **40** 577
[13] Inawashiro S, Thompson C J 1983 *Physics Letters* ,**97A** 245
[14] Inawashiro S, Thompson C J, Honda G 1983 *J. Stat. Phys.* Vol. **33** No. 2 419
[15] Yokoi C S O, Oliveira M J, Salinas S R 1985 *Phys. Rev. Lett.*, **54** 163
[16] Tragtenberg M H R, Yokoi C S O 1995 *Phys. Rev. E*, **52** 2187
[17] Coutinho S, Morgado W A M, Curado E M F, da Silva L 2006 *Physical Review B* **74**(9):094432 1
[18] Ganikhodjaev N N, Mukhamedov F M, Mendes J F F 2006 *Jour. Stat. Mech.* P08012
[19] Ganikhodjaev N N, Mukhamedov F M, Pah C H 2008*Physics Letters A* **373** 33
[20] Katsura S, Takizawa M 1974*Prog. Theor. Phys.* **51** 82
[21] Ganikhodjaev N N, Pah C H, Wahiddin M R B 2004 *J. Math. Phys.* **45** 3645
[22] Ganikhodjaev N N, Rozikov U A 2009 *Mathematical Physics, Analysis and Geometry*,**12**,141
[23] Ganikhodjaev N N, Uuz S 2011 *Physics A* **390**, 4160
[24] Monroe J L, 1992 *J.Stat.Phys.* **67** 1185
[25] Monroe J L 1994 *Phys.Lett. A* **188** 80
[26] Potts R B 1952 *Proc. Cambridge Philos. Soc.* **48** 106
[27] Wu F Y 1982*Rev. Mod. Phys.* **54** 235
[28] Ganikhodjaev N N, Temir S, Akin H, 2009 *J. Stat. Phys.* **137** 701
[29] H. Moraal, *Physica* **92A**:305(1978)
[30] Ganikhodjaev N N and Siti Fatimah Zakaria 2011 *Australian Journal of Basic and Applied Sciences*, **5(8)** 1296
[31] Ganikhodjaev N and Fatimah Abdul Razak 2010 *Mathematical Physics, Analysis and Geometry* **13(1)** 1

2012 iCAST: Contemporary Mathematics, Mathematical Physics and their Applications IOP Publishing
Journal of Physics: Conference Series **435** (2013) 012034 doi:10.1088/1742-6596/435/1/012034

Interaction of Two Pulses in Defocusing Nonlinear Schrodinger Equation

T H Hassan[1], S S Tadjimuratov[2], M R B Wahiddin[3], N Regaa[4] and B A Umarov[4]

[1] Faculty of Engineering, International Islamic University Malaysia P.O. Box 10, 50728, Kuala Lumpur, Malaysia
[2] Physical-Technical Institute of the Uzbek Academy of Sciences, 2-b, Bodomzor str., 100084, Tashkent, Uzbekistan
[3] Faculty of Information and Communication Technology, International Islamic University Malaysia, P.O. Box 10, 50728 Kuala Lumpur, Malaysia
[4]Department of Physics, Faculty of Science, International Islamic University Malaysia, P.O Box 141, 25710 Kuantan, Malaysia

E-mail: bakhram@iiu.edu.my

Abstract. In this work we are using initial conditions in the form of two separated bright pulses with the rectangular shape to analyse the interaction of pulses in a defocusing Nonlinear Scrödinger Equation (NLSE). By exact solution of the direct scattering problem associated with the defocusing NLSE, we have obtained the expressions for long time behavior of the solution. Obtained results interpreted as a nonlinear interference of two pulses.

1. Introduction

Interaction of a beams and pulses is considered as one of the most important problems of the wave theory and has many applications in physics and engineering. This problem is well investigated in a small perturbation approximation, when an equations describing process of interaction are linear, and the developed theory explains the interference phenomena [1]. Recently, the study of an interaction in nonlinear media attracted much attention of researches [2], especially in optics, physics of Bose Einstein condensates, plasma physics, hydrodynamics etc. One of the generic equations of nonlinear wave theory is the Nonlinear Schrodinger equation (NLSE), which describes the propagation of an envelope of modulated wave in weekly nonlinear dispersive media [2]. This equation in one dimensional case is solvable by inverse scattering transform (IST) method, which allows to find different solutions of NLSE. By this method the bright soliton solutions and interaction of pulses were investigated in details for focusing NLSE [3]. The interaction of two pulses of the defocusing NLSE was considered in [4] with the help of Inverse Scattering Transform method. Obtained results were applied to explain some features of the nonlinear interference in Bose-Einstein condensates observed in experiments [5]. Recent development in this area presented in the review paper [6]. In the paper [4] the initial condition is chosen in the Gaussian form. This choice is not allowed to obtain explicit expression for the description of the interference patterns. This circumstance has compelled authors to use the approximate methods in deriving of final results. In this work we are using initial conditions in the form of two separated bright pulses with the rectangular shape. This choice allows us to study the important features of a nonlinear interference. By exact solution of the direct scattering problem associated with the defocusing NLSE, we have obtained the expressions for long time behavior of the solution.

2. Formulation of the problem

Following [4] we will write the defocusing NLSE in scaled form

$$iq_\tau + \frac{1}{2}q_{xx} - g|q|^2 q = 0. \tag{1}$$

Let us first rewrite the Eq.(1) in the more convenient for calculations form

$$iq_\tau + \frac{1}{2}q_{xx} - 2|q|^2 q = 0. \tag{2}$$

where we have used the following change of variables $t \to t/2$, $x \to x$, $q \to \sqrt{g}q$.

The equation (2) is completely integrable [7], that is it can be presented as a compatibility condition of two linear systems $\psi_x = U\psi$, $\psi_t = V\psi$, where U and V are 2×2 matrices.

$$U = i\lambda \begin{pmatrix} -1 & 0 \\ 0 & 1 \end{pmatrix} + \begin{pmatrix} 0 & q \\ q^* & 0 \end{pmatrix}$$

$$V = -2i\lambda^2 \begin{pmatrix} 1 & 0 \\ 0 & -1 \end{pmatrix} - 2\lambda \begin{pmatrix} 0 & q \\ q^* & 0 \end{pmatrix} - \\ i \begin{pmatrix} 0 & q_x \\ -q_x^* & 0 \end{pmatrix} + i \begin{pmatrix} |q|^2 & 0 \\ 0 & -|q|^2 \end{pmatrix}.$$

The eigenvalue Zakharov-Shabat problem for the auxiliary two-component eigenfunction $(\psi_1, \psi_2)^T$ is written as the following system

$$\frac{\partial \psi_1}{\partial x} = -i\lambda\psi_1 + iq(0,x)\psi_2, \tag{3}$$

$$\frac{\partial \psi_2}{\partial x} = i\lambda\Psi_1 + iq^*(0,x)\psi_1. \tag{4}$$

It is well known that the defocusing NLSE (2), does not have the bright soliton solutions, it follows from the fact that the system (3-4) does not have discrete eigenvalues. Any initial localized pulse on zero background will disperse and eventually disappear. Our aim is to consider the interaction of two localized pulses with the objective to discuss the influence of nonlinearity to interference process. The long time asymptotic solution of Eq(2) can be found by calculation of the transmission coefficient $a(\lambda)$ for system (3-4), considering $q(x,0)$ as the potential. The expression for intensity $|q|^2$ in this case has the following form [8]

$$|q|^2 = \frac{}{1/4\pi t} \ln |a(\lambda = -x/4t)|^2. \tag{5}$$

3. Solution for two-box initial conditions

Let us now to concretize the initial conditions which we are going to consider for Eq(2). Initial conditions are two rectangular pulses:

$$q(x,0) = q_{10}B(x,x_1,x_2)\exp(2i\nu_1 x) + q_{20}B(x,x_3,x_4)\exp(2i\nu_2 x), \tag{6}$$

where q_{10} and q_{20} are constant amplitudes, $x_1 < x_2 < x_3 < x_4$, $x_3 - x_2$ is the separation distance, $d_1 = x_2 - x_1$ and $d_2 = x_4 - x_3$ are the widths of each of the pulses , $B(y, y_1, y_2) = [\theta(y - y_1) - \theta(y - y_2)]$, and θ is the Heavyside step function.

We will try now to solve the system (3-4), considering $q(x, 0)$ formally as the external potential. Our approach is to find solutions in different regions and then use the continuity conditions. In the domains where the potential equals to zero the solution is

$$\Psi = \begin{pmatrix} \psi_1 \\ \psi_2 \end{pmatrix} = \begin{pmatrix} A\exp(-i\lambda x) \\ B\exp(i\lambda x) \end{pmatrix}, \tag{7}$$

where A and B are some unknown coefficients.

Let us represent the potential in the nonzero regions in the following general form $q = q_0 \exp(i\nu x)$. To find the solutions in the regions with nonzero potential, we will transform the system (3-4) to the following scalar second order ordinary differential equation for v

$$v_{xx} + \xi^2 v = 0 \tag{8}$$

where $\lambda + \nu = k$, $(\lambda + \nu)^2 - |q|^2 = \xi^2$ and the new variable v defined by the equation $\psi_1 = v\exp(i\nu x)$.

Solution of the Eq.(8) is

$$v = \alpha_1 e^{-i\xi x} + \alpha_2 e^{i\xi x},$$

so the solution for ψ_1 will be

$$\psi_1 = \alpha e^{-i(\xi-\nu)x} + \delta e^{i(\xi+\nu)x}.$$

By the same way we can find solution for ψ_2

$$\psi_2 = \sigma e^{-i(\xi-\nu)x} + \beta e^{i(\xi+\nu)x}.$$

Solutions for ψ_1, ψ_2 are overdetermined there are four constants, instead two. One can determine the first pair through second as follows

$$\sigma = -\frac{k-\xi}{iq_0}\alpha,$$
$$\delta = \frac{k-\xi}{iq_0^*}\beta.$$

So we have

$$\psi_1 = \alpha e^{-i(\xi-\nu)x} + \frac{(k-\xi)\beta}{iq_0^*} e^{i(\xi+\nu)x}, \tag{9}$$

$$\psi_2 = -\frac{(k-\xi)\alpha}{iq_0} e^{-i(\xi-\nu)x} + \beta e^{i(\xi+\nu)x}. \tag{10}$$

Now using formulas (7) and(9-10) one is able to write the solutions in the different domains.

I. We are considering the scattering problem for system (3-4), so in the domain $x < x_1$ we have only the initial plane wave with unit amplitude

$$\Psi = \begin{pmatrix} \exp(-i\lambda x) \\ 0 \end{pmatrix}. \tag{11}$$

II. $x_1 < x < x_2$

$$\psi_1 = \alpha_1 e^{-i(\xi_1-\nu_1)x} + \frac{(k_1-\xi_1)\beta_1}{iq_{10}^*} e^{i(\xi_1+\nu_1)x}, \tag{12}$$

$$\psi_2 = -\frac{(k_1-\xi_1)\alpha_1}{iq_{01}} e^{-i(\xi_1-\nu_1)x} + \beta_1 e^{i(\xi_1+\nu_1)x}. \tag{13}$$

III. $x_2 < x < x_3$

$$\Psi = \begin{pmatrix} C\exp(-i\lambda x) \\ D\exp(i\lambda x) \end{pmatrix}. \tag{14}$$

IV. $x_3 < x < x_4$

$$\psi_1 = \alpha_2 e^{-i(\xi_2-\nu_2)x} + \frac{(k_2-\xi_2)\beta_2}{iq_{20}^*} e^{i(\xi_2+\nu_2)x}, \tag{15}$$

$$\psi_2 = -\frac{(k_2-\xi_2)\alpha_2}{iq_{02}} e^{-i(\xi_2-\nu_2)x} + \beta_2 e^{i(\xi_2+\nu_2)x}. \tag{16}$$

V. $x_4 < x$

$$\Psi = \begin{pmatrix} a\exp(-i\lambda x) \\ b\exp(i\lambda x) \end{pmatrix}. \tag{17}$$

The next step is to use continuity conditions and to fit the solution at the boundary points between different domains. After some simple but lengthly calculations one is able to obtain explicit expressions for the coefficients C, D, α_1, β_1, α_2, β_2, a and b. We will right below only the formula for a

$$a(\lambda) = \exp[i(k_1d_1 + k_2d_2)]\{[\cos(\xi_1 d_1) - i\frac{k_1}{\xi_1}\sin(\xi_1 d_1)][\cos(\xi_2 d_2) - i\frac{k_2}{\xi_2}\sin(\xi_2 d_2)] + \frac{q_{10}q_{20}}{\xi_1\xi_2}\exp[2i(-k_1x_2 + k_2x_3)]\sin(\xi_1 d_1)\sin(\xi_2 d_2)\} \tag{18}$$

4. Discussion and conclusions

The expression for long time asymptotic solutions for intensity of Eq.(1) can be now written by returning to the original variables and substituting expression for $a(\lambda)$ to Eq.(5). Analysis of this expression shows that the asymptotic behavior is as expected dispersing with the clear interference fringes. The account of nonlinearity changes the distribution of the fringes and amplitude of the field, and should be taken into account. Obtained formulas may help in explaining results of interference experiments in one-dimensional nonlinear systems and also can find applications in applied interferometry for determining a materials properties from an interference experiments.

Acknowledgments

This work has been supported by the research endowment grant of IIUM EDW B10-089-0428.

References

[1] Hariharan P, 2007 *Basics of interferometry*, Elsevier
[2] *Encyclopedia of Nonlinear Science* 2005 ed A Scott (New York: Routledge)
[3] Tsoy E N and Abdullaev F Kh 2003 *Phys. Rev.* E **67** 056610
[4] Wu-Ming Liu, Biao Wu, and Qian Nju 2000 *Phys. Rev. Lett.* **84** 2294
[5] Andrews M R, Townsend C G, Miesner H J, Durfee D S, Kurn D M, and Ketterle 1997 *Science* **275** 637
[6] Chaohong Lee et.al 2012 *Front. Phys.* **7** 109
[7] Zakharov V E and Shabat A B 1972 *Sov. Phys. JETP* **34** 62
[8] Segur H and Ablowitz M J 1976 *J. Math. Phys.* **17** 714

On Paramagnetic Phases of the Potts Model on a Bethe Lattice in the Presence of Competing Interactions

Nasir Ganikhodjaev[1], **Anasbek Rahmatullaev**[2]

[1]Department of Computational and Theoretical Sciences, Faculty of Science, IIUM, 25200 Kuantan, MALAYSIA

[2]Institute of Mathematics and Informational Technologies, 100125, Tashkent, UZBEKISTAN.

E-mail: `gnasir@iium.edu.my`, `anasbek@mail.ru`

Abstract. We consider a Potts model on a Bethe lattice with competing nearest-neighbour J and next-nearest-neighbour interactions J_p. The phase diagram of this model was studied by Ganikhodjaev *et al.* In this paper we investigate the problem of phase transition for the considered model and show that for some parameter values of the model there is phase transition.

1. Introduction

The Ising model with competing interactions on the Bethe lattice (or the Cayley tree; see [1] for terminology) has received widespread attention by many authors recently since the appearance of the Vannimenus model [2], [3],[4],[5] etc. A Potts model just as an Ising model with competing interactions has recently been studied extensively because of the appearance of nontrivial magnetic ordering (see [6],[7], [8] and reference therein). The Bethe lattice, that is, an infinite connected tree whose sites have the same coordination number, has a thin structure without closed paths but with infinite dimensionality. By introducing competing interactions between Ising or Potts spin variables, assigned to each site, we enable the system to present a very rich phase diagram with many modulated phases. The Vannimenus model [2], that is, the Ising model on a Bethe lattice of coordination number $q = 3$, with ferromagnetic nearest-neighbor interaction and with an antiferromagnetic next-nearest-neighbor interaction (then latter restricted to the sites belonging to the same branch) is the counterpart of the anisotropic next-nearest-neighbor Ising model (ANNNI) model defined on regular lattices [9]. The three states Potts model on the Bethe lattice tree of coordination number $q = 3$ with nearest-neighbor and next-nearest-neighbor interactions (latter restricted to the sites belonging to the same branch) was considered by Ganikhodjaev *et al* [6]. The Hamiltonian of this model is

$$H(\sigma) = -J \sum_{\langle x,y\rangle \in L} \delta_{\sigma(x)\sigma(y)} - J_p \sum_{\rangle \widetilde{x,y} \langle} \delta_{\sigma(x)\sigma(y)}, \qquad (1)$$

where $J, J_p \in R$ are coupling constants with $J > 0, J_p < 0$ and $\langle x, y\rangle$ stands for nearest-neighbours vertices and $\rangle\widetilde{x,y}\langle$ stands for next-nearest-neighbor interaction restricted to the sites belonging to the same branch.

As usual, one can introduce the notions of the conditional Gibbs measure, translation-invariant and periodic limiting Gibbs distributions (see [10],[11],[12]). In order to produce the recurrent equations, we consider the relation of the partition function on V_n to the partition function on subsets of V_{n-1} (see details in [6].) Let $Z^{(n)}(i_1, i_0, i_2)$ be a partition function on V_n where the spin in the root x^0 is i_0 and the two spins in the proceeding x^1 and x^2 are i_1 and i_2, respectively. As shown in [6] one can select only five independent variables $Z^{(n)}(1,1,1)$, $Z^{(n)}(2,1,2)$, $Z^{(n)}(1,2,1)$, $Z^{(n)}(2,2,2)$, $Z^{(n)}(3,2,3)$ and with the introduction of new variables

$$\begin{aligned} u_1^{(n)} &= \sqrt{Z^{(n)}(1,1,1)},\ u_2^{(n)} = \sqrt{Z^{(n)}(2,1,2)},\\ u_3^{(n)} &= \sqrt{Z^{(n)}(1,2,1)},\ u_4^{(n)} = \sqrt{Z^{(n)}(2,2,2)},\\ u_5^{(n)} &= \sqrt{Z^{(n)}(3,2,3)}, \end{aligned}$$

straightforward calculations (see more detail [6]) show that

$$\begin{aligned} u_1^{(n+1)} &= a(bu_1^{(n)} + 2u_2^{(n)})^2,\\ u_2^{(n+1)} &= (bu_3^{(n)} + u_4^{(n)} + u_5^{(n)})^2,\\ u_3^{(n+1)} &= (u_1^{(n)} + (b+1)u_2^{(n)})^2,\\ u_4^{(n+1)} &= a(u_3^{(n)} + bu_4^{(n)} + u_5^{(n)})^2,\\ u_5^{(n+1)} &= (u_3^{(n)} + u_4^{(n)} + u_5^{(n)})^2, \end{aligned} \tag{2}$$

and the total partition function is given in terms of (u_i) by

$$Z^{(n)} = (u_1^{(n)} + 2u_2^{(n)})^2 + 2(u_3^{(n)} + u_4^{(n)} + u_5^{(n)})^2, \tag{3}$$

where β is the inverse temperature and $a = \exp(\beta J), b = \exp(\beta J_p)$. We note that, in the paramagnetic phase (high symmetry phase), $u_1 = u_4$ and $u_2 = u_3 = u_5$. For discussing the phase diagram, the following choice of reduced variables is convenient:

$$\begin{aligned} x &= \tfrac{2u_2+u_3+u_5}{u_1+u_4}, \qquad y_1 = \tfrac{u_1-u_4}{u_1+u_4},\\ y_2 &= \tfrac{u_2-u_3}{u_1+u_4}, \qquad y_3 = \tfrac{u_2-u_5}{u_1+u_4}. \end{aligned} \tag{4}$$

The variable x is just a measure of the frustration of the nearest-neighbour bonds and is not an order parameter like y_1, y_2 and y_3. In this case the basic equations have following form:

$$\begin{aligned} x' &= \tfrac{1}{2aD}[P(y_1,y_2,y_3) + ((b+1)x + 2 - y_1 - by_2 - y_3)^2],\\ y_1' &= \tfrac{2}{D}(b+x)(by_1 + y_2 + y_3),\\ y_2' &= -\tfrac{1}{aD}[y_1 + by_2 - y_3][2 + (b+1)x - (b-1)(y_2 - y_3)],\\ y_3' &= \tfrac{1}{aD}(b-1)(y_3 - y_2)[2 + (b+1)x - 2y_1 - (b+1)(y_2 + y_3)], \end{aligned} \tag{5}$$

where

$$\begin{aligned} D &= (b+x)^2 + (by_1 + y_2 + y_3)^2,\\ P(y_1,y_2,y_3) &= 3y_1^2 + (4b^2 - 4b + 3)y_2^2 + (3b^2 - 4b + 4)y_3^2 + 2(2b+1)y_1y_2 + 2(b+2)y_1y_3\\ &- (2b^2 - 7b + 2)y_2y_3. \end{aligned}$$

In the next section we consider the problem of phase transition for considered model (1).

2. Translation-invariant Gibbs Measures:Phase Transitions

To investigate the phase transitions of considered model we consider the dynamic system (2) and study its limiting behaviour. Let $\mathbf{u} = (u_1, u_2, u_3, u_4, u_5) \in R_+^5$ and the dynamic system $F : R_+^5 \to R_+^5$ is defined by

$$\begin{aligned} u_1' &= a\,(bu_1 + 2u_2)^2 \\ u_2' &= (bu_3 + u_4 + u_5)^2 \\ u_3' &= (u_1 + (b+1)u_2)^2 \\ u_4' &= a\,(u_3 + bu_4 + u_5)^2 \\ u_5' &= (u_3 + u_4 + bu_5)^2, \end{aligned} \tag{6}$$

Then the recurrent equations (2) one can rewrite as $\mathbf{u}^{(n+1)} = F(\mathbf{u}^{(n)})$, $n \geq 0$. Let us describe fixed points of this dynamical system, i.e., solutions of equation $F(\mathbf{u}) = \mathbf{u}$. Denote $\mathrm{Fix}(F) = \{\mathbf{u} : F(\mathbf{u}) = \mathbf{u}\}$.

As noted in [6], in the paramagnetic phase (high symmetry phase), we have $u_1 = u_4$ and $u_2 = u_3 = u_5$. One can see that a set $M = \{\mathbf{u} = (u_1, u_2, u_3, u_4, u_5) \in R_+^5 :\ u_1 = u_4,\ u_2 = u_3 = u_5\}$ is an invariant set of the transformation F. On this set the reduced variables x, y_1, y_2, y_3 (4) have the following form $x = \frac{2u_2}{u_1}, y_1 \equiv 0, y_2 \equiv 0, y_3 \equiv 0$ and the system of equations (4) is reduced to single equation

$$x' = \frac{1}{2a}\left(\frac{(b+1)x+2}{b+x}\right)^2. \tag{7}$$

In this case the measure of the frustration $x^{(n)}$ one can consider as

$$x^{(n)} = 2\sqrt{\frac{Z^{(n)}(1,1,1)}{Z^{(n)}(2,1,2)}}$$

and since a conditional Cibbs measure of the cylinder set $\{x^0 = i_0, x^1 = i_1, x^2 = i_2\}$ is defined as

$$\mu^{(n)}(\{x^0 = i_0, x^1 = i_1, x^2 = i_2\}) = \frac{Z^{(n)}(i_0, i_1, i_2)}{Z^{(n)}}$$

one can conclude that corresponding limit Gibbs measure is fully determined by fixed points of recurrent equation (7). If we define $f : R^+ \to R^+$ by

$$f(x) = \left(\frac{(b+1)x+2}{b+x}\right)^2. \tag{8}$$

then f is bounded and thus the curve $y = f(x)$ must intersect the line $y = 2ax$, i.e., the recurrent equation (7) has a fixed points (see [13]). Note that if there is more than one positive fixed point, then there is more than one translation-invariant paramagnetic Cibbs measure corresponding to these fixed points. It is thus worthwhile to examine how many solutions the equation $f(x) = 2ax$ has.

Lemma 1 The equation

$$\left(\frac{(b+1)x+2}{b+x}\right)^2 = 2ax \tag{9}$$

(with $x \geq 0, a > 0, b > 0$) has one solution if $b < \frac{\sqrt{73}-1}{2}$. If $b > \frac{\sqrt{73}-1}{2}$, then there exist $\nu_1(b)$ and $\nu_2(b)$ with $0 < \nu_1(b) < \nu_2(b)$ such that the equation (9) has three solutions if $\nu_1(b) < 2a < \nu_2(b)$ and has two solutions if either $2a = \nu_1(b)$ or $2a = \nu_2(b)$. In fact

$$\nu_i(b) = \frac{1}{x_i}\left(\frac{(b+1)x_i+2}{b+x_i}\right)^2,$$

where x_1 and x_2 are the solutions of the equation

$$(b+1)x^2 - (b^2+b-6)x + 2b = 0.$$

Proof As before let

$$f(x) = \left(\frac{(b+1)x+2}{b+x}\right)^2;$$

thus we have

$$f'(x) = 2(b-1)(b+2)\frac{[(b+1)x+2]}{(x+b)^3},$$

$$f''(x) = 2(b-1)(b+2)\frac{[b^2+b-6-2(b+1)x]}{(x+b)^4}.$$

In particular if $b \leq 1$ then f is decreasing and there can only be one solution of $f(x) = 2ax$; thus we can restrict ourselves to the case $b > 1$. We have that f is convex for $x < \frac{b^2+b-6}{2(b+1)}$ and is concave for $x > \frac{b^2+b-6}{2(b+1)}$. If $\frac{b^2+b-6}{2(b+1)} < 0$ there can be only be one solution of $f(x) = 2ax$, i.e., $b < 2$. Thus there are at most three solutions to $f(x) = 2ax$ if $\frac{b^2+b-6}{2(b+1)} > 0$, i.e., $b > 2$. In fact it is quite easy to see that there is more than one solution if and only if there is more than one solution to the equation $xf'(x) = f(x)$, which is the same as

$$(b+1)x^2 - (b^2+b-6)x + 2b = 0.$$

The discriminant of this equation is equal to $(b-1)(b+2)(b^2+b-18)$ and it has two positive roots x_1 and x_2 if $b > \frac{\sqrt{73}-1}{2}$. This completes the proof.
Denote

$$T_c = \frac{J_p}{\ln\frac{\sqrt{73}-1}{2}},$$

where $J_p > 0$. Using this Lemma 1 we obtain following
Theorem 1 Let $J_p > 0$. Then if $T \geq T_c$ the model has unique paramagnetic phase and if $T < T_c$ and the conditions of the Lemma 1 are satisfied then for the model there are three paramagnetic translation-invariant Gibbs measures, i.e., there is phase transition.

3. Periodic Gibbs Measures: Phase Transitions

To describe all phases of considered model, one iterates the recurrence relations (5) and observes their behavior after a large number of iterations. In the simplest situation a fixed point $(x^*, y_1^*, y_2^*, y_3^*)$ is reached. It corresponds to a paramagnetic phase if $y_1^* = 0, y_2^* = 0, y_3^* = 0$ or to a ferromagnetic phase if $y_1^*, y_2^*, y_3^* \neq 0$. Secondary, the system may be periodic with period p, where case $p = 2$ corresponds to antiferromagnetic phase and case $p = 4$ corresponds to so-called antiphase, that denoted $< 2 >$ for compactness. Finally, the system may remain aperiodic. The distinction between a truly aperiodic case and one with a very long period is difficult to make numerically. Below we discuss existing of paramagnetic phases with period $p = 2$. Using numerical method in [6] presented new phase, namely, paramodulated phase. In this section we prove existence of such phases and study the phase transition problem. Let us first describe periodic points of the equation $F(F(u)) = u$ with $p = 2$ on the set M. In this case this equation can be reduced to following equation

$$f(f(x)) = x. \tag{10}$$

To find periodic solutions we consider the equation

$$\frac{f(f(x))-x}{f(x)-x}=0.$$

Simple but tedious algebra gives that the last equation is equivalent to the following

$$[(b+1)^2+2ab]^2x^2+4a[4a^2b^3+a(b^4+2b^3+9b^2+8b-4)+2(b+1)^3]x+4(ab^2+b+1)^2=0 \quad (11)$$

It is evident that the equation $Ax^2+Bx+C=0$ has two positive solutions if $b^4+2b^3+9b^2+8b-4<0$, i.e., $0<b<0.3498$, i.e., $J_p<0$. To prove the existing of two positive roots we need to show that for some values of parameters a and b we have $B<0$ and discriminant $D=B^2-4AC>0$, where

$$\begin{aligned} B &= 4a[4a^2b^3+a(b^4+2b^3+9b^2+8b-4)+2(b+1)^3], \\ D &= 16[4a^3b^2+(b^4+4b^3+9b^2+8b-4)a^2+(b^3+2b^2+4b+1)a+(b+1)^2] \\ &\quad [4a^3b^2+(b^4+9b^2+8b-4)a^2-(b^3+3b^2-2)a-(b+1)^2]. \end{aligned}$$

Consider the quadratic equation (with respect to a) $B=0$, i.e.,

$$4a^2b^3+a(b^4+2b^3+9b^2+8b-4)+2(b+1)^3=0.$$

with discriminant $\Delta_B=(b+1)^2(b-1)^2(b^4+2b^3-11b^2-12b+4)$. For $0<b<0.2702$ we have $\Delta_B>0$ and two positive roots $a_1(b)$ and $a_2(b)$. Then for any $0<b<0.2702$ (since $0<b<0.3498$) and a with $a_1<a<a_2$ we have $B<0$. Similarly one can prove that for any $0<b<0.2702$ there are $a_1^*(b)$ and $a_2^*(b)$ such that for a with $a_1^*(b)<a<a_2^*(b)$ discriminant $D>0$. In fact, equalizing to zero each factor of discriminant D, we consider two equations (with respect to a)

$$4a^3b^2+(b^4+4b^3+9b^2+8b-4)a^2+(b^3+2b^2+4b+1)a+(b+1)^2=0 \quad (12)$$

and

$$4a^3b^2+(b^4+9b^2+8b-4)a^2-(b^3+3b^2-2)a-(b+1)^2=0. \quad (13)$$

Since $0<b<0.2702$ first equation has two positive roots $\bar{a}_1(b)$ and $\bar{a}_2(b)$, such that for any a with $\bar{a}_1(b)<a<\bar{a}_2(b)$, first factor is negative. Second equation has single positive root $a(b)$ such that for any a with $0<a<a(b)$ second factor also is negative. Simple but tedious calculus shows that for any b in $(0,0.2702)$ there are $\tilde{a}_1$ and $\tilde{a}_2$ such that for any a in $(\tilde{a}_1,\tilde{a}_2)$ we have $B<0$ and $D>0$. Thus we have proved the following

Lemma 2 The equation (10) has no solution if $b\geq 0.2702$ and if $b<0.2702$ then there exist $\tilde{a}_1,\tilde{a}_2$ such that the equation has two positive solutions if $\tilde{a}_1<a<\tilde{a}_2$.

Let

$$T_c^{(2)}=\frac{J_p}{\ln 0.2702},$$

where $J_p<0$. From Lemma 2 we get the following

Theorem 2 Let $J_p<0$. Then if $T\geq T_c^{(2)}$ the model has no paramagnetic phase with period 2 and if $T<T_c^{(2)}$ and the conditions of the Lemma 2 are satisfied then for the model there are two paramagnetic Gibbs measures with period 2, i.e., there is phase transition.

Remark The paramagnetic Gibbs measure with period 2 was found in [6] numerically and called paramodulated phase.

4. Conclusion

It is proved the existence of phase transition for translation-invariant paramagnetic Gibbs measures when $J_p > 0$ and for periodic with period 2 paramagnetic Gibbs measures when $J_p < 0$. These results fully consistent with numerical results in [6].

4.1. Acknowledgments

This work is supported by the FRGS Grant Phase 2/2010, FRGS 10-022-0141.

5. References

[1] Baxter R J 1982 *Exactly Solved Models in Statistical Mechanics* (Academic Press, London/New York)
[2] Vannimenus J 1981 *Z. Phys. B* **43** 141
[3] Inawashiro S and Thompson C J 1983 *Phys. Lett.* **97A** 245
[4] Mariz A M, Tsallis C and Albuquerque E L 1985 *J. Stat. Phys.* **40** 577
[5] Pemantle R and Peres Y 2010 *Ann. Probab.* **38** 184
[6] Ganikhodjaev N N, Mukhamedov F M and Pah C H 2008 *Phys. Lett. A* **373** 33
[7] Ganikhodjaev N N, Temir S and Akin H 2009 *J. Stat. Phys.* **137** 701
[8] Coutinho S, Morgado W A M, Curado E M F and da Silva M 2006 *Physical Review B* **74**(9) 094432-1-7
[9] Elliott R J 1961 *Phys.Rev* **124** 346
[10] Georgii H O 1988 *Gibbs measures and phase transitions* (Walter de Gruyter, Berlin)
[11] Ganikhodjaev N N and Rozikov U A 1997 *Theor. Math. Phys.* **111** 480
[12] Ganikhodjaev N N and Rozikov U A 2009 *Math. Phys. Anal.Geom.* **12** 141
[13] Preston C 1974 *Gibbs states on countable sets* (Cambridge University Press, London)

Journal of Physics: Conference Series **435** (2013) 012036 doi:10.1088/1742-6596/435/1/012036

On the number of eigenvalues of a model operator in fermionic Fock space

Zahriddin Muminov[1], **Fudziah Ismail**[2], **Zainidin Eshkuvatov**[3]

[1,2,3] Department of Mathematics, Faculty of Science, University Putra Malaysia, Malaysia
[2,3] Institute for Mathematical Research, UPM, Malaysia
[1]Samarkand State University, Samarkand, Uzbekistan

E-mail: [1]zimuminov@mail.ru, [2]fudziah@science.upm.edu.my, [3]ezaini@science.upm.edu.my

Abstract. We consider a model describing a truncated operator H (truncated with respect to the number of particles) acting in the direct sum of zero-, one-, and two-particle subspaces of a fermionic Fock space $\mathcal{F}_a(L^2(\mathbb{T}^3))$ over $L^2(\mathbb{T}^3)$. We admit a general form for the "kinetic" part of the hamiltonian H, which contains a parameter γ to distinguish the two identical particles from the third one.
In this note:
(i) We find a critical value γ^* for the parameter γ that allows or forbids the Efimov effect (infinite number of bound states if the associated generalized Friedrichs model has a threshold resonance) and we prove that only for $\gamma < \gamma^*$ the Efimov effect is absent, while this effect exists for any $\gamma > \gamma*$.
(ii) In the case $\gamma > \gamma*$ we also establish the following asymptotics for the number $N(z)$ of eigenvalues z below E_{min}, the lower limit of the essential spectrum of H :

$$\lim_{z\to E_{min}^-} \frac{N(z)}{|\log|E_{min}-z||} = \mathcal{U}_0(\gamma) \quad (\mathcal{U}_0(\gamma) > 0), \quad \forall\gamma > \gamma^*.$$

1. Introduction

The spectral theory of continuous and lattice three particle Schrödinger operators in $\mathbb{R}^3$ shows the remarkable phenomenon known as *Efimov effect*: if all hamiltonians of all the two-body subsystems are positive and if at least two of them have a zero-energy resonance, then the three-body system has an infinite number of negative eigenvalues accumulating at zero.

This remarkable spectral property was discovered by V. Efimov [11] and has since become the subject of many papers [2, 3, 7, 9, 12, 20, 22, 23, 24, 26]. The first mathematical proof of the existence of this effect was given by Yafaev ([26]) and A. Sobolev estabilished [22] the asymptotics of the number of eigenvalues near the threshold of the essential spectrum.

Recently, Wang [25] has proved the existence of the Efimov effect in the system with $N \geq 4$ particles in $\mathbb{R}^3$ but in this case the properties of the spectrum have not been fully comprehended yet.

In statistical physics Minlos and Shpon [18], Malishev and Minlos [16], solid-state physics Mattis [17], Mogilner [19] and the theory of quantum fields Friedrichs [13], Buhler et.al. [8] some important problems arise where the number of quasi-particles is bounded, but not fixed. In Sigal, Soffer [21] has developed geometric and commutator techniques to find the location of the spectrum and to prove absence of singular continuous spectrum for Hamiltonians without conservation of the particle number.

Journal of Physics: Conference Series **435** (2013) 012036 doi:10.1088/1742-6596/435/1/012036

Notice that the study of systems describing n particles in interaction, without conservation of the number of particles is reduced to the investigation of the spectral properties of self-adjoint operators acting in the *cut subspace* $\mathcal{H}^{(n)}$ of the Fock space, consisting of $r \leq n$ particles [8, 13, 18, 19, 21, 27].

The model operator H, associated with a system describing two bosons and one particle another nature in interaction, without conservation of the number of particles, was considered in Albeverio et.al. [5],[6] and there existence of the Efimov effect was proved. This model described a truncated operator (truncated with respect to the number of particles) that acts in the direct sum of zero-, one-, and two-particle subspaces of the *bosonic Fock space*.

In the present paper we consider a model describing a truncated operator H (truncated with respect to the number of particles) acting in the direct sum of zero-, one-, and two-particle subspaces of a *fermionic Fock space* $\mathcal{F}_a(L^2(\mathbb{T}^3))$ over $L^2(\mathbb{T}^3)$. The model operator H is a lattice analogue of the Hamiltonians of the spin-fermion models for diluted magnetic semiconductors [8].

The main aim of the present paper is to give a thorough mathematical treatment of the spectral properties for a model operator H with emphasis on the asymptotics for the number of infinitely many eigenvalues (Efimov's effect case).

We admit a general form for the "kinetic" part of the hamiltonian H, which contains a parameter γ to distinguish the two identical particles from the third one. Under some smoothness assumptions we obtain the following results:

(i) We find a critical value γ^* for the parameter γ that allows or forbids the Efimov effect and we prove that only for $\gamma < \gamma^*$ the Efimov effect is absent, while this effect exists for any $\gamma > \gamma*$.

(ii) In the case $\gamma > \gamma*$ we also establish the following asymptotics for the number $N(z)$ of eigenvalues z below E_{min}, the lower limit of the essential spectrum of H :

$$\lim_{z \to E_{min}^-} \frac{N(z)}{|\log|E_{min} - z||} = \mathcal{U}_0(\gamma) \quad (\mathcal{U}_0(\gamma) > 0), \quad \forall \gamma > \gamma^*.$$

We notice that the assertion (i) is surprising and similar assertions have not yet been proved for the models in the *bosonic Fock space* ([5][6]).

We remark that for the three-particle discrete Schrödinger operator H the authors Dell'Antonio et.al. [10] found an explicit value of the parameter (which is nicely characterized as the only positive solution of a transcendental equation in one of the appendices) γ, say γ^* , such that only for values of γ below this number, the Efimov effect is absent for the sector of the Hilbert space which contains functions which are antisymmetric with respect to the two identical particles, while it is present for all values of the parameter γ on the symmetric sector. Moreover, in this work, striking similarities between the Efimov and Thomas effects, aside the occurrence of infinitely many bound states are discussed.

The organization of the present paper is as follows. Section 1 is an introduction to the whole work. In Section 2, the model operator is described as a bounded self-adjoint operator H in $\mathcal{H}^{(3)}$ and the main results of the present paper are formulated. Some spectral properties of the corresponding Friedrichs models $h(p)$, $p \in \mathbb{T}^3$, are recalled and the location and structure of the essential spectrum of H are given. Section 3 deals with the review the Birman-Schwinger principle for the operator H. In Section 4, we represent the sketch of the proof of the main results. We follow closely [10] to derive the proof of our main results (Theorem 2.10).

2. The model operator and statement of the main results

Let $\mathbb{T}^3$ be the three – dimensional torus, the cube $(-\pi, \pi]^3$ with appropriately identified sides. We remark that the torus $\mathbb{T}^3$ will always be considered as an abelian group with respect to the addition and multiplication by the real numbers regarded as operations on $\mathbb{R}^3$ modulo $(2\pi\mathbb{Z})^3$. Denote by $L^2_{as}((\mathbb{T}^3)^2)$ the subspace of antisymmetric functions of the Hilbert space $L^2((\mathbb{T}^3)^2)$.

Set

$$\mathcal{H}_0 = \mathbb{C}, \quad \mathcal{H}_1 = L^2(\mathbb{T}^3), \quad \mathcal{H}_2 = L^2_{as}((\mathbb{T}^3)^2).$$

The Hilbert space $\mathcal{H}^{(3)} = \mathcal{H}_0 \oplus \mathcal{H}_1 \oplus \mathcal{H}_2$ is called the *direct sum of zero-, one-, and two-particle subspaces* of a fermionic Fock space $\mathcal{F}_a(L^2(\mathbb{T}^3))$ over $L^2(\mathbb{T}^3)$.

Let H_{ij} be annihilation (creation) operators [13] defined in the Fock space for $i < j$ $(i > j)$. We note that in physics, an annihilation operator is an operator that lowers the number of particles in a given state by one, a creation operator is an operator that increases the number of particles in a given state by one, and it is the adjoint of the annihilation operator.

In this paper, we consider the case, where the number of annihilations and creations of the particles of the considering system equal to 1. It means that $H_{ij} \equiv 0$ for all $|i-j| > 1$. So, a model describing a truncated operator H acts in the Hilbert space $\mathcal{H}^{(3)}$ as a matrix operator:

$$H = \begin{pmatrix} H_{00} & H_{01} & 0 \\ H_{10} & H_{11} & H_{12} \\ 0 & H_{21} & H_{22} \end{pmatrix},$$

where the operators $H_{ij} : \mathcal{H}_j \to \mathcal{H}_i, i, j = 0, 1, 2$ are defined by the forms:

$$(H_{00}f_0)_0 = u_0 f_0, \quad (H_{01}f_1)_0 = \int_{\mathbb{T}^3} b(q')f_1(q')dq', \quad (H_{10}f_0)_1 = b(p)f_0,$$

$$(H_{11}f_1)_1 = u(p)f_1(p), \quad (H_{12}f_2)_1 = \int_{\mathbb{T}^3} b(q')f_2(p,q')dq',$$

$$(H_{21}f_1)_2 = \frac{1}{2}(b(q)f_1(p) - b(p)f_1(q)), \quad (H_{22}f_2)_2 = E_\gamma(p,q)f_2(p,q).$$

Here $f_i \in \mathcal{H}_i$, $i = 0, 1, 2$, u_0- fixed real number, $b(\cdot), u(\cdot)$ - real-valued analytic functions on $\mathbb{T}^3$, $E_\gamma(\cdot,\cdot) : (\mathbb{T}^3)^2 \to \mathbb{R}$, $\gamma > 0$, is given in the form

$$E_\gamma(p,q) = \gamma\varepsilon(p+q) + \varepsilon(p) + \varepsilon(q), \quad p, q \in \mathbb{T}^3,$$

where the function $\varepsilon(\cdot)$ is real valued conditionally negative definite three times differentiable function on $\mathbb{T}^3$ with a unique non-degenerate minimum at the origin.

Recall that a complex-valued bounded function $\varepsilon : \mathbb{T}^m \to \mathbb{C}$ is called conditionally negative definite if $\varepsilon(p) = \overline{\varepsilon(-p)}$ and

$$\sum_{i,j=1}^{n} \varepsilon(p_i - p_j) z_i \bar{z}_j \le 0$$

for all $p_1, p_2, \ldots, p_n \in \mathbb{T}^m$ and all $\mathbf{z} = (z_1, z_2, \ldots, z_n) \in \mathbb{C}^n$ satisfying $\sum_{i=1}^n z_i = 0$.

2.1. The Friedrichs model operator

To formulate the main results of the paper we introduce a family of Friedrichs models $h(p)$, $p \in \mathbb{T}^3$ which act in $\mathcal{H}^{(2)} \equiv \mathcal{H}_0 \oplus \mathcal{H}_1$ with the entries

$$(h_{00}(p)f_0)_0 = u(p)f_0, \; h_{01} = \frac{1}{\sqrt{2}} H_{01},$$
$$h_{10} = h_{01}^*, \; (h_{11}(p)f_1)_1(q) = w_p(q)f_1(q),$$

where $w_p(q) = E_\gamma(p,q)$.

Let the operator $h_0(p)$, $p \in \mathbb{T}^3$, act in $\mathcal{H}^{(2)}$ as

$$h_0(p)\begin{pmatrix} f_0 \\ f_1(q) \end{pmatrix} = \begin{pmatrix} 0 \\ w_p(q)f_1(q) \end{pmatrix}.$$

The perturbation $h(p) - h_0(p)$ of the operator $h_0(p)$ is a self-adjoint operator of rank 2. Therefore in accordance with the invariance of the essential spectrum under finite rank perturbations, the essential spectrum $\sigma_{ess}(h(p))$ of $h(p)$ fills the following interval on the real axis:

$$\sigma_{ess}(h(p)) = [m(p), M(p)],$$

where

$$m(p) = \min_{q\in\mathbb{T}^3} w_p(q), \quad M(p) = \max_{q\in\mathbb{T}^3} w_p(q).$$

Definition 2.1 *Let $u(0) \neq E_{\min}$. The operator $h(0)$ is said to have a threshold resonance if the number 1 is an eigenvalue of the operator*

$$(\mathrm{G}\psi)(q) = \frac{b(q)}{2(u(0) - E_{\min})} \int_{\mathbb{T}^3} \frac{b(t)\psi(t)dt}{w_0(t) - E_{\min}}, \quad \psi \in C(\mathbb{T}^3),$$

and the associated eigenfunction ψ (up to a constant factor) satisfies the condition $\psi(0) \neq 0$.

Remark 2.2 *The spectrum and resonances of this Friedrichs model are studied in [5, 6].*

For any $p \in \mathbb{T}^3$ we define an analytic function $\Delta(p, z)$(the Fredholm determinant associated with the operator $h(p)$) in $\mathbb{C} \setminus [m(p), M(p)]$ by

$$\Delta(p, z) = u(p) - z - \frac{1}{2} \int_{\mathbb{T}^3} \frac{b^2(t)dt}{E_\gamma(p, t) - z}.$$

Since $\Delta(0, \cdot)$ is continuous in $z \leq E_{\min}$ the following finite limit exists

$$\Delta(0, E_{\min}) = \lim_{z\to E_{\min}^-} \Delta(0, E_{\min}).$$

The lemmas below play crucial role in the proof of the main results when the operator $h(0)$ has a threshold resonance, and the proof of the lemmas were given in [5] and [6].

Lemma 2.3 *(i) If $u(0) \leq E_{\min}$, then the operator $h(0)$ has no a threshold resonance.*
(ii) If $u(0) > E_{\min}$, then the operator $h(0)$ has a threshold resonance iff $\Delta(0, E_{\min}) = 0$ and $b(0) \neq 0$.
(iii) Assume that $u(0) > E_{\min}$ and $\Delta(0, E_{\min}) = 0$.
a) If $b(0) \neq 0$, then the operator $h(0)$ has a threshold resonance and the vector $f = (f_0, f_1)$, where

$$f_0 = const \neq 0, \; f_1(q) = -\frac{b(q)f_0}{\sqrt{2}(w_0(q) - E_{\min})} \in L^1(\mathbb{T}^3) \setminus L^2(\mathbb{T}^3), \tag{1}$$

obeys the equation $h(0)f = E_{\min}f$.
b) If $b(0) = 0$, then the number $z = E_{\min}$ is an eigenvalue of the operator $h(0)$ and the vector $f = (f_0, f_1)$, where $f_0 \in \mathbb{C}^1$ and $f_1 \in L^2(\mathbb{T}^3)$ defined by (1), is the corresponding eigenvalue.

Let W be the 3×3-matrix of the second order partial derivatives of function $\varepsilon(\cdot)$ at the point $p = 0$.

Lemma 2.4 *Let the operator $h(0)$ has a threshold resonance. Then for any $p \in U_\delta(0), \delta > 0$ sufficiently small, and $z \leq E_{\min}$ the following decomposition*

$$\Delta(p, z) = \frac{4\sqrt{2}\pi^2 b^2(0)}{(1+\gamma)^{\frac{3}{2}} \det(W)^{\frac{1}{2}}} \sqrt{m(p) - z} + \Delta^{(02)}(m(p) - z) + \Delta^{(20)}(p, z)$$

holds, where $\Delta^{(02)}(m(p) - z)$ (resp. $\Delta^{(20)}(p, z)$) is a function behaving like $O((m(p) - z)^{\frac{1+\theta}{2}})$ (resp. $O(|p|^2)$) as $|m(p) - z| \to 0$ (resp. $p \to 0$ uniformly in $z \leq E_{\min}$).

Lemma 2.5 *Let the operator $h(0)$ has a threshold resonance. Then there exist positive numbers c, C and δ such that*

$$c|p| \leq \Delta(p, E_{\min}) \leq C|p| \quad \textit{for any} \quad p \in U_\delta(0),$$

and

$$\Delta(p, E_{\min}) \geq c \quad \textit{for any} \quad p \in \mathbb{T}^3 \setminus U_\delta(0).$$

2.2. *The essential spectrum of the operator H*
The following theorem describes the essential spectrum of the operator H.

Theorem 2.6 *For the essential spectrum $\sigma_{ess}(H)$ of the operator H the equality*

$$\sigma_{ess}(H) = \cup_{p\in\mathbb{T}^3}\sigma_d(h(p)) \cup [E_{\min}, E_{\max}]$$

holds, where $\sigma_d(h(p))$ is the discrete spectrum of the operator $h(p), p \in \mathbb{T}^3$.

Assumption 2.7 *Let $u(\cdot), b(\cdot)$ be even functions and $u(p) > u(0), \quad 0 \neq p \in \mathbb{T}^3$.*

The following lemma describes the location of the essential spectrum of the operator H.

Lemma 2.8 *Let Assumption 2.7 be fulfilled and $h(0)$ has a threshold resonance at $E_{\min}$. Then*

$$\sigma_{ess}(H) = [E_{\min}, E_{\max}].$$

2.3. *Statement of the main results*
Henceforth we assume that $h(0)$ has a threshold resonance.

By $N(z)$ we denote the number of eigenvalues and counted according their multiplicities of H lying below $z \le E_{\min}$.

Let γ^* be a solution of the equation

$$\frac{2(1+\gamma)^2}{\pi\gamma\sqrt{1+2\gamma}} - \frac{2(1+\gamma)^2}{\pi\gamma^2}\arcsin\frac{\gamma}{1+\gamma} = 1, \quad \gamma > 0.$$

Remark 2.9 *This equation has a unique positive solution (see [10, Appendix A]).*

In the following main theorem, we precisely describe the dependence of the number of eigenvalues of H on the parameters $\gamma > 0$.

Theorem 2.10 *Let Assumption 2.7 be fulfilled and $h(0)$ has a threshold resonance at $E_{\min}$. Then*

(i) For any $0 < \gamma < \gamma^*$ the operator H has a finite number of eigenvalues lying below the bottom $E_{\min}$ of the essential spectrum.

(ii) For any $\gamma > \gamma^*$ the operator H has infinitely many eigenvalues lying below $E_{\min}$. The function $N(z)$ obeys the relation

$$\lim_{z\to E_{\min}^-} \frac{N(z)}{|\log|E_{\min} - z||} = \mathfrak{U}_0(\gamma) \quad (\mathfrak{U}_0(\gamma) > 0). \tag{2}$$

Remark 2.11 *In [10], a result analogue to Theorem 2.10 has been proven for the three-particle Schrödinger operators on the lattice $\mathbb{Z}^3$.*

3. The Birman-Schwinger principle
For a bounded self-adjoint operator B, we define $n(\lambda, B)$ by

$$n(\lambda, B) = sup\{dimF : (Bu, u) > \lambda,\ u \in F,\ ||u|| = 1\}.$$

$n(\lambda, B)$ is equal to infinity if λ is in the essential spectrum of B and if $n(\lambda, B)$ is finite, it is equal to the number of the eigenvalues of B larger than λ. By the definition of $N(z)$ we have

$$N(z) = n(-z, -H),\ -z > -E_{\min}.$$

Journal of Physics: Conference Series **435** (2013) 012036 doi:10.1088/1742-6596/435/1/012036

In our analysis of the spectrum of H the crucial role is played by the self-adjoint compact Faddeev-Newton type integral operator $T(z), z < E_{\min}$, in the space $\mathcal{H}^{(2)}$ with the entries

$$(T_{00}(z)f_0)_0 = (1 - u_0 - z)f_0, \quad (T_{01}(z)f_1)_0 = -\int_{\mathbb{T}^3} \frac{b(q')f(q')dq'}{\sqrt{\Delta(q',z)}},$$

$$T_{10}(z) = T_{01}^*(z), \quad (T_{11}(z)f_1)_1(p) = -\frac{b(p)}{2\sqrt{\Delta(p,z)}} \int_{\mathbb{T}^3} \frac{b(q')f(q')dq'}{\sqrt{\Delta(q',z)}(E_\gamma(p,q') - z)}.$$

The following lemma follows from the well known Birman-Schwinger principle for the operator H (see [4, 22, 24]). In a system describing three particles in interaction, without conservation of the number of particles, case the Birman-Schwinger principle was obtained in [5], [6]. We refer to these paper for the proof.

Lemma 3.1 *The operator $T(z)$ is compact and continuous in $z < E_{\min}$ and*

$$N(z) = n(1, T(z)).$$

3.1. The Birman-Schwinger principle at the threshold

It should be noted that the operator $T(z)$ can be defined as a bounded operator even for the point $z = E_{\min}$ by

$$(T_{00}(E_{\min})f_0)_0 = (1 - u_0 - E_{\min})f_0, \quad (T_{01}(E_{\min})f_1)_0 = -\int_{\mathbb{T}^3} \frac{b(q')f(q')dq'}{\sqrt{\Delta(q',z)}},$$

$$T_{10}(E_{\min}) = T_{01}^*(E_{\min}), \quad (T_{11}(E_{\min})f_1)_1(p) = -\frac{b(p)}{2\sqrt{\Delta(p,E_{\min})}} \int_{\mathbb{T}^3} \frac{b(q')f(q')dq'}{\sqrt{\Delta(q',E_{\min})}(E_\gamma(p,q') - E_{\min})}.$$

Remark 3.2 *The operator $T(z)$ converges strongly (but not uniformly) as $z \to E_{\min} - 0$ to $T(E_{\min})$. Here we do not give the proof of this convergence. The convergence of the these type of operators was shown in [26],[14].*

Next lemma is an analogue of Lemma 3.1, and we omit its proof.

Lemma 3.3 *For any $z, z \leq E_{\min}$, the inequality*

$$N(z) \leq n(1, T(E_{\min}))$$

occurs.

4. The sketch of the proof of the main results

By the definition of $\varepsilon(\cdot)$ we get

$$E_\gamma(p,q) = E_{\min} + \frac{1}{2}\big((1+\gamma)(Wp,p) + 2\gamma(Wp,q) + (1+\gamma)(Wq,q)\big) + O(|p|^{3+\theta} + |q|^{3+\theta}), \tag{3}$$

as $p, q \to 0$ and

$$m(k) = E_{\min} + \frac{1}{2}\frac{1+2\gamma}{1+\gamma}(Wk,k) + O(|k|^{3+\theta}) \quad \text{as} \quad k \to 0.$$

Applying the asymptotics for $m(p)$ and using Lemma 2.4 we have

$$\Delta(p,z) = \frac{4\pi^2 b^2(0)}{(1+\gamma)^{3/2}\det(W)^{\frac{1}{2}}}\,[n(Wp,p) + 2(E_{\min} - z)]^{\frac{1}{2}} + O\big((|p|^2 + |E_{\min} - z|)^{\frac{1+\theta}{2}}\big) \quad \text{as } p, |E_{\min} - z| \to 0, \tag{4}$$

where

$$n = \frac{1+2\gamma}{1+\gamma}.$$

4.1. The infiniteness of the discerete spectrum of H

In this subsection we shall derive the asymptotics (2) for the number of eigenvalues of H.

We recall that in the subsection we closely follow A. Sobolev's method (see [22]) to derive the asymptotics for the number of eigenvalues of the operator H (see Theorem 2.10).

Let $T(\delta;|z-E_{\min}|)$ be the operator in $\mathcal{H}^{(2)}$ defined by

$$T(\delta;|z-E_{\min}|)=\begin{pmatrix}0 & 0\\ 0 & T_{11}(\delta;|z-E_{\min}|)\end{pmatrix},$$

where the $T_{11}(\delta;|E_{\min}-z|)$ is the integral operator in $\mathcal{H}_1$ with the kernel

$$T(\delta,|E_{\min}-z|;p,q)=$$
$$-\,\mathrm{d}_0\frac{\hat\chi_\delta(p)\hat\chi_\delta(q)(n(Wp,p)+2|E_{\min}-z|)^{-\frac14}(n(Wq,q)+2|E_{\min}-z|)^{-\frac14}}{(1+\gamma)(Wp,p)+2\gamma(Wp,q)+(1+\gamma)(Wq,q)+2|E_{\min}-z|},$$

where $\hat\chi_\delta(\cdot)$ is the characteristic function of the region $\hat U_\delta(0)=\{p\in\mathbb{T}^3:\ |W^{\frac12}p|<\delta\}$ and

$$\mathrm{d}_0=\frac{\det W^{\frac12}}{2\pi^2}(1+\gamma)^{\frac32}.$$

Lemma 4.1 *Let the conditions of Theorem 2.10 be fulfilled. The operator $T(z)-T(\delta;|E_{\min}-z|)$ belongs to the Hilbert-Schmidt class and is continuous in $z\le E_{\min}$.*

Proof. Applying asymptotics (3), (4) one can estimate the kernel of the operator $T(z)-T(\delta;|E_{\min}-z|)$, $z\le E_{\min}$, by the square-integrable function

$$C\Big(\frac{|p|^{1+\theta}+|q|^{1+\theta}}{|p|^{\frac12}(p^2+q^2)|q|^{\frac12}}+\frac{|E_{\min}-z|^{\frac{\theta}{2}}(p^2+q^2)^{-1}}{(|p|^2+|E_{\min}-z|)^{\frac14}(|q|^2+|E_{\min}-z|)^{\frac14}}+1\Big).$$

Hence the operator $T(z)-T(\delta;|E_{\min}-z|)$ belongs to the Hilbert-Schmidt class for all $z\le E_{\min}$. In combination with the continuity of the kernel of the operator in $z<E_{\min}$ this gives the continuity of $T(z)-T(\delta;|E_{\min}-z|)$ in $z\le E_{\min}$. □

Let $\hat T_0(\delta;|E_{\min}-z|)$ be the restriction of the integral operator $T(\delta;|E_{\min}-z|)$ to the subspace $L^2(\hat U_\delta(0))$. One verifies that the operator $\hat T_0(\delta;|E_{\min}-z|)$ is unitarily equivalent to the integral operator $T_1(r)$ acting in $L^2(U_r(0))$, where $r=|E_{\min}-z|^{-\frac12}$ and $U_r(0)=\{p\in\mathbb{R}^3:|p|<r\}$, with the kernel

$$T_1(r;p,q)=-\mathrm{d}_1\frac{(np^2+2)^{-1/4}(nq^2+2)^{-1/4}}{(1+\gamma)p^2+2\gamma(p,q)+(1+\gamma)q^2+2},$$

where

$$\mathrm{d}_1=\frac{(1+\gamma)^{3/2}}{2\pi^2}.$$

The equivalence is given by the unitary dilation

$$B:L^2(\hat U_\delta(0))\to L^2(U_r(0)),\quad (B_rf)(p)=(\frac{r}{\delta})^{-3/2}f(\frac{\delta}{r}W^{\frac12}p).$$

Further, we may replace

$$(np^2+2)^{-1/4},\ (nq^2+2)^{-1/4}\quad\text{and}\quad(1+\gamma)p^2+2\gamma(p,q)+(1+\gamma)q^2+2$$

by

$$(np^2)^{-1/4}(1-\chi_1(p)),\ (nq^2)^{-1/4}(1-\chi_1(q)) \quad \text{and} \quad (1+\gamma)p^2+2\gamma(p,q)+(1+\gamma)q^2,$$

respectively, since the error will be a Hilbert-Schmidt operator continuous up to $z=E_{\min}$.

We have denoted by $\chi_1(\cdot)$ the characteristic function of the ball $U_1(0)$. By the the replacement we obtain the integral operator $T_2(r)$ in $L^2(U_r(0)\setminus U_1(0))$ with a kernel

$$T_2(r;p,q)=-\frac{\mathrm{d}_1}{n^{\frac{1}{2}}}\frac{|p|^{-1/2}|q|^{-1/2}}{(1+\gamma)p^2+2\gamma(p,q)+(1+\gamma)q^2}.$$

By the dilation

$$M: L^2(U_r(0)\setminus U_1(0))\to L^2((0,\mathbf{r}),\mathbb{S}^2),\ \mathbf{r}=1/2\big|\log|E_{\min}-z|\big|,$$

where $\mathbb{S}^2$ is the unit sphere in $\mathbb{R}^3$, $(M\,f)(x,w)=e^{3x/2}f(e^x w),\quad x\in(0,\mathbf{r}),\ w\in\mathbb{S}^2$, one sees that the operator $T_2(r)$ is unitarily equivalent to the integral operator $\mathbf{S_r}$ with the kernel $S_\gamma(x-x';<\xi,\eta>)$, $\xi,\eta\in\mathbb{S}^2,\ x,x'\in\mathbb{R}^+$, where

$$S_\gamma(x;t)=-(2\pi)^{-2}\frac{u}{\cosh x+st},$$
$$u=\frac{1+\gamma}{\sqrt{1+2\gamma}},\quad s=\frac{\gamma}{1+\gamma},\quad t=<\xi,\eta>.$$

Recall the lemma in [22].

Lemma 4.2 *Let $A(z)=A_0(z)+A_1(z)$, where A_0 (A_1) is compact and continuous in $z<0$ $(z\leq 0)$. Assume that for some function $f(\cdot)$, $f(z)\to 0,\ z\to -0$ the limit*

$$\lim_{z\to -0} f(z)n(\lambda,A_0(z))=l(\lambda),$$

exists and is continuous in $\lambda>0$. Then the same limit exists for $A(z)$ and

$$\lim_{z\to -0} f(z)n(\lambda,A(z))=l(\lambda).$$

The following theorem is important for the proof of the asymptotics (2) and can be proved in similar way as Theorem 6.4 in [10].

Theorem 4.3 *Let the conditions of the part (ii) of Theorem 2.10 are fulfilled. The following equalities*

$$\lim_{|E_{\min}-z|\to 0}\frac{n(1,T(z))}{\big|\log|E_{\min}-z|\big|}=\lim_{\mathbf{r}\to\infty}\frac{1}{2}\mathbf{r}^{-1}n(1,\mathbf{S_r})=\mathfrak{U}_0(\gamma),\quad \mathfrak{U}_0(\gamma)>0,$$

hold.

This theorem, together with Lemmas 3.1 and 4.2 completes the proof of part *(ii)* of Theorem 2.10.

4.2. The finiteness of the discerete spectrum of H

The following lemma and Lemma 3.3 completes the proof of part *(i)* of Theorem 2.10.

Lemma 4.4 *Let $\gamma<\gamma^*$ and the hypothesis of part (i) of Theorem 2.10 be fulfilled. Then there exists the number $\epsilon=\epsilon_\gamma$ depending on γ so that*

$$\sup\sigma_{ess}(T(E_{\min}))<1-\epsilon_\gamma.$$

Proof. See [10, appendix B]. □

Acknowledgments
This work was partially supported by University Putra Malaysia under research university grant scheme (RUGS), and project code is 9190883. The first named author thanks S.N. Lakaev for useful discussions. This work is also partially supported by the Fundamental Science Foundation of Uzbekistan.

References

[1] Albeverio S., Gesztesy F. and Høegh-Krohn R. 1982 *Ann. Inst. H. Poincaré Sect. A (N.S.)* **37**, 1–28.
[2] Albeverio S., Høegh-Krohn R., and Wu T.T. 1971 *Phys. Lett. A* **83**, 105–109.
[3] Albeverio S., Lakaev S.N., Makarov K.A. 1998 *Letters in Math. Phys.* **43**, 73-85.
[4] Albeverio S., Lakaev S.N., Muminov Z.I. 2004 *Ann. Henri Poincaré* **5**, 743–772.
[5] Albeverio S., Lakaev S.N., and Rasulov T.H. 2007 *J. Stat. Phys.* **127**, 191220.
[6] Albeverio S., Lakaev S.N., and Rasulov T.H. 2007 *Methods Funct. Anal. Topology* **13**, no.1, 116.
[7] Amado R.D. and Noble J.V. 1971 *Phys. Lett.* **B.35**. no.1, 25-27.
[8] Buhler Ch., Yunoki S., and Moreo A. 2000 *Phys. Rev. Lett.* 84(12):26902693.
[9] Dell'Antonio G.F., Figari R., Teta A. 1994 *Ann. Inst. H. Poincaré Phys. Théor.* **60** no.3, 253-290.
[10] Dell'Antonio G.F., Muminov Z.I., Shermatova Y.M. 2011 *J. Phys. A: Math. Theor.* no.44 315302 doi: 10.1088/1751-8113/44/31/315302.
[11] Efimov V. 1973 *Nucl. Phys. A* **210**, 157–158.
[12] Faddeev L.D., Merkuriev S.P. 1993 *Quantum scattering theory for several particle systems.* (Kluwer Academic Publishers) p 424.
[13] Friedrichs K.O. 1965 *Perturbation of spectra in Hilbert space* (AMS., Providence, Rhode Island) p 178.
[14] Lakaev S.N. 1993 *Funkcionalnii analiz i ego priloj.* **27**, No. 3, 15–28 (translation in Funct. Anal.Appl.).
[15] Lakaev S.N. and Rasulov T.Kh. 2003 *Mathematical Notes* **73**, No.3, 521–528.
[16] Malishev V.A. and Minlos R.A. 1995 *Linear infinite-particle operators. Translations of Mathematical Monographs, 143* (American Mathematical Society, Providence, RI) p 298.
[17] Mattis D.C. 1986 *Rev.Modern Phys.* **58** No.2, 361-379.
[18] Minlos R.A. and Spohn H. 1996 *Amer. Math. Soc. Transl.(2)* **177** 159–193.
[19] Mogilner A.I. 1991 *Advances in Sov. Math.* **5** 139–194.
[20] Ovchinnikov Yu.N. and Sigal I.M. 1989 *Ann. Physics* **123** 274–295.
[21] Sigal I.M., Soffer , Zielinski L. 2002 *J. Math. Phys.* **43** No.4, 1844–1855.
[22] Sobolev A.V. 1993 *Commun. Math. Phys.* **156** 127–168.
[23] Tamura H. 1991 *J. Funct. Anal.* **95** 433–459.
[24] Tamura H. 1994 *Adv. Stud. Pure Math., Math. Soc. Japan, Tokyo* **23** 311–322.
[25] Wang X.P. 2004 *J. Func. Anal.* **209** 137161.
[26] Yafaev D.R. 1974 *Math. USSR-Sb.* **23** 535-559.
[27] Zhukov Yu., Minlos R. 1995 *Theoret. and Math. Phys.* **103** No.1, 398–411.

Journal of Physics: Conference Series **435** (2013) 012037 doi:10.1088/1742-6596/435/1/012037

Performance of selected imputation techniques for missing variances in meta-analysis

NRN Idris[1,4], MH Abdullah[2], SMTolos[3]

[1,2,3]Department of Computational and Theoretical Sciences, Faculty of Science, International Islamic University Malaysia, Bandar Indera Mahkota, 25200 Kuantan, Pahang, MALAYSIA

Email : ruzni@iium.edu.my

Abstract. A common method of handling the problem of missing variances in meta-analysis of continuous response is through imputation. However, the performance of imputation techniques may be influenced by the type of model utilised. In this article, we examine through a simulation study the effects of the techniques of imputation of the missing SDs and type of models used on the overall meta-analysis estimates. The results suggest that imputation should be adopted to estimate the overall effect size, irrespective of the model used. However, the accuracy of the estimates of the corresponding standard error (SE) is influenced by the imputation techniques. For estimates based on the fixed effects model, mean imputation provides better estimates than multiple imputations, while those based on the random effects model responds more robustly to the type of imputation techniques. The results showed that although imputation is good in reducing the bias in point estimates, it is more likely to produce coverage probability which is higher than the nominal value.

1. Introduction

A meta-analysis is a statistical technique for integrating quantitative results of the same research question from several sources. Theoretically, combining the results from multiple trials should enhance the precision and accuracy of any pooled results. In practice however, there are a number of potential problems that may affect the validity of such results. One widely debated controversy related to meta-analysis concerns the choice between a fixed and random effects model for providing an overall estimate of the effect size [1]. When the differences in the effect sizes across the studies is due only to sampling error, they are considered homogeneous. This source of variation can be accommodated in meta-analysis by using the fixed effect model. However, if the variability in the effect size estimates exceeds those from sampling error alone, then a random effects model would be more appropriate as it takes into account the unexplained heterogeneity attributed to systematic differences between studies [2].

In meta-analysis of continuous data, variability measures, particularly the standard deviations (SDs), are often not reported in the published report of the trials [3]. A popular

[4] Any correspondences should be addressed to ruzni@iium.edu.my

approach in handling this problem is through imputation of the missing SDs [4]. Earlier studies which examined the effects of imputing the missing SDs on the overall meta-analysis estimates [5]-[7] concluded that imputation recovers most of the lost information in the estimate of effect size and the corresponding SE. These studies did not examine the effects of different techniques of imputation used to estimate the missing variances.

This article investigates empirically, the effects of imputing the missing SDs and the choice of meta-analysis model on the overall meta-analysis estimates. Meta-analysis estimates were based on the fixed and random effect models obtained from three sets of simulated data, namely, (1) complete data, where all studies are assumed to report the SDs; (2) incomplete data, where studies with missing SDs were excluded; and (3) the imputed data, where the missing SDs were imputed, and the studies with imputed SDs were included in the analysis.

The imputation methods considered in this study are the weighted mean imputation and the multiple imputations (MI). These techniques of imputation are the most common techniques in meta-analyses, particularly in psychology and social science [3]. The mean imputation has some popularity among the analysts due to its simplicity and ease of understanding and interpretation. This technique is recommended if less than 10% of the data is missing [6] and known to perform better if the data is approximately normal, and the missing values are missing completely at random (MCAR) [8]. The MI is a popular choice of imputation technique, and the main advantage is that it allows for the computation of the uncertainty due to imputation, in addition to those due to sampling error [9]. In this study we assumed that the study level SDs is missing completely at random (MCAR).

2. Material and Method

2.1. Simulation of data

The data for each meta-analysis was simulated at individual levels and summarized into study or aggregate levels. The following model was used to generate the individual level data,

$$y_{ij} = \beta_{0i} + \beta_{1i} T_{ij} + \epsilon_{ij}$$

where y_{ij} is the response of the jth patient in the ith study, β_{0i} is the random study effect, T_{ij} represents the dummy covariates for treatment, which takes two values (namely, 0 for the control and 1 for the treatment arm), β_{1i} is the random treatment effect which is assumed to vary across the studies, but would take a fixed value under the fixed effect model, and ε_{ij} are the random error terms. y_{ij} , $\beta_{0i,}$ β_{1i} and ε_{ij} are assumed to be independent and normally distributed with $\varepsilon_i \sim N(0, \sigma_\varepsilon^2)$; $\sigma_\varepsilon^2 = 1$; $\beta_{0i} \sim N(\beta_0, \sigma_{\upsilon 0}^2)$; $\beta_0 = 0$, $\sigma_{\upsilon 0}^2 = 1$; $\beta_{1i} \sim N(\beta_1, \sigma_{\upsilon 1}^2)$; $\beta_1 = 1$, $\sigma_{\upsilon 1}^2 = 1$. Additionally, the number of patients in each study for each N are assumed to be equal, i.e., $n_i = n$ for $i = 1,2,\ldots,N$, with an equal number of patients undergoing the two treatments , i.e $n_{0i} = n_{1i} = 1/2\ n$. Total responses in each simulation run was Σn_{ij}. The parameters varied across the simulation are the number of studies in each meta-analysis (N) at 10, 20, 30; the number of patients within each study in a meta-analysis (n) at 20, 60, 100; and the percentage of studies with missing SDs (x %) at 10, 30, 50. The combinations of the number of studies, N, the sample size, n, and the percentage of missing studies, x %, generated 27 meta-analyses. Each of these combinations was replicated 500 times.

2.2. Creation of missing SD's
To create the SDs missing completely at random, x % (x = 10, 30, 50) studies were selected at random from N studies, and excluded from the data.

2.3. Imputation techniques
Mean imputation: The missing variances were replaced by the weighted mean of the available variances, where the weight is the number of patients in each study (n).

Multiple imputations: The missing variance was replaced by a randomly selected value from the available variances and estimates based on this data were recorded. This process was repeated 250 times in each of the 500 simulation runs, and the mean and standard deviations of the estimates were computed. This allows the uncertainty induced by the imputation to be incorporated into the overall estimate of the variance of the estimate. [8].

2.4. Measures of performance
The performance of the two imputation techniques were evaluated using the percentage relative bias (PRB) for the point estimates and the coverage probability for the interval estimates. The PRB is computed between the estimates from complete data (all studies reported the SDs) and the estimates from the imputed data and the incomplete data. To estimate the coverage probability, a random interval was computed based on estimates from the imputed data and incomplete data in each simulation run for each meta-analysis. The coverage probability for $\hat{\theta}_{true}$ is estimated as the proportion of the random intervals, out of 500 replications, which contains $\hat{\theta}_{true}$. The nominal level used was 0.95.

3. Results

3.1. Fixed effects model
The percentage relative bias (PRB) for the estimates based on the fixed effects model is tabulated in Table 1. Clearly the PRB in the SE of the estimates were much higher compared to those of the effect size. Furthermore, the PRB was generally smaller when the missing SDs is imputed, compared to the approach of excluding the studies with missing SDs. The trends of the PRB in the effect estimates for the different values of x% are illustrated in Figure 1. There was not much difference in the magnitude of the PRB when the missing SDs were imputed using the two techniques of imputations. The mean imputation performed only slightly better than the MI. On the other hand, the PRB in the SE of the estimates (Figure 2) were much higher when the studies with missing SDs were excluded compared to when the missing SDs were imputed (> 300 %). Furthermore, unlike the trends for the effect estimates, there were significant differences in the PRB for the two imputation techniques. The mean imputation performed far better than MI, particularly for the larger percentages of missing SDs (> 30 %). Additionally, the effects of x on PRB in SE was very notable, i.e. the bias increases with x%.

Table 1. Fixed effects model : Percentage relative bias in effect size and the SE of the effect size (n=60)

x %	N	% Relative bias of effect size			% Relative bias of SE of effect size		
		Not imputed	MI	Mean imp	Not imputed	MI	Mean imputed
10	10	0.16	0.02	0.05	-5.5	-2.40	-0.09

	20	-0.45	0.02	0.01	-5.4	-2.64	-0.11
	30	-0.21	0.03	-0.01	-5.4	-2.70	0.03
30	10	-0.14	0.03	-0.17	-19.7	-6.01	-0.21
	20	-0.01	-0.07	0.14	-19.5	-7.77	-0.20
	30	0.62	0.02	-0.06	-19.5	-7.90	-0.23
50	10	0.57	0.34	-0.01	-41.5	-10.5	-0.36
	20	-0.83	-0.06	0.02	-41.5	-12.3	-0.39
	30	-0.82	0.08	0.06	-41.3	-12.6	-0.27

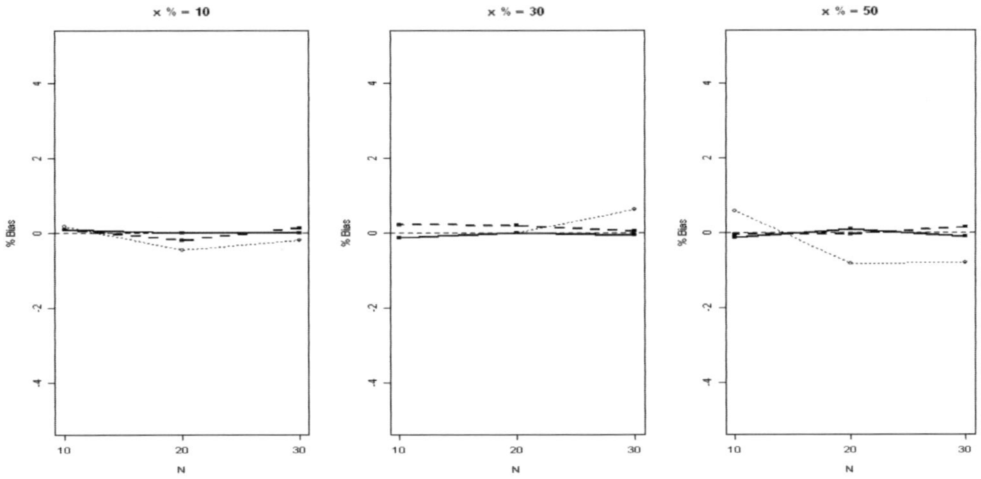

Figure 1. Fixed effects model : percentage relative bias in effect size (*n* = *60*). Legend : thin dotted-line - studies with missing SDs omitted ; thick break-line - mean imputation ; thick solid-line - multiple imputation.

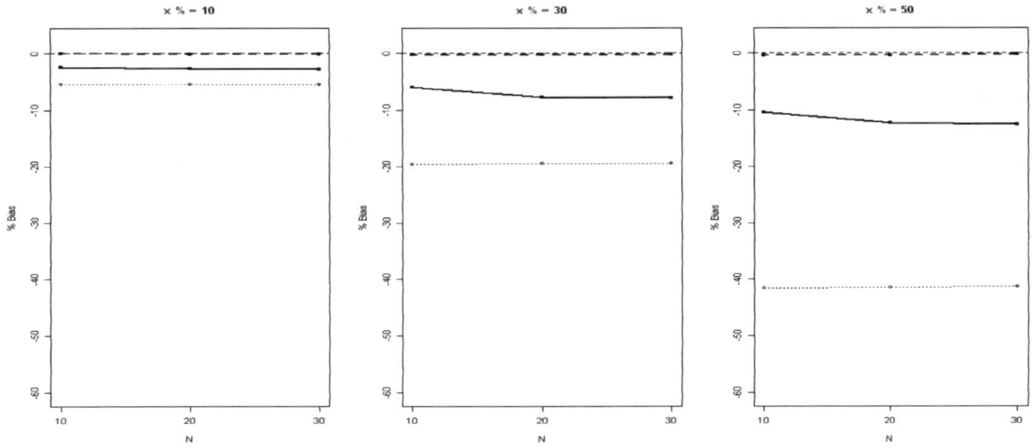

Figure 2. Fixed effect model : percentage relative bias in the SE of the effect size (*n* = *60*). Legend: thin dotted-line - studies with missing SDs omitted ; thick break-line - mean imputation ; thick solid-line - multiple imputation.

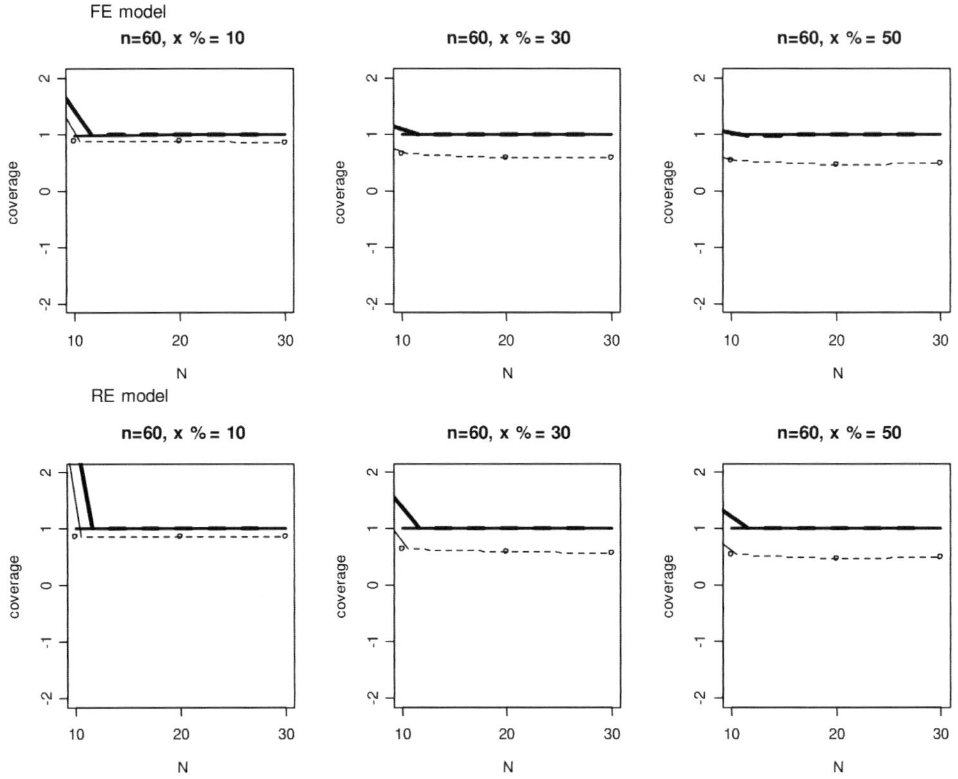

Figure 3. The coverage probability for the true effect sizes based on fixed and random effects model.

The estimated coverage probability for $\hat{\theta}_{true}$ is illustrated in Figure 3. The coverage probability from both the mean imputed and MI data were above the nominal value for all the meta-analyses, while in incomplete data, the coverage were mostly well below the nominal level.

3.2. Random effects model

The PRB for the estimates based on the random effect model are presented in Table 2. Consistent with the results for the fixed effect model, the PRB were much smaller in the effect sizes compared to those in the SE of the effect size. Additionally the PRBs were much larger if the studies with missing SDs were excluded. However there were no notable differences in the magnitude of the PRB in the effect size using either the mean imputation or the MI (Figure 4).

As in the case for the fixed effect model, for estimates of effect size, the percentage of missing SDs and the sample sizes did not have much effect on the magnitude of the relative bias when imputation was used. By contrast, when studies with missing SDs were excluded, the PRB in the SE of the effect size increased significantly with increasing x%. In this case, the PRB increased up to 40% when half of the studies were excluded. Imputing the missing SDs appears to recover most of the information, as illustrated in Figure 5. The relative biases were all very close to zero for both techniques of imputation. This was different from the results obtained from estimates based on the fixed effect model, where mean imputation seemed to be more superior in recovering the information on the SE of the estimates compared to the MI imputation. Additionally, when the SDs were imputed, the study sizes *n* and the percentage of missing SDs, x %, had little effect on the bias in the SE.

Table 2. Random effect model : percentage relative bias in the effect size and the SE of the effect size

X %	N	% Relative bias of effect size			% Relative bias of SE of effect size		
		Not imputed	MI	Mean imp	Not imputed	MI	Mean imputed
10	10	-0.19	0.012	0.007	-4.9	0.12	0.01
	20	0.30	-0.003	0.003	-5.4	0.01	0.01
	30	-0.29	0.0001	0.0002	-5.4	0.03	0.02
30	10	-0.13	-0.022	-0.007	-17.5	0.01	0.09
	20	0.54	-0.008	0.001	-19.0	0.01	0.02
	30	-0.83	-0.0001	-0.004	-19.1	0.02	0.02
50	10	-0.25	0.082	0.002	-35.8	0.03	0.40
	20	0.60	0.005	-0.005	-39.6	0.04	0.12
	30	-0.46	0.002	-0.006	-39.7	0.09	0.08

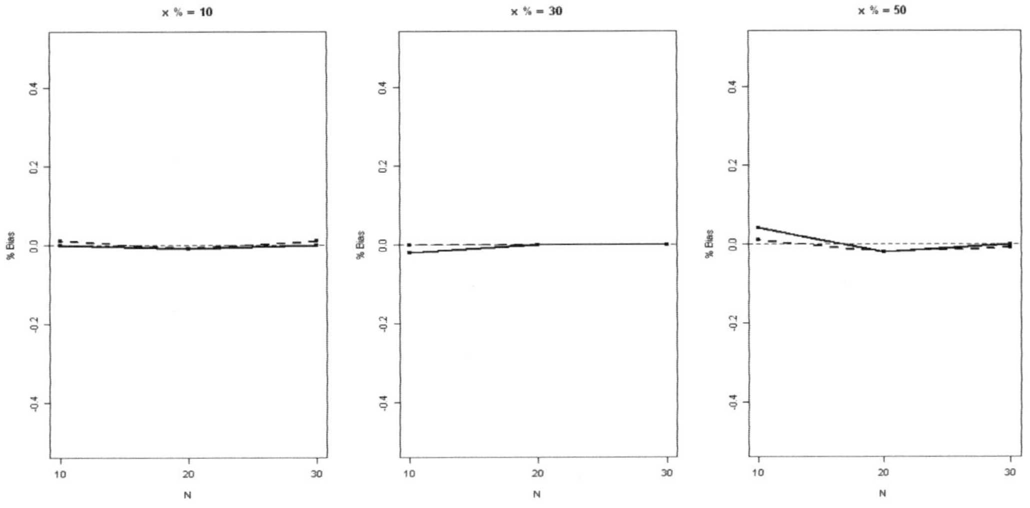

Figure 4. Random effects model : percentage relative bias in effect size *(n = 60)*.
Legend : thick break-line - mean imputation ; thick solid-line - multiple imputation.

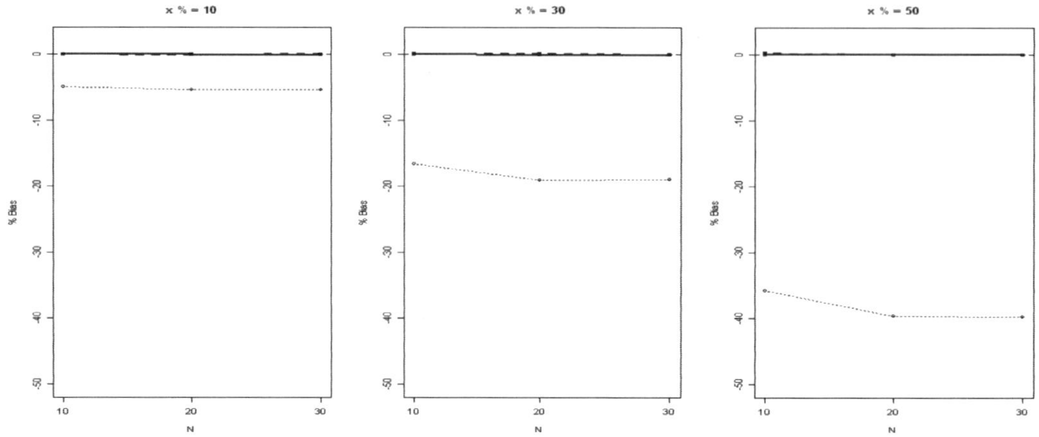

Figure 5. Random effects model : percentage relative bias in SE of the effect size *(n=60)*.
Legend : thin dotted-line - studies with missing SDs omitted ; thick break-line - mean imputation ; thick solid-line - multiple imputation.

4. Discussion

This paper investigates the influence of the imputation techniques and the choice of a meta-analysis model on the estimates of meta-analysis parameters—namely, the treatment effect size and the corresponding SE—in continuous response data with missing SDs in some of the studies. Comparisons in PRBs and the coverage probability were made for estimates based on the random and fixed effect model for the data imputed using the mean imputation and MI. The main conclusions had drawn from this project support many of the findings in previous literature. We have additionally illustrated that irrespective of whether the estimates of overall effect size is based on the fixed or random effect model, imputation is a good approach in handling the problem of missing study-level SDs. The PRB produced using this approach is relatively small compared to the studies with missing SDs. In fact, it was shown that the expected bias in this case will tend to be zero if the SDs across the studies is assumed to be completely homogeneous [10]. Imputation is therefore always recommended to estimate the SE of the effect size in data with missing SDs, as other serious bias may be introduced. The results show that if the random effect model is used to estimate the SE of the estimate for data with some missing SDs, both the MI and mean imputation will give equally good estimates (no difference in PRB; $p < 0.337$). On the other hand, if the estimate of the SE is based on the fixed effect model, the techniques of imputation adopted will have some impact on the PRB introduced into the estimate of the SE. In this particular study, it was observed that if the fixed effect model is used, then mean imputation is expected to produce smaller PRB compared to those using the MI (difference; $p < 0.001$). These results are clearly illustrated in Figure 2.

The results also showed that coverage probability for estimates based on excluded studies are mostly well below the nominal level despite a large percentage of overestimation in the SE. On the other hand, the coverage is mostly above nominal level for the data with imputed SDs due to overestimation of the effect sizes. This suggests that the bias in the effects size has more influence on the coverage probability compared to the bias in the SE of the estimates. For estimates based on the fixed effect model, high coverage level using both the MI and mean imputed data suggests that the intervals are too wide. Despite a small bias in the effect sizes (from 0.01% – 1.7 %), the intervals are wider due to overestimation in the SE effect sizes. In the incomplete data, low coverage is mainly due to underestimation in the effect sizes. Although the SEs in this data are overestimated and much larger than those in imputed data, their influences are relatively small compared to those of the effect sizes. Similarly, for the random effect model, the coverage probability is consistently above the nominal value for both methods of imputation. This is expected, as the intervals are generally wider for the estimates using the random effect model due to incorporation of the between-study variances. For MI, the intervals are even wider, with the inclusion of additional uncertainty due to imputation in the estimate of the variances. Generally the coverage is more affected by the bias in effect size itself than the precision. We found no superior performance using the two techniques of imputation for the two models with respect to coverage probability. Both models generated wider intervals than the nominal level. Thus imputation in general may produce intervals which are on average too wide. Nonetheless, this approach is better than exclusion of studies with missing SDs, which produces coverage that is too low.

When selecting the imputation technique to compute missing SDs, an analyst should also consider the type of meta-analysis model that the estimate is based on in order to minimise the bias. The random effect model responds more robustly to the type of imputations used. Additionally, in the random effect model, both the MI and mean imputation produced smaller PRB compared to those from the fixed effect model.

Although the random effects model appears to be a safer choice, there are some concerns regarding its general application in practice, such as the assumptions of normally distributed random effects that pose problems with both validity and in our ability to check the validity for meta-analyses based on small number of studies [11].

In this study, the simulation of data was carried out under the assumption that the SDs is homogeneous across the studies. Further work should consider this limitation and that the extension of analyses is also possible on the effects of other imputation techniques, such as imputation by regression or the hot deck. These analyses are not intended to provide a specific guide for the model and imputation techniques to be utilized, but to investigate their influences on the estimates of meta-analysis parameters. However, we may suggest that when there are missing SDs, an analyst is therefore advised to exercise caution in choosing the technique of imputation which is best suited for the type of meta-analysis model used. This study shows that if the fixed effect model is used, then mean imputation is recommended. The random effect model is more vigorous and either the mean or MI technique is suitable.

References

[1] Brockwell S.E. and Gordon I.R., 2001. "A comparison of statistical method for meta-analysis," Stat Med, **20** , 825-840.

[2] Whitehead A., 2002. *Meta-analysis of controlled clinical trials*, London : John Wiley.

[3] Rothstein H.R., Sutton A.J., Borenstein M., 2005. *Publication bias in meta-analysis; prevention, assessment and adjustments.* London : John Wiley & Sons.

[4] Wiebe N., Vandermeer B., Platt R.W., Klassen T.P., Moher D., Barrowman N.J., 2006. "A systematic review identifies a lack of standardization in methods for handling missing variance data," J. Clin. Epidemiol, **59,** 342-353.

[5] Furukawa TA., Barbui C., Cipriani A., Brambilla P., Watanabe N., 2006. "Imputing missing standard deviations in meta analyses can provide accurate results," J. Clin. Epidemiol, **59** 7-10

[6] Idris, N.R.N., Robertson C., 2009. "The effects of imputing the missing standard deviations on the standard error of the meta-analysis estimates," Commun Stat – Simul C, **38,** 513-526

[7] Thiessen, P.H., Barrowman, N., Garg, A.X., 2007. "Imputing variance estimates do not alter the conclusions of a meta-analysis with continuous outcomes: a case study of changes in renal function after living kidney donation," J. Clin. Epidemiol. **60**: 3, 228-240.

[8] Little, J.A. and Rubin, D.B., 1987. *Statistical analysis with missing data.* London:John Wiley.

[9] Marshall, A., Altman, D.G., Holder, R.L., 2010. "Comparison of imputation methods for handling missing covariate data when fitting a Cox proportional hazards model: a resampling study," *BMC Medical Research Methodology*,**10** 112.

[10] Idris N.R., 2006. *Estimating meta-analysis parameters in non-standard data.* PhD Thesis, University of Strathclyde.

[11] Sutton, A.J., Abrams, K.R., Jones, D.R., Sheldon, T.A., Song, F., 2000. "Missing data," In : *Methods for meta-analysis in medical research*, New York : John Wiley, 199-204.

An investigation of the relationship between extrapolation parameters and approximate solutions in solving fuzzy linear systems using refinement of jacobi over relaxation method

Lazim Abdullah[1] and Nurhakimah Ab Rahman[2]
Department of Mathematics, Faculty of Science and Technology
Universiti Malaysia Terengganu, 21030 Kuala Terengganu, Terengganu.

E-mail: lazim_m@umt.edu.my[1], hamikahrun@yahoo.com[2]

Abstract. Extrapolation parameters in Refinement of Jacobi over Relation method in solving fuzzy linear systems were hypothesized to influence the rate of convergence. This paper aims to investigate the relationship between extrapolation parameters and approximate solutions of fuzzy linear systems when solved using Refinement of Jacobi over Relaxation method. The Algorithms of Refinement of Jacobi over Relaxation method are proposed based on the Jacobi method that specifically used in solving the system. A five by five fuzzy linear system is given to investigate the convergence to exact solution with three different values of extrapolation parameters. The numerical results show that there is a positive correlation between extrapolation parameter and convergence to exact solution. The three extrapolation parameters suggest that convergence to exact solution can be increased in line with the increase in the values of extrapolation parameters.

1.Introduction

One of the most important discussions in applied mathematics is system of linear equations. The systems are germane for modelling and solving in many branches of knowledge areas such as physics, engineering, health sciences, economics, finance and even social sciences. In most of the applications, parameters of the system are represented by crisp. However, there are cases where parameters represented by fuzzy numbers. This situation is true especially when estimation of the system coefficients is imprecise and only some vague knowledge about the actual value of the parameters is available. It has been suggested that the system may be convenient to represent some or all of them with fuzzy numbers. The concept of fuzzy numbers and arithmetic operations with these numbers were first introduced and investigated by Zimmermann [11]. Fuzzy numbers and its arithmetic operations can be used to generalize crisp mathematical concept to fuzzy sets [6].

Fuzzy numbers were further developed in 1990s when Buckley and Qu [3] and Buckley [4] investigated the system and subsequently Buckley [5] proposed a general model for solving a fuzzy $n \times n$ linear system whose coefficient matrix is crisp and the right-hand side column is an arbitrary fuzzy number vector. They used the parametric form of fuzzy numbers and replaced the original fuzzy $n \times n$ linear system by a crisp $2n \times 2n$ linear system. Years after that, there were many researchers proposed various methods to solve fuzzy linear system of equations. For example, Friedman et al. [9] developed a method for solving a fuzzy linear system, while Asady et al. [2] solved rectangular fuzzy linear system of equations based on matrix decomposition method. They dealt with Householder process to obtain a QR-decomposition for the coefficient matrix of the extended linear system of the fuzzy linear system equation. Another decomposition method that is Adomian decomposition method was used by Matinfar et al. [10] to solve fuzzy linear equation. Hence it is immensely important to further develop numerical procedures that would appropriately treat general fuzzy linear systems and solve them.

Another class of methods for solving fuzzy linear systems is iterative methods. The Jacobi method for solving fuzzy linear system was introduced by Allahviranloo [1] for the first time. Dafchahi [7] builds a refinement to the Jacobi method and proposed the Refinement of Jacobi (RJ) method for solution of linear system equations. This method is presented to increase the rate of convergence of Jacobi. Besides the Jacobi method, there are some discussions about others iterative methods. They are the Jacobi over Relaxation (JOR) and the Successive over Relaxation (SOR) methods [8]. What is uniquely in the JOR and the SOR methods is that, there is an extrapolation parameter (ω) in their formulas. The JOR and SOR are the two most popular methods that used an extrapolation parameter. Extrapolation parameter used in both methods is presented by ω where $0 \le \omega \le 1$.Note that the value of extrapolation parameter influences the convergence of the system. Therefore, it motivates to extend the RJ method to be the Refinement of Jacobi over Relaxation (RJOR) method. The extension is done by adding an extrapolation parameter into the formula of RJ method. However, the convergence of the system depends on the extrapolation parameter. It was hypothesized that when smaller value of ω is used, the system converges very slowly. Therefore, this paper investigates the relationship between extrapolation parameters and convergence to the exact solution when fuzzy linear system is solved using Refinement of Jacobi over Relaxation method.

2. Preliminaries

In this section, some necessary backgrounds and notions of fuzzy linear systems, Jacobi over Relaxation, Successive over Relaxation, and Refinement of Jacobi method are reviewed.

Definition 2.1: Fuzzy linear systems

Friedman et al. [9] have provided definitions of fuzzy linear systems.

$$\begin{aligned} a_{11}x_1 + a_{12}x_2 + \dots + a_{1n}x_n &= y_1 \\ a_{21}x_1 + a_{22}x_2 + \dots + a_{2n}x_n &= y_2 \\ \vdots \quad \vdots \quad\quad \vdots \quad &\quad \vdots \\ a_{n1}x_1 + a_{n2}x_2 + \dots + a_{nn}x_n &= y_n \end{aligned} \tag{2.1}$$

The $n \times n$ matrix form of the above equations is $AX = Y$, where the coefficient matrix $A = (a_{ij}), 1 \le i, j \le n$ is a crisp $n \times n$ matrix and $y_i \in E^1, 1 \le i \le n$. This system is called a fuzzy linear system (FLS).

Definition 2.2: Jacobi over relaxation (JOR) method

Based on Dehgan and Hashemi [8], Jacobi over Relaxation method is an extrapolated of Jacobi method.

$$\begin{bmatrix} \underline{X} \\ \overline{X} \end{bmatrix} = \begin{bmatrix} \omega D_1^{-1}\underline{Y} + (I_n - \omega D_1^{-1}B)\underline{X} + \omega D_1^{-1}C\overline{X} \\ \omega D_1^{-1}\overline{Y} + (I_n - \omega D_1^{-1}B)\overline{X} + \omega D_1^{-1}C\underline{X} \end{bmatrix}$$

So the JOR iterative method will be

$$\begin{cases} \underline{X^{(m+1)}} = \omega D_1^{-1}\underline{Y} - \omega D_1^{-1}\left[\left(1 - \dfrac{1}{\omega}\right)D_1 + L_1 + U_1\right]\underline{X^{(m)}} + \omega D_1^{-1}C\overline{X^{(m)}} \\ \overline{X^{(m+1)}} = \omega D_1^{-1}\overline{Y} - \omega D_1^{-1}\left[\left(1 - \dfrac{1}{\omega}\right)D_1 + L_1 + U_1\right]\overline{X^{(m)}} + \omega D_1^{-1}C\underline{X^{(m)}} \end{cases}$$

Hence we have the JOR method in matrix form:

$$X^{(m+1)} = D^{-1}(D-\omega S)X^{(m)} + \omega D^{-1}Y \quad (2.2)$$

where D is a diagonal matrix, S is a singular matrix and ω is an extrapolation parameter.

Definition 2.3 Successive over relaxation (SOR) method
There are forward and backward Successive over Relaxation method [8].
The forward SOR iterative method will be

$$\begin{cases} \underline{X}^{(m+1)} = \omega(D_1+\omega L_1)^{-1}\underline{Y} + (D_1+\omega L_1)^{-1}[(1-\omega)D_1 - \omega U_1]\underline{X}^{(m)} - \\ \quad \omega(D_1+\omega L_1)^{-1}C\overline{X}^{(m)} \\ \\ \overline{X}^{(m+1)} = -\omega^2(D_1+\omega L_1)^{-1}C(D_1+\omega L_1)^{-1}\underline{Y} + \omega(D_1+\omega L_1)^{-1}\overline{Y} + \\ \quad (D_1+\omega L_1)^{-1}\left[(1-\omega)D_1 - \omega U_1 + \omega^2 C(D_1+\omega L_1)^{-1}C\right]\overline{X}^{(m)} - \\ \quad \omega(D_1+\omega L_1)^{-1}C(D_1+\omega L_1)^{-1}[(1-\omega)D_1 - \omega U_1]\underline{X}^{(m)} \end{cases} \quad (2.3)$$

Similarly, the backward SOR iterative method will be

$$\begin{cases} \underline{X}^{(m+1)} = \omega(D_1+\omega U_1)^{-1}\underline{Y} + (D_1+\omega U_1)^{-1}[(1-\omega)D_1 - \omega L_1]\underline{X}^{(m)} - \\ \quad \omega(D_1+\omega U_1)^{-1}C\overline{X}^{(m)} \\ \\ \overline{X}^{(m+1)} = -\omega^2(D_1+\omega U_1)^{-1}C(D_1+\omega U_1)^{-1}\underline{Y} + \omega(D_1+\omega U_1)^{-1}\overline{Y} + \\ \quad (D_1+\omega U_1)^{-1}\left[(1-\omega)D_1 - \omega L_1 + \omega^2 C(D_1+\omega U_1)^{-1}C\right]\overline{X}^{(m)} - \\ \quad \omega(D_1+\omega U_1)^{-1}C(D_1+\omega U_1)^{-1}[(1-\omega)D_1 - \omega L_1]\underline{X}^{(m)} \end{cases} \quad (2.4)$$

Based on the above equations, D is diagonal for the system, L is strict lower part of the system, U is strict upper part of the systems, C is absolute value of the negative entries of matrix A and X are the variables of the system, where they are lower variables and upper variables.

Definition 2.4: Refinement of jacobi method
Based on the previous study by Buckley [4], stated that Refinement of Jacobi method is an extended of Jacobi method.

$$\begin{aligned} AX &= b \\ (L+D+U)X &= b \\ DX &= -(L+U)X + b \\ DX &= (D-A)X + b \\ DX &= DX + (b-AX) \\ X &= X + D^{-1}(b-AX) \end{aligned}$$

From above form, the iterative refinement of formula in matrix form is obtained.

$$\overline{X}^{(k+1)} = X^{(k+1)} + D^{-1}\left(b - AX^{(k+1)}\right) \quad (2.5)$$

From Jacobi formula, Refinement of Jacobi method in matrix form is proposed as follows:

$$X^{(k+1)} = \left(D^{-1}(L+U)\right)^2 X^{(k)} + \left(I - D^{-1}(L+U)\right)D^{-1}b \quad (2.6)$$

Hence, formula (2.6) can be used to solve fuzzy linear systems. Based on the formula, D is diagonal for the system, L is strict lower part of the system, U is strict upper part of the systems, b is fuzzy vector for the system which including fuzzy lower bound and fuzzy upper bound, and X are the variables of the system, where they are also lower variables and upper variables.

3. Refinement of jacobi over relaxation method and extrapolation parameters

In this study, Refinement of Jacobi over Relaxation method is proposed to solve fuzzy linear systems. This method is chosen because Dehgan and Hashemi [8] and Allahviranloo [1] find out that Jacobi over Relaxation and Successive over Relaxation method are iterative methods in fuzzy linear systems. This motivates us to make an extension of Refinement of Jacobi method to Refinement of Jacobi over Relaxation method as an iterative method. Previous studies by Allahviranloo [1] have shown that Successive over Relaxation method is an extension of Gauss Seidel method. Similarly, Jacobi over Relaxation method is an extension of Jacobi method.

Based on the formula (2.2), there is an extrapolation parameter in formula of Jacobi over Relaxation method. Therefore, Refinement of Jacobi method is extending by adding an extrapolation parameter in the formula (2.6). Hence, we have the Refinement of Jacobi over Relaxation method in matrix form:

$$X^{(k+1)} = \left(D^{-1}(L+\omega U)\right)^2 X^{(k)} + \left(I - D^{-1}(L+U)\right)\omega D^{-1} b \tag{3.1}$$

The above formula also can be written as

$$X^{(k+1)} = \left(D^{-1}(L+\omega U)\right)^2 X^{(k)} + \left(I - D^{-1}(L+\omega U)\right)\omega D^{-1} b \tag{3.2}$$

It is interesting to note that when the both formulas are used to solve fuzzy linear systems, the solution is same even though there are only two extrapolation parameters in formula (3.1) while three extrapolation parameters in formula (3.2). A possible explanation for this might be related to $\omega = 1$. However in this study, three different extrapolation parameters will be used in Refinement of Jacobi over Relaxation method to see how of convergence effects in the systems when $\omega \neq 1.0$. Buckley [5] uses $\omega = 0.87$ for SOR method and $\omega = 0.90$ for JOR method in his numerical examples. Both extrapolation parameters are close to $\omega = 1.0$. So, in order to see how convergence of systems when $0 \leq \omega \leq 1$ are used, we choose a middle point between 0 and 1, and two more points are at two ends between 0 and 1. Therefore, three extrapolation parameters will be used are $\omega = 0.25, 0.5, \text{and} 0.9$.

In order to understand clearly how to get the solution fuzzy linear system by Refinement of Jacobi over Relaxation method, an algorithm is summarized as follows:

Step 1: Determine L, U, D^{-1} and b of the systems.

Step 2: Multiply extrapolation parameter ω with U.

Step 3: Sum up L and ωU and multiply them with the inverse of D.

Step 4: Evaluate $((L+\omega U)D^{\ 1})^2$.

Step 5: Identify the number of variables (X) of the systems. Make sure they are including lower bound and upper bound variables.

Step 6: Multiply the variables with the $((L+\omega U)D^{-1})^2$.

Step 7: Identify an identity matrix for the systems and then subtract it to $((L+U)D^{\ 1})$.

Step 8: Multiply b with the inverse of diagonal matrix of the systems and extrapolation parameter.

Step 9: Multiply $\omega D^{\ 1}b$ with the $I \quad (L+U)D^{\ 1}$.

Step 10: Substitute all the terms to the formula (3.1).

Step 11: When $\frac{x_i^{(k+1)} \quad x_i(k)}{x_i^{(k+1)}} < \varepsilon$, stop the calculation. Note that, the decision to stop the iterations were based on the criterion.

4. Numerical applications

To illustrate the proposed method and demonstrate the relationship, the 5×5 system retrieved from Dehgan and Hashemi [8] is presented.

Example

Consider the 5×5 fuzzy system

$$
\begin{aligned}
&8x_1+2x_2+x_3-3x_5=(r,2-r)\\
&-2x_1+5x_2+x_3-x_4+x_5=(4+r,7-2r)\\
&x_1-x_2+5x_3+x_4+x_5=(1+2r,6-3r)\\
&-x_3+4x_4+2x_5=(1+r,3-r)\\
&x_1-2x_2+3x_5=(3r,6-3r)
\end{aligned}
$$

The proposed algorithm (refer to Section 3) is implemented to solve the system. Three extrapolation parameters $\omega=0.25, 0.5$, and 0.9 are tested. The solutions for each extrapolation parameters are represented graphically in triangular fuzzy numbers.

The exact solution for the above system is given as

$$
\begin{aligned}
&x_1=(0.729\text{-}0.331{*}r,\ 0.044+0.354{*}r)\\
&x_2=(0.614+0.166{*}r,\ 1.077-0.297{*}r)\\
&x_3=(0.126+0.291{*}r,\ 0.918-0.501{*}r)\\
&x_4=(0.242-0.331{*}r,\ -0.416+0.326{*}r)\\
&x_5=(0.475+0.912{*}r,\ 2.395-1.007{*}r)
\end{aligned}
$$

The exact solutions can also be depicted graphically. It is shown in Figure 1.

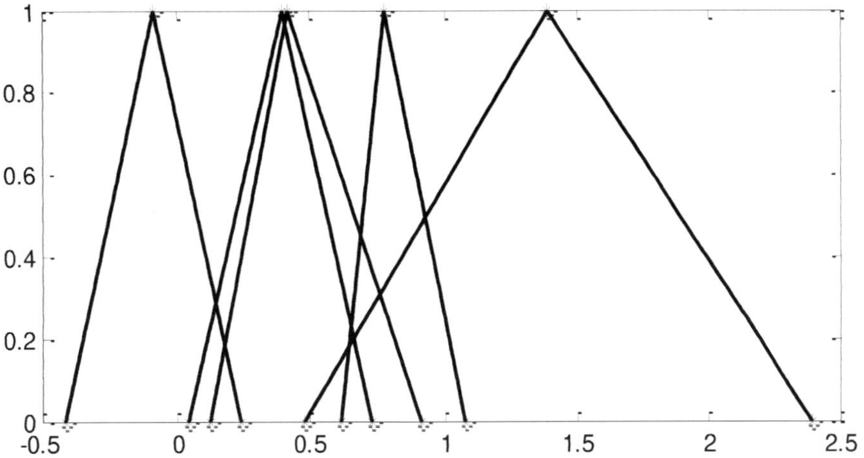

Figure 1: Exact solution in triangular fuzzy numbers.

The tests are extended to obtain approximation solutions using the algorithm for three different extrapolation parameters. The solutions are summarized in the Table 1-3.

Table 1: Approximation solutions and triangular fuzzy numbers for $\omega = 0.25$

Extrapolation parameter	Solutions
$\omega = 0.25$	$x_1 = (0.153 - 0.095 * r,\ -0.074 + 0.121 * r)$ $x_2 = (0.260 - 0.039 * r,\ 0.200 - 0.030 * r)$ $x_3 = (0.105 + 0.013 * r,\ 0.163 - 0.081 * r)$ $x_4 = (0.142 - 0.105 * r,\ -0.085 + 0.100 * r)$ $x_5 = (0.231 + 0.172 * r,\ 0.571 - 0.223 * r)$ 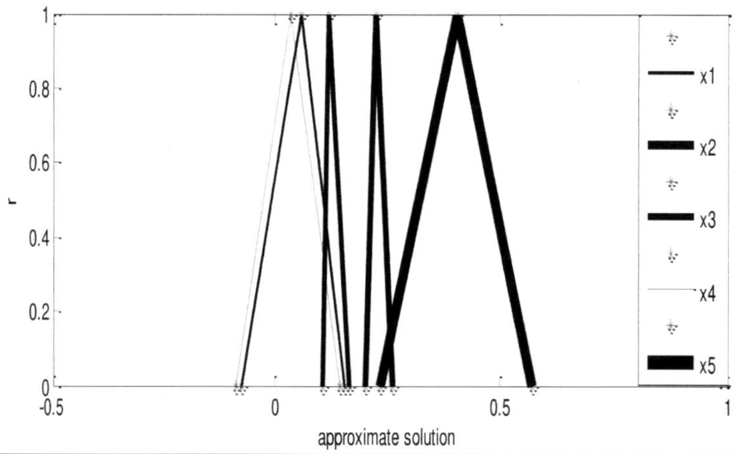

Table 2: Approximation solutions and triangular fuzzy numbers for $\omega = 0.5$

Extrapolation parameter	Solutions
$\omega = 0.5$	$x_1 = (0.355 - 0.203 * r,\ -0.128 + 0.245 * r)$ $x_2 = (0.509 - 0.057 * r,\ 0.403 - 0.027 * r)$ $x_3 = (0.199 + 0.037 * r,\ 0.324 - 0.165 * r)$ $x_4 = (0.276 - 0.219 * r,\ -0.221 + 0.211 * r)$ $x_5 = (0.440 + 0.350 * r,\ 1.119 - 0.440 * r)$ 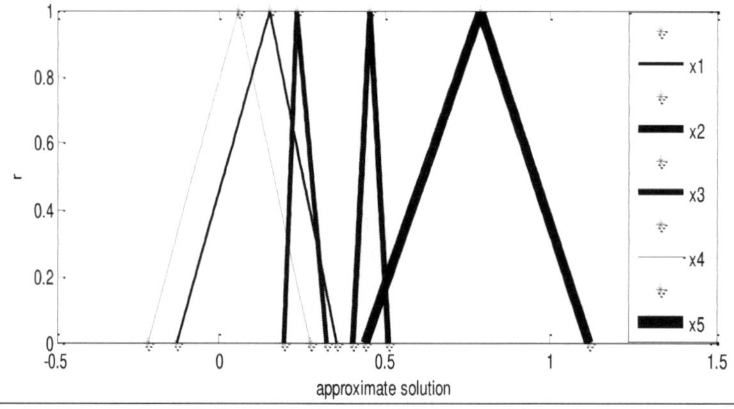

Table 3: Approximation solutions and triangular fuzzy numbers for $\omega = 0.9$

Extrapolation parameter	Solutions
$\omega = 0.9$	$x_1 = (0.719 - 0.354 * r,\ -0.079 + 0.384 * r)$ $x_2 = (0.717 + 0.044 * r,\ 0.842 - 0.175 * r)$ $x_3 = (0.232 + 0.175 * r,\ 0.707 - 0.377 * r)$ $x_4 = (0.346 - 0.362 * r,\ -0.447 + 0.353 * r)$ $x_5 = (0.588 + 0.729 * r,\ 2.050 - 0.835 * r)$ 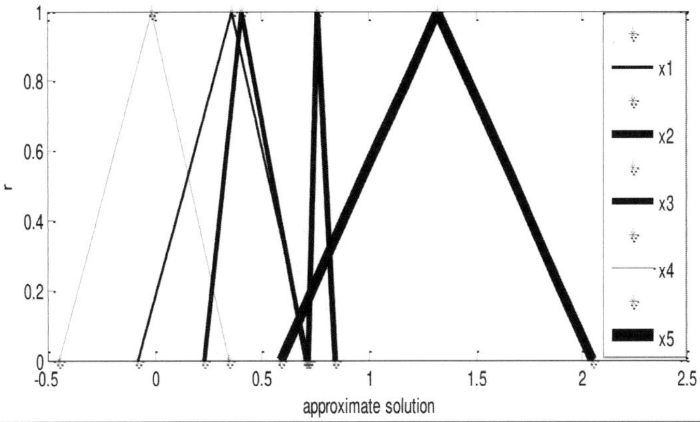

From these results, it can be seen that there are relationships between an exact solutions and the approximation solutions when three different extrapolation parameters are used. When extrapolation parameter used is small, the rate of convergence is very low. On the other hand, when bigger extrapolation parameter is used, the rate of convergence is increased. The system seems converged properly to the exact solution as the values of extrapolation parameters increases. The approximate solutions indicate the importance of extrapolation parameters in solving the system using Refinement of Jacobi over Relaxation. The exact solution possible is achieved when extrapolation parameter is 1. This investigation concludes that the rate of convergence is influenced by the values of extrapolation parameters in solving fuzzy linear systems with Refinement of Jacobi over Relaxation method.

5. Conclusions

This paper was presented to investigate the effect of extrapolation parameters to convergence rate when the fuzzy linear system is solved using Refinement of Jacobi over Relaxation method. Using the numerical applications, it can be concluded that there was a relationship between extrapolation parameter and approximate solutions. Convergence to the exact solution seems to be improved by increasing the values of extrapolation parameters. It would suggest that there is inefficiently to choose other extrapolation parameters other than $\omega = 1$. The most probable explanation is the Refinement of Jacobi over Relaxation method is reduced to Refinement of Jacobi method when $\omega = 1$. Therefore, fuzzy linear systems may be solved by this proposed method and converge to the exact solution when the appropriate value of extrapolation parameter is selected.

6. References

[1] Allahviranloo T 2005 The Adomian decomposition method for fuzzy system of linear equations *Appl. Math. Comput.* **163**, pp. 553–63

[2] Asady B, Abbasbandy S and, Alavi M 2005 Fuzzy general linear systems *Appl. Math. and Comput.* **169** 34-40

[3] Buckley J J and Qu Y 1991 Solving systems of linear fuzzy equations *Fuzzy Sets and Systems* **43** 33–43

[4] Buckley J J 1992 Solving fuzzy equations in economics and finance *Fuzzy Sets and Systems* **48** 289–96

[5] Buckley J J 1992 Solving fuzzy equations *Fuzzy Sets and Systems* **50** 1–14

[6] Chang S L and Zadeh L A 1972 On fuzzy mapping and control *IEEE Trans., Syst. Man Cyb.* **2**

[7] Dafchahi F N 2008 A new refinement of Jacobi method for solution of linear system equations $AX = b$ *Int.J.Contemp.Math.Sciences* **3** **17** 819-27

[8] Dehgan M. and Hashemi B 2006 Iterative solution of fuzzy linear systems, *Applied Mathematics and Computation* **175**, 645-74

[9] Friedman M, Ming M and, Kandel A 1998 Fuzzy linear systems *Fuzzy Sets and Systems* **96**, 201-09

[10] Matinfar M, Nasseri S H and Sohrabi M 2008 Solving fuzzy linear system of equations by using householder decomposition method *Appl. Math. Sci.* **2** **52** 2569-75

[11] Zimmermann H J 1996 *The Fuzzy Set Theory and its Applications* (Kluwer, Dordrecht)

The Finite and Moving Order Multinomial Universal Portfolio

Choon Peng Tan and Sook Theng Pang

Faculty of Engineering and Science,
Department of Mathematical and Actuarial Sciences,
Universiti Tunku Abdul Rahman, Kuala Lumpur Campus,
Jalan Genting Kelang, 53300 Kuala Lumpur, Malaysia

E-mail: tancp@utar.edu.my,pangst@utar.edu.my

Abstract. An upper bound for the ratio of wealths of the best constant -rebalanced portfolio to that of the multinomial universal portfolio is derived. The finite- order multinomial universal portfolios can reduce the implementation time and computer-memory requirements for computation. The improved performance of the finite-order portfolios on some selected local stock-price data sets is observed.

1. Introduction

The uniform universal portfolio[1] and its genenalization, the Dirichlet universal portfolio[2] have been studied by Cover and Ordentlich. One of the advantages of using a universal portfolio is that it does not depend on the underlying probability distribution of the stock prices or the stcohastic model of the stock prices. A measure of its performance is the ratio of wealths achieved by the best constant rebalanced portfolio to the universal portfolio as a function of the number of trading days. In [2] it was shown that this ratio is upper bounded by a polynomial in the number of trading days for the Dirichlet$(\alpha_1, \alpha_2, ..., \alpha_m)$ universal portfolio where $(\alpha_1, \alpha_2, ..., \alpha_m)$ is (1 , 1, ..., 1) or $(\frac{1}{2}, \frac{1}{2}, ..., \frac{1}{2})$. The general polynomial bound was obtained by Tan [3] for any positive vector $(\alpha_1, \alpha_2, ..., \alpha_m)$. The conjectured asymptotic behavior of the ratio of wealths for the Dirichlet universal portfolio was discussed in [4].

In this paper , the discrete-distribution equivalent of the Dirichlet universal portfolio, namely, the finite and moving order multinomial universal portfolios are introduced and studied. An upper bound for the ratio of wealths of the best-constant-rebalanced portfolio to that of the multinomial universal portfolio is established. To derive this upper bound, two inequalities in [2] and [5] are exploited.

2. Main Results

Let the random variables $Y_1, Y_2, ..., Y_m$ have a joint multinomial distribution with parameters $N, p_1, p_2, ..., p_{m-1}$ where $0 < p_i < 1$ for $i = 1, 2, ..., m-1$ and $0 < p_m = 1-p_1-p_2-.....-p_{m-1} < 1$; N is a positive integer bigger than m, which is the number of stocks in the market. The joint probability function $f(y_1, y_2, ..., y_m)$ is given by :

$$f(y_1, y_2, ..., y_m) = \begin{pmatrix} & N & & \\ y_1 & y_2 & .... & y_m \end{pmatrix} p_1^{y_1} p_2^{y_2} ....p_m^{y_m} \tag{1}$$

where the multinomial coefficient $\begin{pmatrix} & N & & \\ y_1 & y_2 & .... & y_m \end{pmatrix}$
is defined by :

$$\begin{pmatrix} & N & & \\ y_1 & y_2 & .... & y_m \end{pmatrix} = \frac{N!}{y_1!y_2!....y_m!} \tag{2}$$

and $y_i = 0, 1, 2, ..., N$ for $i = 1, 2, ..., m$ subject to $y_1 + y_2 + ... + y_m = N$. The joint moment generating function $M(\tau_1, \tau_2, ...\tau_m)$ of the multinomial distribution (1) is given by :

$$M(\tau_1, \tau_2, ...\tau_m) = (p_1 e^{\tau_1} + p_2 e^{\tau_2} + ... + p_m e^{\tau_m})^N \tag{3}$$

Let $M^{n_1,n_2,....,n_m}(\tau_1, \tau_2, ..., \tau_m)$ denote the partial derivative $\frac{\partial^{n_1+n_2+....n_m} M(\tau_1,\tau_2,....,\tau_m)}{\partial\tau_1^{n_1}\partial\tau_1^{n_2}....\partial\tau_m^{n_m}}$ and

$M^{n_1,n_2,....,n_m}(0, 0, ..., 0)$ will denote the value of the partial derivative evaluated at $\tau_1 = \tau_2 = ... = \tau_m = 0$. It is known that

$$E[Y_1^{n_1} Y_2^{n_2} .....Y_m^{n_m}] = M^{n_1,n_2,.n_m}(0, 0, ..., 0) \tag{4}$$

If n_k in (4) is increased by 1 , then we adopt the notation

$$E[Y_1^{n_1 \cdots} Y_k^{n_k+1} .....Y_m^{n_m}] = M^{n_1,n_2,..,n_k+1,..n_m}(0, 0, ..., 0) \tag{5}$$

for $k = 1, 2, ..., m$.

Let $\mathbf{x}_n = (x_{ni}) \geq 0$ be the nonnegative price-relative or asset-relative vector on the nth trading day for $n = 1, 2, ...,$, where the price-relative of a stock/asset is defined to be the ratio of the closing price to its opening price [2]. We adopt the notation $\mathbf{x}^n = (\mathbf{x}_1, \mathbf{x}_2, ..., \mathbf{x}_n)$ to denote a sequence of price-relative vectors from first to the nth trading day.

Let the domain D of the function $f(\mathbf{y})$given by (1) be :

$$D = \{\mathbf{y} : f(\mathbf{y}) > 0, \sum_{i=1}^{m} y_i = N, y_i = 0, 1, ..., N fori = 1, 2, ..., N\} \tag{6}$$

Where the vector $\mathbf{y} = (y_1, y_2, ..., y_m)$. Given a positive integer ν and $\mathbf{x}$, the multinomial universal portfolio of order ν is defined by the sequence of portfolio vectors $\hat{\mathbf{b}}_{n+1,k} = (b_{n+1,k})$ given by :

$$\hat{\mathbf{b}}_{n+1,k} = \frac{\sum_{\mathbf{y}\in D} y_k (\mathbf{y}^t\mathbf{x}_n)(\mathbf{y}^t\mathbf{x}_{n-1})...(\mathbf{y}^t\mathbf{x}_{n-(\nu-1)})f(\mathbf{y})}{N\sum_{\mathbf{y}\in D}(\mathbf{y}^t\mathbf{x}_n)(\mathbf{y}^t\mathbf{x}_{n-1})....(\mathbf{y}^t\mathbf{x}_{n-(\nu-1)})f(\mathbf{y})} \tag{7}$$

for $k = 1, 2, ..., m$ and $n = 1, 2....;$ $f(\mathbf{y})$ and D are given by (1) and (6) respectively. We shall omit D whenever the domain of summation is clear in the context.Note that (7) can be rewritten in the form

$$\hat{\mathbf{b}}_{n+1,k} = \zeta_{n,\nu}\{\sum_{i_1=1}^{m}\sum_{i_2=1}^{m} .....\sum_{i_\nu=1}^{m}(x_{n_{i_1}} x_{n-1,i_2}....x_{n-\nu+1,i_\nu}) \times E[Y_1^{n_1(k;\mathbf{i})} Y_2^{n_2(k;\mathbf{i})} .....Y_m^{n_m(k;\mathbf{i})}]\} \tag{8}$$

for $k = 1, 2, ..., m$ where the constant $\zeta_{n,\nu}$ is given by :

$$\zeta_{n,\nu} = \{N\sum_{i_1=1}^{m}\sum_{i_2=1}^{m} .....\sum_{i_\nu=1}^{m}(x_{n_{i_1}} x_{n-1,i_2}....x_{n-\nu+1,i_\nu}) \times E[Y_1^{n_1(\mathbf{i})} Y_2^{n_2(\mathbf{i})} .....Y_m^{n_m(\mathbf{i})}]\}^{-1}, \tag{9}$$

the vector $\mathbf{i} = (i_1, i_2, ..., i_\nu); 1 \leq i_j \leq m$ for $j = 1, 2, ..., m; n_j(k; \mathbf{i})$ is the number of $y_j^{'}$s in the product $(y_k y_{i_1} y_{i_2} .....y_{i_\nu})$; $n_j(\mathbf{i})$ is the number of $y_j^{'}$ s in the product$(y_{i_1} y_{i_2} ....y_{i_\nu})$; $\sum_{j=1}^{m} n_j(k; \mathbf{i}) = \nu + 1$ and $\sum_{j=1}^{m} n_j(\mathbf{i}) = \nu$. To see this, consider the numerator of $\hat{b}_{n+1,k}$ in (7):

$$\sum_{\mathbf{y}} y_k (\sum_{i_1=1}^{m} y_{i_1} x_{ni_1})(\sum_{i_2=1}^{m} y_{i_2} x_{n-1,i_2}) ....(\sum_{i_\nu=1}^{m} y_{i_\nu} x_{n-\nu+1,i_\nu}) f(\mathbf{y})$$

$$= \sum_{\mathbf{y}} y_k [\sum_{i_1=1}^{m} \sum_{i_2=1}^{m} ... \sum_{i_\nu=1}^{m} (y_{i_1} y_{i_2} ....y_{i_\nu})(x_{ni_1} x_{n-1,i_2} .....x_{n-\nu+1,i_\nu}] f(\mathbf{y})$$

$$= \sum_{i_1=1}^{m} \sum_{i_2=1}^{m} ... \sum_{i_\nu=1}^{m} (x_{ni_1} x_{n-1,i_2} .....x_{n-\nu+1,i_\nu}][\sum_{\mathbf{y}} (y_k y_{i_1} y_{i_2} ....y_{i_\nu}) f(\mathbf{y})]$$

which is the same as the numerator of (8) by counting the number of $y_1^{'}$s, $y_2^{'}$s.....$y_m^{'}$s in the product $(y_k y_{i_1} y_{i_2} ....y_{i_m})$. A similar derivation shows that the denominators of (7) and (8) are the same.

Startifng with an initial wealth of 1 unit, the wealth or capital $\hat{S}(\mathbf{x}^n)$ of the investor after n trading days using the universal portfolio $\{\hat{\mathbf{b}}_n\}$ is given by :

$$\hat{S}(\mathbf{x}^n) = \prod_{i=1}^{n} \hat{\mathbf{b}}_n^t \mathbf{x}_n \tag{10}$$

A constant-rebalanced portfolio is a portfolio $\mathbf{b} = (b_i)$ that is constant over the trading days and the wealth at the end of n trading days is

$$S(\mathbf{x}^n) = \prod_{i=1}^{n} \mathbf{b}^t \mathbf{x}_n \tag{11}$$

The best constant-rebalanced -portfolio wealth is

$$S^*(\mathbf{x}^n) = \prod_{i=1}^{n} \mathbf{b}^{*t} \mathbf{x}_n \tag{12}$$

where the best constant-rebalanced portfolio $\mathbf{b}^*$ maximizes $S(\mathbf{x}^n)$ in (11) over all portfolio vectors $\mathbf{b}$ is the simplex of portfolio vectors B given by :

$$B = \{\mathbf{b} : \sum_{i=1}^{m} b_i = 1, 0 \leq b_i \leq 1, i = 1, 2, ..., m\} \tag{13}$$

To derive an upper bound for $\frac{S_n^*(\mathbf{x}^n)}{\hat{S}(\mathbf{x}^n)}$, we need the following two lemmas. Lemma 1 is due to [2] and [5] whereas Lemma 2 is due to [2].

Lemma 1. If $\zeta_1, \zeta_2, ..., \zeta_l \geq 0$ and $\eta_1, \eta_2, ..., \eta_l \geq 0$, then

$$\min_j(\frac{\zeta_j}{\eta_j}) \leq \frac{\sum_{i=1}^{l} \zeta_i}{\sum_{i=1}^{l} \eta_i} \leq \max_j(\frac{\zeta_j}{\eta_j}). \tag{14}$$

Lemma 2. For $r = 1, 2, ..., m$, let $n_r(j^n)$ be the number of times r occur in the string of indices $j^n \in \{1, 2, ..., m\}^n$, where $j^n = \prod_{i=1}^{n} j_i$ and in the sequel, the letters j^n in $n_r(j^n)$ is suppressed for the sake of simplicity. Then for any constant rebalanced portfolio $\mathbf{b}_m$,

$$\prod_{i=1}^{n} b_{j_i} \leq \frac{\prod_{i=1}^{m} n_i^{n_i}}{\nu^\nu} \leq \frac{\prod_{i=1}^{m} \Gamma(n_i + \frac{1}{2})}{\Gamma(\nu + \frac{1}{2})\Gamma(\frac{1}{2})^{m-1}} \tag{15}$$

where $\sum_{j=1}^{m} n_j = \nu$ and $\Gamma(\alpha) = \int_0^\infty u^{\alpha-1} e^{-u} du$ is the gamma function for $\alpha > 0$.
We state the following result,

Proposition. Let $\hat{S}_n(\mathbf{x}^n)$ be the universal wealth due to the multinomial $(N, p_1, ..., p_{m-1})$ universal portfolio of order ν given by (8). Then the ratio of wealths $\frac{S_n^*(\mathbf{x}^n)}{\hat{S}_n(\mathbf{x}^n)}$ satisfies the following bound:

$$\frac{S_n^*(\mathbf{x}^n)}{\hat{S}_n(\mathbf{x}^n)} \leq \left[\frac{\Gamma(\nu + \frac{1}{2})}{\Gamma(\frac{1}{2})}\right]^{m-1} \left[\frac{N}{\min\limits_{\substack{k,n_1,....,n_m \\ \sum_r n-r=\nu}} \frac{M^{n_1,....,n_k+2,....,n_m}(0,0,...,0)}{M^{n_1,....,n_m}(0,...0)}}\right]^n \tag{16}$$

for $\nu \geq 3$ and for the sepecial cases $\nu = 1, 2$, replace $\{\frac{\Gamma(\nu+\frac{1}{2})}{\Gamma(\frac{1}{2})}\}^{m-1}$ in the bound (16) by $\{\frac{\Gamma(\frac{1}{2})}{\Gamma(\nu+\frac{1}{2})}\}$.
Proof. Using Lemma 1,

$$\begin{aligned} &\frac{\sum_{i_1=1}^{m}\sum_{i_2=1}^{m}.....\sum_{i_\nu=1}^{m} x_{ni_1}x_{n-1,i_2}......x_{n-\nu+1,i_\nu} E\{Y_1^{n_1(k;\mathbf{i})}....Y_m^{n_m(k;\mathbf{i})}\}}{\sum_{i_1=1}^{m}\sum_{i_2=1}^{m}.....\sum_{i_\nu=1}^{m} x_{ni_1}x_{n-1,i_2}......x_{n-\nu+1,i_\nu} E\{Y_1^{n_1(\mathbf{i})}....Y_m^{n_m(\mathbf{i})}\}} \\ &\geq q = \min_{\substack{k,n_1,...,n_m, \\ \sum_r n_r=\nu}} \left\{\frac{E\{Y_1^{n_1(k;\mathbf{i})}....Y_m^{n_m(k;\mathbf{i})}\}}{E\{Y_1^{n_1(\mathbf{i})}....Y_m^{n_m(\mathbf{i})}\}}\right\} \end{aligned} \tag{17}$$

From the definition of $n_j(k;\mathbf{i})$, we obtain equivalently,

$$q = \min_{\substack{k,n_1,...,n_m \\ \sum_r n_r=\nu}} \{\frac{E\{Y_1^{n_1}....Y_k^{n_k+1}....Y_m^{n_m}\}}{E\{Y_1^{n_1}...Y_m^{n_m}\}} = \min_{\substack{k,n_1,....,n_m \\ \sum_r n_r=\nu}} \left\{\frac{M^{n_1,...,n_k+1,....n_m}(0,....,0)}{M^{n_1,...,n_m}(0,...,0)}\right\} \tag{18}$$

The universal single-day wealth at the end of day $(n+1)$ satisfies the following inequality from (8) and (17):

$$\hat{\mathbf{b}}_{n+1}^t \mathbf{x}_{n+1} = \sum_{k=1}^{m} \hat{b}_{n+1,k} x_{n+1,k} \geq \frac{1}{N} \sum_{k=1}^{m} x_{n+1,k} \times q \tag{19}$$

for $n = 1, 2...$

Hence,from (19) and Lemma 2,

$$\begin{aligned}
\frac{S^*_{n+1}(\mathbf{x}^n)}{\hat{S}_{n+1}(\mathbf{x}^n)} &= \prod_{i=1}^{n+1}\left(\frac{\mathbf{b}^{*t}\cdot\mathbf{x}_i}{\hat{b}_i^t\cdot\mathbf{x}_i}\right)\\
&= \frac{\sum_{i_1}\sum_{i_2}\cdots\cdots\sum_{i_{n+1}} x_{n+1,i_1}x_{ni_2}\cdots\cdot x_{1i_{n+1}}b_{i_1}b_{i_2}\cdots\cdot b_{i_{n+1}}}{\sum_{i_1}\sum_{i_2}\cdots\sum_{i_{n+1}} x_{n+1,i_1}x_{ni_2}\cdots\cdot x_{1i_{n+1}}\hat{b}_{i_1}\hat{b}_{i_2}\cdots\cdot\hat{b}_{i_{n+1}}}\\
&\le N^{n+1}\max_{i^n=\prod_{j=1}^{n+1} i_j}\left\{\frac{\prod_{j=1}^{n+1} b_{i_j}}{q^{n+1}}\right\}\\
&\le \frac{N^{n+1}}{q^{n+1}}\frac{\max_{0\le n_j\le\nu}\prod_{j=1}^{m}\Gamma(n_j+\frac{1}{2})}{\Gamma(\nu+\frac{1}{2})\Gamma(\frac{1}{2})^{m-1}}
\end{aligned}$$

which reduces to (16) by replacing $n+1$ by n and noting that

$$\max_{0\le n_j\le\nu}\Gamma(n_j+\frac{1}{2}) = \begin{cases}\Gamma(\nu+\frac{1}{2}) & \text{for } \nu\ge 3\\ \Gamma(\frac{1}{2}) & \text{for } \nu = 1,2.\end{cases}$$

Remarks

(i) Note that

$$\sum_{k=1}^{m}\frac{E[Y_1^{n_1}\cdots\cdot Y_k^{n_k+1}\cdots\cdot Y_m^{n_m}]}{E[Y_1^{n_1}\cdots\cdot Y_M^{n_m}]} = N$$

implies that

$$q \le \min_k\frac{E[Y_i^{n_1}\cdots\cdot Y_k^{n_k+1}\cdots\cdot Y_m^{n_m}]}{E[Y_1^{n_1}\cdots Y_m^{n_m}]} \le \frac{N}{m}$$

and the geometric rate $\frac{N}{q}$ in the bound (16) is larger than m.

(ii) For $\nu = n$, the multinomial universal portfolio of order ν is known as the moving order universal portfolio because n changes with time.

3. Empirical Results

The implementation of the Dirichlet and moving-order multinomial universal portfolios requires a substantial amount of computer memory growing exponentially fast as a function of the number of trading days. To overcome this problem, the memory-saving Dirichlet universal portfolios of finite order is introduced in [6]. Using the finite-order multinomial universal portfolios, the implementation time and computer- memory requirements are reduced substantially.

The finite-order portfolios are run on three stock-price data sets A, B and C selected from the Kuala Lumpur Stock Exchange. The data sets consists of a total of 500 trading days each, covering the period of January 1, 2003 until November 30, 2004. Each set consists of three Malaysian companies listed in the stock exchange. Set A consists of the stocks of Malayan Banking, Genting and Amway(M) Holdings. Set B consists of the stocks of Public bank, Sunrise and YTL Corporation. The last set C consists of the stocks of Hong Leong Bank, RHB Capital and YTL Corporation.

The wealths S_{500} achieved by selected values of the multinomial (N, p_1, p_3) universal portfolio of orders 1,2 and 3 over 500 trading days on data sets A,B and C are listed in Tables 1, 2 and 3 respectively. An initial starting wealth of 1 unit is assumed. The maximum wealths of 2.0362, 4.0761 and 4.0761 shown in the three tables for data sets A,B and C respectively are comparable to the wealths achieved by the Mahalanobis universal portfolios in different study by the first author [6] [7]. The implementation times of the finit-order portfolios are of the order of 0,01, 0,02 and 0,05 seconds for the first, second and third order respectively. On the other hand, the implementation times of the moving-order Dirichlet universal portfolios due to Cover and ordentlich are of the order of 394.77, 394.98 and 394.38 seconds with the maximum wealths of 1.870, 4.0761 and 4.0761 shown in table 4 for data sets A,B and C . Thus, the wealths achieved by the finite-order multinomial universal portfolios are comparable to that of the moving-order Dirichlet universal portfolio with substantial savings in computer memory and implementation time.

Table 1. The wealth S_{500} achieved by the (p_1, p_2, N) universal portfolio together with the final portfolio (b_1, b_2, b_3) and the implementation time for data Set A

	(p_1, p_2, N)	(b_1, b_2, b_3)	S_{500}	Implementation Time (Seconds)
order 1	(0.8,0.01,600)	(0.8,0.01,0.19)	2.0238	0.014 seconds
order 2	(0.8,0.01,600)	(0.8,0.01,0.19)	2.0362	0.022 seconds
order 3	(0.8,0.01,600)	(0.8,0.01,0.19)	2.0315	0.052 seconds

Table 2. The wealth S_{500} achieved by the (p_1, p_2, N) universal portfolio together with the final portfolio (b_1, b_2, b_3) and the implementation time for data Set B

	(p_1, p_2, N)	(b_1, b_2, b_3)	S_{500}	Implementation Time (Seconds)
order 1	(0.000001,0.000001,600)	(0.00,0.00,1.00)	4.0761	0.014 seconds
order 2	(0.000001,0.000001,600)	(0.00,0.00,1.00)	4.0354	0.022 seconds
order 3	(0.000001,0.000001,600)	(0.00,0.00,1.00)	4.0217	0.052 seconds

Table 3. The wealth S_{500} achieved by the (p_1, p_2, N) universal portfolio together with the final portfolio (b_1, b_2, b_3) and the implementation time for data Set C

	(p_1, p_2, N)	(b_1, b_2, b_3)	S_{500}	Implementation Time (Seconds)
order 1	(0.000001,0.000001,600)	(0.00,0.00,1.00)	4.0761	0.014 seconds
order 2	(0.000001,0.000001,600)	(0.00,0.00,1.00)	4.0490	0.022 seconds
order 3	(0.000001,0.000001,600)	(0.00,0.00,1.00)	4.0218	0.052 seconds

Table 4. The wealth S_{500} achieved by the Dirichlet universal portfolio together with the final portfolio (b_1, b_2, b_3) and the implementation time for data Set A, B and C

	$(\alpha_1, \alpha_2, \alpha_3)$	(b_1, b_2, b_3)	S_{500}	Implementation Time (Seconds)
Set A	(0.01,0.01,8000)	(0.00,0.00,1.00)	1.8750	394.77 seconds
Set B	(0.01,0.01,8000)	(0.00,0.00,1.00)	4.0761	394.98 seconds
Set C	(0.01,0.01,8000)	(0.00,0.00,1.00)	4.0761	394.38 seconds

References

[1] Cover T M 1991 *Math. Finance* **1** 1–29

[2] Cover T M and Ordentlich E 1996 *IEEE Trans. Inform. Theory* **42** 348–363

[3] Tan C P 2002 *The 2002 Internation Symposium on Information Theory and Its Applications* (Xi'an China) pp 587–590

[4] Tan C P and Pang S T 2011 *Proceedings of the International Conference in Mathematics and Applications* (Bangkok,Thailand) p 399

[5] Ordentlich E and Cover T M 1998 *Mathematics of Operations Research* **23** 960–982

[6] Tan C P, Chu S Y and Lim W X 2012 *The 16th Internation Congress on Insurance: Mathematics and Economic* (Hong Kong) p 113

[7] Tan C P and Lim W X 2011 *Proceedings of the International Conference in Mathematics and Applications* (Bangkok, Thailand) p 391

Journal of Physics: Conference Series **435** (2013) 012040 doi:10.1088/1742-6596/435/1/012040

Sustainable Multi-Product Seafood Production Planning Under Uncertainty

Ruth Simanjuntak[1], Monalisa Sembiring[2], Rani Sinaga[3], Endang J. Pakpahan[4], Herman Mawengkang[5]

[1,2,3,4]Graduate School of Mathematics, University of Sumatera Utara
[5]FMIPA USU Kampus USU, Medan, Indonesia 20155

Email: mawengkang@usu.ac.id

Abstract. A multi-product fish production planning produces simultaneously multi fish products from several classes of raw resources. The goal in sustainable production planning is to meet customer demand over a fixed time horizon divided into planning periods by optimizing the trade-off between economic objectives such as production cost, waste processed cost, and customer satisfaction level. The major decisions are production and inventory levels for each product and the number of workforce in each planning period. In this paper we consider the management of small scale traditional business at North Sumatera Province which performs processing fish into several local seafood products. The inherent uncertainty of data (e.g. demand, fish availability), together with the sequential evolution of data over time leads the sustainable production planning problem to a nonlinear mixed-integer stochastic programming model. We use scenario generation based approach and feasible neighborhood search for solving the model.

2000 Mathematics Subject Classification : 90B30 secondary 90C15

1. Introduction

Generally, the production planning problem aims to fulfill production and sourcing decisions so as to meet customer demand subject to production capacity, workforce availability and inventory restrictions and is inherently an optimization problem. The objective of the problem is to minimize the total cost or to maximize profit.

In production planning problems, mathematical models can be broadly classified into two classes: deterministic models and stochastic models. Deterministic models assume that the data are known and typically model the uncertainty using "best guesses" of uncertain values. Although various human judgments based and quantitative models have been developed to forecast these variables with uncertainty such as demand, these deterministic models typically end up solving "mean-value" or "worst-case" problems. The solution to such "worst-case" or "mean-value" problems are often inadequate–large error bound arise when one solves "mean-value" problems and "worst-case" formulations that can produce very conservative and expensive solutions ([2]). Without considering uncertainty, the deterministic production planning models, though widely studied in the literature, are less acceptable and deployed in practice.

In this paper we consider production planning problem which arises in marine fisheries industry in Indonesia. Marine fisheries play an important role in the economic development of Indonesia. This industry could also provide employment to people who live at coastal areas, to increase the financial gain of local government, and to conserve sustainability. Fisheries industrial sector can be classified into three different parts, i.e., open sea fishing, fish cultivation and processed fish. This paper is focusing on the latter sector.

Generally the processed fish industry in Indonesia can be found at the coastal area. There are a lot of variety of fish processed can be produced, such as smoked fish, salted fish, crunchy bashed of fish, fish bowl, terrain (preserved fish), etc. The management of processed fish industry is still dominated by the small local traditional business, using conventional management strategy. Consequently, they do not have enough information regarding the market demand and price. In this production planning situation, the current information may be certain, but future events are inevitably stochastic.

The proposed model explicitly permits the incorporation of uncertain parameters. Most of the references concerning optimization problems in the presence of the uncertainty appear under the heading of stochastic programming. See ([1], [5], [8]), for a good presentation of the basics. Two-stage stochastic programs with recourse typify a particularly important class of models. In such models, the objective function commonly corresponds to the minimization of expected costs or to the maximization of expected benefits (linear or nonlinear), although it can also refer to the expected value of the absolute or quadratic deviations of certain specific goals or the variance of the second-stage recourse function. Two kinds of decision variables exist. Those determined before the random variables have been revealed are called *first-stage* or *here-and-now* decision variables and represent proactive decisions; in this paper, they correspond to the production cost and workforce of the first period. Those determined after the realization of the random variables are called *second-stage* or *recourse* decision variables and represent reactive decisions made in recourse or response to the uncertainty factor. See also ([6], [7], [9]) for more information.

A review of some of the existing literature of production planning under uncertainty is provided in [15]. Stochastic programming ([2], [3], [7]) and robust optimization [16] have made several successful applications in production planning. In [4] a multi-stage stochastic programming approach was used for addressing a multi-product production planning model with random demand. Bakir and Byrune [1] developed a stochastic LP model based on the two-stage deterministic equivalent problem to incorporate demand uncertainty in a multi-period multi-product production planning model. Huang [5] proposed multi-stage stochastic programming models for production and capacity planning under uncertainty. Kazemi et al. ([8], [10]) proposed a two-stage stochastic model for addressing multi-product production planning with uncertain yield. [12] proposed a robust optimization model for stochastic aggregate production planning. In [11] a robust optimization model was developed to address a multi-site aggregate production planning problem in an uncertain environment. Wu [22] applied the robust optimization approach to uncertain production loading problems with import quota limits under the global supply chain management environment. [9] proposed two robust optimization models with different recourse cost variability measures to address multi-product production planning with uncertain yield.

In the case of discrete random variables, the resulting two-stage recourse models are usually large and complex, and thus must be solved numerically using suitable algorithmic strategies. Most of these algorithms apply decomposition strategies that break the model down by scenario or stage in an iterative scheme, allowing the resolution of smaller models (smaller in comparison to the deterministic equivalent model in its extensive form which gave rise to the original two-stage model). In this paper, a feasible neighborhood search method is proposed to solve an extended deterministic equivalent model, in which each first-stage variable is replicated for each scenario, with an imposed equality for the new variable values (non-anticipatively constraints).

This article is organized as follows: section two briefly reviews the two-stage recourse model. Section three presents the problem background. The stochastic programming model of the problem can be found in section four. In section five we present the solution basic approach. Computational results and conclusions, respectively, in section six and seven.

2. Framework of Two-Stage Recourse Model

In the following, the framework of two-stage stochastic integer programming model is described briefly. For detail, the reader is referred to van der Vlerk and Haneveld (1999). The stochastic linear programming model is expressed as follows:

$$\min \; c_x^T x + \sum_{s=1}^{S} p_s (q^T y^s) \tag{1}$$

$$s.t. \;\; Ax = b \tag{2}$$

$$T^S x + W y^S = h^S \quad s = 1,..,S \tag{3}$$

$$x, y^S \geq 0 \quad s = 1,..,S \tag{4}$$

Equations (2) represent the first-stage model and (3) represent the second-stage model. $\boldsymbol{x}$ is the vector of first-stage decision variables which is scenario-independent. The optimal value of x is not conditional on the realization of the uncertain parameters. $\boldsymbol{c_x}$ is the vector of cost coefficient at the first-stage. A is the first-stage coefficient matrix and $\boldsymbol{b}$ is the corresponding right-hand-side vector. $\boldsymbol{y}$ is the vector of second-stage (recourse) decision variables. $\boldsymbol{q}$ is the vector of cost (recourse) coefficient matrix and $\boldsymbol{h}^s$ is the corresponding right-hand-side vector and T^s is the matrix that ties the two stages together where $s \in \Omega$ represents scenarios in future and p_s is the probability that scenario s occurs. In the second-stage model, the random constraint defined in (3), $\boldsymbol{h}^s - \boldsymbol{T}^s\boldsymbol{x}$, is the goal constraint: violations of this are allowed, but the associated penalty cost, $\boldsymbol{q}^T\boldsymbol{y}$, will influence the choice of $\boldsymbol{x}$. $\boldsymbol{q}^T\boldsymbol{y}$ is the recourse penalty cost or second-stage value function and $\sum_{s=1}^{S} p_s(q^T y^s)$ denotes the expected value of recourse penalty cost (second-stage value function).

3. Problem Background

Fish and its processed products are the most affordable source of animal protein in the diet of most people. In Indonesia, most of the processed fish industries are found at the coastal area. In these industries fish are processed traditionally. There are eight kinds of fish product to be produced by the community, namely, dried fish, salted fish, BBQ fish, pindang fish, smoked fish, preserved fish, pressed fish, and fish bowl.

Model parameter and decision variables used throughout this paper are defined as follows.

Sets

- T = number of periods
- N = set of products
- M = set of resources
- S = set of scenarios

Variables

- X_{jt} = Quantity of product $j \in N$ in period $t \in T$ (ton)
  - $J = 1$ for dried fish
  - $J = 2$ for salted fish
  - $J = 3$ for BBQ fish
  - $J = 4$ for pindang fish
  - $J = 5$ for smoked fish
  - $J = 6$ for preserved fish
  - $J = 7$ for pressed fish
  - $J = 8$ for fish bowl
- u_{it} : Additional amount of resource $i \in M$ to purchase in $t \in T$ (unit)
- k_t : Number of workers required in period $t \in T$ (man-period)
- k_t^- : Number of workers laid-off in period $t \in T$ (man-period)
- k_t^+ : Number of additional workers in period $t \in T$ (man-period)
- I_{jt} : Quantity of product $j \in N$ to be stored in period $t \in T$ (units)
- B_{jt} : Under-fulfillment of product $j \in N$ in period $t \in T$ (units)

Parameters

- $\alpha, \beta, \gamma, \delta, \mu, \rho, \lambda, \eta$ are all costs
- D_{jt} : Demand for product $j \in N$ in period $t \in T$ (units)
- U_{jt} : Upper bound on u_{jt}
- r_{ij} : Amount of resource $i \in M$ needed to produce one unit of product $j \in N$
- f_{it} : Amount of resource $i \in M$ available at time $t \in T$ (units)
- a_j : Number of worker needed to produce one unit of product $j \in N$
- w_{jt}^p : Waste of fish product $j \in N$ in period $t \in T$ (units)

4. The Stochastic Programming Model

Minimize

$$\begin{aligned}
&\sum_{j\in N}\sum_{t\in T}\alpha_{jt}x_{jt} + \sum_{i\in M}\sum_{t\in T}\beta_{it}u_{it} + \sum_{t\in T}\mu_t k_t + \sum_{t\in T}\gamma_t k_t^- + \sum_{t\in T}\delta_t k_t^+ \\
&+ \sum_{j\in N}\sum_{t\in T}\eta_{jt}w_{jt}^p + \sum_{s\in S}p_s\sum_{j\in N}\sum_{t\in T}\rho_{jt}^s I_{jt}^s + \sum_{s\in S}p_s\sum_{j\in N}\sum_{t\in T}\lambda_{jt}^s B_{jt}^s \\
&+ \sum_{s\in S}p_s\left\{\sum_{j\in N}\sum_{t\in T}\left(\rho_{jt}^{s'}I_{jt}^{s'}\right)^2\right\} + \sum_{s\in S}p_s\left\{\sum_{j\in N}\sum_{t\in T}\left(\lambda_{jt}^s B_{jt}^s - \sum_{s'\in S}\lambda_{jt}^{s'}B_{jt}^{s'}\right)^2\right\}
\end{aligned} \tag{5}$$

Subject to

$$\sum_{j\in N} r_{ji}x_{jt} \le f_{it} + u_{it} \qquad \forall i \in M,\ \forall t \in T \tag{6}$$

$$u_{it} \le U_{it} \qquad \forall i \in M,\ \forall t \in T \tag{7}$$

$$\sum_{j\in N} a_j x_{jt} \le k_t \qquad \forall t \in T \tag{8}$$

$$0.10x_{jt} \le w_{jt}^{p} \le 0.20x_{jt}, \quad \forall j \in N, \ \forall t \in T \tag{9}$$

$$\sum_{j \in N}\sum_{t \in T} w_{jt}^{p} \le C^{p} \tag{10}$$

$$\varepsilon_1 \le \frac{\sum_{s \in S} P_s \sum_{j \in N} \sum_{t \in T} I_{jt}^{s} - \sum_{s' \in S} P_{s'} \sum_{j \in N} \sum_{t \in T} I_{jt}^{s'}}{\sum_{s \in S} P_s \sum_{j \in N} \sum_{t \in T} I_{jt}^{s}} \le \varepsilon_2 \tag{11}$$

$$\varepsilon_3 \le \frac{\sum_{s \in S} P_s \sum_{j \in N} \sum_{t \in T} B_{jt}^{s} - \sum_{s' \in S} P_{s'} \sum_{j \in N} \sum_{t \in T} B_{jt}^{s'}}{\sum_{s \in S} P_s \sum_{j \in N} \sum_{t \in T} I_{jt}^{s}} \le \varepsilon_4 \tag{12}$$

$$k_t = k_{t-1} + k_t^{+} - k_t^{-} \quad t = 2, \ldots T \tag{13}$$

$$x_{jt} + B_{jt-1}^{s} + I_{jt}^{s} - B_{jt}^{s} = D_{jt}^{s} \quad \forall j \in N, \forall t \in T, \forall s \in S \tag{14}$$

$$x_{jt}, u_{it}, k_t, k_t^{-}, k_t^{+}, I_{jt}^{s}, B_{jt}^{s} \ge 0 \ \forall j \in N, \forall i \in M, \forall t \in T, \forall s \in S \tag{15}$$

In this production planning problem we will decide:

- The quantity of each processed fish product to be produced in each period
- The additional resource to be used
- The number of regular additional and laying-off workers in each period
- To minimize the cost of processing fish waste, and
- To minimize the variability

The demand for each period is not known with certainty. Under the random demand in each period we should decide the number of each product to be stored in inventory or to fulfill the under fulfillment for each period.

All of these decisions are formulated in expression (5) of the model as an objective function. Constraint (6) expresses that the amount of resource $i \in M$ needed to produce product $j \in N$ at least should have the same amount of resources available at time $t \in T$ together with the additional resource needed. However, the additional resource needs to have an upper bound (expression (7)). In constraint (8), we have the number of workers needed to produce one unit product $j \in N$. The amount of fish waste where it should be between 10% - 20% can be found in (9). Constraint (10) expresses that the process of fish waste should be width in the capacity C^p. Ranges for the variability can be found in constraints (11) and (12). Constraint (13) ensures that the available workers in any period equal the number of worker from the previous period plus any change in the number of worker level during the current period. The change in the number of worker level may be due to either adding extra workers or laying-off redundant workers. Constraint (14) determines either the quantity of product to be stored in inventory or to purchase from outside to fulfill the shortfall in meeting market demand.

Model formulated in expression (5) through to (15) is in deterministic equivalent form, due to the fact that, the random form has been represented by scenario and in the objective function of these random terms have been premultiplied by the corresponding probabilities p_s. The method for transforming a stochastic programming model to its deterministic equivalent model was addressed in Mawengkang et al. (2006) and Irvan and Mawengkang (2007).

Journal of Physics: Conference Series **435** (2013) 012040 doi:10.1088/1742-6596/435/1/012040

5. The Algorithm

Cycle 1.

After solving the relaxed problem, the procedure for searching a suboptimal but integer-feasible solution from an optimal continuous solution can be described as follows.

Let

$$x = [x] + f, \quad 0 \le f \le 1$$

be the (continuous) solution of the relaxed problem, $[x]$ is the integer component of non-integer variable x and f is the fractional component.

Step 1. Get row i^* the smallest integer infeasibility, such that

$$\delta_{i^*} = \min\{f_i, 1 - f_i\}$$

Step 2. Calculate

$$v_{i^*}^T = e_{i^*}^T B^{-1}$$

this is a pricing operation

Step 3. Calculate $\sigma_{ij} = v_{i^*}^T a_j$

With j corresponds to $\min_j \left\{ \left| \frac{d_j}{\sigma_{ij}} \right| \right\}$

I. *For nonbasic j at lower bound*

If $\sigma_{ij} < 0$ and $\delta_{i^*} = f_i$ calculate $\Delta = \frac{(1-\delta_{i^*})}{-\sigma_{ij}}$

If $\sigma_{ij} > 0$ and $\delta_{i^*} = 1 - f_i$ calculate $\Delta = \frac{(1-\delta_{i^*})}{\sigma_{ij}}$

If $\sigma_{ij} < 0$ and $\delta_{i^*} = 1 - f_i$ calculate $\Delta = \frac{\delta_{i^*}}{-\sigma_{ij}}$

If $\sigma_{ij} > 0$ and $\delta_{i^*} = f_i$ calculate $\Delta = \frac{\delta_{i^*}}{\sigma_{ij}}$

II. For *nonbasic j at upper bound*

If $\sigma_{ij} < 0$ and $\delta_{i^*} = 1 - f_i$ calculate $\Delta = \frac{(1-\delta_{i^*})}{-\sigma_{ij}}$

If $\sigma_{ij} > 0$ and $\delta_{i^*} = f_i$ calculate $\Delta = \frac{(1-\delta_{i^*})}{\sigma_{ij}}$

If $\sigma_{ij} > 0$ and $\delta_{i^*} = 1 - f_i$ calculate $\Delta = \frac{\delta_{i^*}}{\sigma_{ij}}$

If $\sigma_{ij} < 0$ and $\delta_{i^*} = f_i$ calculate $\Delta = \frac{\delta_{i^*}}{-\sigma_{ij}}$

Otherwise go to next non-integer nonbasic or superbasic j (if available). Eventually the column j^* is to be increased form LB or decreased from UB. If none, go to next i^*.

Step 4. Calculate

$$\alpha_{j^*} = B^{-1}\alpha_{j^*}$$

i.e. solve $B\alpha_{j^*} = \alpha_{j^*}$ for α_{j^*}

Step 5. Ratio test; there would be three possibilities for the basic variables in order to stay feasible due to the releasing of nonbasic j^* from its bounds.

If j^* lower bound

Let

$$A = \min_{i' \neq i^* | \alpha_{ij^*} > 0} \left\{ \frac{x_{B_{i'}} - l_{i'}}{\alpha_{ij^*}} \right\}$$

$$B = \min_{i' \neq i^* | \alpha_{ij^*} < 0} \left\{ \frac{u_{i'} - x_{B_{i'}}}{-\alpha_{ij^*}} \right\}$$

$$C = \Delta$$

the maximum movement of j^* depends on:

$$\theta^* = \min(A, B, C)$$

If j^* upper bound

Let

$$A' = \min_{i' \neq i^* | \alpha_{ij^*} < 0} \left\{ \frac{x_{B_{i'}} - l_{i'}}{\alpha_{ij^*}} \right\}$$

$$B' = \min_{i' \neq i^* | \alpha_{ij^*} > 0} \left\{ \frac{u_{i'} - x_{B_{i'}}}{-\alpha_{ij^*}} \right\}$$

$$C' = \Delta$$

the maximum movement of j^* depends on:

$$\theta^* = \min(A', B', C')$$

Step 6. Exchanging basis for the three possibilities

1. If A or A'
   - $x_{B_{i'}}$ becomes nonbasic at lower bound $l_{i'}$
   - x_{j^*} becomes basic (replaces $x_{B_{i'}}$)
   - x_{i^*} stays basic (non-integer)
2. If B or B'
   - $x_{B_{i'}}$ becomes nonbasic at upper bound $u_{i'}$
   - x_{j^*} becomes basic (replaces $x_{B_{i'}}$)
   - x_{i^*} stays basic (non-integer)
3. If C or C'
   - x_{j^*} becomes basic (replaces x_{i^*})

- x_{i*} becomes superbasic at integer-valued

repeat from step 1.

Cycle2, pass1: adjust integer-infeasible superbasics by fractional steps to reach complete integer-feasibility.

Cycle2, pass2: adjust integer feasible superbasics.

This phase aims to conduct a highly-localized neighbourhood search.

6. Computational Result

The processed fish industry under investigation is located at the eastern coastal area of North Sumatra province of Indonesia. The industry run by the community of that area has to make a production plan for these eight processed fish products to fulfill market demand over each period of time t, $t = 1, \ldots, T$. In this case each period equals to three months. Therefore there will be four periods in a year, i.e. $T = \{1, 2, 3, 4\}$.

After we had a survey to the location, then we found out that the market situation for the eight processed fish products could come out within three possible situation i.e., good, fair and poor, with associated probabilities of 0.30, 0.50 and 0.20 respectively. Nevertheless, the method addressed in Mawengkang and Suherman (2007) could be used in order to get an efficient number of scenarios. The data of the problem can be found in Mawengkang [13].

Table 1. The Number of Each Product to be Produced (ton)

Product	**Period**			
	1	*2*	*3*	*4*
1	250	250	250	300
2	900	900	900	950
3	200	200	200	300
4	200	200	200	300
5	200	200	200	300
6	200	200	200	300
7	200	200	200	300
8	200	200	200	300

Table 2. Additional Resources to be used (ton)

Resources	**Period**			
	1	***2***	***3***	***4***
Machine 1	12.20	12.20	12.20	16.95
Machine 2	9.80	9.80	9.70	13.80
Machine 3	8.65	8.75	8.65	12.55

Table 3. Workforce Plan

Policy	**Period**			
	1	***2***	***3***	***4***
Reg. workforce	38	35	35	47

Add. workforce	35	0	0	12
Lay off	0	0	0	0

Table 1 reports the optimal production plan for each product in each period. It can be seen that the number for all product in period 4 increase. Therefore, correspondingly there is an increase in additional resources to be used and the need of workforce in period 4, as shown in Table 2 dan Table 3, repectively.

7. Conclusions

In this paper, we develop a two-stage recourse model for production planning problem of a processed fish industry at coastal area of North Sumatra province of Indonesia with stochastic demand. The model is adequate for solving the planning problem faced by the management of the industry. The model includes the computation of worker which is very useful for the industry in order they will be able to schedule a number of local people, and to conserve sustainability. We also propose an algorithm for solving the mixed integer stochastic programming problem. It is observed that the proposed model would be able to provide a credible and effective methodology for real-world sustainable production planning problems under uncertainty. However, there is still room for further investigation, if scheduling situation is imposed in the model. In this case the proposed algorithm needs to explore particularly in CYCLE 2 of the algorithm.

References

[1] M. A. Bakir, and M. D. Byrune, "Stochastic linear optimization of an MPMP production planning model", International Journal of Production Economics 1998; 55:87-96

[2] J. R. Birge, and F. V. Louveaux, "Introduction to stochastic programming". New York: Springer; 1997.

[3] G. B. Dantzig, "Linear programming under uncertainty", Management Science 1955; 1: 197-206

[4] L. F. Escudero, P. V. Kamesam, A. J. King, and R...J-B. Wetts, "Production planning via scenarios", Annals of Operations Research 1993; 34:311-335

[5] K. Huang, "Multi-stage stochastic programming models for production planning", Thesis(PhD), School of Industrial and Systems Engineering,Georgia Institute of Technology;2005

[6] Irvan, H. Mawengkang, "Characteristics of deterministic equivalent model for multi-stage integer stochastic programs", Mathematics Journal, Special Edition Part II, Universiti Teknologi Malaysia; 2008

[7] P. Kall, and S. Wallace, Stochastic programming, Chichester: Wiley; 1994.

[8] M. K. Zanjani, M. Nourelfath, and D. Ait-Kadi, "A stochastic programming approach for production planning in a manufacturing environment with random yield", CIRRELT; 2007(58), working document, Quebec (Canada)

[9] M. K. Zanjani, D. Ait-Kadi, M. Nourelfath, "Robust production planning in a manufacturing environment with random yield: a case in Sawmill production planning", CIRRELT; 2008(52), working document, Quebec (Canada).

[10] M. K. Zanjani, M. Nourelfath, and D. Ait-Kadi, "A multi-stage stochastic programming approach for production planning with uncertainty in the quality of raw materials and demand", CIRRELT;2009(09), working document, Quebec (Canada).

[11] S. C. H. Leung, S. O. S Tsang, Ng W.L, and Y. Wu, "A robust optimization model for multi-site production planning problem in an uncertain environment", European Journal of Operational Research 2007;181(1):224-238.

[12] S. C. H Leung, and Y. Wu, "A robust optimization model for stochastic aggregate production planning", Production Planning & Control 2004; 15(5):502-514.

[13] H. Mawengkang (2010). Production Planning of Fish Processed Product under Uncertainty, Anziam J. Vol. 51, pp C715 – C733.

[14] H. Mawengkang, S. Suwilo, O. S. Sitompul, "Revision modeling of two-stage stochastic programming problem", Journal of Industrial System 2006;7(4):6-10.

[15] H. Mawengkang, Suherman, "A heuristic method of scenario generation in multi-stage decision problem under uncertainty", Journal of Industrial System 2007; 8(2): 98-105.

[16] J. Mula, R. Poler, J. P. Garcia-Sabater, and F. C. Lario, "Models for production planning under uncertainty: a review", International Journal of Production Economics 2006; 103:271-285.

[17] J. M. Mulvey, R. Van derbei, and S. Zenios, "Robust optimization of large scale systems", Operations research 1995; 43 (2): 264-281.

[18] Rico-V. Ramirez, "Two-stage stochastic linear programming": a tutorial. SIAG/OPT Views-and-News 2002; 13 (1): 8-14.

[19] A. Ruszczynski, A. Shapiro, editors, "Stochastic programming". Handbooks in operations research and management science, vol. 10. New York; North-Holland; 2003.

[20] S. Sen, J. l. Higle, "Introductory tutorial on stochastic linear programming models", Interfaces 1999; 29 (2): 33-61.

[21] M. H. Van der Vlerk, and W. K. Haneveld, "Stochastic integer programming: General models and algorithms", Annals of Operations Research, 85:39-57, 1999.

[22] Wu Yue. "Robust optimization applied to uncertain production loading problems with import quota limits under global supply chain management environment". International Journal of Production Research 2006; 44(5):849-882.

Empirical Power Comparison Of Goodness of Fit Tests for Normality In The Presence of Outliers

Mayette Saculinggan[1], Emily Amor Balase[2]
Zone 1, Cogon El Salvador, Misamis Oriental, 9017 Philippines
[1] Electronics Engineering Department, Capitol University, Corrales Extension, Cagayan de Oro, Philippines
[2] Department of Mathematical Sciences, Mindanao University of Science and Technology, Cagayan de Oro City,Philippines

E-mail :maia0812@yahoo.com[1], stat_emily@yahoo.com[2]

Abstract. Most statistical tests such as t-tests, linear regression analysis and Analysis of Variance (ANOVA) require the normality assumptions. When the normality assumption is violated, interpretation and inferences may not be reliable. Therefore it is important to assess such assumption before using any appropriate statistical test. One of the commonly used procedures in determining whether a random sample of size n comes from a normal population are the goodness-of-fit tests for normality. Several studies have already been conducted on the comparison of the different goodness-of-fit(see, for example [2]) but it is generally limited to the sample size or to the number of GOF tests being compared(see, for example [2] [5] [6] [7] [8]). This paper compares the power of six formal tests of normality: Kolmogorov-Smirnov test (see [3]), Anderson-Darling test, Shapiro-Wilk test, Lilliefors test, Chi-Square test (see [1]) and D'Agostino-Pearson test. Small, moderate and large sample sizes and various contamination levels were used to obtain the power of each test via Monte Carlo simulation. Ten thousand samples of each sample size and contamination level at a fixed type I error rate α were generated from the given alternative distribution. The power of each test was then obtained by comparing the normality test statistics with the respective critical values. Results show that the power of all six tests is low for small sample size(see, for example [2]). But for $n = 20$, the Shapiro-Wilk test and Anderson – Darling test have achieved high power. For $n = 60$, Shapiro-Wilk test and Liliefors test are most powerful. For large sample size, Shapiro-Wilk test is most powerful (see, for example [5]). However, the test that achieves the highest power under all conditions for large sample size is D'Agostino-Pearson test (see, for example [9]).

1. Introduction

Statistical tests require the normality assumption. Parametric statistical procedures assume a normal distribution of the data therefore it is important to assess such assumption before using any appropriate statistical test.

In any research, the use of different shapes of probability distributions plays a big role in achieving the objective of the study. The actual data must be checked to determine if the observed frequencies in an experiment correspond to the probabilities in a model of the experiment. This is called best fit.

There are different goodness of fit tests available in the literature. The most common is the chi - square goodness of fit test. For Ronneu (2003), this type of test must be used for large sample data. The idea behind the chi-square goodness-of-fit test is to see if the sample comes from the population with the claimed distribution. Another way of looking at it is to ask if the frequency distribution fits a specific pattern.

Another method of checking goodness of fit is the Anderson–Darling Test. This is used to test if a sample of data came from a population with a specific distribution. It is a modification of the Kolmogorov-Smirnov (K-S) test and gives more weight to the tails than does the K-S test. The Anderson-Darling test makes use of the specific distribution in calculating critical values. This has the advantage of allowing a more sensitive test and the disadvantage that critical values must be calculated for each distribution.

Another Goodness of Fit Tests is the Shapiro–Wilk Test, and it compares a set of measures against the Normal distribution. It may be used before performing parametric tests to ensure that the data being used follow a Normal distribution. Results on the study of Keskin(2006) showed that Shapiro-Wilk test is the most powerful tests and be used for testing normality of data. The Shapiro-Wilk is an improvement on the more general Kolmogorov-Smirnov curve-fitting algorithm.

One of the most well-known modifications of the Kolmogorov-Smirnov test for goodness of fit is generally referred to as the Lilliefors test for normality. This test was developed independently by Lilliefors (1967) and by Van Soest (1967). The null hypothesis for this test is that the error is normally distributed; that is there is no difference between the observed distribution of the error and a normal distribution. The alternative hypothesis is that the error is not normally distributed.

Different tests of normality often produce different results. Some tests reject while others fail to reject the null hypothesis of normality. The contradicting results are misleading and often confuse practitioners. Several studies have already been conducted on comparing tests for normality. The study of Razali et al. (2011) compared the power of four formal tests of normality. The power comparison was obtained via Monte Carlo simulation of sample data generated from alternative distributions that would follow symmetric and asymmetric distributions. The power of each test was then obtained by comparing the test of normality statistics with the respective critical values. Results in the study showed that Shaprio – Wilk test was the most powerful normality test, followed by Anderson – Darling test, Lilliefors test and Kolmogorov-Smirnov test. However, the power of all four tests was still low for small sample size.

Another study conducted by Oztuna et al. (2006) aimed to compare the four different normality tests. Results showed that the most powerful test for normal distributions was given by Jarqua-Bera and for non – normal distributions, Shapiro – Wilk test yielded the most powerful results. In that study, Oztuna et al. (2006) concluded that Jarqua-Bera test was superior for normal and standard normal distributions whereas for non normal distributions Shapiro – Wilk achieved sufficient power at the smaller sample size.

An additional study by Mendes et al. (2003) compared Shapiro-Wilks, Lilliefors and Kolmogorov-Smirnov Tests for Type I error and also the power of the three tests. For all different sample sizes and distributions, Shapiro-Wilks revealed the most powerful results, followed by the Lilliefors test. Kolmogorov-smirnov test results were the weakest among all three tests. Finally, results showed that all three tests were the most powerful when ran on data with exponential distribution.

Furthermore, the study of Shahabuddin (2009) has found out that in Simple Random Sampling (SRS), the FKS which is a proposed modification of the Kolmogorov-Smirnov test found to perform almost as powerful as Anderson Darling tests (AD) in SRS. However, in all the cases considered, FKS has always outperformed the original Kolmogorov-Smirnov and Cramer-von-Mises (CV) tests.

Moreover, Keskin (2006) conducted a study where it was found out that Shapiro-Wilk test was the most powerful tests and could be suggested for testing normality of data. It was also concluded that performance of the normality tests was greatly affected by the distribution type and sample size.

Another study by D'Agostino (1990) showed that D'Agostino – Pearson K^2 statistics are powerful and informative tests. The tests used in the study were based on $\sqrt{b_1}$ and b_2 which showed that the two are excellent and powerful tests. The study recommended that for all sample sizes $\sqrt{b_1}$ and b_2 should be computed and examined as descriptive statistics. For all sample sizes $n \geq 9$, tests of hypotheses could be based on these two tests. In particular, for n > 50 Shapiro- Wilk test should not be used instead D' Agostino-Pearson K^2 test would be used as the test of choice. This justification is not only because of their fine power but also because of the information supplied on non-normality.

The different studies on goodness-of-fit comparisons have included at most only four tests and have limited the coverage of the research to uncontaminated data with one level of significance.

In this study, the power of the six tests for various degrees of contamination and sample sizes in the presence of outliers is assessed and compared via Monte Carlo simulations. This method is used to determine how contamination of outliers in the tested data affects the sensitivity, performance and reliability of the six tests in identifying if it is normal or non normal.

2. Methods

In this study, Monte Carlo procedure was utilized to assess power of goodness-of-fit tests for various sample sizes and contamination levels. The following were the steps adopted to achieve the objectives of the study.

a. Generate data from normal distributions with a certain proportion of contaminants called outliers for various sample sizes.
b. Set the contamination level and the level of significance.
c. Evaluate the power of chi-square, Kolmogorov-Smirnov, Anderson-Darling, Shapiro-Wilk, Lilliefors and D'Agostino-Pearson test statistics in testing if a random sample of n independent data came from a normal distribution with mean μ and variance σ^2 in the context of a Tukey's contaminated normal model as follows:

$$F(x) = (1 - \varepsilon)N(\mu, \sigma^2) + \varepsilon N(\mu^*, \sigma^{2^*}) \text{ and } 0 < \varepsilon < 1 \tag{2.1}$$

Where $\mu = \sigma^2 = \mu^* = \sigma^{2*}$

For this study the null and alternative hypotheses were:

H_o: The distribution is standard normal

H_1: The distribution is not standard normal

d. Determine the best normality tests under different conditions considering the extent of contamination of outliers. In order to obtain the simulated power of the six normality tests, at α = 1%, 5%, and 10% for each sample size, a total of 10,000 samples will be drawn from normal distribution with specified mean μ and variance σ^2 and contaminated with outliers fixed at ε.

2.1. Power Study Process

A Monte Carlo simulation was accomplished with 10000 iterations, considering various sample sizes and different contamination levels. The steps in the process were:

2.1.1. Random data were generated from standard normal distribution as uncontaminated samples. Samples were then contaminated with contaminant given by Normal (5, 100) where ε = .01, .05, .10, .15 and .20 for various sample sizes classifieds small (n=10, 20), moderate (n=30,60) and large (n=100, 1000).

2.1.2. The statistic of all goodness-of-fit tests considered in this study was calculated for every iteration.

2.1.3. The value of the test statistic was then compared to the critical value at different levels of significance α = .10, .05 and .01.

2.1.4. The number of times the value of the test statistic exceeds the critical value was counted for at α level of significance. Exceeding the critical value is equivalent to rejecting H_o at the corresponding significance level. The process was repeated at various sample sizes and contamination level.

3. Highlights of findings and discussion/data presentation

The result of the power of Goodness-of-Fit tests was presented according to various sample sizes with different degrees of contamination. The degree of contamination was generated based on Tukey's contaminated normal model:

$$F(x) = (1 - \varepsilon)N\left(0, \sigma^2\right) + \varepsilon\, N\left(\mu^*, \sigma^{2^*}\right) \; and \; 0 < \varepsilon < 1 , \tag{3.1}$$

Where $\varepsilon = 0.01, 0.05, 0{,}10, 0{,}15$ *and* 0.20 using n = 10, 20, 30, 60, 100 and 1000 to represent small, moderate and large sample sizes. Every sample size was tested 1000 times with the applicable GOF tests for each degree of contamination with test of hypotheses carried out at the 5% level of significance. The figure below showed the behavior of each test for every significance level.

The power of each Goodness-of-Fit tests for various sample sizes and degree of contamination were evaluated to identify the behavior of each tests. The KS test as shown in Figure 1, provides low power for all sample sizes. It happened because KS test is a distance test and by research it is considered an insensitive test for power. Although it increased power as sample size n increases but still it must reached the 80% level to be considered statistically powerful test for normality.

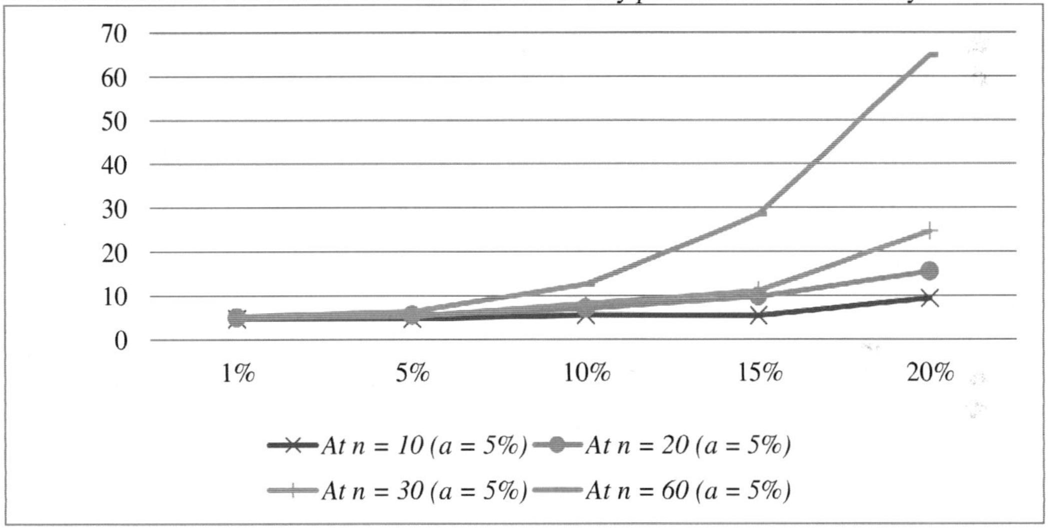

Figure 1. Power versus Degree of Contamination for Kolmogorov-Smirnov Test.

All the six tests were tested at various contamination level to determine the behavior of each test. At a significance level α = 5%, the Kolmogorov-Smirnov test still failed to show power of the test for all conditions. As shown in the figure above, the Kolmogorov-Smirnov test showed power only at n = 60 with contamination of 20 percent. This would mean that for KS test to be sensitive to the tested data, it must be used for large sample size and also must be highly contaminated with outliers.

The Anderson-Darling test at a significance level of .01 performed well especially on moderate sample size. The test showed to be statistically powerful at small samples when the data were highly contaminated with outliers. These results also showed that the test would only have its highest power for very large sample size at n = 1000. The power of Anderson-Darling test increased when sample size and contamination was also increased.

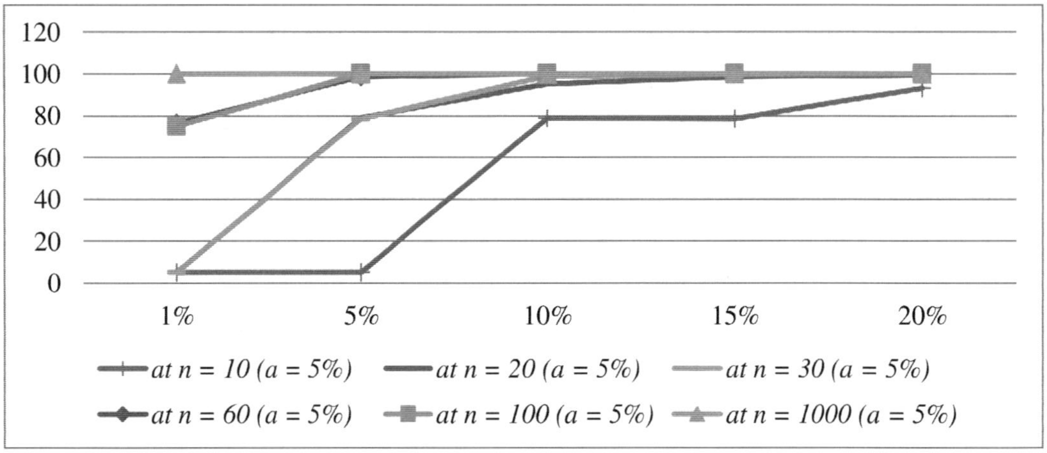

Figure 2. Power versus Degree of Contamination for Anderson-Darling test.

The test was able to show power starting at n = 60 at 0.05 contamination level while for smaller sample size power was shown only when the data were highly contaminated with outlier starting at 10 percent and higher. At n = 1000 the test is already statistically powerful even with 0.01 contamination. For small and moderate samples, the Anderson-Darling test showed power when it was contaminated with at least 5% outliers and showed 100% power only when n = 1000 for all contamination levels.

Shapiro-wilk test was able to detect non normality of the data even at very low contamination level. The result in the figure below showed that Shapiro-Wilk is a powerful test for moderate and large sample sizes. For small sample size, this test would only show power when the data was contaminated with 5% outliers. The Shapiro-Wilk at α = 5%, also showed similar behavior as with Anderson-Darling although at smaller samples like when n = 20 the test was statistically powerful when contaminated with at least 5% of outlier.

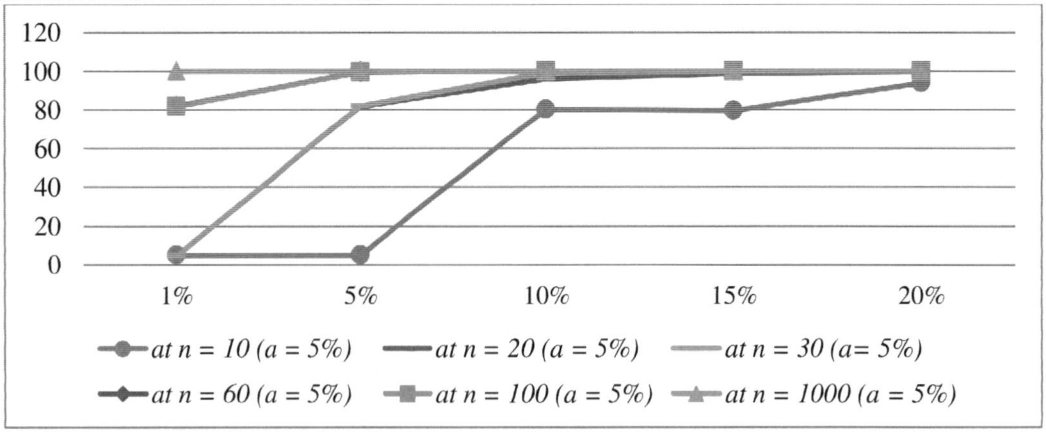

Figure 3. Power versus Degree of Contamination for Shapiro-Wilk Test.

As recommended in other studies, Liliefors test was used only for moderate to large sample data. The result presented in the figure below showed that the test has power when used for moderate samples and only when the data is at least contaminated with 5% outliers. However, with n = 1000, Liliefors could detect non-normality of the data 100% of the time even when it was contaminated with 1 percent outliers.

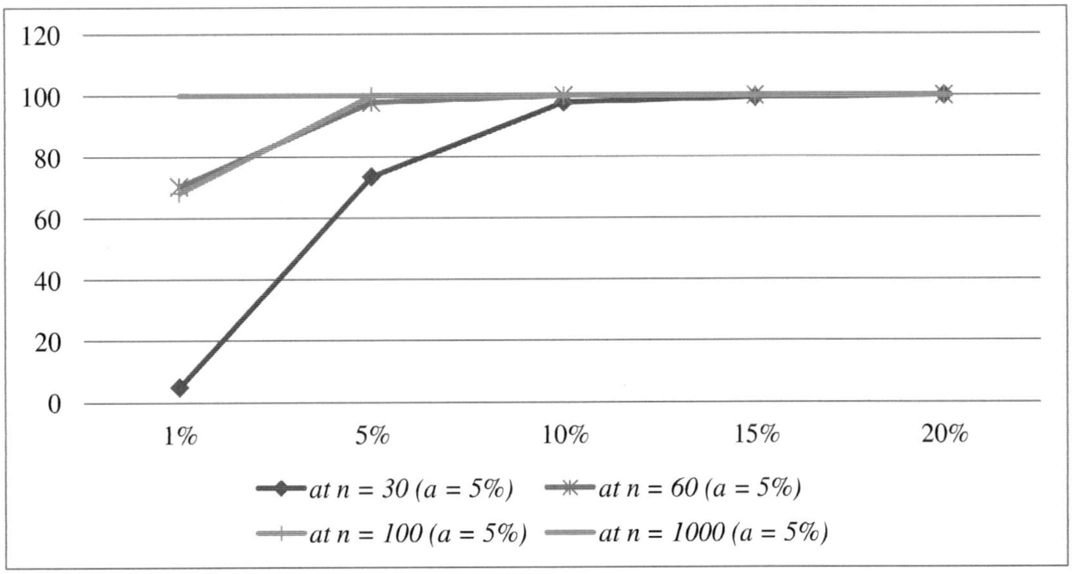

Figure 4. Power versus Degree of Contamination for Liliefors test.

Power of Liliefors test to detect non-normality of data for moderate and large sample sizes was when the data was contaminated with at least 5% outlier. It also revealed power of 100% when sample size is very large like $n = 1000$ for all contamination levels.

As shown in Figure 5 below, power of Chi- Square test increased as sample size and contamination were also increased. Still, the power of this test was low and could not be considered statistically powerful test for normality. This result just shows that Chi-Square had the lowest performance in terms power for large sample data.

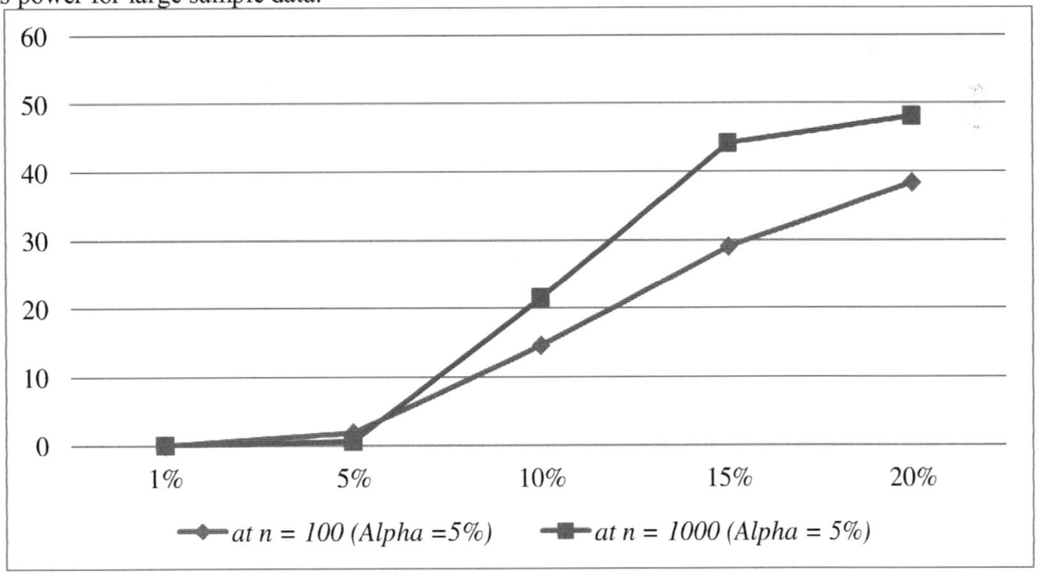

Figure 5 . Power versus Degree of Contamination for Chi-Square test.

Chi-Square test had low power levels and none of the conditions reached a power of 80%. It can be seen in Figure 5 above that as the sample size and contamination is increased, power of the test also increased. But still, Chi-Square was not powerful enough to detect non-normality in the data tested and cannot be considered as omnibus test for normality.

The D'Agostino which was also tested for large sample size has shown high power even at $n = 100$. At 1% significance level, the D'Agostino-Pearson test showed to be a powerful test for large sample data. Starting at 1% contamination up to 20% contamination level for $n = 100$ and $n = 1000$ the simulation result showed that for almost all conditions this test could detect non-normality 100% of the time.

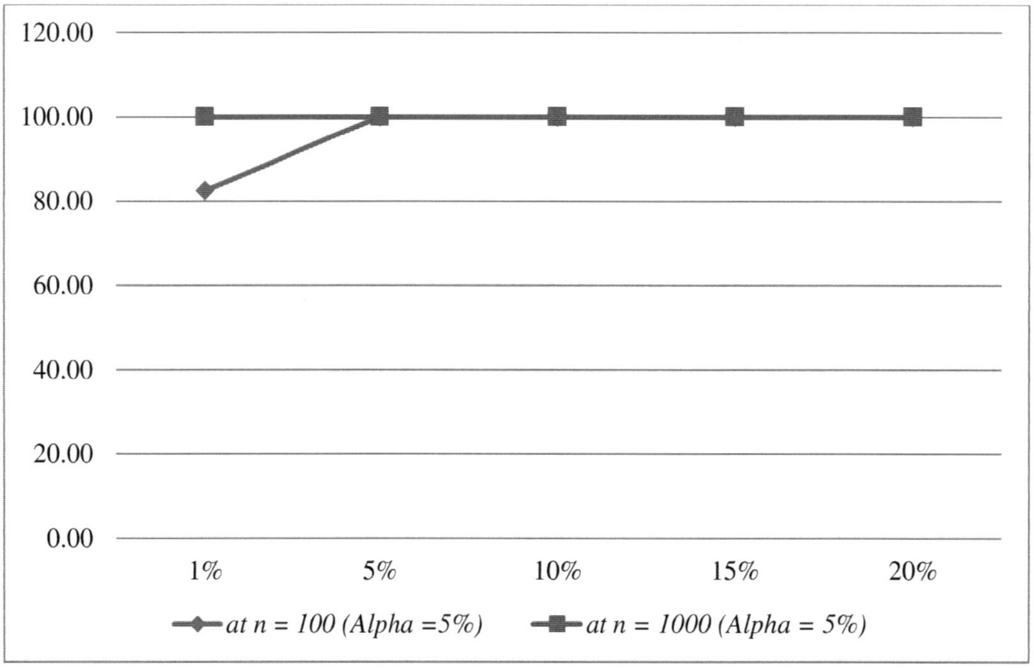

Figure 6. Power versus Degree of Contamination for D'Agostino-Pearson Test.

As shown in the figure above, the D'Agostino-Pearson test has performed well for all contamination levels and at all sample size n. Almost all results showed 100% power except for $n = 100$ at 0.01 level of contamination which could detect only 82.44% of the time on non-normality of data. Opposite to Chi-Square, the D'Agostino-Pearson has shown to be statistically powerful test for all conditions which makes it a good choice for large sample data.

The Kolmogorov-Smirnov and Chi-Square test gave poor power to detect non-normality of data which was consistent with the results of the study conducted by Agostino (1986), Mendes et al. (2003) and Shahabuddin et al. (2009). A modification of the Kolmogorov-Smirnov test was studied by Drezner et al.(2008) in order to enhance the behavior of the test. The result of his study showed that the modfied KS test was able to detect the difference from normality at $n = 200$.

The result of this study showed that, for small and moderate sample sizes, the Shapiro – Wilk gave the most powerful results. Shapiro-Wilk was also most powerful test based on the result of the study conducted by Shapiro, et al. (1986), Mendes et al. (2003), Oztuna et al. (2006), Guner et al. (2009) and Razali et al. (2011). For large sample size data, the D'Agostino-Pearson test outperformed the rest of the test because it was able to give 100% power on almost all sample size and all contamination levels and even at different significance level. This has been similar to the study conducted by D'Agostino (1990), which showed that D'Agostino-Pearson test is an omnibus test for normality.

To be able to show the behavior of each tests per sample, the different GOF tests used for each sample size were compared and discussed below. As shown in figure 7, at $\alpha = 5\%$ small samples with 1% contaminated data, none of the tests were able to show high power.

For moderate sample size, only Shapiro-Wilk test showed power at $n = 60$. D'Agostino-Pearson and Shapiro-Wilk test were able to show power at $n = 100$. For $n = 1000$, all tests except for Chi-Square were able to detect non-normality of the data 100% of the time.

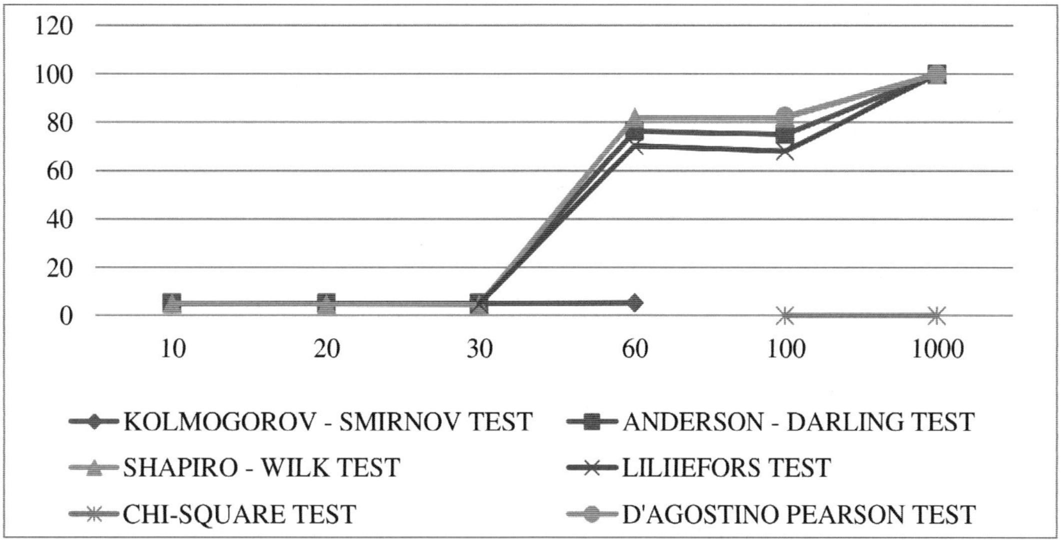

Figure 7. Comparison of Power of Goodness-of-Fit Tests for various Sample Sizes ($\alpha = 5\%$, $\varepsilon = 1\%$).

When the data were contaminated with 5% outlier at $\alpha = 5\%$, only Shapiro-Wilk showed power for small and moderate sampled data at $n = 20$ and $n = 30$ respectively. However, for $n = 60$, three tests showed power except for Kolmogorov-Smirnov test.

All tests showed power and even showed 100% power for very large data except for Chi-Square test.

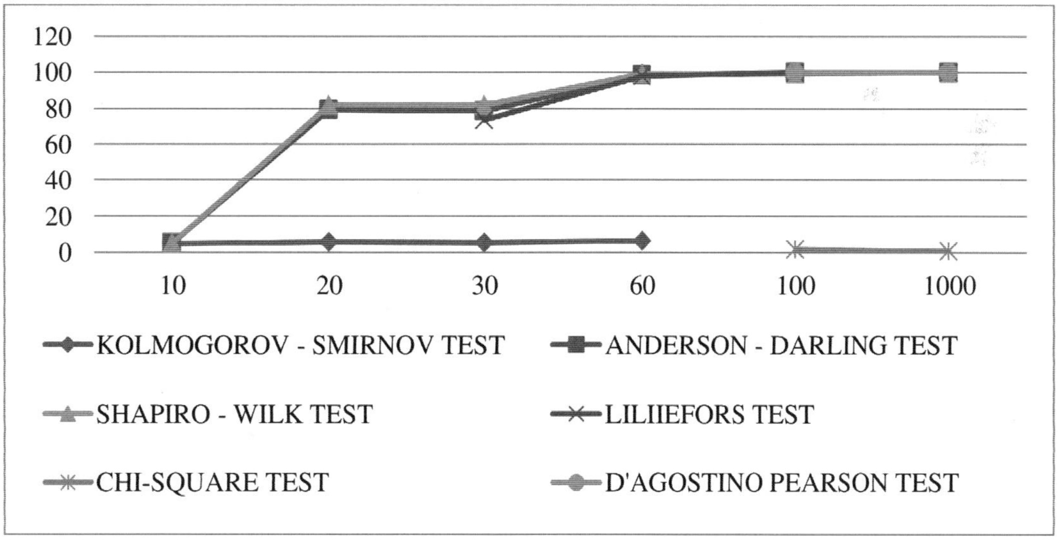

Figure 8. Comparison of Power of Goodness-of-Fit Tests for various Sample Sizes ($\alpha = 5\%$, $\varepsilon = 5\%$).

At 10% contamination of the data, small samples at 5% significance level have shown that for $n = 10$ only Shapiro-Wilk showed power. When the sample size was increased to $n = 20$, only Kolmogorov-Smirnov showed low power for the tested data out of the three tests. The same also for moderate sampled data, only Kolmogorov-Smirnov did not show power of the test as shown in the

Figure below. It also observed that power of all three tests increased as sample size was increased to 60.

All tests showed power for large sample data excluding Chi-Square test that only showed 14.58 and 21.54 power respectively for each sample size as shown in figure 9.

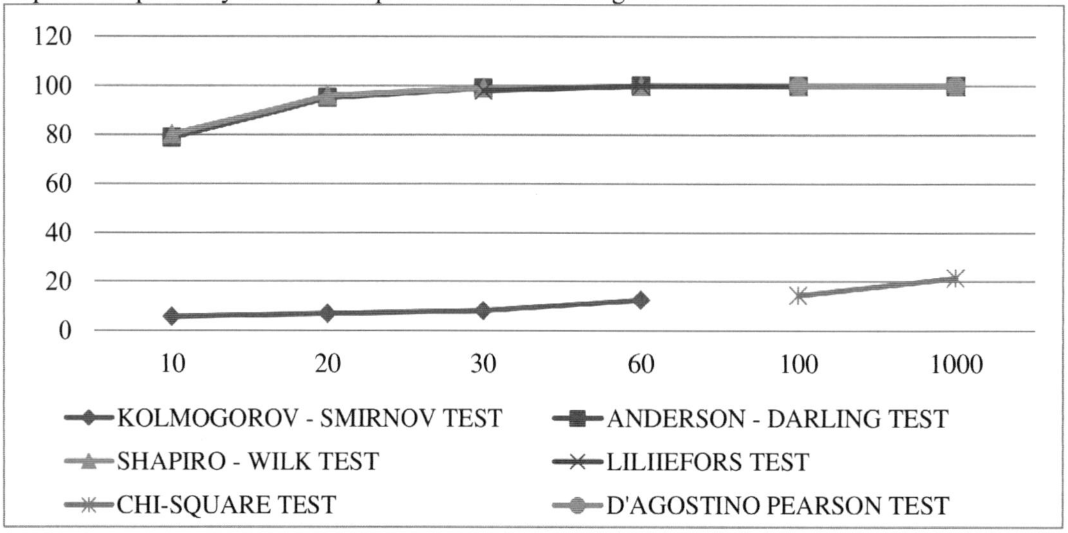

Figure 9. Comparison of Power of Goodness-of-Fit Tests for various Sample Sizes ($\alpha = 5\%$, $\varepsilon = 10\%$).

When the contamination was increased to 15%, at n = 10 all tests failed to show high power to be considered statistically powerful. But when n was increased to 20, the result showed that all three test showed power. For moderate sampled data, only Kolmogorov-Smirnov did not show power of the test. All tests have shown to be statistically powerful by providing 100% power at both sample size except for Chi-Square which could detect only 44.17% of the time to detect non-normality of the tested sample.

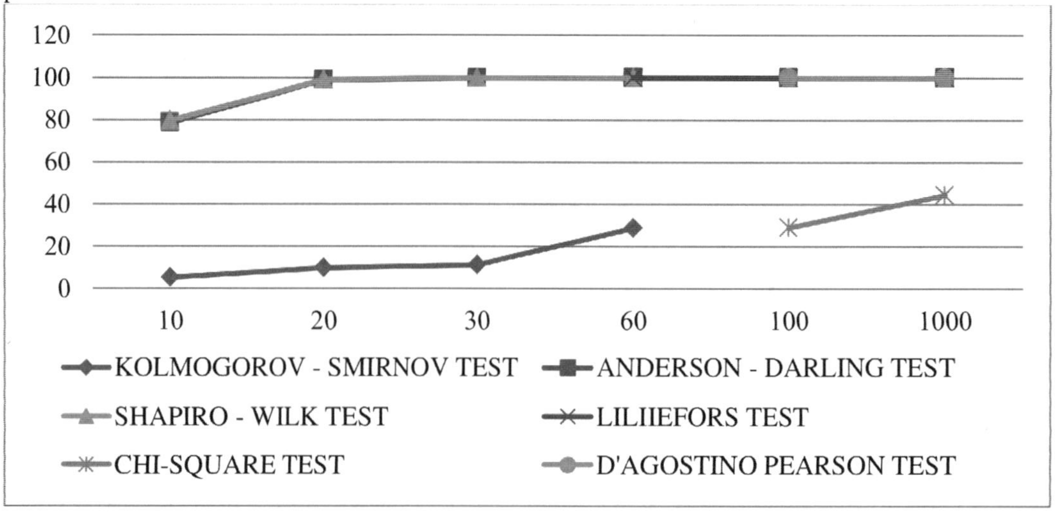

Figure 10. Comparison of Power of Goodness-of-Fit Tests for various Sample Sizes ($\alpha = 5\%$, $\varepsilon = 15\%$)

At n = 1000, only Chi-Square test did not show power of the tested data but all four GOF tests showed 100% power.

When the contamination was increased to 20%, almost all tests had high power even at n = 10 except for Kolmogorov-Smirnov test.

The same with small samples, moderate sample data have shown that three GOF tests showed power of the tests except for Kolmogorov-Smirnov test and would even have 100% power at n = 60.

Only Chi-Square test showed low power of the test for large sample size at 20% contamination as shown in the figure below.

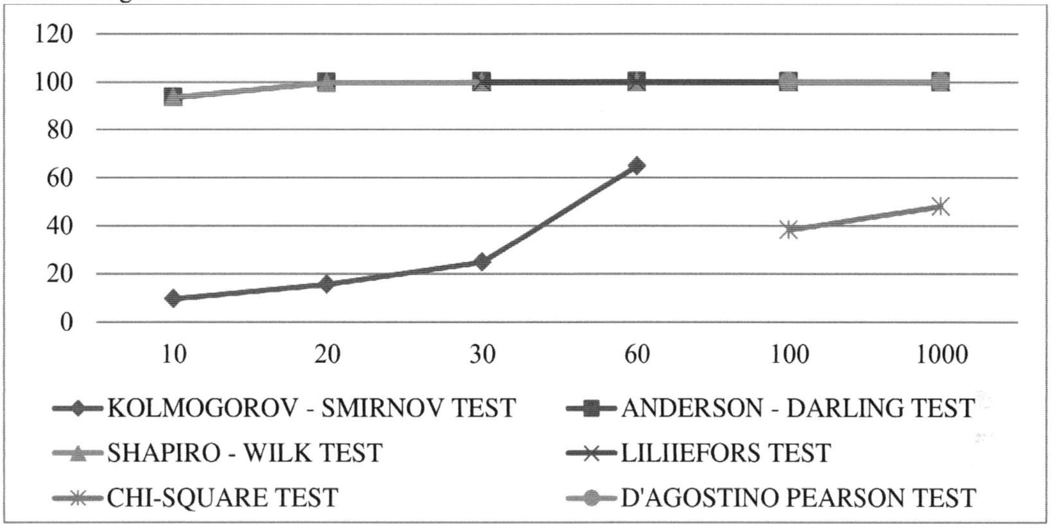

Figure 11. Comparison of Power of Goodness-of-Fit Tests for various Sample Sizes (α = 5%, ε = 20%.

4. Conclusions

Based on the generated results, the different types of Goodness-of-Fit Tests showed different power in various contaminations. For large data samples, the D'Agostino Pearson test has shown to be the appropriate test as it would provide sensitivity for all conditions. For small sample data, Shapiro-Wilk had power at n = 20 for a contamination of 0.05. When the number of samples is moderate, Shapiro-Wilk or Liliefors test has shown more power as compared to Anderson-Darling and Kolmogorov-Smirnov test. Also as evident in the results, when contamination level and sample size were increased the power had also increased.

In this study, Shapiro-Wilk and D'Agostino-Pearson tests have shown to be a powerful test for normality. There were two tests which have shown poor performance in terms of power. These tests were Kolmogorov-Smirnov test and Chi-Square tests. This is because both tests were distance tests and were considered to be inefficient in identifying non normality in the data. Even with outliers introduced into the tested data, the results of the study were supported by the conclusion also made by Agostino and Razali that D'Agostino-Pearson test has been a powerful test and Kolmogorov-Smirnov test to have poor power for normality (see, for example [5] [9]).

References

[1] Romeu J.L. 2003 Selected Topics in Assurance Related Technologies: The Chi-Square : A Large-Sample Goodness of Fit Test **10** 4

[2] Keskin S. 2006 Comparison of Several Univariate Normality Tests
Regarding Type I Error Rate and power of the Test in Simulation based Small Sample *Journal of Applied Science Research*

[3] Lilliefors H. 1969 On the Kolmogorov-Smirnov Test for the Exponential Distribution with Mean Unknown *Journal of American Statistical Association*

[4] Abdi et. al. "Lilliefors/Van Soest's test of normality." Information on http://www.utdallas.edu/~herve/Abdi-Lillie2007-pretty.pdf

[5] Razali N., Wah Y 2011 Power Comparison of Shapiro-Wilk, Kolmogorov-Smirnov, Lilliefors and Anderson Darling tests *Journal of Statistical Modeling and Analytics*

[6] Oztuna D. et. al. 2006 Investigation of Four Different Normality Tests in Terms of Type I Error Rate and Power under Different Distributions *TUBITAK Turk J Medical Science*

[7] Mendes M. , Pala A. 2003 Type I Error Rate and Power of Three Normality Tests Pakistan *Journal of Information and Technology*

[8] Shabuddin A., Azna F., Ibrahim K. and Jemain A.A. 2009 On the Comparison of Several Goodness of Fit tests under Simple Random Sampling and Ranked Set Sampling *World Academy of Science and Engineering Technology*

[9] Agostino et. al, in: Tests for Normal Distribution in Goodness-of-fit techniques (Mercel Dekker Inc.1986)

Journal of Physics: Conference Series **435** (2013) 012042 doi:10.1088/1742-6596/435/1/012042

Numerical Integration Of The Integrals Based On Haar Wavelets

Anvarjon A. Ahmedov[1,2] **Mohammad Hasan bin Abd Sathar** [2]

[1]Department of Process and Food Engineering, Faculty of Engineering, University Putra Malaysia

[2]Institute of Mathematical Research, Universiti Putra Malaysia

E-mail: anvar@eng.upm.edu.my, hanaz_3704@yahoo.com

Abstract. In this work, we present a computational method for solving double and triple integrals with variable limits of integrations which is based on Haar wavelets. This approach is the generalization and improvement of the methods [3]. The advantage of this new methods is its more efficient and simple applicability than the previous methods. Error analysis for the case of two dimension are considered. Finally, we also give some numerical examples to compared with existing methods.

1. Introduction

Integrals and derivatives are basic tools of calculus, with numerous applications in science and engineering. Many problems in mathematics, physics, and engineering involve integration where an explicit formula for the integral is desired. The integrals encountered in a basic calculus course are deliberately chosen for simplicity; those found in real applications are not always so accommodating. Some integrals cannot be found exactly, some require special functions which themselves are a challenge to compute, and others are so complex that finding the exact answer is too slow. This motivates the study and application of numerical methods for approximating integrals, which today use floating-point arithmetic on digital electronic computers. Therefore numerical integration is one of the convenient ways to find approximate values of the functions, see for example [1,2]. In recent years, wavelets become one of the most promising, powerful and reliable tools in this area. There are many kinds of wavelets with different properties, some examples are the Haar wavelets, Daubechies' orthonormal of compact support and Meyer's wavelets [3,4,5]. Haar wavelets which are mathematically the most simple orthonormal wavelets with compact support and they have been used in different numerical approximation problems, for example the Haar wavelets methods for solving linear and nonlinear integral equations was proposed in [6,7,8].In this paper we have modified the method suggested in [3] and the advantage of this methods is its more efficient and simple applicability.The organization of this paper is as follows. In Section 2, numerical integration using Haar wavelets is described and in Section 3 new techniques for solving the double and triple integral equations. Error analysis for Haar Wavelets is given in Section 4 and numerical results are reported in Section 5. Some conclusions are drawn in Section 6.

Journal of Physics: Conference Series **435** (2013) 012042 doi:10.1088/1742-6596/435/1/012042

2. Numerical integration based on Haar wavelets

A wavelet is a wave-like oscillation with an amplitude that starts out at zero, increases, and then decreases back to zero. Generally, wavelets are purposefully crafted to have specific properties that make them useful for signal processing. Wavelets can be combined, using a "reverse, shift, multiply and sum" technique called convolution, with portions of an unknown signal to extract information from the unknown signal. As a mathematical tool, wavelets can be used to extract information from many different kinds of data, including - but certainly not limited to - audio signals and images. Sets of wavelets are generally needed to analyze data fully. A set of "complementary" wavelets will deconstruct data without gaps or overlap so that the deconstruction process is mathematically reversible. Thus, sets of complementary wavelets are useful in wavelet based compression/decompression algorithms where it is desirable to recover the original information with minimal loss. In formal terms, this representation is a wavelet series representation of a square-integrable function with respect to either a complete, orthonormal set of basis functions, or an overcomplete set or frame of a vector space, for the Hilbert space of square integrable functions.

2.1. Haar wavelets

The mother wavelet for the Haar wavelets family is defined on the interval [a,b), as:

$$H\left(2^j\frac{x-a}{b-a}-k\right)=\begin{cases}1 & for\, x\in\left[a+k\frac{(b-a)}{2^j},a+\left(k+\frac{1}{2}\right)\frac{(b-a)}{2^j}\right)\\ -1 & for\, x\in\left[a+\left(k+\frac{1}{2}\right)\frac{(b-a)}{2^j},a+(k+1)\frac{(b-a)}{2^j}\right)\\ 0 & elsewhere,\end{cases}$$

where $n = 1, 2, 3, .$ and write $n = 2^j + k$ with $j = 0, 1, 2$ and $k = 0, 1, , 2^j - 1$ and defined

$$h_n(x)=H\left(2^j\frac{x-a}{b-a}-k\right).$$

All the functions in the Haar wavelet family are defined on subintervals of $[a, b)$ and are generated from by the operations of dilation and translation. The integer j indicates the level of the wavelet and k is the translation parameter. The Haar wavelet functions form the orthogonal functions to each on [a,b)

$$\int_a^b h_n(x)h_\ell(x)\,dx=\begin{cases}(b-a)2^{-j} & n=\ell=2^j+k\\ 0 & n\neq\ell.\end{cases}$$

Also we defined the scalar wavelet $h_0(x) = 1$ for all x. It can be shown that the sequence is a complete orthogonal system in $L_2[a, b]$ and for $f \in C[a, b]$, the series $\sum_n \langle f, h_n\rangle h_n$ converges to f [10], where $\langle f, h_n\rangle = \int_a^b f(x)h_n(x)\,dx$.

2.2. Numerical integration based on Haar wavelets

Here, we consider numerical integration for single, double and triple integrals using Haar wavelets.

2.2.1. Numerical technique for single integrals.

We consider the integral

$$\int_a^b f(x)\,dx$$

over the interval [a,b). Thus any function $f(x)$ which is square integrable in the interval [a,b) may be expanded as:

$$f(x) = \sum_{n=0}^{\infty} a_n h_n(x).$$

In practice, only the first $2M-1$ term of the sum is considered, where $M = 2^{J_1}$ and J_1 is the maximum level of Haar wavelets that is,

$$f(x) \approx \sum_{n=0}^{2M-1} a_n h_n(x). \tag{1}$$

Lemma 1. The approximate value of the integral is

$$f(x) \approx a_0(b-a).$$

Proof. Since

$$\int_a^b h_n(x)\,dx = 0, \quad n = 1, 2, 3, ...,$$

and

$$\int_a^b h_0(x)\,dx = (b-a),$$

$$\int_a^b f(x)\,dx \approx \sum_{n=0}^{2M-1} a_n \int_a^b h_n(x)\,dx \approx a_0(b-a). \tag{2}$$

It is clear from equation (2) that, Haar approximation involves only one coefficient in the evaluation of the definite integral. To calculate the Haar coefficient a_0 we consider the nodal points

$$x_k = a + (b-a)\frac{k+0.5}{2M}, \quad k = 0, 1, ..., 2M-1.$$

The discretized form of (1) can be written as

$$f(x_k) \approx \sum_{n=0}^{2M-1} a_n h_n(x_k), \quad k = 0, 1, ..., 2M-1. \tag{3}$$

The second advantage of Haar wavelet approximation is that we do not need to solve the above system which is computationally expensive for large values of M. The next lemma gives us an easy formula to calculate the value of the Haar coefficient a_0.

Lemma 2. The solution in of the system (3) for a_0 is given by

$$a_0 = \frac{1}{2M}\sum_{k=0}^{2M-1} f(x_k).$$

For the proof see [9].

We apply Lemma 2 to the integral (2) to obtain the following formula for numerical integral equation

$$\int_a^b f(x)\,dx \approx \frac{(b-a)}{2M}\sum_{k=0}^{2M-1} f(x_k) = \frac{(b-a)}{2M}\sum_{k=0}^{2M-1} f\left(a + (b-a)\frac{k+0.5}{2M}\right).$$

Journal of Physics: Conference Series **435** (2013) 012042 doi:10.1088/1742-6596/435/1/012042

2.2.2. Numerical technique for double integrals and triple integrals

The formula derived for numerical integration of double integrals with constant limit in [9] given by

$$\int_c^d \int_a^b f(x,y)\,dx\,dy \approx a_{00}(b-a)(d-c) = \frac{(b-a)(d-c)}{4M^2} \sum_{l=0}^{2M-1} \sum_{k=0}^{2M-1} f(x_k, y_l),$$

where

$$f(x,y) \approx \sum_{j=0}^{2M-1} \sum_{i=0}^{2M-1} a_{ij} h_i(x) h_j(y),$$

and

$$a_{00} = \frac{1}{4M^2} \sum_{l=0}^{2M-1} \sum_{k=0}^{2M-1} f(x_k, y_l). \tag{4}$$

This formula can be extended to triple integral, and given by

$$\int_e^f \int_c^d \int_a^d f(x,y)\,dx\,dy \approx \frac{(b-a)(d-c)(f-e)}{8M^3} \sum_{l=0}^{2M-1} \sum_{k=0}^{2M-1} \sum_{j=0}^{2M-1} f(x_j, y_k, z_l).$$

3. Present method (PM) for double and triple integral with variable limits

3.1. Numerical technique for double integrals with variable limits.

Consider double integral with variable limits of the type

$$\int_a^b \int_{f_1(y)}^{f_2(y)} F(x,y)\,dx\,dy, \qquad f_1(y) \le x \le f_2(y), \;\; a \le y \le b.$$

After substitution $x = X\left(f_2(y) - f_1(y)\right) + f_1(y)$ in integral equation, we obtain

$$\int_a^b \int_0^1 F\left(X\left(f_2(y) - f_1(y)\right) + f_1(y), y\right)\left(f_2(y) - f_1(y)\right) dX\,dy, \quad 0 \le X \le 1, \;\; a \le y \le b$$

We apply formula (4) for a_{00} to approximate the integral equation

$$\begin{aligned} &\textstyle\int_a^b \int_0^1 F\left(X\left(f_2(y) - f_1(y)\right) + f_1(y), y\right)\left(f_2(y) - f_1(y)\right) dX\,dy \approx a_{00}(b-a) \\ &= \tfrac{(b-a)}{4MN} \sum_{l=0}^{2N-1} \sum_{k=0}^{2M-1} F\left(X_k\left(f_2(y_l) - f_1(y_l)\right) + f_1(y_l), y_l\right)\left(f_2(y_l) - f_1(y_l)\right) \end{aligned}$$

where

$$X_k = \frac{k+0.5}{2M}, \quad k = 0, 1, ..., 2M-1,$$

$$y_l = a + (b-a)\frac{l+0.5}{2N}, \quad l = 0, 1, ..., 2N-1.$$

Consider triple integral with variable limits of the type

$$\int_a^b \int_{f_1(z)}^{f_2(z)} \int_{g_1(y,z)}^{g_2(y,z)} F(x,y,z)\,dx\,dy\,dz.$$

The formula for triple integrals with variable limits can be derived in a similar way and is given by substitution$x = X\,(g_2(y,z) - g_1(y,z)) + g_1(y,z)$ and $y = Y\,(f_2(z) - f_1(z)) + f_1(z)$ in the equation , we obtain

$$\int_a^b \int_0^1 \int_0^1 F\,(X(g_2 - g_1) + g_1, Y\,(f_2 - f_1) + f_1, z)\,(f_2 - f_1)(g_2 - g_1) dX\,dY\,dz \quad .$$

Therefore, the formula for approximating triple integrals is given by

$$\begin{aligned}
&\textstyle\int_a^b \int_0^1 \int_0^1 F\,(X(g_2 - g_1) + g_1, Y\,(f_2 - f_1) + f_1, z)\,(g_2 - g_1)\,(f_2 - f_1)\,dX\,dY\,dz \\
&\approx \tfrac{(b-a)}{8M^3} \sum_{l=0}^{2P-1} \sum_{k=0}^{2N-1} \sum_{j=0}^{2M-1} F \left(\begin{array}{c} X_j \left[g_2 \left(\begin{array}{c} Y_k\,(f_2(z_l) - f_1(z_l)) \\ +f_1(z_l), z_l \end{array} \right) - g_1 \left(\begin{array}{c} Y_k\,(f_2(z_l) - f_1(z_l)) \\ +f_1(z_l), z_l \end{array} \right) \right] \\ +g_1 \left(\begin{array}{c} Y_k\,(f_2(z_l) - f_1(z_l)) \\ +f_1(z_l), z_l \end{array} \right), Y_k \left(\begin{array}{c} (f_2(z_l) - f_1(z_l)) \\ +f_1(z_l), z_l \end{array} \right) \end{array} \right) . \\
&\qquad (f_2(z_l) - f_1(z_l)) \left(g_2 \left(\begin{array}{c} Y_k\,(f_2(z_l) - f_1(z_l)) \\ +f_1(z_l), z_l \end{array} \right) - g_1 \left(\begin{array}{c} Y_k\,(f_2(z_l) - f_1(z_l)) \\ +f_1(z_l), z_l \end{array} \right) \right)
\end{aligned}$$

4. Error Analysis

In this section we assume that $f(x,y) \in C^2([a,b] \times [a,b])$ and

$$\exists K > 0; \left| \frac{\partial^2 f(x,y)}{\partial x \partial y} \right| \leq K, \ \forall x, y \in [a,b] \times [a,b].$$

We may proceed as follows, suppose $f_{k\ell}(x,y)$ is the following approximation of $f(x,y)$

$$f_{k\ell}(x,y) = \sum_{n=0}^{k-1} \sum_{m=0}^{\ell-1} a_{nm} h_n(x) h_m(y),$$

where, $k = 2^{\alpha+1}$, $\alpha = 0, 1, 2, ...$ and $\ell = 2^{\beta+1}$, $\beta = 0, 1, 2, ...$ then

$$f(x,y) - f_{k\ell}(x,y) = \sum_{n=k}^{\infty} \sum_{m=\ell}^{\infty} a_{nm} h_n(x) h_m(y) + \sum_{n=k}^{\infty} \sum_{m=0}^{\ell-1} a_{nm} h_n(x) h_m(y) + \sum_{n=0}^{k-1} \sum_{m=\ell}^{\infty} a_{nm} h_n(x) h_m(y).$$

By Parceval's formula we have

$$\begin{aligned}
\|f(x,y) - f_{k\ell}(x,y)\|^2 &= \textstyle\int_a^b \int_a^b (f(x,y) - f_{k\ell}(x,y))^2 dx\,dy \\
&= \sum_{p=k}^{\infty} \sum_{s=\ell}^{\infty} \sum_{n=k}^{\infty} \sum_{m=\ell}^{\infty} a_{nm} a_{ps} \textstyle\int_a^b h_n(x) h_p(x)\,dx \int_a^b h_m(y) h_s(y)\,dy \\
&+ \sum_{p=k}^{\infty} \sum_{s=0}^{\ell-1} \sum_{n=k}^{\infty} \sum_{m=0}^{\ell-1} a_{nm} a_{ps} \textstyle\int_a^b h_n(x) h_p(x)\,dx \int_a^b h_m(y) h_s(y)\,dy \\
&+ \sum_{p=0}^{k-1} \sum_{s=\ell}^{\infty} \sum_{n=0}^{k-1} \sum_{m=\ell}^{\infty} a_{nm} a_{ps} \textstyle\int_a^b h_n(x) h_p(x)\,dx \int_a^b h_m(y) h_s(y)\,dy \\
&= \sum_{n=k}^{\infty} \sum_{m=\ell}^{\infty} a_{nm}^2 + \sum_{n=k}^{\infty} \sum_{m=0}^{\ell-1} a_{nm}^2 + \sum_{n=0}^{k-1} \sum_{m=\ell}^{\infty} a_{nm}^2,
\end{aligned}$$

where,$a_{nm} = \langle h_n(x), \langle f(x,y), h_m(y)\rangle\rangle = \left\langle h_n(x), \int_a^b f(x,y)h_m(y)dy \right\rangle$.
Consider $h_n(x) = H(2^j \frac{x-a}{b-a} - k)$, $k = 0, 1, ..., 2^j - 1$, $j = 0, 1, ...$, we obtain

$$\begin{aligned} a_{nm} &= \left\langle H(2^j \tfrac{x-a}{b-a} - k), \int_a^b f(x,y)H(2^i \tfrac{y-a}{b-a} - k)\, dy \right\rangle \\ &= \int_a^b \left(\int_a^b f(x,y)H(2^i \tfrac{y-a}{b-a} - k)\, dy \right) H(2^j \tfrac{x-a}{b-a} - k)\, dx \\ &= \int_a^b \left(\int_{a+k\frac{b-a}{2^i}}^{a+\left(k+\frac{1}{2}\right)\frac{b-a}{2^i}} f(x,y)dy - \int_{a+\left(k+\frac{1}{2}\right)\frac{b-a}{2^i}}^{a+(k+1)\frac{b-a}{2^i}} f(x,y)dy \right) H(2^j \tfrac{x-a}{b-a} - k)\, dx \end{aligned}$$

Using the mean value theorem we have:

$$a + k\frac{b-a}{2^j} \le y_1 \le a + \left(k + \frac{1}{2}\right)\frac{b-a}{2^j}, a + \left(k + \frac{1}{2}\right)\frac{b-a}{2^j} \le y_2 \le a + (k+1)\frac{b-a}{2^j}$$

Hence,

$$a_{nm} = (b-a)\int_a^b \left(f(x,y_1)2^{-i-1} - f(x,y_2)2^{-i-1}\right) H(2^j \frac{x-a}{b-a} - k)\, dx$$

Using the mean value theorem

$$\begin{aligned} a_{nm} &= 2^{-i-1}(b-a) \int_a^b \left(\tfrac{\partial f(x,y_0)}{\partial y}(y_1 - y_2) \right) H(2^j \tfrac{x-a}{b-a} - k)\, dx, \quad y_1 \le y_0 \le y_2 \\ &= 2^{-i-1}(b-a) \left[\int_{a+k\frac{b-a}{2^j}}^{a+\left(k+\frac{1}{2}\right)\frac{b-a}{2^j}} \tfrac{\partial f(x,y_0)}{\partial y}(y_1 - y_2)dx - \int_{a+\left(k+\frac{1}{2}\right)\frac{b-a}{2^j}}^{a+(k+1)\frac{b-a}{2^j}} \tfrac{\partial f(x,y_0)}{\partial y}(y_1 - y_2)dx \right] \end{aligned}$$

Again we use the mean value theorem as the above step

$$\begin{aligned} a_{nm} &= 2^{-i-1}(b-a)^2(y_1 - y_2)\left[2^{-j-1}\left(\tfrac{\partial f(x_1,y_0)}{\partial y} - \tfrac{\partial f(x_2,y_0)}{\partial y}\right)\right] \\ &= 2^{-i-j-2}(b-a)^2(y_1 - y_2)(x_1 - x_2)\left(\tfrac{\partial^2 f(x_0,y_0)}{\partial x \partial y}\right), x_1 \le x_0 \le x_2 \\ &\le 2^{-i-j-2}(b-a)^2 \cdot 2^{-i}(b-a) \cdot 2^{-j}(b-a)K \\ &= 2^{-2i-2j-2}(b-a)^4 K \\ a_{nm}^2 &= 2^{-4i-4j-4}(b-a)^8 K^2 \end{aligned}$$

Therefore

$$\|f(x,y) - f_{k\ell}(x,y)\|^2 = \sum_{n=k}^{\infty}\sum_{m=\ell}^{\infty} a_{nm}^2 + \sum_{n=k}^{\infty}\sum_{m=0}^{\ell-1} a_{nm}^2 + \sum_{n=0}^{k-1}\sum_{m=\ell}^{\infty} a_{nm}^2$$

$$\begin{aligned} \sum_{n=k}^{\infty}\sum_{m=\ell}^{\infty} a_{nm}^2 &= \sum_{n=2^{\alpha}+1}^{\infty}\sum_{m=2^{\beta}+1}^{\infty} 2^{-4i-4j-4}(b-a)^8 K^2 \\ &= \sum_{n=2^{\alpha}+1}^{\infty}\left((b-a)^8 K^2 \sum_{i=\beta+1}^{\infty}\left(\sum_{m=2^i}^{2^{i+1}-1} 2^{-4i-4j-4}\right)\right) \\ &= \sum_{n=2^{\alpha}+1}^{\infty}\left((b-a)^8 K^2 \sum_{i=\beta+1}^{\infty} 2^{-3i-4j-4}\right) \\ &= \sum_{n=2^{\alpha}+1}^{\infty}\left((b-a)^8 K^2 2^{-4j-4} \sum_{i=\beta+1}^{\infty} 2^{-3i}\right) \\ &= \tfrac{(b-a)^8}{14\ell^3} K^2 \sum_{j=\alpha+1}^{\infty}\left(\sum_{n=2^j}^{2^{j+1}-1} 2^{-4j}\right) \\ &= \tfrac{(b-a)^8}{14k^3} K^2 \sum_{j=\alpha+1}^{\infty} 2^{-3j} \\ &= \tfrac{4(b-a)^8}{49k^3\ell^3} K^2 \end{aligned}$$

$$\begin{aligned}
\sum_{n=k}^{\infty}\sum_{m=0}^{\ell-1} a_{nm}^2 &= \sum_{n=2^{\alpha+1}}^{\infty}\sum_{m=0}^{\ell-1} 2^{-4i-4j-4}(b-a)^8K^2 \\
&= \sum_{n=2^{\alpha+1}}^{\infty}\left((b-a)^8K^2\sum_{i=0}^{\beta}\left(\sum_{m=2^i-1}^{2^{i+1}-1} 2^{-4i-4j-4}\right)\right) \\
&= \sum_{n=2^{\alpha+1}}^{\infty}\left((b-a)^8K^2 2^{-4j-4}\sum_{i=0}^{\beta} 2^{-4i}\left(2^i+1\right)\right) \\
&\le \sum_{n=2^{\alpha+1}}^{\infty}\left((b-a)^8K^2 2^{-4j-4}\sum_{i=0}^{\infty} 2^{-4i}\left(2^i+1\right)\right) \\
&= \sum_{n=2^{\alpha+1}}^{\infty}\left(\tfrac{29}{210}(b-a)^8K^2 2^{-4j}\right) \\
&= \frac{29(b-a)^8}{210}K^2\sum_{j=\alpha+1}^{\infty}\left(\sum_{n=2^j}^{2^{j+1}-1} 2^{-4j}\right) \\
&= \frac{(b-a)^8}{14k^3}K^2\sum_{j=\alpha+1}^{\infty} 2^{-3j} \\
&= \frac{116(b-a)^8}{735k^3}K^2
\end{aligned}$$

$$\sum_{n=0}^{k-1}\sum_{m=\ell}^{\infty} a_{nm}^2 = \frac{116(b-a)^8}{735\ell^3}K^2$$

$$\|f(x,y)-f_{k\ell}(x,y)\|^2 \le 4(b-a)^8K^2\left(\frac{1}{49k^3\ell^3}+\frac{29}{735k^3}+\frac{29}{735\ell^3}\right)$$

Therefore it is obvious that the error bound will approaches 0 as$k \to \infty$and $\ell \to \infty$.

5. Numerical examples

Example 1

$$\int_0^1\int_0^y e^{|x+y-1|}\,dx\,dy = \int_0^1\int_0^1 ye^{|yX+y-1|}\,dX\,dy$$

Table 1. Absolute Errors of Example 1.

Haar	Absolute errors (PM)	Absolute errors (NM)	NM-PM
M = 4, N = 5	0.0016425210	0.0016425205	$-5.0000000017 \times 10^{-10}$
M = 5, N = 10	0.0004951200	0.0004951195	$-5.0000000006 \times 10^{-10}$
M = 10, N = 15	0.0001980110	0.0001980118	$8.0000000000 \times 10^{-10}$
M = 16, N = 16	0.0001185685	0.0001185682	$-3.0000000000 \times 10^{-10}$

The errors are shown in Table 1. From table 1 it is clear that the new method (NM) performs very well then the previous method (PM) [3].

Journal of Physics: Conference Series **435** (2013) 012042 doi:10.1088/1742-6596/435/1/012042

Example 2

$$\int_0^{\frac{\pi}{4}}\int_0^{\sin y}\frac{1}{\sqrt{1-x^2}}\,dx\,dy=\int_0^1\int_0^1\frac{\sin(y)}{\sqrt{1-(X\sin(y))^2}}\,dX\,dy$$

Table 2. Absolute Errors of Example 2.

Haar	Absolute errors (PM)	Absolute errors (NM)	NM-PM
$M=3, N=3$	0.0001711541	0.0001711538	$-2.9999999999\times10^{-10}$
$M=4, N=3$	0.0000978111	0.0000966500	-0.000001161099
$M=5, N=4$	0.0000630685	0.0000627182	$-3.50299999999\times10^{-7}$
$M=6, N=5$	0.0000439774	0.0000438437	$-1.33699999999\times10^{-7}$

Example 3

$$\int\int_R (x+y)^{\frac{-1}{2}}\,dx\,dy \approx 3.549613026789713$$

where R is a quadrilateral region connecting the points (-1,2),(2,1),(3,3) and (1,4). Here divide the region R into three subregions so that the given integrals can be written as

$$\int\int_R (x+y)^{\frac{-1}{2}}\,dx\,dy=\int_1^2\int_{(5-3y)}^{\frac{y+3}{2}}(x+y)^{\frac{-1}{2}}\,dx\,dy+\int_2^3\int_{(y-3)}^{\frac{y+3}{2}}(x+y)^{\frac{-1}{2}}\,dx\,dy$$
$$+\int_3^4\int_{(y-3)}^{(9-2y)}(x+y)^{\frac{-1}{2}}\,dx\,dy.$$

to

$$\int\int_R (x+y)^{\frac{-1}{2}}\,dx\,dy=\int_1^2\int_0^1\frac{7y-7}{2\sqrt{\frac{1}{2}(7y-7)X-2y+5}}\,dX\,dy+\int_2^3\int_0^1\frac{9-y}{2\sqrt{\frac{1}{2}(9-y)X+2y-3}}\,dX\,dy$$
$$+\int_3^4\int_0^1\frac{12-3y}{\sqrt{(12-3y)X+2y-3}}\,dX\,dy.$$

Table 3. Absolute Errors of Example 3.

Haar	Absolute errors (PM)	Absolute errors (NM)	NM-PM
$M=2$, $N=2$	$1.149699028971\times10^{-2}$	$1.149698928971\times10^{-2}$	$-1.0000000827\times10^{-9}$
$M=4$, $N=4$	$2.987381789712\times10^{-3}$	$2.987385289712\times10^{-3}$	$3.4999998455\times10^{-9}$
$M=8$, $N=8$	$7.550162897129\times10^{-4}$	$7.550119897126\times10^{-4}$	$-4.3000003557\times10^{-9}$
$M=16$, $N=16$	$1.892872897131\times10^{-4}$	$1.892849897129\times10^{-4}$	$-2.3000001903\times10^{-9}$
$M=32$, $N=32$	$4.734928971261\times10^{-5}$	$4.735778971287\times10^{-5}$	$8.5000002592\times10^{-9}$
$M=64$, $N=64$	$1.183178971286\times10^{-5}$	$1.184078971272\times10^{-5}$	$8.9999998564\times10^{-9}$
$M=128$, $N=128$	$2.950289712533\times10^{-6}$	$2.928289712933\times10^{-6}$	$-2.1999999599\times10^{-8}$

Example 4

$$\int_0^{\pi}\int_0^{z}\int_0^{zy}\frac{1}{y}\sin\left(\frac{x}{y}\right)dxdydz=\int_0^{\pi}\int_0^{1}\int_0^{1}z^2\sin\left(\frac{X}{Y}\right)dXdY\,dz$$

Table 4. Absolute Errors of Example 4.

Haar	Absolute errors (PM)	Absolute errors (NM)	NM-PM
$\mathrm{M}=4,\mathrm{N}=4$	0.033053436	0.033053436	0
$\mathrm{M}=8,\mathrm{N}=8$	0.030061100	0.030061104	$4.00000000136\times10^{-9}$
$\mathrm{M}=16,\mathrm{N}=16$	0.000773945	0.000773939	$-5.99999999998\times10^{-9}$
$\mathrm{M}=32,\mathrm{N}=32$	0.011996531	0.011996550	$1.90000000004\times10^{-8}$
$\mathrm{M}=64,\mathrm{N}=64$	0.001599302	0.001599289	$-1.29999999998\times10^{-8}$
$\mathrm{M}=128,\mathrm{N}=128$	0.001430520	0.001430510	$-9.99999999994\times10^{-9}$

6. Conclusion

In this paper a new method based on Haar wavelets is applied for numerical integration of double and triple integrals with variable limits. This approach is the generalization and improvement of the methods [3]. From Table 1-4, it can be noticed that the approximate value is better compared to method [3] . Example 1-3 are taken exactly same from [3]. Finally error analysis for the case of two dimensional shows that the approximation become more accurate when k and ℓ are increased. Therefore for better results, using a large value of k and ℓ are recommended.

Acknowledgments

This work has been supported by UPM under the Research University Grant (RUGS), project number 05-01-12-1630RU.

References

[1] Burden R L and Faires J D 1993 *Numerical Analysis*(USA: PWS-KENT).
[2] Ferziger J H 1998 *Numerical Methods for Engineering Application*(New York:John Wiley and Sons Inc).
[3] Imran Aziz, Siraj-ul-Islam and Wajid Khan 2011 *comput.Math.Appl.* **61** 2770.
[4] Babolian E and Shahsavaran A 2009 *J. Comput. Appl.Math.* **225** 87.
[5] Daubechies I 1992 *Ten Lectures on Wavelets* (Philadelphia:SIAM).
[6] Maleknejad K and Mirzaee F 2003 *Int. J. Comput. Math.* **8 (11)** 1397.
[7] Lepik U and Tamme E 2004 *Proc* (Turkey:Antalaya) p494.
[8] Lepik U and Tamme E 2007 *Proc. Estonian Acad.Sci.Phys.Math.* **56** 17.
[9] Imran Aziz, Siraj-ul-Islam and Wajid Khan 2010 *comput.Math.Appl.* **59** 2026.
[10] Wojtaszczyk P 1997, *A Mathematical Introduction to Wavelets* (Cambridge: University Press).

Journal of Physics: Conference Series **435** (2013) 012043 doi:10.1088/1742-6596/435/1/012043

On pricing futures options on random binomial tree

Kamola Bayram[1] and Nasir Ganikhodjaev[2]

[1,2] Department of Computational and Theoretical Sciences,Faculty of Science, International Islamic University Malaysia,P.O. Box 10, 50728 Kuantan, Malaysia.

Abstract. The discrete-time approach to real option valuation has typically been implemented in the finance literature using a binomial tree framework. Instead we develop a new model by randomizing the environment and call such model a random binomial tree. Whereas the usual model has only one environment *(u, d)* where the price of underlying asset can move by *u* times up and *d* times down, and pair *(u, d)* is constant over the life of the underlying asset, in our new model the underlying security is moving in two environments namely (u_1, d_1) and (u_2, d_2). Thus we obtain two volatilities σ_1 and σ_2. This new approach enables calculations reflecting the real market since it consider the two states of market normal and extra ordinal. In this paper we define and study Futures options for such models.

1. Introduction

The Binomial tree for pricing options was first introduced by Cox, Ross, and Rubinstein and [1, 2]. The method arises from discrete random walk models of the underlying security and has a great advantage of being a practical and a flexible tool that can be used to price a large variety of traded options. Moreover, the binomial model is very elegant for pricing of American calls on dividend paying stocks and for pricing of American put options, in general, in that it allows traders to consider early exercise possibilities at each point in the option's price path.

There are two main assumptions underlying the binomial methods. First is that the continuous random walk may be modelled by a discrete random walk with the following properties:[5]

- The asset price S changes only at the discrete times Δt, $2\Delta t$, $3\Delta t$, $4\Delta t$, ..., up to $N\Delta t = T$, the expiry date of the derivative security.
- If the asset price is S^n at the time $n\Delta t$ then at time $(n+1)\Delta t$ it may take only one of the two possible values, $uS^n > S^n$ or $dS^n < S^n$. Thus the price of the underlying asset can move up by u times or move down by d times at each time step Δt where $0 < d < 1 < u$. (See Fig. 1). Note that the up jump u and the down jump d are the same for all time-steps.
- The probability of S moving to uS and the probability of S moving to dS are p and $1 - p$, respectively, are known.

---

[1] Kamola Bayram, E-mail: kamolaomer@hotmail.com

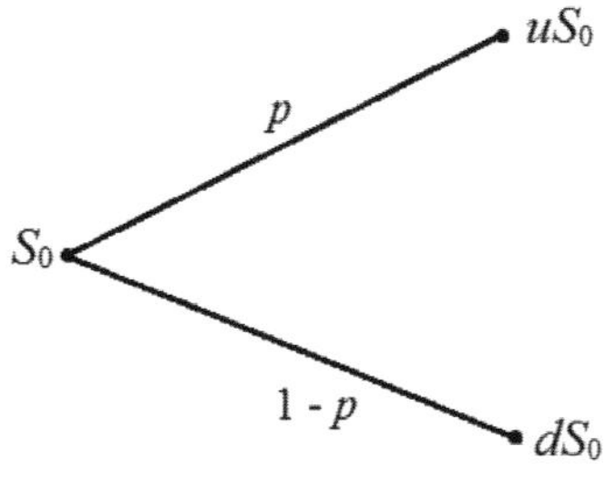

Figure 1. Underlying asset movement in general one-step tree.

Starting with initial value of the underlying asset price S_0 the remaining life-time of derivative security is divided into N time-steps of size $\Delta t = T/N$. The asset price S is assumed to move only at times $n\Delta t$ for $n = 1,2,3,..., N$. Thus a tree of all possible asset price movements could be constructed. At the first step we start with the initial value of the underlying asset $S=S_0$ where we have only two possible asset prices uS_0 and dS_0; see Fig. 2. At the second step we obtain three possible asset prices which are u^2S_0, udS_0 and d^2S_0. And we continue sustaining the same method until the expiry date T of the security is reached.

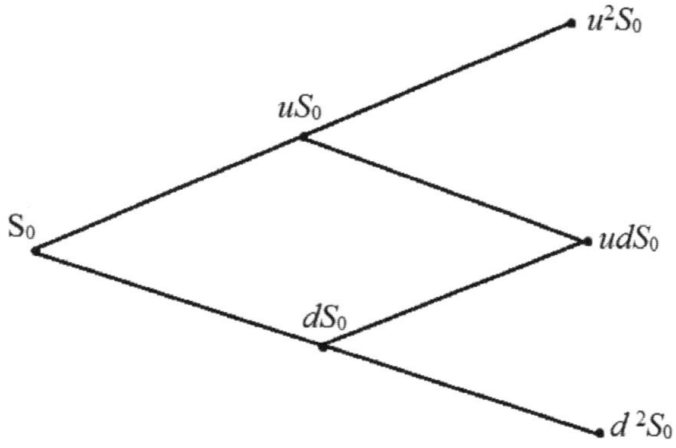

Figure 2. Underlying asset price movement in general two-step tree.

Remark. The value of an asset is the same when an up-jump is followed by a down-jump and a down-jump followed by an up-jump the binomial tree reconnects (that is $udS = duS$), and after n time-steps there are only $n + 1$ possible asset prices.

The second assumption underlying the binomial option pricing model is that no arbitrage opportunity exists.

The no-arbitrage opportunity can be generalized by considering a stock whose price is S_0 and an option on the stock with current price f. As we stated above the price of the underlying stock moves either up to a new level uS_0 or down to a new level dS_0. Suppose that the payoff from the option after up movement of a stock to uS_0 is equals to f_u and to f_d after when it turns to dS_0. Consider a portfolio consisting of a long position in Δ shares and a short position in one option. If the stock price moves up the value of the portfolio at time T is

$$uS_0\Delta - f_u$$

If the price of the underlying stock moves down to the value of the portfolio becomes

$$dS_0\Delta - f_d$$

Two portfolios are equal when the equation

$$uS_0\Delta - f_u = dS_0\Delta - f_d$$

or

$$\Delta = \frac{f_u - f_d}{S_0 u - S_0 d}$$

(1)

The portfolio is riskless in this case, and it must earn the risk-free interest rate that is no arbitrage opportunities could exist. A value Δ in equation (1) shows the ratio of the change in the option price to the change in the stock price.

The binomial tree model is a natural bridge, overture to continuous models for which it is possible to derive the Black-Sholes option pricing formula [7-9].

2. Futures Options

Futures options are options on a futures contract. Here we have only the right to buy a futures contract by a specified price and date whereas in the case of options where underlying asset is commodity or stock share we have the right to buy or sell the underlying asset which is tangible. The very first introduction of futures option market was in the early 1970's and had a short life ended up with scandal. It started with promises of unscrupulous dealers of huge possible rewards with low risk. Their assumption was that the options will never be exercised, so they did not hold the underlying futures contracts but collect millions of dollars from premiums. Their game revealed when adequate number of option buyers whose options end up in the money position wanted to exercise them but faced up with no cashier's checks, disconnected telephones and no forwarding addresses.

In 1982, the futures options have been traded on futures exchanges on an experimental basis, where their financial integrity is warranted by the exchange and an option clearinghouse. In 1987 the permanent trading was approved the popularity of contract has grown tremendously. The main reason why the futures options became such popular is lies under the fact that the futures contracts are more liquid and easier to trade compared to the underlying asset. Furthermore, exercising of futures options usually don't lead to the delivery of the underlying asset, since in most cases the underlying futures contract closes up before two working days, in most cases, to delivery.

Since the option is a contract which gives you the only right to buy or sell a specific underlying asset, the futures contract in our case, has no practical value. So the valuation of an option is the main issue. There are four main factors which influence the options value:

1. *Intrinsic value* – the difference between the options's striking price and the underlying futures price.

   Intrinsic Value = F – K

   where F is the futures price and K is the exercise price of an option.

2. *Time Value* – the value which market participants are willing to pay for the time the option has left to run.

Time Value = Intrinsic Value – Premium

3. *Striking Price – K*

4. *Volatility* – σ

However, the first three factors are known while the last one, volatility, is not concrete but is very important. The buyers enter into options contract with a hope that the underlying futures contract will increase in value whereas the seller has the opposite hope. Thus the value of an option will increase as the volatility increases. An interesting example from 1988 soybean futures market could be given here to show how volatility plays role in option prices. In April,1988, soybeans were traded quietly. Then the spectre of drought appeared, and the market erupted. Therefore, soybean options should have had much more time value in June than they did in May, before the draught occurred.
Lets examine Table 1, to understand the issue.

Table1. Soybean Call Options prices from 5 May to 21 June 1988 [4]

Soybean Call Options					
Date	*Option*	*Underlying Futures Price*	*Option Premium*	*Intrinsic Value*	*Time Value*
May 5	August 675	7.08	40	33	7
June 21	September 1025	10.36	95	11	84

From 5th May to 21st June, the time value of a slightly in-the-money soybean call option with 3 months to expiration increased 12 times from 7 cents to 84 cents. The only factor that is significantly different between the two options is the volatility of the underlying futures contract at the time. In May volatility was normal but six weeks later, soybean futures reached the record high and were swinging wildly between expanded daily price limits [4].

3. Random Binomial Tree Model

3.1 Random Walks in a Random Environment

Random walks in a random environment on the integers Z was introduced and studied by Solomon in 1975 [10]. Let $\{\alpha_n\}$ be a sequence of independent, identically distributed random variables with $0 \le \alpha_n \le 1$ for all n. The random walk in a random environment on the integers Z is the sequence $\{X_n\}$, where $X_0 = 0$ and inductively $X_{n+1} = X_n + 1$, $(X_n - 1)$, with probability $\alpha_{X_n}, (1-\alpha_{X_n})$. Solomon have proved that randomizing the environment in some sense slows down the random walk. Later Menshikov and Petritis [11] studied random walks in a random environment on a regular, rooted, colored tree and the asymptotic behaviour of the walks was classified for ergodicity or transience in terms of the geometric properties of the matrix describing the random environment.

3.2 Random Binomial Tree Model

Since we are dealing with binomial process the stock price can take only two possible values at each node of the tree. Thus during each time step the value of the stock moves either upwards by u times with probability p or downwards by d time with probability $(1-p)$. Let us call the pair (u, d) *environment* of the binomial tree. Note that in this model, pair (u, d) is the same for any moment of the time that is whether this is the first, second or nth step the probability of upward movement u is always equals to p.

Now we define random environment and random binomial tree model as follows. Let $\{U_n\}$ and $\{D_n\}$ be the sequences of independent, identically distributed random variables with $0< D_n < 1 < U_n$ for all n. Then this pair (U_n, D_n) is called a *random environment*.

The random binomial tree model in a random environment (U_n, D_n) is defined as a model such that during n-th moment of time the value of stock either moves up with a certain probability by U_n times or moves down by D_n times with a certain probability for some realization of these random variables.

3.3 Simplest Random Binomial Tree Model

Let $\{U_n\}$ and $\{D_n\}$ be the sequences of independent, identically distributed random variables such that the random variable U_n takes only two values u_1 and u_2 and respectively the random variable D_n takes two values d_1 and d_2 . Thus the pair of random variables (U_n, D_n) describes two possible environments (u_1, d_1) and (u_2, d_2).

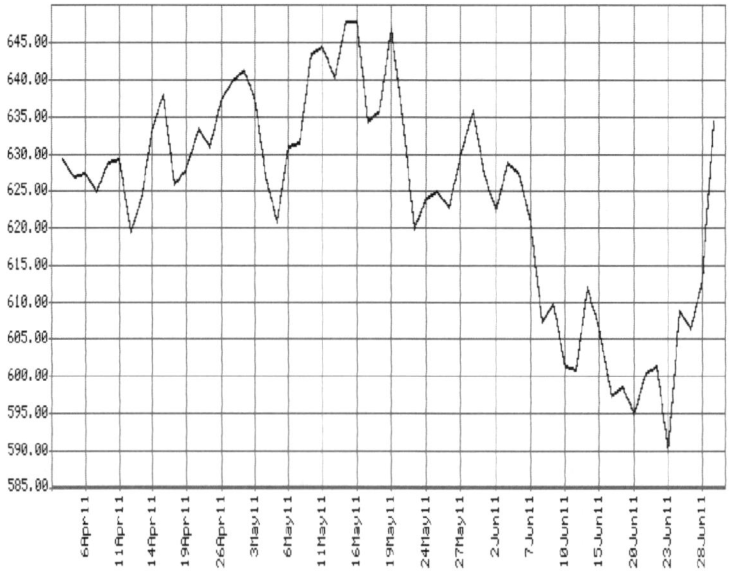

Figure 3. Stock shares of Rolls Royce for period from 6th April to 28th June 2011.(www.rolls-royce.com/investors/detailedshareprice)

To avoid complicated mathematical formulations let introduce and explore simplest example of random environment in binomial tree model using tossing of a coin.

Let us consider two environments (u_1, d_1) and (u_2, d_2). We will repeatedly toss a coin and if the result is a "Head" we will apply the first environment, in the opposite case - the second. That is we will choose environment randomly and we will call such a model *the simplest random binomial tree model*. Our main goal in randomizing the environment is to make the valuation of an option on the underlying stock more accurate. We see that in a usual tree we have only one environment and our tree could move up by u times or down by d times, at each node. However, this assumption is very far from reality, since the stock price movements are very sensitive and react immediately to any kind of information not only domestic but also international. For example, changes in risk free interest rate, conflicts with a neighbour countries, a very good performance of a competitor's company, employment of a new CEO...,etc. This sensitivity implies the randomizing of an environment. We should have at least two environments where one will represent the normal state of market and the second otherwise. Thus one of our environments will be less volatile whereas the second will have comparatively high volatility. And of course the first environment which is less volatile will have a higher probability than the second, hence the extraordinary situations appear once in a while.

To justify the necessity of our model we consider the stock price movements of Rolls Royce for nearly 3 months period given in Figure 3. By examining the graph , one can see that there are two environments instead of one, namely (u_1, d_1) = (1.02;0.98) and (u_2, d_2) = (1.04;0.96). For this case we have two environments and are chosen randomly, namely, *H,T,H,T,H,H,H,T,H,T,T,T, H,H,T,T,T....* In total we get 34 steps. And just 6 of 34 steps represent "Tail", or in other words the second more volatile environment. In a risk-neutral world, it is possible to show that the model parameters must satisfy the following relations:

$$u = e^{\sigma\sqrt{\Delta t}} \tag{2}$$

$$d = e^{-\sigma\sqrt{\Delta t}} \tag{3}$$

By substituting the values of (u_1,d_1) and (u_2,d_2) in equations (2) and (3) one can calculate the volatilities of our two environments as $\sigma_1 = 0.26$ and $\sigma_2 = 0.5$. We see that the second environment is more volatile up to 2 times. That is indicating an extraordinary situations of a company. In this example the probability of getting a "Head" is 0.82 and probability of getting a "Tail" is only 0.18.

One can see that this model is more exactly reflecting the trajectory of stock price.

Theorem 1 The simplest random binomial tree model with two possible environments (u_1 , d_1) and (u_2 , d_2) is arbitrage –free and complete if and only if the following conditions holds

$$\max\{d_1, d_2\} < 1 + r < \min\{u_1, u_2\} \tag{4}$$

where r is the risk free interest rate.

Proof. The proof immediately follows from a similar proposition for binomial tree models with single environment (see for example [3,4]).

4. Futures Options on a Random Binomial Tree

In our paper [12] we have studied European call option. Below we consider futures options.

4.1 Futures Options on a One Step Random Binomial Tree

Firstly, we shall consider European call futures option on a simplest random single-period binomial tree model. In this case N=1 and $\Omega_1=\{H, T\}$ and random environments (u_1, d_1) and (u_2 , d_2) correspond to outcomes of a "Head" and "Tail", respectively. Then we can compute the option prices $f(H)$ and $f(T)$ assuming that condition (4) holds and using the well known formula (see for example [3,4])

$$f = e^{-r\Delta t}[pf_u + (1-p)f_d] \tag{5}$$

where

$$p = \frac{1-d}{u-d} \tag{6}$$

Then we define option price f for a simplest random single-period binomial tree model as expectation of random variable **f**: $\Omega_1 \rightarrow$R, i.e.,

$$f = \alpha f(H) + (1-\alpha) f(T). \tag{7}$$

where α is the probability of getting a "Head".

Example 1.

Let us consider a simplest random binomial tree model on a European call futures option on crude oil with two environments u_1= 1.1, d_1 =0.9 and u_2 = 1.4, d_2=0.7. The time to the options maturity is 6 months, the current futures price is $30, the exercise price is $31, the risk free interest rate is 6% per annum. We consider the probability of turning out the first environment as α = 0.89 and the second as 1-α = 0.11. Since the second environment has high volatility it indicates an extraordinary state of the market which seldom happens.

Using formulas (5-7) one can compute that

$$f(H)=0.9704; \quad f(T)= 4.5752 \text{ and } \quad \mathbf{f}=\alpha f(H) + (1-\alpha)f(T)=1.3670$$

4.2 Futures Options On Two Period Binomial Tree

In this case N=2 and $\Omega_2=\{(H,H);(H,T),(T,H);(T,T)\}$ and random environments (u_1, d_1) and (u_2, d_2) correspond to outcomes of a "Head" and "Tail' respectively.

For outcomes (H,H) and (T,T), the value of corresponding options one can compute directly using well known formula for the case of two time steps of usual binomial tree model.

$$f = e^{-2r\Delta t}[p^2 f_{uu} + 2p(1-p)f_{ud} + (1-p)^2 f_{dd}] \tag{8}$$

where f_{uu} is the value of the futures contract after two up movements, f_{ud} is after one up and one down and f_{dd} is after two down movements. For outcome (H,T) repeated application of equation (5) gives

$$f_{u_1} = e^{-r\Delta t}[p_2 f_{u_1u_2} + (1-p_2)f_{u_1d_2}] \tag{9}$$

$$f_{d_1} = e^{-r\Delta t}[p_2 f_{d_1u_2} + (1-p_2)f_{d_1d_2}] \tag{10}$$

Substituting from equations (9) and (10) into (5), we get

$$f(H,T) = e^{-2r\Delta t}[p_1p_2 f_{u_1u_2} + p_1(1-p_2)f_{u_1d_2} + (1-p_1)p_2 f_{d_1u_2} + (1-p_1)(1-p_2)f_{d_1d_2}]$$

and it is evident that

$$f(H,T)=f(T,H)$$

Then the option price f for a simplest random two-step binomial tree model is defined as expectation of random variable **f**: $\Omega_2 \to R$, i.e.,

$$f = \alpha^2 f(H,H) + 2\alpha(1-\alpha)f(H,T) + (1-\alpha)^2 f(T,T) \tag{12}$$

Example 2.

Now let us look at previous example in a two-step model. Using formulas (8) and (11) one can compute that

$$f(H,H)=1.2478;\ f(H,T) = f(T,H) = 4.4400;\ f(T,T)=4.8094$$

and using formula (12) the value of the option is;

$$f=\alpha^2 f(H,H) + 2\alpha(1-\alpha)f(H,T) + (1-\alpha)^2 f(T,T)=1.9159$$

4.3 Futures Options On Multi Period Random Binomial Tree

Suppose that a tree with N- time steps is used to value a European futures call option with strike price K and life T. Each step is of length T/N. For usual binomial tree model, if there have been j upward movements and N-j downward movements on the tree, the final stock price is $S_0 u^j d^{N-j}$, where u is the proportional up movement, d is the proportional down movement, and S_0 is the initial stock price. Then option price f is computed as follows [3,4]

$$e^{-rT}\sum_{j=0}^{N}\binom{N}{j}p^{j}(1-p)^{N-j}\max(S_0u^{j}d^{N-j}-K,0) \qquad (13)$$

Now consider a simplest random binomial tree model where the randomness of environment is defined by tossing a coin (not necessarily fair) N times. Let Ω_N be a sample space with outcome $\omega = (\omega_1, \omega_2, \ldots, \omega_N)$, where $\omega_i \in \{H,T\}$, $i = 1,2,\ldots, N$, observing "Heads" or "Tails". Let $\Pr\{H\}=\alpha$ and $\Pr\{T\}= 1-\alpha$, where $0\le \alpha \le 1$. For any outcome $\omega\in\Omega_N$ let $M(\omega)$ be the number of tails in tossing of a coin N times. If $M(\omega) =0$ or $M(\omega)=N$, then one can compute option price for European call using of the previous formula. Now we show how to compute the option price for $0< M(\omega)< N$. Let $M(\omega)=1$, then using trivial identity

$$(a+b)^N=(a+b)^{N-1}(a+b),$$

we can find option price f_1 as follows:

$$f_1 = e^{-rT}(\sum_{j=0}^{N-1}\sum_{k=0}^{1}\frac{(N-1)!}{(N-1-j)!\,j!}p_1^{j}(1-p_1)^{N-1-j}p_2^{k}(1-p_2)^{1-k}\max(S_0u_1^{j}d_1^{N-1-j}u_2^{k}d_2^{1-k}-K,0)$$

Similarly for $M(\omega)=m$, where $1\le m \le N$-1, we can find option price f_m as follows:

$$f_m = e^{-rT}(\sum_{j=0}^{N-m}\sum_{k=0}^{m}\frac{(N-m)!}{(N-m-j)!\,j!}p_1^{j}(1-p_1)^{N-m-j}p_2^{k}(1-p_2)^{m-k}\max(S_0u_1^{j}d_1^{N-1-j}u_2^{k}d_2^{m-k}-K,0),$$

(14)

where p_1 and p_2 can be computed using formula (6) for corresponding environments (u_1, d_1) and (u_2, d_2), respectively.

Let

$$A_m=\{\omega\in\Omega_N;\ M(\omega)=m\}$$

It is well known that

$$\Pr(A_m)=\binom{N}{m}\alpha^{N-m}(1-\alpha)^{m}$$

Now option price f for a simplest random N-period binomial tree model we define as expectation of a random variable **f**: $\Omega_N\to$R, i.e

$$\mathrm{f}=\sum_{m=0}^{N}\binom{N}{m}\alpha^{N-m}(1-\alpha)^{m}f_m \qquad (15)$$

5. Conclusion

One way of deriving the famous Black-Scholes-Merton result for valuing a European option on a non-dividend-paying stock is by allowing the number of time steps in a binomial tree to approach infinity. Above we have introduced random binomial tree model and derived formula for computing futures options for such models. Similarly one can define and study other options.

To derive formula for the valuation of a European call option for random binomial tree models that is similar with the Black-Scholes-Merton formula for usual binomial tree models, we have to investigate [12] by allowing the number of time steps to approach infinity. Since in the case of random binomial tree models we have "doubled" randomness that we need to apply and develop methods suggested by Solomon [7] and Menshikov and Peteritis [8]. Also it is an interesting problem to derive the Black-Scholes-Merton differential equation with random coefficients. All these problems will be subject of coming papers.

Acknowledgments

This work was supported by the IIUM Grant EDW B 12-403-0881.

References

[1] Cox J C, Ross S A and Rubinstein M 1979 *J. Fin. Econ.* **7** 229
[2] Rendleman R J and Bartter B J 1979 *J.Fin.* **24** 1093
[3] Hull J C 2011 *Options, Futures, and Other Derivatives* (City Pearson)
[4] Lofton T 2001 *Getting Started in Futures* (New York, John Wiley and Sons Interscience)
[5] Gibson R 1991 *Option Valuation:Analyzing and Pricing Standardized Option Contracts* (Geneva McGraw-Hill Intersience)
[6] Pascucci A 2011 *PDE and Martingale Methods in Option Pricing*. (Bocconi Iniversity Press, Springer)
[7] Black F and Scholes M 1973 *J. Pol. Econ.* **81** 637
[8] Georgiadis, E. 2011.Binomial Options Pricing Has No Closed-Form Solution. *Forthcoming in Algorithmic Finance*, pp. 1-6, http://algorithmicfinance.org.
[9] Merton R C *Bell J. Econ. Man. Sc*, **4** 141
[10] Solomon F 1975 Random walks in a random environment. *The Annals of Probability* 3 **1** 1
[11] Menshikov N and Peteritis D 2001 Random walks in random environment on trees and multiplicative chaos. arXiv:math/0112103v1 [math.PR] pp 1-30
[12] Ganikhodjaev N N and Bayram K 2012 *J.Appl. Scien.* **12(18)** 1978

Journal of Physics: Conference Series **435** (2013) 012044 doi:10.1088/1742-6596/435/1/012044

Novel method for solving singular two-point BVPs

A. Sami Bataineh

Department of Mathematics, Faculty of Science, Al-Balqa Applied University,19117 Al Salt, Jordan

E-mail: a_s_bataineh@yahoo.com

Abstract. In this paper, we present a new algorithm to find exact and approximate analytical solutions of singular two-point boundary-value problems (BVPs) based on Bernstein operational matrix of differentiation. Different from other numerical techniques, Bernstein polynomials and their properties are employed for deriving a general procedure for forming this matrix. The proposed method can be applied to linear and nonlinear problems. The scheme is tested for some examples and the obtained results demonstrate efficiency of the proposed method.
Keywords. Singular two-point boundary value problems; Bernstein polynomials; Operational matrix of differentiation. PACS Numbers: 02.60.Lj; 02.30.Hq; 02.10.Yn; 02.10.Ud; 31.15.B-

1. Introduction

Singular ordinary differential equations arise in the fields of gas dynamics, Newtonian fluid mechanics, fluid mechanics, fluid dynamics, elasticity, reaction-diffusion processes, chemical kinetics and other branches of applied mathematics [1]. The theory of singular boundary-value problems has become a very important area of investigation in recent years [1–5] references therein.

In this work, we consider the singular two-point boundary value problems (BVPs) of the type

$$\frac{1}{p(x)}u''(x) + \frac{1}{q(x)}u'(x) + \frac{1}{r(x)}\left(u(x)\right)^k = g(x), \quad 0 < x \leq 1, \tag{1}$$

subject to the boundary conditions

$$u(0) = \alpha_1, \quad u(1) = \beta, \quad \text{or} \quad u'(0) = \alpha_2, \quad u(1) = \beta, \tag{2}$$

where p, q, r and g are continuous functions on $(0, 1]$ and the parameters $\alpha_1, \alpha_2, \beta$ are real constants.

The aim of the present paper is to apply of Bernstein operational matrix of differentiation for the first time, to propose a reliable numerical technique for solving linear and nonlinear singular two-point BVPs (1)–(2). Several examples are studied to demonstrate the capacity of the method.

2. Bernstein polynomials and its operational matrix of differentiation

The Bernstein polynomials of degree m on $[0, 1]$ are defined by

$$B_{i,m}(t) = \binom{m}{i} t^i (1-t)^{m-i}, \quad i = 0, 1, \ldots, m,$$

where the binomial coefficient is

$$\binom{m}{i} = \frac{m!}{i!\,(m-i)!}.$$

There are $m+1$ nth-degree Bernstein polynomials. For mathematical convenience, we usually set $B_{i,m} = 0$, if $i < 0$ or $i > m$.

In general, we approximate any function $u(x)$ with the first $(m+1)$ Bernstein polynomials as

$$u(x) = \sum_{i=0}^{m} c_i B_{i,m}(x) = \mathrm{C}^{\mathrm{T}} \phi(x), \tag{3}$$

where

$$\begin{aligned} \mathrm{C}^{\mathrm{T}} &= [\mathrm{c}_0, \mathrm{c}_1, \ldots, \mathrm{c}_m], \\ \phi(x) &= [B_{0,m}(x), B_{1,m}(x), \ldots, B_{m,m}(x)]^{\mathrm{T}}. \end{aligned}$$

The derivatives of the vector $\phi(x)$ can be expressed as

$$\frac{\mathrm{d}\phi(x)}{\mathrm{d}x} = \mathrm{D}^1 \phi(x) \tag{4}$$

where D^1 is the $(m+1) \times (m+1)$ operational matrix of derivative given as.

Let,

$$\phi(x) = \begin{pmatrix} B_{0,m} \\ B_{1,m} \\ \vdots \\ B_{m,m} \end{pmatrix} = \begin{pmatrix} a_0 + a_1 x + \cdots + a_m x^m \\ b_0 + b_1 x + \cdots + b_m x^m \\ \vdots \\ z_0 + z_1 x + \cdots + z_m x^m \end{pmatrix} = \begin{pmatrix} a_0 & a_1 & \ldots & a_m \\ b_0 & b_1 & \ldots & b_m \\ \vdots & \vdots & \vdots & \vdots \\ z_0 & z_1 & \ldots & z_m \end{pmatrix} \begin{pmatrix} 1 \\ x \\ \vdots \\ x^m \end{pmatrix},$$

where $a_0, a_1, \ldots, a_m$, $b_0, b_1, \ldots, b_m$ and $z_0, z_1, \ldots, z_m$ are the binomial coefficients. Now,

$$\begin{aligned} \frac{\mathrm{d}\phi(x)}{\mathrm{d}x} &= \begin{pmatrix} a_0 & a_1 & \ldots & a_m \\ b_0 & b_1 & \ldots & b_m \\ \vdots & \vdots & \vdots & \vdots \\ z_0 & z_1 & \ldots & z_m \end{pmatrix} \frac{\mathrm{d}}{\mathrm{d}x} \begin{pmatrix} 1 \\ x \\ \vdots \\ x^m \end{pmatrix} \\ &= \begin{pmatrix} a_0 & a_1 & \ldots & a_m \\ b_0 & b_1 & \ldots & b_m \\ \vdots & \vdots & \vdots & \vdots \\ z_0 & z_1 & \ldots & z_m \end{pmatrix} \begin{pmatrix} 0 \\ 1 \\ \vdots \\ m x^{m-1} \end{pmatrix} \\ &= \begin{pmatrix} a_0 & a_1 & \ldots & a_m \\ b_0 & b_1 & \ldots & b_m \\ . & . & \ldots & . \\ \vdots & \vdots & \vdots & \vdots \\ z_0 & z_1 & \ldots & z_m \end{pmatrix} \begin{pmatrix} 0 & 0 & \ldots & 0 & 0 \\ 1 & 0 & \ldots & 0 & 0 \\ 0 & 2 & \ldots & 0 & 0 \\ \vdots & \vdots & \vdots & \vdots & \vdots \\ 0 & 0 & \ldots & m & 0 \end{pmatrix} \begin{pmatrix} 1 \\ x \\ . \\ \vdots \\ x^m \end{pmatrix} \end{aligned}$$

$$
\begin{aligned}
&= \begin{pmatrix} a_0 & a_1 & \dots & a_m \\ b_0 & b_1 & \dots & b_m \\ . & . & \dots & . \\ \vdots & \vdots & \vdots & \vdots \\ z_0 & z_1 & \dots & z_m \end{pmatrix} \begin{pmatrix} 0 & 0 & \dots & 0 & 0 \\ 1 & 0 & \dots & 0 & 0 \\ 0 & 2 & \dots & 0 & 0 \\ \vdots & \vdots & \vdots & \vdots & \vdots \\ 0 & 0 & \dots & m & 0 \end{pmatrix} \begin{pmatrix} a_0 & a_1 & \dots & a_m \\ b_0 & b_1 & \dots & b_m \\ . & . & \dots & . \\ \vdots & \vdots & \vdots & \vdots \\ z_0 & z_1 & \dots & z_m \end{pmatrix}^{-1} \\
&\times \begin{pmatrix} a_0 & a_1 & \dots & a_m \\ b_0 & b_1 & \dots & b_m \\ . & . & \dots & . \\ \vdots & \vdots & \vdots & \vdots \\ z_0 & z_1 & \dots & z_m \end{pmatrix} \begin{pmatrix} 1 \\ x \\ . \\ \vdots \\ x^m \end{pmatrix} \\
&= \begin{pmatrix} a_0 & a_1 & \dots & a_m \\ b_0 & b_1 & \dots & b_m \\ . & . & \dots & . \\ \vdots & \vdots & \vdots & \vdots \\ z_0 & z_1 & \dots & z_m \end{pmatrix} \begin{pmatrix} 0 & 0 & \dots & 0 & 0 \\ 1 & 0 & \dots & 0 & 0 \\ 0 & 2 & \dots & 0 & 0 \\ \vdots & \vdots & \vdots & \vdots & \vdots \\ 0 & 0 & \dots & m & 0 \end{pmatrix} \begin{pmatrix} a_0 & a_1 & \dots & a_m \\ b_0 & b_1 & \dots & b_m \\ . & . & \dots & . \\ \vdots & \vdots & \vdots & \vdots \\ z_0 & z_1 & \dots & z_m \end{pmatrix}^{-1} \phi(x) \\
&= \mathrm{D}^1\phi(x).
\end{aligned}
$$

Which is satisfy Eq. (4), also it can be generalized Eq. (4) as

$$\frac{\mathrm{d}^2\phi(x)}{\mathrm{d}x^2} = (\mathrm{D}^1)^2\phi(x) = (\mathrm{D}^1)(\mathrm{D}^1), \dots, \frac{\mathrm{d}^n\phi(x)}{\mathrm{d}x^n} = \underbrace{(\mathrm{D}^1)(\mathrm{D}^1)\dots(\mathrm{D}^1)}_{n-\text{times}} \tag{5}$$

3. Applications of the operational matrix of derivative

To solve (1)–(2) by means of the Bernstein operational matrix of derivative, we approximating $(u(x))^k$ as

$$(u(x))^k \simeq \left(\mathrm{C}^{\mathrm{T}}\phi(x)\right)^k. \tag{6}$$

By using Eqs.(4)– (5), we have

$$
\begin{aligned}
u''(x) &\simeq \mathrm{C}^{\mathrm{T}}(\mathrm{D}^1)^2\phi(x), && (7)\\
u'(x) &\simeq \mathrm{C}^{\mathrm{T}}\mathrm{D}^1\phi(x), && (8)
\end{aligned}
$$

Employing Eqs. (6)–(8), then we can defined the residual $\Re(x)$ for Eq. (1) as

$$\Re(x) \simeq \frac{1}{p(x)}\mathrm{C}^{\mathrm{T}}(\mathrm{D}^1)^2\phi(x) + \frac{1}{q(x)}\mathrm{C}^{\mathrm{T}}\mathrm{D}^1\phi(x) + \frac{1}{r(x)}\left(\mathrm{C}^{\mathrm{T}}\phi(x)\right)^k - g(x). \tag{9}$$

Now, to find the solution $u(x)$ given in Eq. (3) we generate $m-1$ linear equations as in a typical tau method [6] by applying

$$\int_0^1 \Re(x)B_{i,m}(x)\,\mathrm{d}x = 0, \quad i = 0, 1, \dots, m-2. \tag{10}$$

Also, by substituting boundary conditions (2) into Eqs. (6) and (8) we have

$$u(0) = \mathrm{C}^{\mathrm{T}}\phi(0) = \alpha_1, \quad u(1) = \mathrm{C}^{\mathrm{T}}\phi(1) = \beta, \tag{11}$$

or

$$u'(0) \;=\; \mathrm{C}^{\mathrm{T}}\mathrm{D}^{1}\phi(0) = \alpha_2, \quad u(1) = \mathrm{C}^{\mathrm{T}}\phi(1) = \beta, \tag{12}$$

Eq. (10), associate with (11) or (12) generate (m+1) set of linear and nonlinear equations respectively. These equations can be solved for unknown coefficients of the vector C^{T}. Consequently, $u(x)$ given in Eq. (6) can be easily calculated.

4. Numerical experiments

To illustrate the effectiveness of the presented method, we shall consider the following examples of singular two-point BVPs.

4.1. Example 1.

We first consider the linear singular two-point BVP [7, 8],

$$u''(x) + \frac{1}{x}u'(x) + u(x) = g(x), \quad 0 < x \leq 1, \tag{13}$$

subject to the boundary conditions

$$u(0) = 0, \quad u(1) = 0. \tag{14}$$

The exact solution of (13) subject to (14) in the case of $g(x) = 4 - 9x + x^2 - x^3$ is

$$u(t) = x^2 - x^3. \tag{15}$$

To solve (13)–(14) we applying the technique described above with $m = 3$, we approximate the solution as

$$u_(x) = c_0\, B_{0,3}(x) + c_1\, B_{1,3} + c_2\, B_{2,3} + c_3\, B_{3,3} = \mathrm{C}^{\mathrm{T}}\phi(x), \tag{16}$$

According to (4), we have

$$\mathrm{D}^1 = \begin{pmatrix} -3 & -1 & 0 & 0 \\ 3 & -1 & -2 & 0 \\ 0 & 2 & 1 & -3 \\ 0 & 0 & 1 & 3 \end{pmatrix}, \quad (\mathrm{D}^1)^2 = \begin{pmatrix} 6 & 4 & 2 & 0 \\ -12 & -6 & 0 & 6 \\ 6 & 0 & -6 & -12 \\ 0 & 2 & 4 & 6 \end{pmatrix}. \tag{17}$$

Therefore using Eq. (10) we obtain

$$-\frac{3}{56} - \frac{79}{280}c_0 + \frac{1}{56}c_1 + \frac{9}{56}c_2 + \frac{43}{280}c_3 \;=\; 0, \tag{18}$$

$$\frac{11}{280} + \frac{1}{56}c_0 - \frac{15}{56}c_1 - \frac{33}{280}c_2 + \frac{131}{280}c_3 \;=\; 0. \tag{19}$$

Now, by applying Eq. (11) we have

$$c_0 \;=\; 0, \tag{20}$$

$$c_3 \;=\; 0. \tag{21}$$

Finally by solving Eqs. (18)–(21) we get

$$c_0 = 0, \quad c_1 = 0, \quad c_3 = \frac{1}{3}, \quad c_4 = 0.$$

Thus

$$u(x) = \begin{pmatrix} 0 & 0 & \frac{1}{3} & 0 \end{pmatrix} \begin{pmatrix} 1 - 3\,x + 3\,x^2 - x^3 \\ 3\,x - 6\,x^2 + 3\,x^3 \\ 3\,x^2 - 3\,x^3 \\ x^3 \end{pmatrix} = x^2 - x^3,$$

which is the same as exact solution (15) and the solution obtained by MHAM [8].

4.2. Example 2.

Consider the linear singular two-point BVPs [8],

$$\left(1 - \frac{x}{2}\right) u''(x) + \frac{3}{2}\left(\frac{1}{x} - 1\right) u'(x) + \left(\frac{x}{2} - 1\right) u(x) = g(x), \quad 0 < x \le 1, \tag{22}$$

subject to the boundary conditions

$$u'(0) = 0, \quad u(1) = 0. \tag{23}$$

The exact solution of (22) subject to (23) in the case

$$g(x) = 5 - \frac{29x}{2} + \frac{13x^2}{2} + \frac{3x^3}{2} - \frac{x^4}{2},$$

is

$$u(x) = x^2 - x^3. \tag{24}$$

By the same manipulations as in the previous example and assuming $m = 3$, we have

$$c_0 = 0, \quad c_1 = 0, \quad c_3 = \frac{1}{3}, \quad c_4 = 0.$$

Thus

$$u(x) = \begin{pmatrix} 0 & 0 & \frac{1}{3} & 0 \end{pmatrix} \begin{pmatrix} 1 - 3\,x + 3\,x^2 - x^3 \\ 3\,x - 6\,x^2 + 3\,x^3 \\ 3\,x^2 - 3\,x^3 \\ x^3 \end{pmatrix} = x^2 - x^3,$$

which is the same as exact solution (15) and the solution obtained by MHAM [8].

4.3. Example 3.

Now, we deal with the following linear singular two-point BVP [8, 9],

$$u''(x) + \frac{1}{x} u'(x) + u(x) = g(x), \quad 0 < x \le 1, \tag{25}$$

subject to the boundary conditions

$$u'(0) = 0, \quad u(1) = \frac{17}{16}. \tag{26}$$

The exact solution of (25) subject to (26) in the case of $g(x) = \frac{5}{4} + \frac{x^2}{16}$ is

$$u(t) = 1 + \frac{x^2}{16}. \tag{27}$$

Again, to solve (25)–(26) we applying the technique described with $m = 2$, where

$$\mathrm{D}^1 = \begin{pmatrix} -2 & -1 & 0 \\ 2 & 0 & -2 \\ 0 & 1 & 2 \end{pmatrix},$$

then we get

$$c_0 = 1, \quad c_1 = 1, \quad c_2 = \frac{17}{16}.$$

Thus

$$u(x) = \begin{pmatrix} 1 & 1 & \frac{17}{16} \end{pmatrix} \begin{pmatrix} 1 - 2x + x^2 \\ 2x - 2x^2 \\ x^2 \end{pmatrix} = 1 + \frac{x^2}{16},$$

which is the same as exact solution (15) and the solution obtained by MHAM [8].

4.4. Example 4.

Finally, consider the nonlinear singular two points BVP [10]

$$u''(x) + \frac{2}{x}u'(x) + (u(x))^5 \;=\; g(x), \quad 0 < x \leq 1 \tag{28}$$

subject to the boundary conditions

$$u(0) \;=\; 1, \qquad u(1) = \frac{\sqrt{3}}{2}. \tag{29}$$

The exact solution of (28) subject to (29) in the case $g(x) = 0$ is

$$u(x) = \frac{1}{\sqrt{1 + \frac{x^2}{3}}}. \tag{30}$$

The values of the unknown matrix C^{T} for $m = 2$, 3 and 4 are given in Table (1). Fig (1) show the absolute errors of $u(x)$ in the interval $x \in [0, 1]$ for different value of m, from this figure we conclude that when we increases the numbers of terms of the Bernstein polynomials the solution become more and more accurate.

5. Conclusions

In this paper, the Bernstein operational matrix of derivative was applied for the first time to solve a class of singular two-point BVPs. Different from other numerical techniques, only small size operational matrix is required to provide the solution at high accuracy. This matrix is used to approximate numerical solution of a class of singular two-point BVPs. It can be clearly seen in the paper that the proposed method is working well even on few numbers of terms of the Bernstein polynomials.

Table 1. The values of the unknown matrix C^T for $m = 2$, 3 and 4 of example 4.

c_i	$m = 2$	$m = 3$	$m = 4$
c_0	1.000000000	1.000000000	1.000000000
c_1	0.9871249039	1.001384632	1.000462557
c_2	0.8660254040	0.9383714768	0.9706859946
c_3		0.8660254040	0.9199858903
c_4			0.8660254040

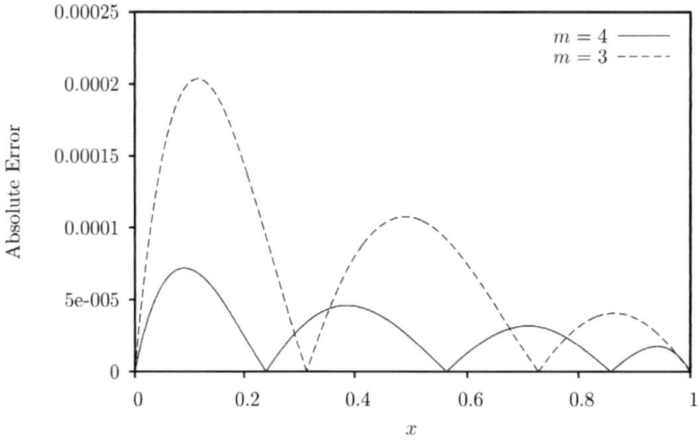

Figure 1. Absolute errors of $u(x)$ in the interval $x \in [0, 1]$ for different value of m of example 4.

6. References

[1] Huanmin Y and Yingzhen L 2009 *J. Comput. Appl. Math.* **223** 703
[2] Zhongli W and Changci P 2001 *Nagoya Math. J.* **162** 127
[3] Zhongli W 2007 *J. Math. Anal. Appl.* **327** 930
[4] Zhongli W 2006 *Appl. Math. Comput.* **182** 1280
[5] Y. Zhang1994 *J. Math. Anal. Appl.* **185** 215
[6] Canuto C Hussaini M Y Quarteroni A and Zang TA 1988 Spectral Methods in Fluid Dynamic (Prentice-Hall, Englewood Cliffs, NJ)
[7] Minggen C and Fazhan G 2007 *J. Comput. Appl. Math.* **205** 6
[8] Bataineh A S Noorani M S M and Hashim I *Phys. Lett. A* **372** 4062
[9] Junfeng L U 2007 *J. Comput. Appl. Math.* **207** 92
[10] Q. Ruibin Q and Agarwal R 1997 *J. Comput. Appl. Math.* **83** 147

AUTHOR INDEX